METHODS IN MOLECULAR BIOLOGY

Series Editor
John M. Walker
School of Life and Medical Sciences
University of Hertfordshire
Hatfield, Hertfordshire, UK

For further volumes:
http://www.springer.com/series/7651

For over 35 years, biological scientists have come to rely on the research protocols and methodologies in the critically acclaimed *Methods in Molecular Biology* series. The series was the first to introduce the step-by-step protocols approach that has become the standard in all biomedical protocol publishing. Each protocol is provided in readily-reproducible step-by-step fashion, opening with an introductory overview, a list of the materials and reagents needed to complete the experiment, and followed by a detailed procedure that is supported with a helpful notes section offering tips and tricks of the trade as well as troubleshooting advice. These hallmark features were introduced by series editor Dr. John Walker and constitute the key ingredient in each and every volume of the *Methods in Molecular Biology* series. Tested and trusted, comprehensive and reliable, all protocols from the series are indexed in PubMed.

Statistical Population Genomics

Edited by

Julien Y. Dutheil

Department of Evolutionary Genetics, Max Planck Institute for Evolutionary Biology, Plön, Germany

OPEN   ⁑ Humana Press

Editor
Julien Y. Dutheil
Department of Evolutionary Genetics
Max Planck Institute for Evolutionary Biology
Plön, Germany

ISSN 1064-3745 ISSN 1940-6029 (electronic)
Methods in Molecular Biology
ISBN 978-1-0716-0201-0 ISBN 978-1-0716-0199-0 (eBook)
https://doi.org/10.1007/978-1-0716-0199-0

This Humana imprint is published by the registered company Springer Science+Business Media, LLC, part of Springer Nature.
The registered company address is: 233 Spring Street, New York, NY 10013, U.S.A.

Preface

With the advent of so-called "next generation" sequencing technologies, the study of genetic variation in populations gained a new dimension, turning *population genetics* into *population genomics*. While a scaling-up in data set size accompanied this shift, population genomics is more than just "big data" population genetics: because the objects of study here are "genomes" and not only "multiple genes," the newly emerging field of population genomics comes along with its specific biological questions and statistical models.

Developments of new statistical methods are linked to the availability of particular data sets. As the first large population genomic initiatives came from primates, the development of many new methods targeted these organisms. Following the generation of increasingly diverse data sets, it is essential to promote the application of these methods to a broader range of organisms and questions.

The goal of this volume is to present the reader with state-of-the-art inference methods in population genomics. It focuses on data analysis based on rigorous statistical techniques. Data set preparation and preprocessing are covered in other volumes such as *Statistical Genomics* (Mathé and Davis eds) and *Data Production and Analysis in Population Genomics* (Pompanon and Bonin eds), while *Evolutionary Genomics* (Anisimova Ed) provides a more general background in evolutionary genomics.

The content of the book is divided into three parts. Part I recalls general concepts related to the biology of genomes and their evolution. Part II covers state-of-the-art methods for the analysis of genomes in populations, allowing to compute basic statistics (Chapters 2 and 3), understand population structure (Chapter 4), study selective processes (Chapters 5 and 6), and uncover the demographic history of populations (Chapters 7 and 8). More advanced tools allowing to simulate evolutionary scenarios (Chapter 9) or possible sample histories of a given data set (Chapter 10) are also presented. Chapters of this part come with practical examples of data analysis, with all necessary material available from the companion website of this book. Finally, part III of this collection offers an overview of the current knowledge that we acquired by applying such methods to a large variety of eukaryotic organisms: plants (Chapters 11 and 12), fungi (Chapters 13 and 14), insects (Chapter 15), fishes (Chapter 16), birds (Chapter 17), rodents (Chapter 18), and primates (Chapter 19). Without pretending to exhaustivity, these chapters highlight the exciting diversity of questions that the study of genome evolution at the population level can address, together with the originality of the model systems and approaches that have been instrumental in answering them.

Plön, Germany   *Julien Y. Dutheil*

Acknowledgments

I am very thankful to all authors who dedicated some of their precious time to contribute to this collection. Additional thanks go to Gustavo Barroso, Alice Feurtey, Annabelle Haudry, Filipa Moutinho, and Eva Stukenbrock for their valuable help in providing feedback on the chapters.

Contents

Contributors

NIKOLAOS ALACHIOTIS • *Institute of Computer Science, Foundation for Research and Technology Hellas, Heraklion, Greece*

GUSTAVO V. BARROSO • *Department of Evolutionary Genetics, Max Planck Institute of Evolutionary Biology, Plön, Germany*

THOMAS BATAILLON • *Bioinformatics Research Center, Aarhus University, Aarhus, Denmark*

DAVID CASTELLANO • *Bioinformatics and Genomics, Centre for Genomic Regulation (CRG), The Barcelona Institute of Science and Technology (BIST), Barcelona, Spain*

CHRISTOPHER C. CHANG • *GRAIL, Inc., Menlo Park, CA, USA*

JADE YU CHENG • *Bioinformatics Research Centre, Aarhus University, Aarhus, Denmark*

FABIEN DE BELLIS • *UMR AGAP, Univ Montpellier, CIRAD, INRA, Montpellier SupAgro, Montpellier, France*

KIRA E. DELMORE • *MPRG Behavioural Genomics, Max Planck Institute for Evolutionary Biology, Plön, Germany*

JULIEN Y. DUTHEIL • *Department of Evolutionary Genetics, Max Planck Institute of Evolutionary Biology, Plön, Germany*

CHRISTOPH J. ESCHENBRENNER • *Environmental Genomics, Christian-Albrechts University of Kiel, Kiel, Germany; Max Planck Institute for Evolutionary Biology, Plön, Germany*

ALICE FEURTEY • *Environmental Genomics, Christian-Albrechts University of Kiel, Kiel, Germany; Max Planck Institute for Evolutionary Biology, Plön, Germany*

PIERRE GLADIEUX • *UMR BGPI, Univ Montpellier, CIRAD, INRA, Montpellier SupAgro, Montpellier, France*

CHRISTOPHER HANN-SODEN • *Department of Plant and Microbial Biology, University of California, Berkeley, Berkeley, CA, USA*

ANNABELLE HAUDRY • *Laboratoire de Biométrie et Biologie Evolutive UMR 5558, CNRS, Université de Lyon, Université Lyon 1, Villeurbanne, France*

MELISSA HUBISZ • *Cornell University, Ithaca, NY, USA*

HANNA JOHANNESSON • *Department of Organismal Biology, Uppsala University, Uppsala, Sweden*

MARTIN KAPUN • *Department of Biology, University of Fribourg, Fribourg, Switzerland; Department of Evolutionary Biology and Environmental Studies, University of Zürich, Zürich, Switzerland; Department of Cell and Developmental Biology, Medical University of Vienna, Vienna, Austria*

JEROME KELLEHER • *Big Data Institute, Li Ka Shing Centre for Health Information and Discovery, University of Oxford, Oxford, UK*

ANGELOS KOROPOULIS • *Institute of Computer Science, Foundation for Research and Technology Hellas, Heraklion, Greece; Computer Science Department, University of Crete, Crete, Heraklion, Greece*

BENJAMIN LAENEN • *Department of Ecology, Environment, and Plant Sciences, Science for Life Laboratory, Stockholm University, Stockholm, Sweden*

KENNETH LANGE • *Department of Computational Medicine, University of California, Los Angeles, CA, USA; Department of Human Genetics, University of California, Los Angeles, CA, USA; Department of Statistics, University of California, Los Angeles, CA, USA*

STEFAN LAURENT • *Department of Comparative Development and Genetics, Max Planck Institute for Plant Breeding Research, Cologne, Germany*

MIRIAM LIEDVOGEL • *MPRG Behavioural Genomics, Max Planck Institute for Evolutionary Biology, Plön, Germany*

CHI-CHUN LIU • *Department of Human Genetics, University of Chicago, Chicago, IL, USA*

KONRAD LOHSE • *Institute of Evolutionary Biology, University of Edinburgh, Edinburgh, UK*

ANNE LORANT • *Department of Plant Sciences, University of California, Davis, Davis, CA, USA; Génétique Quantitative et Evolution—Le Moulon, Institut National de la Recherche Agronomique, Université Paris-Sud, Centre National de la Recherche Scientifique, AgroParisTech, Université Paris-Saclay, Gif-sur-Yvette, France*

THOMAS MAILUND • *Bioinformatics Research Centre, Aarhus University, Aarhus, Denmark*

TIINA M. MATTILA • *Department of Ecology and Genetics, University of Oulu, Oulu, Finland*

ANA FILIPA MOUTINHO • *Department of Evolutionary Genetics, Max Planck Institute of Evolutionary Biology, Plön, Germany*

KASPER MUNCH • *Bioinformatics Research Centre, Aarhus University, Aarhus C, Denmark*

ARNE W. NOLTE • *AG Ökologische Genomik, Institut für Biologie und Umweltwissenschaften, Carl von Ossietzky Universität Oldenburg, Oldenburg, Germany*

JOHN NOVEMBRE • *Department of Human Genetics, Department of Ecology and Evolution, University of Chicago, Chicago, IL, USA*

PAVLOS PAVLIDIS • *Institute of Computer Science, Foundation for Research and Technology Hellas, Heraklion, Greece*

JEFFREY ROSS-IBARRA • *Department of Plant Sciences, University of California, Davis, Davis, CA, USA; Center for Population Biology, University of California, Davis, Davis, CA, USA; Genome Center, University of California, Davis, Davis, CA, USA*

STEPHAN SCHIFFELS • *Department of Archaeogenetics, Max Planck Institute for the Science of Human History, Jena, Germany*

SUYASH SHRINGARPURE • *23andMe Inc., Sunnyvale, CA, USA*

ADAM SIEPEL • *Simons Center for Quantitative Biology, Cold Spring Harbor Laboratory, Cold Spring Harbor, NY, USA*

TANJA SLOTTE • *Department of Ecology, Environment, and Plant Sciences, Science for Life Laboratory, Stockholm University, Stockholm, Sweden*

EVA H. STUKENBROCK • *Environmental Genomics, Christian-Albrechts University of Kiel, Kiel, Germany; Max Planck Institute for Evolutionary Biology, Plön, Germany*

JESPER SVEDBERG • *Department of Organismal Biology, Uppsala University, Uppsala, Sweden*

PAULA TATARU • *Bioinformatics Research Center, Aarhus University, Aarhus, Denmark*

DIETHARD TAUTZ • *Department of Evolutionary Genetics, Max Planck Institute for Evolutionary Biology, Plön, Germany*

JOHN W. TAYLOR • *Department of Plant and Microbial Biology, University of California, Berkeley, Berkeley, CA, USA*

MAUD TENAILLON • *Génétique Quantitative et Evolution—Le Moulon, Institut National de la Recherche Agronomique, Université Paris-Sud, Centre National de la Recherche Scientifique, AgroParisTech, Université Paris-Saclay, Gif-sur-Yvette, France*

KRISTIAN K. ULLRICH • *Department of Evolutionary Genetics, Max Planck Institute for Evolutionary Biology, Plön, Germany*

KE WANG • *Department of Archaeogenetics, Max Planck Institute for the Science of Human History, Jena, Germany*

Part I

Essential Concepts

Chapter 1

A Population Genomics Lexicon

Gustavo V. Barroso, Ana Filipa Moutinho, and Julien Y. Dutheil

Abstract

Population genomics is a growing field stemming from soon a 100 years of developments in population genetics. Here, we summarize the main concepts and terminology underlying both theoretical and empirical statistical population genomics studies. We provide the reader with pointers toward the original literature as well as methodological and historical reviews.

Key words Population genetics, Neutral theory, Coalescent theory, Mutation, Recombination, Selection, Lexicon

1 Genomic Variation

1.1 Loci, Alleles, and Polymorphism

Population genomics studies the evolution of genome variants in populations. A *locus (pl. loci)* refers to a given location in the genome. The particular sequence at a given locus may vary between individuals, each variant being termed an *allele*. We call loci with at least two alleles *polymorphic* and invariant loci *monomorphic*. The term *polymorphism* refers to the presence of multiple alleles but is commonly used as a countable noun as a substitute for "polymorphic locus" (*one polymorphism, several polymorphisms*).

Alleles may differ because of the nucleotide content, but also in length, as a result of nucleotide insertions or deletions (*a.k.a. indels*). Variable loci of length one can have up to four distinct alleles (A, C, G, or T) and are termed *single nucleotide polymorphisms (SNPs)*. SNPs constitute, so far, the majority of the data accounted for by population genetic models.

1.2 Mutations

Molecular events altering the genome are termed *mutations*. Mutations include substitution of a nucleotide into another one, removal or addition of one or several nucleotides, as well as multiplication of some part of the genome. Mutation is the process by which new

---

Authors Gustavo V. Barroso and Ana Filipa Moutinho contributed equally to this work.

Julien Y. Dutheil (ed.), *Statistical Population Genomics*, Methods in Molecular Biology, vol. 2090, https://doi.org/10.1007/978-1-0716-0199-0_1, © The Author(s) 2020

alleles are formed. The *infinite site model* assumes that during the timeframe of evolution modeled, each locus have undergone at most one mutation [1–3]. This model also implies that each mutation creates a new allele in the population and that there is no "backward" or "reverse" mutation. The infinite site model is a generally reasonable assumption as the mutation rate is typically low and genomes are large. It might be locally invalidated, however, in case of mutation hotspots or when larger evolutionary timescales are considered. Under this premise, at most two alleles are expected per locus. Loci with two alleles are termed *diallelic* or *biallelic*, the first term having historical precedence and being more accurate [4], while the second is more commonly used since the 1990s. Furthermore, in a population genomic dataset, a sampled diallelic locus is called a *singleton* if one of the two alleles is present in only one haploid genome, and a *doubleton* if it is present in precisely two haploid genomes.

1.3 The Wright–Fisher Model

The simplest process of allele evolution within a single population is named the *Wright–Fisher model*. It describes the evolution of alleles in a population of fixed and constant size, where all alleles have the same fitness, and therefore the same chance to be transmitted to the next generation (*neutral evolution*). The population is assumed to be *panmictic*, that is, individuals are randomly mating. Time is discretized in *non-overlapping generations* so that the alleles in the current generation are a random sample of the alleles from the previous generation, without new alleles being generated by mutation. Under such conditions, allelic frequencies evolve only because of the stochasticity in the sampling of gametes that will contribute to the next generation, a process termed *genetic drift*. Because populations are of finite size, alleles will be sampled at their actual frequencies on average only and the ultimate fate of any allele is either to reach frequency zero in the population and be lost, when by chance no individual carrying this allele has any descendant in the next generation or to become fixed when all other alleles have been lost. The time until fixation depends on the population size: smaller populations will show a stronger sampling effect and shorter times to fixation. When genetic drift is the only force acting on a population, the number of alleles at a given locus is necessarily decreasing over time.

The *Wright–Fisher model with mutation* extends the Wright–Fisher model by introducing new alleles in the population, at a given rate. As the mutation rate is low, new mutations appear in a single copy, their initial frequency is then $1/2N$ in a diploid population. Mutation and drift act in opposite direction and a *mutation-drift equilibrium* is reached when the rate of allele creation by mutation equals the rate of allele loss by drift. The genetic diversity is then determined by the sole product of the population size N and the mutation rate u. Under the infinite site model, the expected heterozygosity at a locus in a population of diploid individuals is approximated by [1]

$$\hat{h} = \frac{4 \cdot N \cdot u}{4 \cdot N \cdot u + 1}$$

while the expected number of distinct alleles and their respective frequencies can be estimated using *Ewens's sampling formula* [5].

A *substitution* occurs when a new mutation has spread in the population, increasing from frequency $1/(2N)$ to 1 (*see* **Note 1**). Kimura showed that the average time to fixation of a new mutation is $4N$ in a population of diploid individuals [6]. Furthermore, as a neutral mutation has a probability of reaching fixation equal to $1/(2N)$ and given that there are $2N \cdot u$ new mutations per generation, in a purely neutrally evolving population, the expected number of substitutions per generation is equal to $2N \cdot u \cdot 1/(2N) = u$. The substitution rate is therefore independent of the population size and, assuming that the mutation rate is constant in time, the number of substitutions between two populations is a direct measure of the number of generations separating them, a phenomenon termed *molecular clock* [7].

1.4 The Backward Wright–Fisher Model: The Standard Coalescent

While the Wright–Fisher process naturally describes the evolution of sequences within populations one generation after the other, population genetic data typically represent individuals sampled at a given time point. For inference purposes, it is therefore convenient to model the history of the genetic material that gave rise to the sample. The modelization of the ancestry of a sample (also known as the *genealogy*) is typically done backward in time, as every locus find a common ancestor in the past, until the *most recent common ancestor (MRCA)* of the sample. The merging of two lineages in the past is called a *coalescence event*, and the set of mathematical tools describing this process under a variety of demographic models is referred to as the *coalescence theory*. Kingman [8] first described the *standard coalescent*, the genealogical model corresponding to the Wright–Fisher model (but *see* refs. 9 and 10 for a historical perspective). The standard coalescent is, therefore, also referred to as the *Kingman's coalescent*.

2 Beyond the Wright–Fisher Model

The Wright–Fisher model has been extended in several ways to include more realistic assumptions on the underlying evolutionary process. These extensions led to the concept of *Effective population size (Ne)*, originally defined as the number of individuals contributing to the gene pool. When a population deviates from the assumptions of the Wright–Fisher model, Ne is no longer equal to the census population size (N). Often (but not always) in such cases,

Ne can be obtained by a linear scaling of *N* such that it reflects the number of individuals from an idealized Wright–Fisher population that would display the same genetic diversity as the actual population under study [11].

2.1 Demography

A possible deviation from the Wright–Fisher assumptions happens when the population size is not constant across generations. The term *demographic history* generally refers to the collection of demographic parameters (effective sizes, growth rates) that describes the history of the population until its most recent common ancestor [12]. When population size varies in a cyclic manner with relatively small period *n* generations, the resulting genealogies can be modeled by a Wright–Fisher process with a population size equal to the harmonic mean of the historical population sizes, so that

$$Ne = \frac{n}{\sum_i^n \frac{1}{N_i}},$$

where N_i refer to the *i*th population size [13]. More drastic demographic effects include *genetic bottlenecks*, corresponding to a sharp decrease (shrinkage) in population size.

2.2 Population Structure

In the absence of *panmixia*, genetic exchanges occur more often between certain individuals, resulting in *population structure* with several subpopulations. Population structure may occur for different reasons such as overlapping generations, assortative mating, or geographic isolation [12]. *Assortative mating* occurs when individuals choose their mates according to some similarity between their phenotypes. If the phenotype is genetically determined, assortative mating can influence the level of heterozygosity in the population [14].

Gene flow describes the migration of genetic variants between subpopulations under a scenario of population structure. It reduces genetic differentiation among subpopulations [15]. Ultimately, subpopulations can diverge and become genetically isolated, a process called *speciation*. The simplest speciation processes involve spontaneous isolation (*isolation model*) or spontaneous isolation followed by a period of gene flow (*isolation with migration model*) [16].

When speciation events occur in a short timeframe and ancestral population sizes are large, ancestral polymorphism may persist in the ancestral species, a phenomenon called *incomplete lineage sorting (ILS)* [17]. The expected amount of ILS depends on the number of generations between two isolation events (Δ_T) and the ancestral effective population size Ne_A [18]:

$$\Pr(ILS) = \frac{2}{3} e^{\left(-\frac{2 \cdot \Delta_T}{Ne_A} \right)}$$

The term *introgression* is used to depict the transfer of genetic material between diverged populations or species through secondary contact [19]. As a result, extant lineages share a common ancestor that predates the two isolation or speciation events. The resulting genealogy may, therefore, be incongruent with the phylogeny defined by the two splits, depending on the order of coalescence events between lineages [20].

3   Statistics on Nucleotide Diversity

Statistics are needed to infer population genetics parameters from polymorphism data. The *site frequency spectrum (SFS)* describes the empirical distribution of allele frequencies across segregating sites of a given (set of) loci in a population sample. For a sample of n sequences (in n haploid individuals or $n/2$ diploid individuals), the so-called unfolded SFS is the set of counts of derived alleles $X = (X_1, X_2, \ldots, X_{n-1})$, where sample configurations X_i denote the number of sites that have $n - i$ ancestral and i derived alleles. The ancestral state is usually estimated using an outgroup sequence. In cases where we cannot assess the ancestral allele, the folded site frequency spectrum, X', may be calculated instead. X' represents the distribution of the minor allele frequencies, such as $X'_i = X_i + X_{n-i}$ for $i < n/2$ and $X'_{n/2} = X_{n/2}$ [13, 21, 22]. The shape of the SFS is affected by underlying population genetic processes, such as demography and selection, and therefore serves as the input of many population genetics methods [23] (*see* Fig. 1).

Watterson's theta, here noted $\hat{\theta}_S$, is an estimator of the population mutation rate $\theta = 4Ne \cdot u$, where Ne is the (diploid) effective population size and u the mutation rate. It is derived from the number of segregating sites S_n of a sample of size n [25]. Assuming an infinite sites model, S_n is equal to the product of u and the expected time to coalescence, corrected by the sample size:

$$E[Sn] = u \cdot 4 \cdot Ne \sum_{i=1}^{n-1} i.$$

Since $4Ne \cdot u = \theta$ the equation may be written as $E[Sn] = \theta \cdot a_n$, where $a_n = \sum_{i=1}^{n-1} i$. The proposed estimator of θ for the sample is

$$\hat{\theta}_S = \frac{\hat{S}_n}{a_n} = \frac{\hat{S}_n}{\left(1 + \frac{1}{2} + \ldots + \frac{1}{n-1}\right)},$$

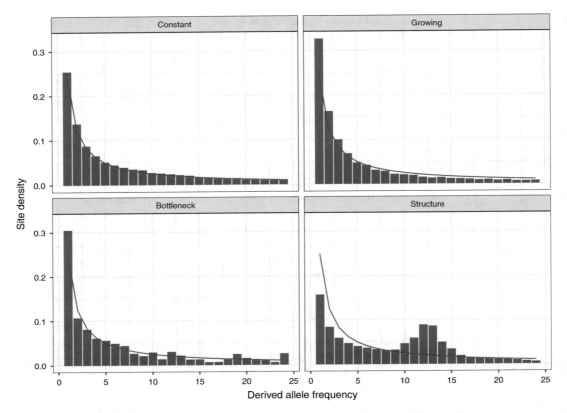

Fig. 1 Effect of demography on the shape of the site frequency spectrum (SFS). The figure depicts four scenarios: constant population size, exponential growth, genetic bottleneck, and population structure. The red curve shows the expectation under a constant population size. In the case of exponential growth or a genetic bottleneck, the SFS displays an excess of low-frequency variants. Population structure, here simulated as two subpopulations exchanging migrants at a low rate, results in an excess of intermediate frequency variant when we reconstruct a single SFS from the two subpopulations. Simulations were performed using the `msprime` software [24] (*see* also Chapter 9 and the online companion material)

where $\hat{S}_n$ is the observed number of segregating sites in the sample. In order to be comparable, values of θ are usually reported per site, and $\hat{\theta}_S$ is then further divided by the sequence length L. This estimator is unbiased when the data is generated from a Wright–Fisher process but is not robust to deviations from it, due to selection or demography [26].

Tajima's π, the *average pairwise heterozygosity* is a measure of nucleotide diversity defined as the number of pairwise differences between a set of sequences [27]. Under the infinite sites model, the number of mutations separating two orthologous chromosomes D_{ij} is equal to the number of nucleotide differences between sequences i and j. As the expectation of the average pairwise nucleotide differences between all pairs of sequences in a sample is equal to $\theta = 4Ne \cdot u$ [28], Tajima's estimator of θ is:

$$\hat{\theta}_\pi = \frac{2}{n(n-1) \cdot L} \sum_{i=1}^{n-1} \sum_{j=i+1}^{n} D_{ij},$$

where L is the total sequence length.

4   Selective Processes

4.1   Protein-Coding Genes

The coding region of a protein-coding gene, also known as *Coding DNA Sequence (CDS)* is the portion of DNA, or RNA, that encodes a protein. A start and stop codons limit the coding region at the five-prime and three-prime end, respectively. In mRNAs, the CDS is bounded by the five-prime untranslated region (5-UTR) and the three-prime untranslated region (3'-UTR), also included in the exons. Mutations within coding regions are expected to be of distinct types: *synonymous mutations* lead to no change of amino-acid at the protein level due to the redundancy of the genetic code, as opposed to *non-synonymous mutations*. Non-synonymous mutations can further be classified as *conservative* and *non-conservative* (= *radical*), whether they replace an amino-acid by a biochemically similar one or not. Because of the structure of the genetic code, the four types of mutations at one site (toward A, C, G, or T) can be in principle both synonymous and non-synonymous. Sites where n out of four possible mutations are synonymous are called *n-fold degenerated*. *Four-fold degenerated* sites only undergo synonymous mutations, while a mutation at a so-called *zero-fold degenerated* site is necessarily non-synonymous. Most of second codon positions are zero-fold degenerated, while many of the third positions are four-fold degenerated.

4.2   Fitness Effect

The resulting change of fitness at the organism level characterizes the type of mutations: neutral mutations have no impact on the fitness, while harmful or deleterious mutations induce a lower fitness. Conversely, advantageous mutations increase the fitness of the organism compared to the wild-type genotype. There is, however, a wide range of selective effects, which extends the categorization of mutations from strongly deleterious, through weakly deleterious, neutral to mildly and highly adaptive mutations. The relative frequencies of these types of mutations represent the distribution of fitness effects [29, 30].

The *selection coefficient (s)* is a measure of differences in fitness, which determines the changes in genotype frequencies that occur due to selection. It is commonly expressed as a relative fitness. If one considers a single locus with two alleles A and a, a standard parametrization is to attribute a fitness of 1 to the homozygote AA and relative fitness of $1 + s$ for the homozygote aa. The heterozygote Aa is attributed a fitness of $1 + h \cdot s$, where h is the so-called *coefficient of dominance*. The s parameter varies between -1 and $+\infty$ (but *see* **Note 2**), wherein values comprised among -1 and 0 are indicative of negative selection, while positive values correspond to positive selection [13, 31]. The efficiency of selection, however, depends on both s and the effective population size, Ne, so that mutations with $Ne \cdot s \ll 1$ behave in effect like neutral mutations, whose fate is determined by genetic drift only [29].

4.3 Types of Selection

Positive selection acts on alleles that increase fitness, raising their frequency in the population over time, while *negative selection* (= *purifying selection*) decreases the frequency of alleles that impair fitness. Both positive and negative selection decrease genetic diversity. Conversely, *balancing selection* acts by maintaining multiple alleles in the gene pool of a population at frequencies higher than expected by drift alone. Three mechanisms are generally acknowledged: *heterozygous advantage*, where heterozygotes have a higher fitness than homozygotes and maintain genetic polymorphism; *frequency-dependent selection*, where the fitness of the genotype is inversely proportional to its frequency in the population; and *environment-dependent fitness* of genotypes (also known as *local adaptation*) [31, 32].

4.4 Inference of Selection in Protein-Coding Sequences

The strength and direction of selection acting on protein-coding regions may be assessed by contrasting the rate of non-synonymous (potentially under selection, dN) to synonymous (assumed to be neutral, dS, but see, for instance, [33]) substitutions between species. In a population of sequences evolving neutrally, all substitutions are neutral and the two rates are equal, leading to a dN/dS ratio equal to one on average. Assuming non-synonymous mutations are either neutral or deleterious while synonymous mutations are always neutral, the rate of non-synonymous substitutions will be lower than the rate of synonymous substitutions, and the dN/dS ratio will be lower than one. Conversely, if non-synonymous mutations are positively selected, their rate of fixation may exceed the rate of synonymous mutation, leading to a higher substitution rate and a dN/dS ratio higher than one.

At the population level, the ratio of non-synonymous (pN) and synonymous (pS) polymorphism is indicative of the strength of purifying selection acting on a protein. Because non-synonymous mutations are more likely to have a negative fitness effect and be counter-selected, they tend to be removed from the population by purifying selection or segregate at low-frequency. We can estimate the synonymous and non-synonymous genetic diversity by computing the average pairwise heterozygosity π separately for non-synonymous and synonymous mutations, noted π_N and π_S, respectively. The π_N/π_S ratio is therefore generally below one, the stronger the purifying selection, the closer the ratio is to zero.

Contrasting the dN/dS and pN/pS ratios allows to test the selection regime acting on the sequences [34]. If mutations are all neutral or deleterious, we expect the ratios dN/dS and pN/pS to be equal. Positively selected mutations will tend to quickly rise to fixation and will not be observed as polymorphism, leading to an increased dN/dS ratio higher than pN/pS. Conversely, balancing selection will lead to an excess of polymorphism detectable as $dN/dS < pN/pS$ [35]. A simple measure of the proportion of amino-acid substitutions resulting from positive selection (α) is given by $1 - (dS \cdot pN/dN \cdot pS)$ [36]. Using the complete synonymous and non-synonymous site frequency spectra, it is further possible to estimate the distribution of fitness effects and account for slightly deleterious and slightly advantageous mutations when estimating the rate of adaptive substitutions (*see* Chapter 5) [37].

5   Linkage and Recombination

5.1   The Coalescent with Recombination

In sexually reproducing species, *recombination* refers to both the shuffling of non-homologous chromosomes and the rearrangement of homologous chromosomes during meiosis. Such cross-over events cause each chromosome to have two parent chromosomes in the previous generation, which are themselves the products of recombination events in the previous generations. Therefore, any chromosome in the current generation can be viewed as a mosaic of chromosomes that existed in the past (*see* Fig. 2) [38]. The collection of coalescence and recombination events that describes the history of sampled chromosomes until the most recent common ancestor of each non-recombining block is reached (*see* Fig. 2) is called the *ancestral recombination graph (ARG)* [39]. Compared to a tree-like genealogy of a sample without recombination, whose complexity depends only on the sample size, the complexity of the ARG grows with the sample size and the number of recombination events in the ancestry of the sample.

Backward-in-time, the *most recent common ancestor (MRCA)* denotes the first individual where the entire sample (population) coalesces for a particular non-recombining block. The *TMRCA* notes the timing of such event. DNA sequences provide no information beyond the MRCA in a sample of genomes since all individuals will share any mutation that happens further back in time [40]. In the presence of recombination, different parts of the genome will have different MRCAs. In this case, all ancestral material is eventually found as a contiguous sequence in the *grand most*

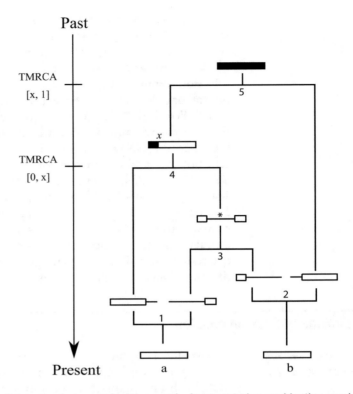

Fig. 2 An ancestral recombination graph. An ancestral recombination graph is a collection of recombination (1–2) and coalescence (3–5) events. In each depicted chromosome, white bars represent segregating ancestral material, black bars represent coalesced ancestral material, and thin lines represent non-ancestral material. The asterisk denotes trapped non-ancestral material. Note that "1" does not impact the sample because the resulting segments are joined back together in "4" before coalescing in "5." There are thus only two relevant TMRCAs in the ARG, separated at position x

recent common ancestor (GMRCA) of the sample (*see* Fig. 2). If the GMRCA is not an MRCA for any nucleotide, this individual does not have any significance for DNA sequences [39].

In the ARG, nucleotide segments that are found both in past chromosomes and in contemporary samples are termed *ancestral genetic material* (*see* Fig. 2). Conversely, *non-ancestral genetic material* refers to segments that are found in past chromosomes but not in contemporary samples. Furthermore, non-ancestral genetic material flanked on both sides by ancestral genetic material is referred to as *trapped genetic material*. In this setting, recombination events that happen in trapped genetic material can affect linkage disequilibrium between present-day nucleotides (*see* Fig. 2). Thus the existence of trapped genetic material introduces long-range correlations between genealogies rendering the coalescent with recombination a non-Markovian process along chromosomes

[41]. The *Sequentially Markov coalescent (SMC)* is an approximation to the coalescent with recombination whereby recombination events are assumed to happen only within ancestral material. This approximation allows the use of efficient algorithms in both simulation and data analysis [42, 43].

5.2 Impact of Linkage on Selection

An excess of linkage between loci compared to a random association is termed *linkage disequilibrium (LD)*. LD arises from genetic drift, population admixture, and selection, but is reduced by recombination each generation. It is, therefore, higher between close loci and decays with increasing physical distance [44].

Linked selection refers to the reduction of diversity at neutral sites that happens as a result of their physical linkage to variants under selection [45]. In the absence of recombination, all variants segregating in a chromosome would undergo the same shift in frequency as the selected variant. However, recombination creates new allelic combinations and reduces this correlation as the physical distance from the selected locus increases (*see* Fig. 3).

Background selection refers to a form of linked selection where the reduction of diversity at neutral loci results from linkage to a locus under purifying selection [46], and *genetic hitchhiking* is commonly used to depict linked selection due to linkage to a locus under positive selection [47], where a new beneficial mutation will rise in frequency in a population. As the new positively selected allele increases its frequency, nearby linked alleles on the chromosome will "hitchhike" along with it, also growing in frequency, thus producing a *selective sweep* of genetic diversity (*see* Fig. 3d). *Hard sweeps* occur when a new mutation is positively selected and is therefore exclusively associated with the genetic background where it arose. Conversely, *soft sweeps* occur when a mutation is already segregating in the population at the onset of selection. This mutation may exist in several genetic backgrounds and therefore does not prompt a complete loss of genetic variation after the selective sweep [47] (*see* Fig. 3a–c).

Linkage of two or more loci can also impair the efficacy of positive selection, a phenomenon termed *Hill–Robertson interference (HRI)* [48]. When two advantageous mutations at distinct loci in distinct individuals segregate in the population, one will be lost unless a recombination event brings them together. In the absence of recombination between the selected loci, only the unlikely event of recurrent mutations can generate the optimal haplotypic combination [49] (*see* Fig. 3e).

A) Incomplete, then complete hard sweep

B) Incomplete, then complete soft sweep from standing genetic variation

C) Incomplete, then complete soft sweep from recurrent mutations

D) Background selection

E) Hill-Robertson interference

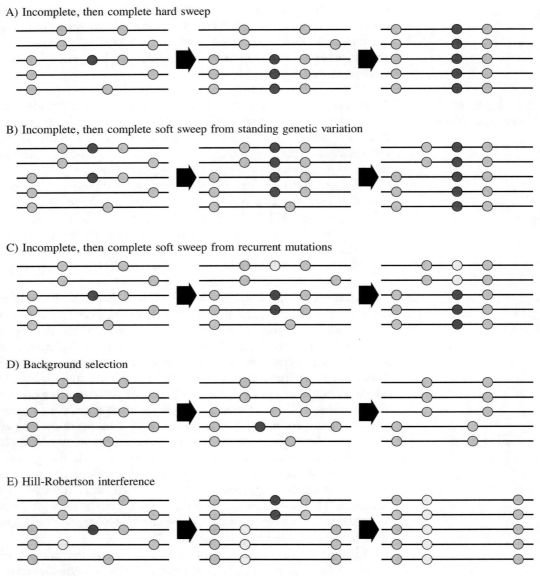

Fig. 3 Impact of selection on genetic diversity. Black lines represent individual genomes. SNP variants are displayed by filled circles. Distinct variants at the same position are depicted with different colors: neutral variants in gray, positive variants in red or yellow, and negative variant in blue. (**a**) A positively selected new variant spreads in the population and removes genetic diversity at linked loci, generating a hard selective sweep. (**b** and **c**) Segregation of several positively selected variants in different genetic backgrounds, either from standing variation or recurrent mutations, resulting in a soft selective sweep. (**d**) Reduction of neutral diversity because of linkage to deleterious mutations (background selection). (**e**) Competitive segregation of positively selected variant at distinct loci, resulting in the loss of advantageous variants (Hill–Robertson interference)

6 Notes

1. The use of the term *substitution* differs in population genetics and molecular biology. In the latter case, it describes a particular type of mutation where a single nucleotide replaces a distinct one (as opposed to insertions/deletions, for instance).

2. In some instances, s is substituted by $-s$, so that the relative fitnesses become $\omega_{AA} = 1$, $\omega_{Aa} = 1 - h \cdot s$ and $\omega_{aa} = 1 - s$.

References

1. Kimura M, Crow JF (1964) The number of alleles that can be maintained in a finite population. Genetics 49:725–738

2. Kimura M (1969) The number of heterozygous nucleotide sites maintained in a finite population due to steady flux of mutations. Genetics 61(4):893–903

3. Crow JF (1989) Twenty-five years ago in genetics: the infinite allele model. Genetics 121(4):631–634

4. Elston RC, Satagopan J, Sun S (2017) Statistical genetic terminology. Methods Mol Biol 1666:1–9. https://doi.org/10.1007/978-1-4939-7274-6_1

5. Ewens WJ (1972) The sampling theory of selectively neutral alleles. Theor Popul Biol 3(1):87–112

6. Kimura M (1970) The length of time required for a selectively neutral mutant to reach fixation through random frequency drift in a finite population. Genet Res 15(1):131–133

7. Kimura M (1983) The neutral theory of molecular evolution. Cambridge University Press, Cambridge. http://ebooks.cambridge.org/ref/id/CBO9780511623486

8. Kingman JFC (1982) The coalescent. Stoch Process Appl 13(3):235–248. https://doi.org/10.1016/0304-4149(82)90011-4

9. Barton NH (2016) Richard Hudson and Norman Kaplan on the coalescent process. Genetics 202(3):865–866. https://doi.org/10.1534/genetics.116.187542

10. Kingman JFC (2000) Origins of the Coalescent: 1974–1982. Genetics 156(4):1461–1463. http://www.genetics.org/content/156/4/1461

11. Sjödin P, Kaj I, Krone S, Lascoux M, Nordborg M (2005) On the meaning and existence of an effective population size. Genetics 169(2):1061–1070. https://doi.org/10.1534/genetics.104.026799

12. Wakeley J (2008) Coalescent theory: an introduction, 1st edn. Roberts and Company Publishers, Reading

13. Wright S (1938) Size of population and breeding structure in relation to evolution. Science 87:430–431

14. Jiang Y, Bolnick DI, Kirkpatrick M (2013) Assortative mating in animals. Am Nat 181(6):E125–138. https://doi.org/10.1086/670160

15. Sousa V, Hey J (2013) Understanding the origin of species with genome-scale data: modelling gene flow. Nat Rev Genet 14(6):404–414. https://doi.org/10.1038/nrg3446

16. Hey J, Nielsen R (2004) Multilocus methods for estimating population sizes, migration rates and divergence time, with applications to the divergence of Drosophila pseudoobscura and D. persimilis. Genetics 167(2):747–760. https://doi.org/10.1534/genetics.103.024182

17. Dutheil JY, Hobolth A (2012) Ancestral population genomics. Methods Mol Biol 856:293–313. https://doi.org/10.1007/978-1-61779-585-5_12

18. Hobolth A, Christensen OF, Mailund T, Schierup MH (2007) Genomic relationships and speciation times of human, chimpanzee, and gorilla inferred from a coalescent hidden Markov model. PLoS Genet 3(2):e7. https://doi.org/10.1371/journal.pgen.0030007

19. Martin SH, Jiggins CD (2017) Interpreting the genomic landscape of introgression. Curr Opin Genet Dev 47:69–74. https://doi.org/10.1016/j.gde.2017.08.007

20. Mailund T, Munch K, Schierup MH (2014) Lineage sorting in Apes. Annu Rev Genet

https://doi.org/10.1146/annurev-genet-120213-092532

21. Bustamante CD, Wakeley J, Sawyer S, Hartl DL (2001) Directional selection and the site-frequency spectrum. Genetics 159 (4):1779–1788

22. Wright S (1968) Evolution and the genetics of populations, vol 2. The theory of gene frequencies. The University of Chicago Press, Chicago

23. Schraiber JG, Akey JM (2015) Methods and models for unravelling human evolutionary history. Nat Rev Genet 16(12):727–740. https://doi.org/10.1038/nrg4005

24. Kelleher J, Etheridge AM, McVean G (2016) Efficient coalescent simulation and genealogical analysis for large sample sizes. PLoS Comput Biol 12(5):e1004842. https://doi.org/10.1371/journal.pcbi.1004842

25. Watterson GA (1975) On the number of segregating sites in genetical models without recombination. Theor Popul Biol 7(2):256–276

26. Tajima F (1989) Statistical method for testing the neutral mutation hypothesis by DNA polymorphism. Genetics 123(3):585–595

27. Nei M, Tajima F (1981) Genetic drift and estimation of effective population size. Genetics 98(3):625–640. http://www.genetics.org/content/98/3/625

28. Tajima F (1983) Evolutionary relationship of DNA sequences in finite populations. Genetics 105(2):437–460

29. Eyre-Walker A, Keightley PD (2007) The distribution of fitness effects of new mutations. Nat Rev Genet 8(8):610–618. https://doi.org/10.1038/nrg2146

30. Orr HA (2009) Fitness and its role in evolutionary genetics. Nat Rev Genet 10(8):531–539. https://doi.org/10.1038/nrg2603

31. Gillespie JH (2004) Population genetics: a concise guide. JHU Press, Baltimore

32. Nielsen R (2005) Molecular signatures of natural selection. Annu Rev Genet 39:197–218. https://doi.org/10.1146/annurev.genet.39.073003.112420

33. Pouyet F, Bailly-Bechet M, Mouchiroud D, Guéguen L (2016) SENCA: a multilayered codon model to study the origins and dynamics of codon usage. Genome Biol Evol 8 (8):2427–2441. https://doi.org/10.1093/gbe/evw165

34. McDonald JH, Kreitman M (1991) Adaptive protein evolution at the Adh locus in Drosophila. Nature 351(6328):652–654. https://doi.org/10.1038/351652a0

35. Parsch J, Zhang Z, Baines JF (2009) The influence of demography and weak selection on the McDonald-Kreitman test: an empirical study in Drosophila. Mol Biol Evol 26(3):691–698. https://doi.org/10.1093/molbev/msn297

36. Smith NGC, Eyre-Walker A (2002) Adaptive protein evolution in Drosophila. Nature 415 (6875):1022–1024. https://doi.org/10.1038/4151022a

37. Keightley PD, Eyre-Walker A (2007) Joint inference of the distribution of fitness effects of deleterious mutations and population demography based on nucleotide polymorphism frequencies. Genetics 177 (4):2251–2261. https://doi.org/10.1534/genetics.107.080663

38. Stumpf MPH, McVean GAT (2003) Estimating recombination rates from population-genetic data. Nat Rev Genet 4(12):959–968. https://doi.org/10.1038/nrg1227

39. Hein J, Schierup MH, Wiuf C (2005) Gene genealogies, variation and evolution: a primer in coalescent theory. Oxford University Press, Oxford

40. Rosenberg NA, Nordborg M (2002) Genealogical trees, coalescent theory and the analysis of genetic polymorphisms. Nat Rev Genet 3 (5):380–390. https://doi.org/10.1038/nrg795

41. Rasmussen MD, Hubisz MJ, Gronau I, Siepel A (2014) Genome-wide inference of ancestral recombination graphs. PLoS Genet 10(5): e1004342. https://doi.org/10.1371/journal.pgen.1004342

42. McVean GAT, Cardin NJ (2005) Approximating the coalescent with recombination. Philos Trans R Soc Lond B Biol Sci 360 (1459):1387–1393. https://doi.org/10.1098/rstb.2005.1673

43. Marjoram P, Wall JD (2006) Fast "coalescent" simulation. BMC Genet 7:16. https://doi.org/10.1186/1471-2156-7-16

44. Slatkin M (2008) Linkage disequilibrium—understanding the evolutionary past and mapping the medical future. Nat Rev Genet 9 (6):477–485. https://doi.org/10.1038/nrg2361

45. Cutter AD, Payseur BA (2013) Genomic signatures of selection at linked sites: unifying the disparity among species. Nat Rev Genet 14

(4):262–274. https://doi.org/10.1038/nrg3425

46. Charlesworth B, Morgan MT, Charlesworth D (1993) The effect of deleterious mutations on neutral molecular variation. Genetics 134 (4):1289–1303

47. Maynard Smith J, Haigh J (1974) The hitchhiking effect of a favourable gene. Genet Res 23(1):23–35

48. Hill WG, Robertson A (1966) The effect of linkage on limits to artificial selection. Genet Res **8**(3):269–294

49. Roze D, Barton NH (2006) The Hill-Robertson effect and the evolution of recombination. Genetics 173(3):1793–1811. https://doi.org/10.1534/genetics.106.058586

Part II

Statistical Methods for Analyzing Genomes in Populations

Chapter 2

Processing and Analyzing Multiple Genomes Alignments with MafFilter

Julien Y. Dutheil

Abstract

As the number of available genome sequences from both closely related species and individuals within species increased, theoretical and methodological convergences between the fields of phylogenomics and population genomics emerged. Population genomics typically focuses on the analysis of variants, while phylogenomics heavily relies on genome alignments. However, these are playing an increasingly important role in studies at the population level. Multiple genome alignments of individuals are used when structural variation is of primary interest and when genome architecture permits to assemble *de novo* genome sequences. Here I describe MafFilter, a command-line-driven program allowing to process genome alignments in the Multiple Alignment Format (MAF). Using concrete examples based on publicly available datasets, I demonstrate how MafFilter can be used to develop efficient and reproducible pipelines with quality assurance for downstream analyses. I further show how MafFilter can be used to perform both basic and advanced population genomic analyses in order to infer the patterns of nucleotide diversity along genomes.

Key words Multiple genome alignment, Synteny, Alignment post-processing, Quality filtering, Multiple alignment format

1 Introduction: Multiple Genome Alignments

Multiple genome alignments (MGAs) record the homology relationships between related genome sequences. While conventional sequence alignments contain information about nucleotide substitutions, insertions, and deletions, MGAs encode evolutionary events occurring at a larger scale. Such events include chromosome fusion, fission, and rearrangements, which break colinearity between sequences (*aka synteny break*). Furthermore, genome

---

The original version of this chapter was revised. The correction to this chapter is available at https://doi.org/10.1007/978-1-0716-0199-0_20

Electronic supplementary material: The online version of this chapter (https://doi.org/10.1007/978-1-0716-0199-0_2) contains supplementary material, which is available to authorized users.

Julien Y. Dutheil (ed.), *Statistical Population Genomics*, Methods in Molecular Biology, vol. 2090,
https://doi.org/10.1007/978-1-0716-0199-0_2, © The Author(s) 2020, Corrected Publication 2021

sequences, as opposed to gene sequences, are generally segmented. The underlying cause of this segmentation may be biological (presence of multiple chromosomes) or technical (genome sequence could only be assembled at the contig or scaffold level).

MGAs are typically used to compare genomes in distinct species (see, for instance, the 99 vertebrate genome alignments from the UCSC Genome Browser [1]). Conversely, population genomic analyses typically focus on micro-variation events—single nucleotide polymorphisms (SNP) and short indels—and assume synteny and karyotype conservation between individual genomes. As a result, genetic variation is stored as *variant calls* with respect to a *reference* genome, often as a file in the Variant Call Format (VCF) [2] or MAP format [3]. Variant files, however, do not usually contain information about invariable positions and need to be combined with additional information for most evolutionary applications (e.g., as a list of "callable positions," that is, positions where enough information was available to detect a SNP if any).

The Multiple Alignment Format (MAF, not to be confounded with the Mutation Annotation Format) describes the homology relationships between several genomes, as flat text files (see https://genome.ucsc.edu/FAQ/FAQformat.html, last accessed 29/08/18). A MAF file is a list of several alignment blocks where the constitutive sequences are in synteny (*see* Fig. 1). While the structure of each block is identical to traditional sequence alignments (as in the Clustal or Phylip formats), where homologous positions in each sequence are on top of each other and form an *alignment column*, sequence names follow a dedicated syntax in order to record genome coordinates. Besides, several annotation lines can be included, including, for instance, sequence quality scores. Genome alignment programs producing MAF files as output include TBA [4], Mugsy [5], ROAST http://www.bx.psu.edu/~cathy/toast-roast.tmp/README.toast-roast.html (last accessed 29/08/18), Last [6], and Mauve [7].

MGAs are also used in population genomic studies, either when complete individual genomes can be obtained (e.g., [8, 9]) or when pseudo-genomes can be generated [10] (*see* **Note 1**). Because they contain information about both variable and invariable positions, MGAs can be directly used for conducting evolutionary analyses, accounting for missing data and structural variation. This, however, comes at the cost of extended computer requirements, in particular in terms of file size. Additional alignment quality checks are also typically required, as full-genome aligners do not include post-processing steps as most variant calling pipelines do.

In this chapter, we will see how to use the `MafFilter` program to conduct population genomic analyses. In the following, we assume that the data is available as a MGA in the MAF format. Conversion to variant call formats will also be discussed.

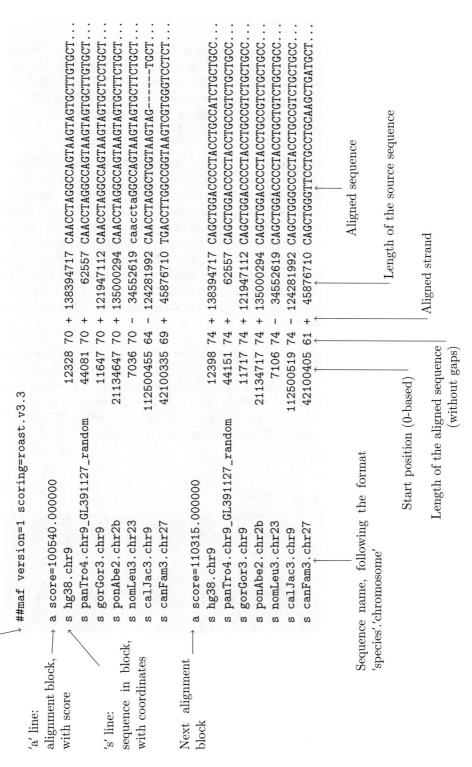

Header line:
format version and scoring scheme used

##maf version=1 scoring=roast.v3.3

'a' line:
alignment block,
with score

a score=100540.000000

's' line:
sequence in block,
with coordinates

s	hg38.chr9	12328	70	+	138394717	CAACCTAGGCCAGTAAGTAGTGCTTGTGCT...
s	panTro4.chr9_GL391127_random	44081	70	+	62557	CAACCTAGGCCAGTAAGTAGTGCTTGTGCT...
s	gorGor3.chr9	11647	70	+	121947112	CAACCTAGGCCAGTAAGTAGTGCTCCTGCT...
s	ponAbe2.chr2b	21134647	70	+	135000294	CAACCTAGGCCAGTAAGTAGTGCTTCTGCT...
s	nomLeu3.chr23	7036	70	-	34552619	caacctaGGCCAGTAAGTAGTGCTTCTGCT...
s	calJac3.chr9	112500455	64	-	124281992	CAACCTAGGCTGGTAAGTAG------TGCT...
s	canFam3.chr27	42100335	69	+	45876710	TGACCTTGGCCGGTAAGTCGTGGGTCCTCT...

Next alignment
block

a score=110315.000000

s	hg38.chr9	12398	74	+	138394717	CAGCTGGACCCCTACCTGCCATCTGCTGCC...
s	panTro4.chr9_GL391127_random	44151	74	+	62557	CAGCTGGACCCCTACCTGCCGTCTGCTGCC...
s	gorGor3.chr9	11717	74	+	121947112	CAGCTGGACCCCTACCTGCCGTCTGCTGCC...
s	ponAbe2.chr2b	21134717	74	+	135000294	CAGCTGGACCCCTACCTGCCGTCTGCTGCC...
s	nomLeu3.chr23	7106	74	-	34552619	CAGCTGGACCCCTACCTGCTGTCTGCTGCC...
s	calJac3.chr9	112500519	74	-	124281992	CAGCTGGGCCCCTACCTGCCGTCTGCTGCC...
s	canFam3.chr27	42100405	61	+	45876710	CAGCTGGGTTCCTGCCTGCCGCAAGCTGATGCT...

Aligned sequence

Length of the source sequence

Aligned strand

Start position (0-based)

Length of the aligned sequence
(without gaps)

Sequence name, following the format
'species'.'chromosome'

Fig. 1 Structure of a MAF file. Data source: UCSC alignment of human chromosome 9 together with 19 Mammals, among which 16 Primates http:/hgdownload. soe.ucsc.edu/goldenPath/hg38/multiz20way/

2 General Principles on `MafFilter` Usage

As MAF files were initially used for multi-species alignments, each input genome is referred to as a *species*. In the following, a *species* can, however, also denote a particular strain or individual in a population. Similarly, the term *chromosome* will be used in a broad sense encompassing scaffolds and contigs, in case of unmapped genome assemblies (*see* Fig. 1).

2.1 Serial Processing of Alignment Blocks: Filters

As MAF files are organized into a series of syntenic blocks, `Maf-Filter` sequentially processes input files one block at a time by applying *filters*. A filter takes a MAF alignment block as input, conducts one or several analyses, and returns a MAF block. Depending on the type of analysis performed, the output block might be identical to the input one or a modified version. In some cases, the filter can compute additional information that can be written to an output file or stored as meta-data (*see* Table 1 for examples). Filters are combined sequentially, the output of one filter serving as input to the next one, allowing to design advanced analysis workflows.

2.2 Option Files and Command Line Arguments

The `MafFilter` program can be controlled by arguments that are passed from the command line or, more conveniently, as a script file. Arguments take the form of 'parameter'='value' statements, which can potentially be nested. Arguments can also be called within the script, allowing to define global variables. Below is a minimalist example demonstrating the syntax:

Table 1
Example types of filters supported by `MafFilter`

Filter name	Filter function	Output
MafStatistic	Compute statistics on a block	Unmodified input block
MinBlockLength	Filter blocks given alignment length	Unmodified input block if its length is larger than a given threshold, otherwise the block is discarded
Subset	Keep only a subset of species	A block with sequences from the specified set of species
WindowSplit	Split a block into smaller blocks of a given size	Multiple smaller blocks
DistanceEstimation	Compute an evolutionary distance matrix from all sequences in the block	Unmodified input block with a distance matrix attached as meta-data

```
1  # maffilter param=MinimalistExample.bpp DATA=chr9
2  input.file=../Primates/$(DATA).maf.gz
3  input.file.compression=gzip
4  input.format=Maf
5  output.log=$(DATA).maffilter.log
6  maf.filter=\
7    MinBlockLength(min_length=1000),\
8    Output(file=$(DATA).min1kb.maf.gz, compression=gzip)
```

Line 1 is a comment line, which will not be parsed. Bash style comments (starting with #), C style (surrounded by /* and */) and C++ style (starting with //) are recognized. The script uses a global variable named "DATA" that is set via the command line and whose value is called using the Makefile syntax $(DATA). The script can be run using the command

```
maffilter param=MinimalistExample.bpp DATA=chr9
```

It will parse the input alignment (here human chromosome 9 aligned with 19 other Mammals, downloaded from the UCSC genome browser), keep only blocks that are at least 1 kb in length, and write the result to a new MAF file. Line 2 indicates the path to the input MAF file; line 3 specifies that the file was compressed using gzip; line 4 indicates that the file is in the MAF format. While MafFilter is dedicated to the analysis of MAF files, it can also take as input a Fasta file for a single species, with one sequence per chromosome. Line 5 indicates the path to a log file, where information about the analysis will be written. Line 6 shows the main argument, maf.filter, which contains a comma-separated list of options, one per filter. Filters will be applied in their order of specification, so that the output of filter 1 will be the input of filter 2, etc. As the line can be rather long, it is split using the "\" character. In this most simple example, there are two filters specified: **MinBlockLength**, which discards blocks below 1 kb, and **Output**, which writes the resulting alignment to a new gzip-compressed MAF file.

In the following, we will see more advanced examples of filters and how they can be combined to conduct genomic analyses.

3 MafFilter as a Data Processor

3.1 Extracting Data of Interest

A MAF alignment contains information about all genomic regions in a set of species, and some analyses can focus on a subset of such species. Besides, certain types of analyses involve only a subset of positions, such as protein-coding sites. MafFilter allows to process a MAF alignment and restrict it both to a subset of species and positions. In the first case, selected species are specified as an argument of a filter, while in the second, a file describing which positions to keep is provided, as a feature file (such as a BED or GFF-like file, see https://genome.ucsc.edu/FAQ/FAQformat. html, last accessed 29/08/18).

The following example illustrates these aspects. The pipeline filters block to keep only the ones with sequences in Human, Chimpanzee, Bonobo, Gorilla, and Orangutan. Additional sequences for other species, if any, are discarded. In a second step, coding regions are extracted and written as a separate alignment file.

```
1  # maffilter param=ExtractingData.bpp DATA=chr9
2  # Note: need to create subdirectory Alignments
3  # before running this script
4
5  input.file=../Primates/$(DATA).maf.gz
6  input.file.compression=gzip
7  input.format=Maf
8  output.log=$(DATA).maffilter.log
9  SPECIES=(hg38, panPan1, panTro4, gorGor3, ponAbe2)
10 maf.filter=                                        \
11   Subset(                                          \
12     species=$(SPECIES),                            \
13     strict=yes, keep=no,                           \
14     remove_duplicates=yes),                        \
15   Merge(                                           \
16     species=$(SPECIES),                            \
17     dist_max=0),                                   \
18   ExtractFeature(                                  \
19     ref_species=hg38,                              \
20     feature.file=../Primates/chr9.CDS1kb.gtf,      \
21     feature.format=GTF,                            \
22     complete=yes,                                  \
23     ignore_strand=no),                             \
24   OutputAlignments(                                \
25     format=Clustal,                                \
26     file=Alignments/FivePrimates%i-%c-%b-%e.aln,   \
27     reference=hg38)                                \
```

The **Subset** filter (line 11) extracts blocks where certain species are aligned (given as a list, here provided as a global variable set line 9). The strict and keep arguments can be combined to obtain various behaviors: with strict set to "yes" and keep set to "no", we only keep blocks where the five selected species are all present and discard sequences from putative additional species. The remove_ duplicates argument further removes blocks where any of the selected species might be present more than once (paralogous sequences). The **Merge** filter (line 15) subsequently fuses consecutive blocks in complete synteny, which might have been split apart because of a synteny break in one of the non-selected species.

The position extraction is done by the **ExtractFeature** filter (line 18), which retains regions specified in a file in the Gene Transfer Format (GTF). The GTF file contains only Coding DNA Sequences (CDS) with at least 1 kb in length. We further specify to only extract regions that are fully covered in the alignment (`complete` argument, line 23). The `ignore_strand` argument, line 24, tells whether regions on the negative strand should be reverse-complemented ("no" option) or kept as is ("yes" option).

Finally, the writing of the extracted blocks is done by the **OutputAlignments** filter (line 24). Each block is written in the Clustal alignment format [11] into a file with path `Alignments/FivePrimates% i-% c-% b-% e.aln`, where %i will be replaced by the index of the block. As a result, each block will be written in a separate file. If the special %i code is omitted, all alignments will be appended to a single output file. The additional special codes %c, %b, and %e can be optionally used in combination with %i and correspond to the coordinates of the block (chromosome, begin and end, respectively) according to one "reference" species specified by the `reference` argument. Further note that `MafFilter` cannot create directories, only files. In case the provided output path is not valid, no output will be generated.

3.2 Statistics with MafFilter

The effect of each data extraction step can be visualized using statistics filters. The **SequenceStatistics** filter is a powerful and generic way of computing and reporting measures for each block. It takes as input a list of statistics names and generates a table file with computed statistics as columns, and each block as a row. The table also contains the coordinates of the block according to one *reference* species.

The following pipeline is a modification of the one presented in Subheading 3.1. After each step, a **SequenceStatistics** filter is added to report the length (number of alignment columns) and size (number of sequences) of each block. This creates four files, summarized in Fig. 2.

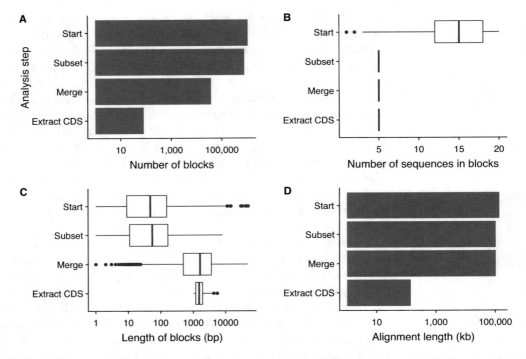

Fig. 2 Effect of data extraction filters, as measured with statistics filters. Four steps are plotted: before filtering ("Start"), after subsetting to five primate species ("Subset"), after merging synteny blocks ("Merge") and after extracting CDS regions ("Extract CDS"). (**A**) Number of blocks after each step. (**B**) Distribution of block sizes, that is, the number of species represented in each block. (**C**) Distribution of block lengths, that is, number of alignment columns in each block. (**D**) Total alignment length, that is, the sum of all block lengths

```
1   # maffilter param=Statistics.bpp DATA=chr9
2
3   input.file=../Primates/$(DATA).maf.gz
4   input.file.compression=gzip
5   input.format=Maf
6   output.log=$(DATA).maffilter.log
7   SPECIES=(hg38, panPan1, panTro4, gorGor3, ponAbe2)
8   maf.filter=                                          \
9     SequenceStatistics(                                \
10      statistics=(BlockLength,BlockSize),              \
11      ref_species=hg38,                                \
12      file=$(DATA).statistics1.txt),                   \
13    Subset(                                            \
14      species=$(SPECIES),                              \
15      strict=yes, keep=no,                             \
16      remove_duplicates=yes),                          \
17    SequenceStatistics(                                \
18      statistics=(BlockLength,BlockSize),              \
19      ref_species=hg38,                                \
20      file=$(DATA).statistics2.txt),                   \
```

```
21 │ Merge(                                                  \
22 │    species=$(SPECIES),                                  \
23 │    dist_max=0),                                         \
24 │ SequenceStatistics(                                     \
25 │    statistics=(BlockLength,BlockSize),                  \
26 │    ref_species=hg38,                                    \
27 │    file=$(DATA).statistics3.txt),                       \
28 │ ExtractFeature(                                         \
29 │    ref_species=hg38,                                    \
30 │    feature.file=../Primates/chr9.CDS1kb.gtf,            \
31 │    feature.format=GTF,                                  \
32 │    complete=yes,                                        \
33 │    ignore_strand=no),                                   \
34 │ SequenceStatistics(                                     \
35 │    statistics=(BlockLength,BlockSize),                  \
36 │    ref_species=hg38,                                    \
37 │    file=$(DATA).statistics4.txt)                        \
```

After filtering, 81 alignment blocks are created. This is less than the 146 entries in the GTF file, the difference being due to CDS that are (at least partially) missing or not in synteny in any of the five selected species. When only the human and chimpanzee genomes are considered, for instance, the number of complete CDS present in the alignment becomes 118.

3.3 Pre-Processing the Data for Quality Insurance

Comparative evolutionary analyses of sequences require high-quality input data, as any error at this stage is likely to propagate in the downstream analyses. Such errors may occur both at the individual sequence level (sequencing and assembly errors) and at the alignment level (wrong orthology inference, alignment errors). In some cases, we also want to discard regions (e.g., protein-coding positions) that are likely to violate the prior assumptions of a given analysis (e.g., neutral evolution).

The MAF format allows storing position-specific scores. Using **QualFilter**, it is possible to remove regions with a low score in a given set of species. The filter further allows computing the average score in a sliding window with user-specified size. Windows with an average score below a given threshold are discarded, and the corresponding block split accordingly. Similarly, **MaskFilter** can be used to clean blocks according to the proportion of masked positions in a given set of sequences. Masked regions are coded as lowercase nucleotides and are typically used to annotate low-complexity regions.

The local quality of the alignment can be assessed via the distribution of gaps in sliding windows. **AlnFilter** and **AlnFilter2** both slide windows along the alignment and discard regions with too many gaps. They differ by their scoring criteria: **AlnFilter** computes the global frequency of gap characters, while **AlnFilter2** estimates the number of indel events, independently of the length of the insertion or deletion track. **EntropyFilter** can also be used to remove highly variable regions in the alignment.

Finally, **FeatureFilter** can be used to exclude regions from the alignment. Features to exclude can be specified as an annotation file, in GFF, GTF, or BedGraph format. When GFF or GTF annotation files are provided, it is further possible to exclude only a given subset of features.

Most filters allow writing the filtered regions in a separate file optionally. This feature enables to finely tune the filtering criteria by visually assessing which regions are kept or removed. Using the **SequenceStatistics** filter is also convenient to monitor the proportion of alignment discarded. In the following sections, concrete example analyses will demonstrate the use of these filters. Before getting there, however, we will introduce a last set of filters enabling inter-operability between analysis tools: format conversion filters.

3.4 Conversion to Other Formats

When MGAs store genomes from individuals of the same species, they can be exported as variants. This requires that a reference genome is specified, usually implying that any synteny break will be further ignored, together with parts of the alignment that do not include the chosen reference species. When exporting to variant formats, it is generally recommended to first project the alignment on the reference species, so that the variants are sorted (see, for instance, program `maf_project` in Subheading 5). MafFilter can export in three distinct variant formats: the widely used VCF [2] (**VcfOutput** filter), Plink ped and map files [3] (**PlinkOutput** filter), and MSMC [12] (*see* Chapter 7, **MsmcOutput** filter).

Synteny block can also be exported into standard alignment format with the **OutputAlignments** filter, as seen in Subheading 3.1. The **OutputAlignments** filter further accepts a `ldhat_header` argument allowing to export alignments readable by the `convert` program from the `LDhat` package [13].

Meta-data associated with alignment blocks can be exported using dedicated filters. The **OutputDistanceMatrices** filter exports all matrices into a file in the Phylip format. Similarly, the **OutputTrees** filter exports trees in Newick format. Both require the specification of a tag name used to attach the meta-data to each block (e.g., `MLdistance` or `BioNJ`).

4 Examples of Advanced Analyses

4.1 Example Analysis 1: Computing Nucleotide Diversity Along the Genome

This section describes the first complete analysis example. We use the publicly available Drosophila Population Genomics Project phase 3 (DPGP3, *see* Chapter 13) [10], containing 197 genomes from a single African ancestral population. We restrict our analysis to one chromosome arm (2L) and ten individuals. The corresponding dataset has been combined into a single MAF file (see online Supplementary Information). The following script first uses **AlnFilter** to process the data in 10 bp windows slid by one bp in order to remove regions with too many gaps, which discards 10% of the alignment (*see* Fig. 3A). This leads to many more blocks (*see* Fig. 3B), of shorter length (*see* Fig. 3C). The resulting split blocks are then merged if not further apart than a 100 bp, using the **Merge** filter. When merged, missing regions are filled by unresolved characters ("N"). The resulting blocks are split into non-overlapping windows of 10 kb, and smaller blocks are discarded (**MinBlock-Length** and **WindowSplit** filters). As a result, 32% of the original alignment is lost (*see* Fig. 3A). Two statistics are used to compute population genetics quantities: **SiteFrequencySpectrum**, which counts minor allele frequencies (*see* Chapter 1 in this volume) and **DiversityStatistics**, which computes various diversity estimators (*see* Table 2).

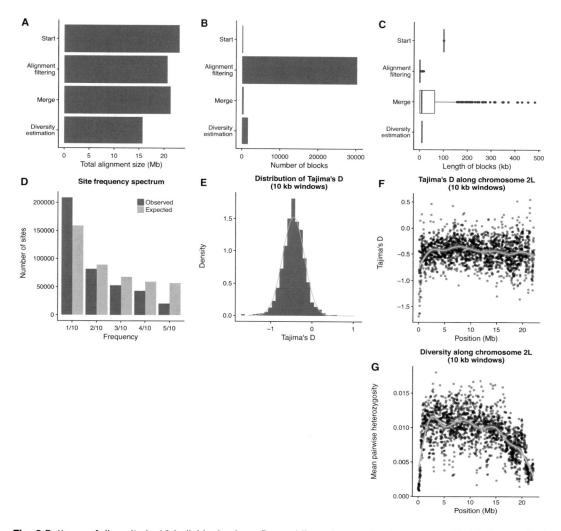

Fig. 3 Patterns of diversity in 10 individuals along *Drosophila melanogaster* chromosome 2L. First row: effect of data filtering. (**A**) Total alignment length. (**B**) Number of blocks. Second and third rows: diversity statistics. (**C**) Distribution of block lengths. (**D**) Site frequency spectrum, observed and expected under a Wright-Fisher model with identical global diversity. (**E**) Distribution of Tajima's D in 10 kb windows (histogram) and normal distribution fit (curve). (**F**) Tajima's D in 10 kb windows plotted along the chromosome. The smooth line was computed using a generalized additive model (GAM). (**G**) Average pairwise heterozygosity in 10 kb windows plotted along the chromosome. The smooth line was computed using a generalized additive model (GAM)

Table 2
Available statistics for the `SequenceStatistics` filter

Statistic name	Statistic function	Output
`BlockSize`	Report the number of sequences	Numerical value
`BlockLength`	Report the number of alignment columns	Numerical value
`SequenceLength`	Report the number of nucleotides for a given sequence	Numerical value
`AlnScore`	Report the alignment score for the block, as encoded in the input MAF file	Numerical value
`BlockCounts`	Report the frequencies of each nucleotide	Numerical values, one for each character state
`SiteStatistics`	Compute the number of sites with missing data/gaps, numbers of mono, di, tri, and quadri-allelic sites, number of parsimony-informative sites	Numerical values, one for each statistic
`PairwiseDivergence`	Compute the percentage of mismatches between two user-specified species	Numerical values
`SiteFrequencySpectrum`	Count sites based on their minor allele frequency, according to user-specified bins	Numerical values, one per bin
`PolymorphismStatistics`	Compare two sets of sequences and compute the number of fixed and polymorphic positions in both sets.	Numerical values, for all combination of fixed and polymorphic sites (e.g., fixed in one set and polymorphic in the other)
`DiversityStatistics`	Compute the number of segregating sites, Watterson's theta, Tajima's pi, and Tajima's D.	Four numerical values
`ModelFit`	Given a phylogenetic tree, fit a nucleotide substitution model using maximum likelihood and report the parameter estimates. A large variety of models from Jukes-Cantor to General Time Reversible are available, including rate across sites variation models.	Numerical values, one per estimated parameter

The `MafFilter` script starts by defining a few variables: dataset (line 3), list of individual sequences used (lines 4–5), reference sequence used for the output of coordinates (line 6) and size of the window for which estimators are computed (here 10 kb, line 7).

The script generates a text file with all computed statistic per 10 kb windows, together with their coordinates in the reference genome. Besides, simpler statistics files are generated at each step of the analysis to summarize the data used. The actual alignment is also output as a new MAF file for further assessment.

```
1   # maffilter param=Maffilter-Diversity.bpp
2
3   DATA=dpgp3_Chr2L_10indv
4   INDV=ZI152,ZI173,ZI190,ZI199,ZI211,ZI219,\
5         ZI253,ZI344,ZI374,ZI490
6   REF=ZI152
7   WSIZE=10000
8
9   input.file=../Drosophila/$(DATA).split.maf.gz
10  input.file.compression=gzip
11  input.format=Maf
12  output.log=$(DATA).maffilter-diversity.log
13  maf.filter=                                         \
14    SequenceStatistics(                               \
15      statistics=(BlockLength,BlockSize),             \
16      ref_species=$(REF),                             \
17      file=$(DATA).statistics1.txt),                  \
18    AlnFilter(                                        \
19      species=($(INDV)),                              \
20      window.size=10,                                 \
21      window.step=1,                                  \
22      missing_as_gap=yes,                             \
23      max.gap=0.3,                                    \
24      max.ent=-1,                                     \
25      relative=yes,                                   \
26      file=$(DATA).trash_aln.maf,                     \
27      compression=none,                              \
28      verbose=yes),                                   \
29    SequenceStatistics(                               \
30      statistics=(BlockLength,BlockSize),             \
31      ref_species=$(REF),                             \
32      file=$(DATA).statistics2.txt),                  \
33    Merge(species=($(INDV)), dist_max=100),          \
34    SequenceStatistics(                               \
35      statistics=(BlockLength,BlockSize),             \
36      ref_species=$(REF),                             \
37      file=$(DATA).statistics3.txt),                  \
38    Output(                                           \
39      file=$(DATA).filtered.maf.gz,                   \
40      compression=gzip),                             \
41    MinBlockLength(min_length=$(WSIZE)),              \
42    WindowSplit(                                       \
43      preferred_size=$(WSIZE),                        \
44      align=center,                                   \
45      keep_small_blocks=no),                          \
46    SequenceStatistics(                               \
47      statistics=(BlockLength,BlockSize,              \
```

```
48          DiversityStatistics(,                              \
49            ingroup=($(INDV))),                              \
50          SiteFrequencySpectrum(                             \
51            ingroup=($(INDV)),                               \
52            bounds=(-0.5,0.5,1.5,2.5,3.5,4.5,5.5))           \
53          ),                                                 \
54        ref_species=$(REF),                                 \
55        file=$(DATA).diversity_statistics.txt)              \
```

This example demonstrates the use of **AlnFilter**: lines 20–21 specify the size of the window and the amount by which it is slid (10 nucleotides slid by 1). Line 22 further tells the filter that missing nucleotides ("N") should be counted as gaps. The maximal proportion of gaps allowed in the window is set to 0.3 (line 23). Absolute numbers of gaps can also be specified by changing line 25 to "no." In this example, we do not filter according to the site variability, and the maximal entropy is set to − 1 (line 24). Alternatively, windows will be discarded if they both display a number of gaps and entropy higher than the specified thresholds. Discarded regions are output to a separate MAF file (lines 26–27), for further assessment. Finding the optimal alignment filtering criteria requires to compare both the retained and rejected regions. Multiple **Aln-Filter** can be combined in order to achieve the desired quality.

Diversity estimators are computed as standard statistics (*see* Subheading 3.2). **DiversityStatistics** takes only one input arguments, the list of individuals to use (line 49, in this case, all of them). **SiteFrequencySpectrum** requires, in addition, specifying boundaries for the frequencies to compute (line 52). As we have ten genomes, the possible SNPs minor frequencies are 0, 1, 2, 3, 4, and 5 out of 10. We therefore specify as boundaries − 0.5, 0.5, 1.5, 2.5, 3.5, 4.5, and 5.5. Note that it is possible to specify fewer boundaries to pull two or more categories. Each category generates one column in the output statistic file. Besides, positions with unresolved characters or more than two alleles are counted separately and excluded from the site frequency spectrum calculation.

The computed site frequency spectrum reveals an excess of low-frequency variants (*see* Fig. 3d), resulting in a globally negative Tajima's D value (*see* Fig. 3e). The effect is relatively constant along the chromosome, except for the most telomeric region (*see* Fig. 3f), suggesting that this population underwent a demographic expansion. Patterns of heterozygosity, on the other hand, show a substantial reduction at the telomere, and a positive correlation with the distance to the centromere, at the right end of the alignment (Kendall's tau = 0.28, p-value $< 2.2 \cdot 10^{-16}$, *see* Fig. 3g).

**4.2   Example
Analysis 2: Inferring
Phylogenetic
Relationships**

In this example, we infer the phylogenetic relationships of five great apes. We use the UCSC 20-way genome alignment, containing 16 Primates genomes. For the sake of computational efficiency, we restrict the analysis to chromosome 9 only. We implement the following pipeline:

1. extract the genome alignment for human, chimpanzee, bonobo, gorilla, and orangutan,

2. filter the alignment to remove ambiguously aligned regions,

3. split the resulting filtered alignment into non-overlapping 10 kb windows,

4. compute a pairwise distance matrix using maximum likelihood and estimate a BioNJ tree for each window,

5. root each tree using the orangutan sequence as an outgroup,

6. write the resulting trees to a file,

7. fit a model of sequence evolution on the human, bonobo, chimpanzee, and gorilla ingroup using maximum likelihood and output parameters to a file.

This results in the following `MafFilter` option file:

```
1   # maffilter param=MafFilter-Phylogeny.bpp
2
3   DATA=chr9
4   SPECIES=(hg38, panPan1, panTro4, gorGor3, ponAbe2)
5   WSIZE=10000
6
7   input.file=../Primates/$(DATA).maf.gz
8   input.file.compression=gzip
9   input.format=Maf
10  output.log=$(DATA).maffilter.log
11  maf.filter=                                              \
12    Subset(                                                \
13      species=$(SPECIES),                                  \
14      strict=yes, keep=no,                                 \
15      remove_duplicates=yes),                              \
16    XFullGap(species=$(SPECIES), verbose=no),              \
17    MinBlockLength(min_length=10),                         \
18    AlnFilter2(verbose=no,                                 \
19      species=$(SPECIES),                                  \
20      window.size=10, window.step=1,                       \
21      missing_as_gap=yes,                                  \
22      max.gap=2, max.pos=2, relative=no,                   \
23      file=None,                                           \
24      compression=none),                                   \
25    Merge(                                                 \
26      species=$(SPECIES),                                  \
27      dist_max=100,                                        \
28      rename_chimeric_chromosomes=yes),                    \
```

```
29    Output(                                                    \
30      file=$(DATA).filtered.maf.gz,                            \
31      compression=gzip),                                       \
32    MinBlockLength(min_length=$(WSIZE)),                        \
33    WindowSplit(                                                \
34      preferred_size=$(WSIZE),                                 \
35      align=center,                                            \
36      keep_small_blocks=no),                                   \
37    DistanceEstimation(verbose=no,                             \
38      method=ml,                                               \
39      model=K80(kappa=2),                                      \
40      rate=Gamma(n=4, alpha=0.5),                              \
41      parameter_estimation=initial,                            \
42      max_freq_gaps=0.33,                                      \
43      gaps_as_unresolved=yes,                                  \
44      profiler=none,                                           \
45      message_handler=none,                                    \
46      extended_names=yes),                                     \
47    DistanceBasedPhylogeny(verbose=no,                         \
48      method=bionj, dist_mat=MLDistance),                      \
49    NewOutgroup(                                                \
50      tree_input=BioNJ,                                        \
51      tree_output=BioNJ_rooted,                                \
52      outgroup=ponAbe2),                                       \
53    OutputTrees(                                                \
54      tree=BioNJ_rooted,                                       \
55      file=$(DATA).trees.dnd,                                  \
56      compression=none,                                        \
57      strip_names=yes),                                        \
58    DropSpecies(                                                \
59      tree_input=BioNJ_rooted,                                 \
60      tree_output=BioNJ_subtree,                               \
61      species=ponabe2),                                        \
62    SequenceStatistics(                                         \
63      statistics=(                                             \
64        BlockCounts(suffix=.all),                              \
65        BlockCounts(species=hg38   , suffix=.hs),              \
66        BlockCounts(species=panPan1, suffix=.pp),              \
67        BlockCounts(species=panTro4, suffix=.pt),              \
68        BlockCounts(species=gorGor3, suffix=.gg),              \
69        BlockCounts(species=ponAbe2, suffix=.pa),              \
70        ModelFit(                                              \
71          model=HKY85(kappa=1, theta=0.5,                      \
72                      theta1=0.5, theta2=0.5),                 \
73          rate_distribution=Gamma(n=4, alpha=0.5),             \
74          root_freq=Full,                                      \
75          tree=BioNJ_subtree,                                  \
76          parameters_output=(HKY85.theta_1,                    \
77              HKY85.theta1_1, HKY85.theta2_1,                  \
78              HKY85.kappa_1, Gamma.alpha,                      \
79              Full.theta1, Full.theta2, Full.theta),           \
80          fixed_parameters=(),                                 \
81          reestimate_brlen=no,                                 \
82          max_freq_gaps=0.3,                                   \
83          gaps_as_unresolved=yes)),                            \
84        ref_species=hg38,                                      \
85        file=$(DATA).model-statistics.csv),                    \
```

This rather large option file starts with the selection of the species of interest, which we store as a list in the SPECIES variable (line 4). The **Subset** filter (lines 12–15) extracts the corresponding species for each block, excluding blocks where not all five species are present (strict=yes), and removing any additional species that might be present (keep=no). Besides, we discard any block where a species might be present more than once because of paralogy (remove_duplicates=yes). As a result, after this step, all blocks contain exactly five sequences, one for each species.

We then proceed with alignment filtering (starting line 16). We first remove all alignment columns containing a gap in all kept sequences, due to putative indels with more distant species, which have now been discarded. This is achieved via the **xFullGap** filter (line 16). We then slide a 10 bp window in order to exclude regions with a least two indel events, independent of their size. Only indel events involving at least two species are counted (**AlnFilter2**, with arguments max.pos=2 and max.gap=2). The number of gaps is specified as a number of occurrences (relative=no); it can also be specified as a proportion of the number of sequences. As we are sliding 10 bp windows, we first discard alignment blocks with less than ten columns (**MinBlockLength** filter, line 17). The resulting alignment is spread into numerous, potentially small blocks. In order not to discard too much data in subsequent steps of the analysis, we perform a merging step (lines 25–28). With the specified configuration, consecutive blocks will only be merged if all input species are syntenic, that is, the sequences in the two blocks are colinear (same chromosome, same strand, same distance between the start of the new block and end of the previous one). By specifying a subset of species only, it is possible to merge according to some focus species, resulting in coordinates being lost for other species. We further consider a maximum distance of 100 bp in order to merge consecutive blocks (line 27). When two blocks are merged, so are the sequence names, which may result in excessively long names. Using the rename_chimeric_chromosomes argument, we tell the program to arbitrarily give new names to merged sequences, which will be called chimtigXX, XX being a unique number. When merged, missing positions will be replaced by "N" characters, allowing to preserve coordinates. In effect, the combination of the **AlnFilter2** and **Merge** filters result in a masking of the discarded positions.

We analyze the resulting filtered dataset in windows of 10 kb. The focus window size is specified as a global variable (line 5) and can be changed in order to assess the impact of the window size on the results. The **WindowSplit** filter breaks each block into non-overlapping blocks of a given size (lines 33–36). Input arguments allow specifying how to cut a block when its size exceeds the specified window size: either start from the left, center on the block while discarding start and end regions or adjust the size in order not to lose any data. Note that in the latter case, the input window size w will be the minimum size. The resulting window size can, therefore, be comprised between w and $2 \times w - 1$. When the `keep_small_blocks` option (line 36) is set to yes, and the window size is not adjusted, out-of-window alignment parts and block smaller than the specified window size will be kept as separate blocks. Otherwise, they will be discarded.

Phylogenetic reconstruction in each window is performed using a distance method (BioNJ, [14]), which requires first to estimate a pairwise distance matrix. We use a maximum likelihood method, with a K80 substitution model (line 39) and a discrete gamma distribution of rates across sites (line 40). For computational efficiency purpose, we only estimate distances and keep other parameters fixed to realistic values (transition/transversion ratio equal to 2 and gamma shape parameter equal to 0.5). We further consider gaps as unknown characters in the modeling and discard positions with more than one-third of unresolved characters (lines 42 and 43). For each windowed block, the resulting matrix is stored as meta-data, with label `MLDistance`. This distance matrix is then given as input to the **DistanceBasedPhylogeny** filter, which reconstructs a tree using the BioNj method (line 50) and stores it under the label `BioNJ`. Further processing includes rooting each tree using the Orangutan sequence (**NewOutgroup** filter, lines 49–52) and removal of the outgroup branch (**DropSpecies** filter, lines 58–61). Rooted trees are saved into a text file using the **Output-Trees** filter for further analysis.

The final step of the analysis consists in the estimation of substitution parameters for each window. This is done via a **SequenceStatistics** filter, and two dedicated statistics: **Block-Counts** and **ModelFit**. The **BlockCounts** statistics is rather straightforward, as it computes nucleotide frequencies in a given set of species. We use a combination of six calls to this statistic to compute averaged (line 64) as well as species-specific frequencies (lines 65–69). Input arguments include the set of species to use in the calculation, as well as suffix strings to distinguish the different output results. The **ModelFit** statistic is more complex and requires to specify a substitution model, similar to the **DistanceEstimation** filter. As the model is being fitted to the full tree using Felsenstein's dynamic algorithm [15], a more parameter-rich model can be employed (HKY85, [16]). In particular, we use a non-stationary model, allowing us to estimate the observed and equilibrium frequencies separately. Under such a model, the ancestral nucleotide frequencies are different from the equilibrium ones and are fully parameterized (line 74). In order to reduce computational time, we do not reestimate branch lengths and keep them to the values resulting from the BioNJ algorithm (line 81). Enabling branch length reestimation does not change the results significantly (see companion material). Further parameters can be fixed to their initial or default value using the `fixed_parameters` argument (line 80). Finally, the `output_parameters` argument allows specifying which estimated parameters should be output to the result file. As the nomenclature of parameter names can be complicated, `MafFilter` outputs the list of available parameters when run. A two-step run might, therefore, be needed in order to fit the desired model.

The results of this analysis are summarized in two files: a spreadsheet file containing numerical values, one statistic per column and one 10 kb window per line (file `chr9.model-statistics.csv`), and one text file containing a list of trees, one line per window (file `chr9.trees.dnd`). R scripts are provided as companion material in order to analyze these output files. The analysis led to 883 trees. A majority rule consensus tree leads to a topology compatible with the well-established phylogeny of the species (*see* Fig. 4A) [17]. This topology is supported by a majority of windows (Fig. 4B), but four other "minor" topologies are also inferred: topology C and D are supported by 55 and 54 windows, and group human with gorilla and chimpanzee + bonobos with gorilla, respectively. Such topologies result from incomplete lineage sorting in the humans-chimpanzees-bonobos ancestral populations

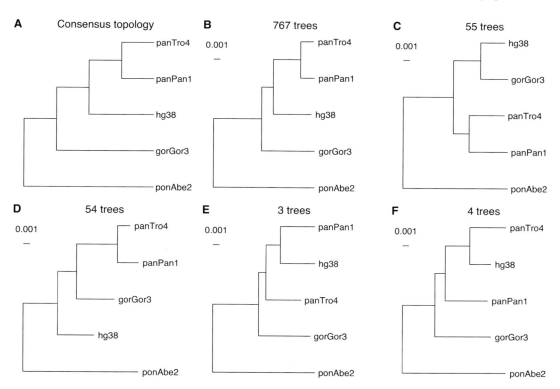

Fig. 4 Incomplete lineage sorting along chromosome 9 of the great apes. A) Consensus topology. B-F) Topologies with mean branch lengths sorted by frequency of occurrence

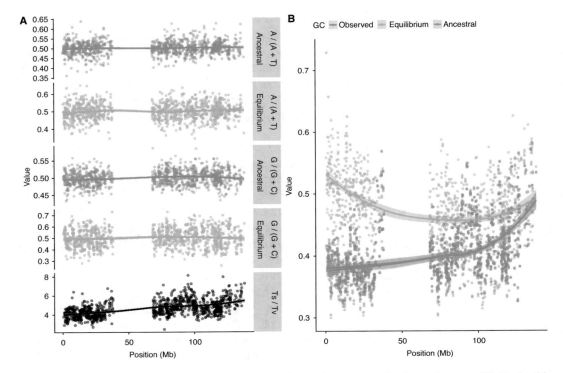

Fig. 5 Characteristics of the substitution process along chromosome 9 of great apes. (**A**) Nucleotide frequencies and substitution rates. Ts: transitions. Tv: transversions. (**B**) GC content. The smooth lines were computed using a generalized additive model (GAM)

[18]. The two last topologies, E and F, are supported by three and four windows and group humans with bonobos and humans with chimpanzees, respectively, revealing incomplete lineage sorting in the common ancestor of bonobos and chimpanzees [19].

Having inferred the underlying genealogy for each window, we could fit a model of sequence evolution and estimate parameters related to the underlying substitution process. We find that the proportions of A vs. T and G vs. C nucleotide are constant over the chromosome and equals $(A/(A + T) \simeq G/(G + C) \simeq 0.5)$. The ratio of transitions over transversions increases along the chromosome, from ~4 on the left end to ~5 on the right end (*see* Fig. 5A). Observed GC content is highly conserved between all species (see companion material), and increases at the right end of the chromosome (ancestral GC, *see* Fig. 5B). Equilibrium GC content, on the other hand, is higher in the two telomeric regions, mirroring the recombination rate. Such relationships between recombination and equilibrium GC content are expected when GC-biased gene conversion is occurring [20]. In the online companion material, an extended version of this script is provided, which removes protein-coding regions in addition to filtering the alignment. This increases the total execution time but does not significantly affect the results.

4.3 Example Analysis 3: Running External Software

MafFilter can integrate external tools within its analysis pipeline. Programs can be run on each alignment block, and their result parsed and further processed. Two types are currently supported, based on the nature of the output.

The **SystemCall** filter exports each block as a standard alignment file, runs a program generating a new alignment, and subsequently replaces the original alignment block with the new alignment. This procedure allows improving the genome alignment by running standard gene alignment programs on synteny block. The following script demonstrates this ability using the MAFFT aligner [21] on the Primates chromosome 9 alignment:

```
1   DATA=chr9
2   SPECIES=(hg38, panPan1, panTro4, gorGor3, ponAbe2)
3
4   input.file=../Primates/$(DATA).maf.gz
5   input.file.compression=gzip
6   input.format=Maf
7   output.log=$(DATA).maffilter-realign.log
8   maf.filter=                                               \
9     Subset(                                                 \
10       species=$(SPECIES),                                  \
11       strict=yes, keep=no,                                 \
12       remove_duplicates=yes),                              \
13     XFullGap(species=$(SPECIES), verbose=no),              \
14     SequenceStatistics(                                    \
15       statistics=(BlockSize, BlockLength),                 \
16       ref_species=hg38,                                    \
17       file=$(DATA)_subset.statistics.csv),                 \
18     WindowSplit(                                           \
19       preferred_size=10000,                                \
20       align=adjust,                                        \
21       keep_small_blocks=yes),                              \
22     SystemCall(                                            \
23       name=MAFFT,                                          \
24       input.file=blockIn.fasta,                            \
25       input.format=Fasta,                                  \
26       output.file=blockOut.fasta,                          \
27       output.format=Fasta,                                 \
28       call=./runMafft.sh),                                 \
29     Merge(                                                 \
30       species=$(SPECIES),                                  \
31       dist_max=0,                                          \
32       rename_chimeric_chromosomes=yes),                    \
33     SequenceStatistics(                                    \
34       statistics=(BlockSize, BlockLength),                 \
35       ref_species=hg38,                                    \
36       file=$(DATA)_windows_realigned.statistics.csv),\
37     Output(                                                \
38       file=$(DATA)_realigned.maf.gz,                       \
39       compression=gzip,                                    \
40       verbose=1)                                           \
```

As in Subheading 4.2, the pipeline starts by extracting sequences for five species and removing full gap positions (lines 9–13). The **SystemCall** filter runs a wrapper script named run-Mafft.sh, located in the current directory (lines 22–28). The script reads a file named blockIn.fasta and writes the realigned sequences into a new file blockOut.fasta, which will be parsed by MafFilter. It further checks that the input block has at least two sequences:

```
1  #! /bin/bash
2  if [ `grep '>' blockIn.fasta | wc -l` == "1" ];
3  then
4    #Only one sequence in block, we pass...
5    cp blockIn.fasta blockOut.fasta
6  else
7    mafft --fft --nomemsave --maxiterate 2 --thread -1 \
8        blockIn.fasta > blockOut.fasta 2> mafft.log
9  fi
```

For computational efficiency, we ensure that input alignments are no longer than 10,000 sites and split long blocks using the **Window-Split** filter (lines 18–21). The keep_small_blocks option is set to yes, so that smaller blocks are kept unsplit and not discarded. Realigned blocks are subsequently re-assembled using the **Merge** filter (lines 29–32). However, note that this script will typically take ca. 1 day to complete on a standard desktop computer. The final alignment is exported to a new Maf file (lines 37–40), and statistics are computed before and after realignment. The total alignment length (number of aligned positions) slightly shrinks from 89.245 to 89.223 Mb after realigning with MAFFT.

The **ExternalTreeBuilding** filter enables running an external phylogeny reconstruction software on each alignment block and import the resulting tree. As done in Subheading 4.2, we filter the alignment of chromosome 9 and reconstruct the phylogenetic tree in 10 kb windows, this time using the PhyML program [22]:

```
1  # maffilter param=MafFilter-Phylogeny.bpp
2
3  DATA=chr9_realigned
4  SPECIES=(hg38, panPan1, panTro4, gorGor3, ponAbe2)
5  WSIZE=10000
6
7  input.file=$(DATA).maf.gz
8  input.file.compression=gzip
9  input.format=Maf
10 output.log=$(DATA).maffilter-phylogeny.log
11 maf.filter=                                          \
12   MinBlockLength(min_length=10),                     \
13   AlnFilter2(verbose=no,                             \
```

```
14    species=$(SPECIES),                              \
15    window.size=10, window.step=1,                   \
16    missing_as_gap=yes,                              \
17    max.gap=2, max.pos=2, relative=no,               \
18    file=$(DATA).trash_aln.maf,                      \
19    compression=none),                               \
20  Merge(                                             \
21    species=$(SPECIES),                              \
22    dist_max=100,                                    \
23    rename_chimeric_chromosomes=yes),                \
24  Output(                                            \
25    file=$(DATA)_filtered.maf.gz,                    \
26    compression=gzip),                               \
27  MinBlockLength(min_length=$(WSIZE)),               \
28  WindowSplit(                                       \
29    preferred_size=$(WSIZE),                         \
30    align=center,                                    \
31    keep_small_blocks=no),                           \
32  ExternalTreeBuilding(                              \
33    input.file=blockIn.phy ,                         \
34    input.format=Phylip(                             \
35      order=sequential, type=extended),             \
36    output.file=blockIn.phy_phyml_tree,             \
37    output.format=Newick,                           \
38    property_name=PhyML,                            \
39    call=./runPhyml.sh),                            \
40  NewOutgroup(                                       \
41    tree_input=PhyML,                               \
42    tree_output=PhyML_rooted,                        \
43    outgroup=ponAbe2),                              \
44  OutputTrees(                                       \
45    tree=PhyML_rooted,                              \
46    file=$(DATA).trees.dnd,                         \
47    compression=none,                               \
48    strip_names=yes)                                \
```

The **ExternalTreeBuilding** filter exports the current block as an alignment file (lines 32–39) and the runPhyml.sh script launches phyml:

```
1  #! /bin/bash
2  phyml -i blockIn.phy -d nt -q -m HKY85 -f m -t e -c 4 \
3    -a e -s BEST -o tlr -b 0 > phyml.log 2> phyml.err
```

An HKY85 model of nucleotide substitutions is used with a four-class discrete gamma distribution of rate fitted to the data. The best tree from both nearest neighbor interchange (NNI) and subtree pruning and regrafting (SPR) algorithms for topology search is selected and further read by MafFilter. After rerooting (line 40), the trees for every block are collected and output. This pipeline produces exactly 1000 trees, compared to 883 when no realignment was performed, demonstrating that the realignment step substantially increased the quality of the alignment. Results are consistent with the BioNJ analysis, 852 blocks supporting the well-established phylogeny of the species. 85 and 58 trees cluster human and gorilla or chimpanzee and bonobo with gorilla, respectively, consistent with the occurrence of incomplete lineage sorting. Interestingly, these analyses reveal a dissymmetry in the frequency of the two ILS topologies, the one grouping human and gorilla being more frequent than the one grouping gorilla and chimpanzee. This was previously observed [23] and shown to be due to a higher rate of sequencing errors in the chimpanzee genome [18].

4.4 Example Analysis 4: Coordinates Translation from One Species to Another

Many evolutionary analyses require inter-specific comparisons. When the compared species are closely related enough, it is possible to perform a joint genome alignment in order to work with a single, common reference genome. This may not always be the preferable option, however, in particular when species are divergent and/or have undergone substantial structural variation and the patterns under study are intrinsically dependent on the genome position (e.g., linkage disequilibrium [9]). In such cases, analyses are conducted independently in each species, and coordinates are then converted into a common reference for comparison.

The LiftOver utility, available at UCSC, can be used to convert genome coordinates from one genome assembly to another, but should not be used to map genomes of distinct species. MafFilter, however, has a function allowing to perform such task, providing a genome alignment of the two species is available. Such an alignment can be obtained with software like BlastZ and LastZ [24], TBA [4], or Mummer [25]. The following example shows how to convert human gene coordinates into their gorilla homologs using MafFilter and the 20-way genome alignment from UCSC:

```
1   DATA=chr9
2   SPECIES=(hg38, gorGor3)
3
4   input.file=../Primates/$(DATA).maf.gz
5   input.file.compression=gzip
6   input.format=Maf
7   output.log=$(DATA).maffilter-liftover.log
8     maf.filter=                                          \
9       Subset(                                            \
10        species=$(SPECIES),                              \
11        strict=yes, keep=no,                             \
12        remove_duplicates=yes),                          \
13      XFullGap(species=$(SPECIES), verbose=no),          \
14      Merge(species=$(SPECIES)),                         \
15      LiftOver(                                          \
16        ref_species=hg38,                                \
17        target_species=gorGor3,                          \
18        target_closest_position=yes,                     \
19        feature.file=../Primates/chr9.CDS1kb.gtf,        \
20        feature.file.compression=none,                   \
21        feature.format=GTF,                              \
22        file=hg38_to_gorGor3.tln,                        \
23        compression=none)                                \
```

The option-file starts by reading the genome alignment and specifying the path to the log file (lines 4–7). It then selects the two species to compare, as shown in Subheading 3.1. The **LiftOver** filter specifies the path towards the feature file to translate (lines 19–21), here in GTF format (MafFilter currently supports translation form GFF, GTF, and BedGraph files, eventually compressed). Lines 16 and 17 allow setting the reference and target species, respectively. Argument target_closest_position set the behavior in case the matching position in the target genome is a gap. If set to yes, the closest non-gap position will be returned. Original and translated positions will be returned in a tabular file, specified at lines 22 and 23. Note that for the **LiftOver** filter to work correctly, the alignment should be projected on the reference genome (in this case hg38), for instance using the maf_project program from the TBA package. Besides, feature coordinates will only be translated if they are wholly contained in an alignment block, that is if the feature does not overlap with a synteny break. It is therefore essential, for optimal efficiency, that the alignment blocks reflect the synteny structure of the reference and target species only, which will be the case if the two species have been pairwise aligned. When the two species are from a multiple genome alignment, the **Subset** and **Merge** filters should be used to combine syntenic blocks.

The output file recalls the query coordinates and their translation, for the features that could be translated. It is often convenient to merge this translation file with the original query, which can be done in R:

```
1  anno <- read.table("../Primates/chr9.CDS1kb.gtf",
2                      sep = "\t")
3  tln <- read.table("hg38_to_gorGor3.tln", header = TRUE)
4  tln$begin.ref <- tln$begin.ref + 1
5  tln$begin.target <- tln$begin.target + 1
6  anno2 <- merge(anno, tln, by.x = c(1,4,5),
7                 by.y = c(1,3,4), all = TRUE)
```

The first 5 lines read the original GTF file as a table. GTF annotations are 1-based, inclusive $[a, b]$, while MafFilter uses 0-based, exclusive coordinates $[a, b[$. GTF coordinates are automatically converted when reading the file, but MafFilter outputs results in its coordinate system. We convert them back at lines 4 and 5, before merging the two tables, lines 6 and 7. Furthermore, the strand column in the translation table does not match the strand column in the input GTF file. In the feature file, this column indicates on which strand the feature is to be found, information that is not used in the translation step. The "strand" column in the translation file indicates which strand of the sequence was present in the genome alignment. Since the alignment was projected on the reference genome, the corresponding value is always positive. In the target genome, the value will be positive if the genomes are colinear, and negative in the case of a genomic inversion.

5 Other Useful Tools

MafFilter provides tools to analyze a MAF file sequentially. These tools primarily focus on processing data for statistical analyses. It has a limited formatting capacity, in particular when long-range operations are involved, such as reordering alignment blocks. The TBA [4] and Last [6] packages contain several useful tools for that purpose, which can be used in combination with MafFilter.

From the TBA package:

- The maf_order program permits to select and order sequence according to their species names.

- The maf_project program order alignment blocks according to a reference genome. Blocks where the reference genome is on the negative strand will be reversed. All blocks that do not contain the reference species will be discarded.

From the Last package:

- The maf-join program allows combining several (sorted) multiple alignments.

- The maf-sort program permits to sort alignments according to sequence names.

6 Conclusion

The MafFilter program allows to efficiently process multiple genome alignment files, by sequentially analyzing synteny blocks. It features a flexible and extensible syntax permitting the design of reproducible pipelines for the post-processing of genome data. Beyond filtering and quality assessment, MafFilter can be used to analyze patterns of diversity along genomes, within and between species.

7 Note

Note 1: pseudo-genomes

1. Pseudo-genomes are obtained by applying a set of inferred variants to the corresponding reference genome. All positions for which a variant could not be called (whether there is one or not) will, therefore, be identical to the reference genome in the resulting pseudo-genome.

References

1. Casper J, Zweig AS, Villarreal C, Tyner C, Speir ML, Rosenbloom KR, Raney BJ, Lee CM, Lee BT, Karolchik D, Hinrichs AS, Haeussler M, Guruvadoo L, Navarro Gonzalez J, Gibson D, Fiddes IT, Eisenhart C, Diekhans M, Clawson H, Barber GP, Armstrong J, Haussler D, Kuhn RM, Kent WJ (2018) The UCSC Genome Browser database: 2018 update. Nucleic Acids Res 46(D1): D762–D769. https://doi.org/10.1093/nar/gkx1020

2. Danecek P, Auton A, Abecasis G, Albers CA, Banks E, DePristo MA, Handsaker RE, Lunter G, Marth GT, Sherry ST, McVean G, Durbin R, 1000 Genomes Project Analysis Group (2011) The variant call format and VCFtools. Bioinformatics 27(15):2156–2158. https://doi.org/10.1093/bioinformatics/btr330

3. Chang CC, Chow CC, Tellier LC, Vattikuti S, Purcell SM, Lee JJ (2015) Second-generation PLINK: rising to the challenge of larger and richer datasets. Gigascience 4:7. https://doi.org/10.1186/s13742-015-0047-8

4. Blanchette M, Kent WJ, Riemer C, Elnitski L, Smit AFA, Roskin KM, Baertsch R, Rosenbloom K, Clawson H, Green ED, Haussler D, Miller W (2004) Aligning multiple genomic sequences with the threaded blockset aligner. Genome Res 14(4):708–715. https://doi.org/10.1101/gr.1933104

5. Angiuoli SV, Salzberg SL (2011) Mugsy: fast multiple alignment of closely related whole genomes. Bioinformatics 27(3):334–342. https://doi.org/10.1093/bioinformatics/btq665

6. Kiełbasa SM, Wan R, Sato K, Horton P, Frith MC (2011) Adaptive seeds tame genomic sequence comparison. Genome Res 21(3):487–493. https://doi.org/10.1101/gr.113985.110

7. Darling AE, Mau B, Perna NT (2010) progressiveMauve: multiple genome alignment with gene gain, loss and rearrangement. PLoS ONE 5(6):e11147. https://doi.org/10.1371/journal.pone.0011147

8. Stukenbrock EH, Christiansen FB, Hansen TT, Dutheil JY, Schierup MH (2012) Fusion of two divergent fungal individuals led to the recent emergence of a unique widespread pathogen species. Proc Natl Acad Sci USA 109 (27):10954–10959. https://doi.org/10.1073/pnas.1201403109

9. Stukenbrock EH, Dutheil JY (2018) Fine-scale recombination maps of fungal plant pathogens reveal dynamic recombination landscapes and intragenic hotspots. Genetics 208 (3):1209–1229. https://doi.org/10.1534/genetics.117.300502

10. Lack JB, Cardeno CM, Crepeau MW, Taylor W, Corbett-Detig RB, Stevens KA, Langley CH, Pool JE (2015) The Drosophila genome nexus: a population genomic resource of 623 Drosophila melanogaster genomes, including 197 from a single ancestral range population. Genetics 199(4):1229–1241. https://doi.org/10.1534/genetics.115.174664

11. Higgins DG, Sharp PM (1988) CLUSTAL: a package for performing multiple sequence alignment on a microcomputer. Gene 73 (1):237–244

12. Schiffels S, Durbin R (2014) Inferring human population size and separation history from multiple genome sequences. Nat Genet 46 (8):919–925. https://doi.org/10.1038/ng.3015

13. Myers S, Bottolo L, Freeman C, McVean G, Donnelly P (2005) A fine-scale map of recombination rates and hotspots across the human genome. Science 310(5746):321–324. https://doi.org/10.1126/science.1117196

14. Gascuel O (1997) BIONJ: an improved version of the NJ algorithm based on a simple model of sequence data. Mol Biol Evol 14 (7):685–695. https://doi.org/10.1093/oxfordjournals.molbev.a025808

15. Felsenstein J (1981) Evolutionary trees from DNA sequences: a maximum likelihood approach. J Mol Evol 17(6):368–376

16. Hasegawa M, Kishino H, Yano T (1985) Dating of the human-ape splitting by a molecular clock of mitochondrial DNA. J Mol Evol 22 (2):160–174

17. Hasegawa M, Kishino H, Yano T (1987) Man's place in Hominoidea as inferred from molecular clocks of DNA. J Mol Evol 26 (1–2):132–147

18. Scally A, Dutheil JY, Hillier LW, Jordan GE, Goodhead I, Herrero J, Hobolth A, Lappalainen T, Mailund T, Marques-Bonet T, McCarthy S, Montgomery SH, Schwalie PC, Tang YA, Ward MC, Xue Y, Yngvadottir B, Alkan C, Andersen LN, Ayub Q, Ball EV, Beal K, Bradley BJ, Chen Y, Clee CM, Fitzgerald S, Graves TA, Gu Y, Heath P, Heger A, Karakoc E, Kolb-Kokocinski A, Laird GK, Lunter G, Meader S, Mort M, Mullikin JC, Munch K, O'Connor TD, Phillips AD, Prado-Martinez J, Rogers AS, Sajjadian S, Schmidt D, Shaw K, Simpson JT,

Stenson PD, Turner DJ, Vigilant L, Vilella AJ, Whitener W, Zhu B, Cooper DN, de Jong P, Dermitzakis ET, Eichler EE, Flicek P, Goldman N, Mundy NI, Ning Z, Odom DT, Ponting CP, Quail MA, Ryder OA, Searle SM, Warren WC, Wilson RK, Schierup MH, Rogers J, Tyler-Smith C, Durbin R (2012) Insights into hominid evolution from the gorilla genome sequence. Nature 483 (7388):169–175. https://doi.org/10.1038/nature10842

19. Prüfer K, Munch K, Hellmann I, Akagi K, Miller JR, Walenz B, Koren S, Sutton G, Kodira C, Winer R, Knight JR, Mullikin JC, Meader SJ, Ponting CP, Lunter G, Higashino S, Hobolth A, Dutheil J, Karakoç E, Alkan C, Sajjadian S, Catacchio CR, Ventura M, Marques-Bonet T, Eichler EE, André C, Atencia R, Mugisha L, Junhold J, Patterson N, Siebauer M, Good JM, Fischer A, Ptak SE, Lachmann M, Symer DE, Mailund T, Schierup MH, Andrés AM, Kelso J, Pääbo S (2012) The bonobo genome compared with the chimpanzee and human genomes. Nature 486(7404):527–531. https://doi.org/10.1038/nature11128

20. Duret L, Galtier N (2009) Biased gene conversion and the evolution of mammalian genomic landscapes. Annu Rev Genomics Hum Genet 10:285–311. https://doi.org/10.1146/annurev-genom-082908-150001

21. Katoh K, Misawa K, Kuma K, Miyata T. (2002), MAFFT: a novel method for rapid multiple sequence alignment based on fast Fourier transform. Nucleic Acids Res 30 (14):3059–3066

22. Guindon S, Dufayard JF, Lefort V, Anisimova M, Hordijk W, Gascuel O (2010) New algorithms and methods to estimate maximum-likelihood phylogenies: assessing the performance of PhyML 3.0. Syst Biol 59 (3):307–321. https://doi.org/10.1093/sysbio/syq010

23. Slatkin M, Pollack JL (2008) Subdivision in an ancestral species creates asymmetry in gene trees. Mol Biol Evol 25(10):2241–2246. https://doi.org/10.1093/molbev/msn172

24. Schwartz S, Kent WJ, Smit A, Zhang Z, Baertsch R, Hardison RC, Haussler D, Miller W (2003) Human-mouse alignments with BLASTZ. Genome Res 13(1):103–107. https://doi.org/10.1101/gr.809403

25. Kurtz S, Phillippy A, Delcher AL, Smoot M, Shumway M, Antonescu C, Salzberg SL (2004) Versatile and open software for comparing large genomes. Genome Biol 5(2):R12. https://doi.org/10.1186/gb-2004-5-2-r12

Chapter 3

Data Management and Summary Statistics with PLINK

Christopher C. Chang

Abstract

PLINK is a versatile program which supports data management, quality control, and common statistical computations on matrices of genomic variant calls, in a computationally efficient manner. In population genomics, it is frequently used to take care of the "basics," so they do not need to be reimplemented when a new type of analysis needs to be performed on such a matrix. I describe several of these basic operations, and discuss uses and pitfalls.

Key words Allele frequency, Hardy–Weinberg equilibrium, Linkage disequilibrium, Principal component analysis, Relationship inference, Sex inference, Variant call format

1 Introduction

Genotyping chips and sequencing machines produce data in a wide variety of formats. However, they are all trying to measure the same thing: what are the genome sequences of these organisms? These sequences will tend to be 99%+ -identical between different organisms of the same species, of course, but thanks to mutation and sexual reproduction, interesting variation will remain, and it is this variation that is the primary object of study for population genomics.

The most commonly studied type of variation is the "single nucleotide polymorphism" (SNP), an isolated position in the genome that noticeably varies between organisms of the same species while adjacent positions remain identical. They account for a large fraction of total variation, they are relatively easy to detect with modern technologies, and they introduce fewer analytical difficulties than, e.g., genomic rearrangements. Thus, many population-genomic datasets are in the form of a SNP × sample

The original version of this chapter was revised. The correction to this chapter is available at https://doi.org/10.1007/978-1-0716-0199-0_20

Electronic supplementary material: The online version of this chapter (https://doi.org/10.1007/978-1-0716-0199-0_3) contains supplementary material, which is available to authorized users.

Julien Y. Dutheil (ed.), *Statistical Population Genomics*, Methods in Molecular Biology, vol. 2090, https://doi.org/10.1007/978-1-0716-0199-0_3, © The Author(s) 2020, Corrected Publication 2021

matrix. PLINK 1.0 [1] introduced a simple and efficient binary encoding for biallelic-SNP × sample matrices which has become a de facto standard (*see* **Note 1**).

PLINK itself also supports a variety of common data management and quality control operations on such matrices, along with some useful summary statistics; and the wider ecosystem of software directly supporting the PLINK 1 binary format can handle much more (see, e.g., the ADMIXTURE software discussed in Chapter 4). In this chapter, I will cover the following:

- How to convert data into the PLINK 1 binary format.
- Filtering out samples and SNPs with too much missing data.
- Selecting a sample subset without very close relatives.
- Minor allele frequency reporting, and using MAFs to filter out SNPs.
- Hardy–Weinberg equilibrium statistics, and filtering applications.
- Selecting a SNP subset in approximate linkage equilibrium.
- Principal component analysis.
- Sex validation and imputation.
- Reporting linkage disequilibrium statistics.
- Exporting data to other file formats.

2   Materials

You will want PLINK 1.9 and 2.0 [2] for everything that follows. If they are already installed on your system, typing `plink` or `plink2` with no additional command-line arguments into a `bash` (or Windows `cmd`) shell should cause version information and partial lists of supported commands to be printed:

```
~$ plink
PLINK v1.90b6.4 64-bit (7 Aug 2018)          www.cog-genomics.org/plink/1.9/
(C) 2005-2017 Shaun Purcell, Christopher Chang   GNU General Public License v3

plink [input flag(s)...] {command flag(s)...} {other flag(s)...}
plink --help {flag name(s)...}

Commands include --make-bed, --recode, --flip-scan, --merge-list,
--write-snplist, --list-duplicate-vars, --freqx, --missing, --test-mishap,
--hardy, --mendel, --ibc, --impute-sex, --indep-pairphase, --r2, --show-tags,
--blocks, --distance, --genome, --homozyg, --make-rel, --make-grm-gz,
--rel-cutoff, --cluster, --pca, --neighbour, --ibs-test, --regress-distance,
--model, --bd, --gxe, --logistic, --dosage, --lasso, --test-missing,
--make-perm-pheno, --unrelated-heritability, --tdt, --dfam, --qfam, --tucc,
--annotate, --clump, --gene-report, --meta-analysis, --epistasis,
--fast-epistasis, and --score.

'plink --help | more' describes all functions (warning: long).
```

If either is not already installed on your system, prebuilt binaries for Linux, OS X, and Windows can be downloaded from
https://www.cog-genomics.org/plink/1.9/ and
https://www.cog-genomics.org/plink/2.0/
Alternatively, you can download the source code from
https://github.com/chrchang/plink-ng
and build the programs yourself. On some systems, you can also easily install PLINK 1.9 from a package manager such as APT, Bioconda, or Homebrew.

PLINK is designed to interoperate well with R, and this chapter includes an R plotting command. On the off chance you do not have R installed, it can be downloaded from
https://cran.r-project.org/mirrors.html
Sample datasets (based on 1000 Genomes phase 1 [3]) and PLINK outputs for many operations described in this chapter can be downloaded from the companion website.

3   Methods

3.1   Getting Started: Importing and Merging Data

A PLINK 1 binary fileset contains three files:

- A text file with the ".fam" extension, containing sample IDs and possibly some pedigree information (sex, parental IDs). PLINK sample IDs normally have two components: a family ID ("FID") in the first column and an individual ID ("IID") in the second column of the .fam file.

- A text file with the ".bim" extension, containing biallelic-variant IDs, positions, and the two observed alleles for each variant. (Alleles can contain more than one nucleotide; PLINK is designed to work with SNP-like data, but it is not restricted to just SNPs.) Usually, the less common allele is in the 5th column and the more common one is in the 6th, but if your organism has already been sequenced enough to have an official "reference genome," the 6th column may always contain the reference-genome allele.

   PLINK 1.9 refers to the 5th-column allele as "A1," and the 6th-column allele as "A2."

- A binary file with the ".bed" (*see* **Note 2**) extension, containing a compact representation of the variant × sample matrix. Logically, for an autosome in a diploid genome, each matrix entry is either "0 copies of A2 allele," "1 copy," "2 copies," or "NA."

Sometimes, you will be given data in a different format, and/or multiple filesets which should be merged into a single PLINK binary fileset for more convenient analysis.

Genome sequencing data is frequently represented in Variant Call Format (VCF) [4] or its binary counterpart BCF. In addition to variants and genotype calls, these files may contain genotype and mapping quality statistics, and many other tidbits of information. However, in population genomics, you usually only need the variants and genotype calls. These can be converted to PLINK-format with a command like

```
plink --vcf original_data.vcf.gz \
    --keep-allele-order \
    --make-bed \
    --out converted_data
```

which generates a {converted_data.bed, converted_data.bim, converted_data.fam} fileset with the same genotype calls as original_data.vcf.gz.

Let us walk through the pieces of the command line above.

- "–vcf original_data.vcf.gz" specifies that the primary source of input data for this run is original_data.vcf.gz, and it is in VCF format. The .gz at the end of the filename indicates that the file is gzip-compressed; PLINK automatically decompresses the file in this case.

- The backslash at the end of the first line indicates that we are not done typing the command. It is not strictly necessary: you could put all four flags on a single line instead. But the one-flag-per-line style is usually more readable.

- "–keep-allele-order" tells PLINK 1 to keep the allele ordering in the input file, because VCF files explicitly specify which allele is in the species "reference genome" and which allele deviates from it, and that information is useful at times. A2 is set to the reference allele, and A1 is set to the alternate allele.

  Without this flag, PLINK 1 automatically reorders the alleles such that A2 is the most common ("major") allele *in the immediate dataset*, and A1 is the least common ("minor"). Thus, it is normally unsafe to make assumptions about which allele is A1 and which is A2: if a particular allele is present at 48% frequency, it is easy to imagine that allele having 51% frequency in one dataset (and thus being "major" there) and 45% in another. However, there are several ways to impose a specific ordering when necessary; –ref-from-fa (PLINK 2 only) and –a2-allele are generally the most useful.

- "–make-bed" is the command to generate a new PLINK 1 binary fileset. This is used a lot.

- "–out" lets you set the output filename prefix for the current run. Without it, all output filenames would be of the form plink. {extension} instead.

You can find more documentation on each of these flags by going to

https://www.cog-genomics.org/plink/1.9/

and typing the flag name into the "Quick index search" box at the bottom of the left sidebar. Or you can run "plink –help [flag name]" to get a quick summary. Both sources of documentation also discuss some quality filters PLINK can apply, if very-low-quality genotype calls remain in the VCF file; the information on how PLINK sample IDs are generated from VCF sample IDs may also be relevant.

The command line

```
plink2 --vcf original_data.vcf.gz \
    --make-bed \
    --out converted_data
```

has the same effect. Note the lack of "–keep-allele-order": PLINK 2's default assumption is that the 6th .bim column contains reference-genome alleles, and you need to add "–maj-ref" to tell PLINK 2 to reorder the alleles the same way PLINK 1 does. This reflects the fact that reference genomes are more widely agreed on and more stable than they were in 2007.

For brevity, we will ignore –keep-allele-order, –a2-allele, and the like in the rest of this chapter, but be aware that allele order matters sometimes. We will also just give the PLINK 1.9 form of a command when either PLINK version can be used.

3.1.2 PLINK text ({.ped, .map})

A significant number of older datasets are in PLINK's original text fileset format, where the .map file contains variant IDs and positions, and the .ped file stores both sample IDs/pedigree info and genotype calls. These can be imported with

```
plink --file original_data \
    --make-bed \
    --out converted_data
```

Replace –file with –tfile to import a "transposed text" fileset ({. tped, .tfam}).

3.1.3 Other Formats

PLINK is capable of importing several other commonly used genotype data formats. The necessary command lines are all very similar; you usually just need to replace –vcf/–file with another flag. Refer to

https://www.cog-genomics.org/plink/1.9/input and
https://www.cog-genomics.org/plink/2.0/input
for details.

3.1.4 Alternate Chromosome/Contig Sets

Unless you specify otherwise, PLINK interprets chromosome codes as if you were working with human data: 1–22 (or "chr1"–"chr22") refer to the respective autosomes, 23 refers to the X chromosome, 24 refers to the Y chromosome, 25 refers to pseudoautosomal regions, and 26 refers to mitochondria.

If you are working with another species, use the "–chr-set" flag to indicate the number of autosomes. A positive parameter indicates a diploid genome, while a negative number tells PLINK to treat the genome as haploid; so "–chr-set 38" is appropriate for dog data, while "–chr-set -19" is correct for some bog moss species.

Some datasets also contain unplaced contigs. As long as their names start with non-numeric characters (e.g., "contig35" is ok, "35" is not), PLINK will accept them when you include "–allow-extra-chr" (–aec for short) on the command line.

(Unfortunately, if you need –chr-set or –aec once, you will usually need to include it *every* time you run PLINK on data for that species.)

3.1.5 Missing Variant IDs

Sometimes, your input files contain lots of variants which have not been assigned a unique ID (e.g., a VCF file where lots of "ID" column entries are "."). This can prevent some PLINK commands from working properly (such as the fileset merge operation we will discuss next), so it is best to address this during or immediately after data import.

PLINK's –set-missing-var-ids flag provides one solution. Given a template string where "@" represents the chromosome code and "#" represents the base-pair coordinate on the chromosome, it replaces every "." variant ID with a template-based ID. For example,

```
plink --bfile converted_data \
    --set-missing-var-ids @:# \
    --make-bed \
    --out idfilled_data
```

would assign the ID "3:5331691" to an unnamed variant at chromosome 3, position 5331691. This is good enough for SNP-only data.

PLINK 2 also allows the template string to include REF/ALT allele codes, supports several ways of handling very long allele codes, and lets you rewrite all variant IDs with the template instead of just missing IDs; see https://www.cog-genomics.org/plink/2.0/data#set_all_var_ids.

3.1.6 Merging

So you have converted all your data to PLINK 1 binary format, but it is split across multiple filesets. To merge them into a single fileset,

- Create a text file with each of the input filename prefixes, one per line. Call this input_sources.txt.

- Run

```
plink --merge-list input_sources.txt \
    --out merged_data
```

Note that, if you are merging a thousand single-sample input filesets which all have the same sample ID, PLINK will assume all genotype calls are for the same individual (and, as a consequence, most or all genotype calls in the merged dataset will be missing; PLINK's merger normally only keeps a genotype call when all input files agree on it). Assign a different sample ID to each individual before merging.

3.1.7 Filling in Missing Pedigree Information

VCF files normally do not contain pedigree information. However, you will often know at least the sexes of most of the samples anyway. Here is one way to integrate them with your merged PLINK fileset.

- Create a text file where the first two columns are PLINK sample IDs, and the third column indicates sex (1 or M = male, 2 or F = female) (*see* **Note 3**). Call this file sex_info.txt.
- Run

```
plink --bfile merged_data \
    --update-sex sex_info.txt \
    --make-bed \
    --out merged_data_with_sex
```

You can import parental IDs in a similar manner; see the documentation on –update-parents. (Or you can write your own script to manipulate the .fam file; there is more than one way to do these things.)

3.2 Missingness Filters

As of this writing, genotyping chips and genome sequencers are not perfect. For this and other reasons, many datasets contain a bunch of "NA" entries in the variant × sample genotype call matrix.

The usual practice is to filter out the samples and variants with high missing-entry frequencies; these tend to be caused by mistakes in the lab, bad SNP probes, variant calling limitations, and similar issues where throwing out the entire row/column is an appropriate solution. (You still have lots of other rows and columns to work with, so at least in population genomics it usually is not worth the effort to try to salvage it.) PLINK's –mind (for **ind**ividual **m**issingness) and –geno flags provide a way to do this:

```
plink --bfile merged_data \
    --geno 0.1 \
    --mind 0.1 \
    --make-bed \
    --out missingness_filtered_data
```

Walking through this command line,

- –bfile specifies that the primary source of input data for this run is {merged_data.bed, merged_data.bim, merged_data.fam}.

- "–geno 0.1" tells PLINK to throw out every variant where more than 10% of the genotype calls are "NA"s.

- "–mind 0.1" tells PLINK to throw out every sample where more than 10% of the genotype calls are "NA"s. This happens *before* – geno, not simultaneously with it (or after it; command-line flag order is ignored). You can see PLINK 1.9's full order of operations at

  https://www.cog-genomics.org/plink/1.9/order

- –make-bed generates a new PLINK 1 fileset, with the high-missingness samples and variants removed. When using PLINK, you generally do not "edit" data in-place; instead you generate a new fileset with a different name whenever you apply filters or other data transformations.

- –out causes the new files to be written to {missingness_filtered_data.bed, missingness_filtered_data.bim, missingness_filtered_data.fam} instead of {plink.bed, plink.bim, plink.fam}.

  Most PLINK runs also generate a .log file which includes the original command line as well as other information printed to the console; in this case, it would be named missingness_filtered_data.log. As long as you do not delete these .log files or reuse the same –out arguments at inappropriate times, this makes it easy to see how each PLINK output file in a directory was generated.

3.3 Selecting a Sample Subset Without Very Close Relatives

Many population-genomic statistics (such as the allele frequencies we are about to discuss) and analyses are distorted when there are lots of very close relatives in the dataset; you are generally trying to make inferences about the population as a whole, rather than a few families that you oversampled. PLINK 2 includes an implementation of the KING-robust [5] pairwise relatedness estimator, which can be used to prune all related pairs (*see* **Note 4**). For example, to get rid of all first-degree relations (parent–child and sibling–sibling), you could run

```
plink2 --bfile missingness_filtered_data \
       --king-cutoff 0.177 \
       --make-bed \
       --out relpruned_data
```

If you want to do fancier things based on this relatedness estimator, such as trying to reconstruct the entire pedigree, take a look at the KING program, which can be downloaded from http://people.virginia.edu/~wc9c/KING/Download.htm

3.4 Minor Allele Frequency Reporting, Filtering

Allele frequencies are a primary object of study in population genetics and genomics. PLINK's —freq command reports empirical allele frequencies, and its —maf filter removes all variants with minor allele frequency below the given threshold.

For example,

```
plink --bfile relpruned_data \
      --freq \
      --out allele_freqs
```

writes an empirical allele frequency report to allele_freqs.frq, where the first column contains the chromosome, the second column contains the variant ID, columns 3–4 contain the minor and major allele codes in that order, column 5 contains the minor allele frequency, and column 6 contains the number of allele observations. (This file format is spelled out at

https://www.cog-genomics.org/plink/1.9/formats#frq,

and all other PLINK 1.9 output file formats are described elsewhere on the page.)

If you are interested in the allele frequency spectrum (refer to Chapter 1 for some of its applications), PLINK 2 —freq has an additional convenience that may come in handy:

```
plink2 --bfile relpruned_data \
       --freq alt1bins=0.01,0.02,0.03,0.04,0.05,0.1,0.2,0.3,0.4 \
       --out allele_spectrum
```

This generates an additional allele_spectrum.afreq.alt1.bins file which reports the number of variants with MAF in $[0, 0.01)$, $[0.01, 0.02)$, ..., and $[0.4, 0.5]$. (And the main allele frequency file is formatted slightly differently from PLINK 1.9, with a more VCF-like header row, and single-tab delimiters instead of multiple spaces. These changes are small enough that it should not be difficult to switch an old codebase from PLINK 1 to PLINK 2, but be aware that the latter is not a drop-in replacement for the former.)

To filter out all variants with minor allele frequency below 5%, you would run

```
plink --bfile relpruned_data \
      --maf 0.05 \
      --make-bed \
      --out maf_filtered_data
```

3.5 Hardy–Weinberg Equilibrium Statistics

Checking for deviation from Hardy–Weinberg equilibrium is useful for multiple reasons:

- If there are far more heterozygous calls than would be expected under Hardy–Weinberg equilibrium, that is usually due to a systematic variant calling error. Any such variants should be removed from the dataset.

- Population stratification can cause large violations of Hardy–Weinberg equilibrium, in the fewer-hets-than-expected direction. Natural selection, migration, and some mating patterns can drive smaller violations in either direction.

- Several statistical methods assume approximate Hardy–Weinberg equilibrium. You can use a more-stringently-filtered subset of your data for just these methods.

PLINK includes functions for computing Hardy–Weinberg equilibrium exact test p-values, based on the method of Wigginton et al. [6] The –hwe flag filters out variants with p-values more extreme than a given threshold; the following PLINK 2 command line is appropriate during initial quality control:

```
plink2 --bfile maf_filtered_data \
    --hwe 1e-25 keep-fewhet \
    --make-bed \
    --out hwe_filtered_data
```

The "keep-fewhet" modifier causes this filter to be applied in a one-sided manner (so the fewer-hets-than-expected variants that one would expect from population stratification would *not* be filtered out by this command), and the 1e-25 threshold is extreme enough that we are unlikely to remove anything legitimate. (Unless the dataset is primarily composed of F1 hybrids from an artificial breeding program.) After you have a good idea of population structure in your dataset, you may want to follow up with a round of two-sided –hwe filtering, since large (*see* **Note 5**) violations of Hardy–Weinberg equilibrium in the fewer-hets-than-expected direction *within* a subpopulation are also likely to be variant calling errors; with multiple subpopulations, the –write-snplist and –extract flags can help you keep just the SNPs which pass all subpopulation HWE filters.

The –hardy command can be used to dump all of the p-values.

3.6 Selecting a SNP Subset in Approximate Linkage Equilibrium

Some analyses, such as the PCA we will discuss next, work best on a genome-spanning subset of SNPs which are in approximate linkage equilibrium. PLINK 2's –indep-pairwise command (*see* **Note 6**) is a computationally efficient method of identifying a reasonable subset. For example,

```
plink2 --bfile hwe_filtered_data \
    --indep-pairwise 200kb 1 0.5 \
    --out ldpruned_snplist
```

removes SNPs so that no pair within 200 kilobases have squared-allele-count-correlation (r^2) greater than 0.5, and saves the IDs of the remaining SNPs to ldpruned_snplist.prune.in. (There is nothing magical about the $r^2 = 0.5$ threshold; it is useful to adjust it depending on the number of SNPs you want to keep. The lower the threshold you use, the larger your kilobase window should be.)

You can then create a fileset with just this SNP subset by combining –extract (which lets you select an arbitrary subset of variants, by ID) with –make-bed:

```
plink --bfile hwe_filtered_data \
    --extract ldpruned_snplist.prune.in \
    --make-bed \
    --out ldpruned_data
```

3.7   Principal Component Analysis

Once you have LD-pruned and MAF-filtered your dataset, PLINK 2's –pca command has a good shot of revealing large-scale population structure. For example,

```
plink2 --bfile ldpruned_data \
    --pca 5 \
    --out pca_results
```

writes a tab-delimited table to pca_results.eigenvec, with one sample per row and one principal component per later column. (Eigenvalues are written to pca_results.eigenval.)

You will usually want to sanity-check the output at this point, and verify that the top principal components do not correlate too strongly with, e.g., sequencing facility or date. (A full discussion of "batch effects" and how to deal with them could take up an entire chapter; worst case, you may have to analyze your batches separately, or even redo all genotyping/sequencing from scratch. I will be optimistic here and suppose that no major problem was uncovered by PCA, but be aware that this is frequently your best chance to catch data problems that would otherwise sink your entire analysis.)

It is also a good idea to throw out gross outliers at this point; any sample which is more than, say, 8 standard deviations out on any top principal component is likely to have been genotyped/sequenced improperly; you can remove such samples by creating a text file with the bad sample IDs, and then using –remove + – make-bed:

```
plink --bfile ldpruned_data \
    --remove bad_samples.txt \
    --make-bed \
    --out ldpruned_data2
```

If you do this, follow it up by repeating the PCA, since the bad samples might have distorted the principal components:

```
plink2 --bfile ldpruned_data2 \
    --pca 5 \
    --out pca_results2
```

(Occasionally, the new principal components will reveal *another* bad sample, and you have to repeat these two steps, etc. EIGEN-SOFT [7, 8] has some additional built-in principal component analysis options, including automated iterated outlier removal, and a top-eigenvalue-based test for significant population structure.)

Anyway, once there is nothing obviously wrong with the PCA results, you can load the table in R and plot the top pairs of principal components against each other:

```
> pca_table <- read.table("pca_results.eigenvec", header=TRUE, comment.char="")
> plot(pca_table[, c("PC1", "PC2", "PC3", "PC4", "PC5")])
```

Results are shown in Fig. 1.

If there are obvious clusters in the first few plots, I recommend jumping ahead to Chapter 4 (on ADMIXTURE) and using it to label major subpopulations before proceeding.

3.8 Sex Validation and Imputation

If you have X-chromosome population-genomic data, you can employ PLINK's –check-sex command to sanity-check the sex information in your .fam file. (The method is based on chrX heterozygosity rates.) Similarly, –impute-sex uses chrX heterozygosity rates to fill in missing sex entries when appropriate.

This is typically a two- or three-step process:

0. If your species has pseudoautosomal regions, ensure they are not encoded as part of the X chromosome. If they are, PLINK 1.9's –split-x or PLINK 2's –split-par flag can be used to change the chromosome codes of the relevant variants before proceeding.

1. Run –check-sex once without additional parameters, just to see the distribution of F (inbreeding) coefficients.

```
plink --bfile ldpruned_data \
    --check-sex \
    --out f_distribution
```

Then plot the distribution of values in the sixth column of f_distribution.sexcheck. If both genders are well-represented in the dataset, you should see a big tight clump near 1 (corresponding to the males), and a more widely dispersed set of values centered near 0 (corresponding to the females).

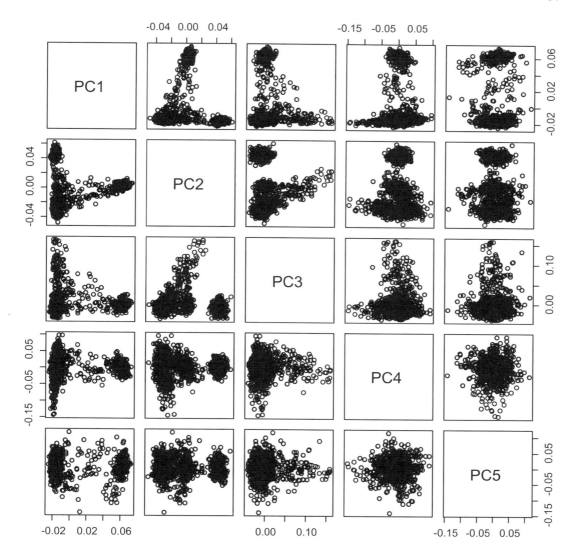

Fig. 1 PCA bi-plot generated by R

2. Select a pair of decision boundaries (where an F coefficient below the first boundary causes the sample to be classified as female, above the second boundary causes the sample to be classified as male, and in between causes the sample to be labeled as unknown-sex), and run –impute-sex (or just –check-sex again) with the boundaries:

```
plink --bfile ldpruned_data \
    --impute-sex 0.7 0.8 \
    --make-bed \
    --out sexfilled_data
```

If you also have Y-chromosome genotype calls and can expect males to have fewer missing chrY calls than females, you can take this into account in the sex inference process; refer to the –check-sex/–impute-sex documentation at

https://www.cog-genomics.org/plink/1.9/basic_stats#check_sex

3.8.1 Subpopulation Allele Frequencies, and –read-freq

By default, –check-sex/–impute-sex and several other PLINK commands assume that your dataset is representative of a single population, and all samples are members of that population; population allele frequencies and F coefficients are estimated under these assumptions.

Thus, if you identified multiple subpopulations in the previous section, you should perform sex validation/imputation on one subpopulation at a time. PLINK's –filter flag provides one way to do this; put sample IDs in the first two columns of subpops.txt and subpopulation IDs in the third column, then

```
plink --bfile ldpruned_data \
    --filter subpops.txt AFR \
    --make-bed \
    --out afr_data

plink --bfile afr_data \
    --check-sex \
    ...
```

You can then use –update-sex to copy the inferred sexes back to your main datasets.

Also, you sometimes have access to more accurate (sub)population allele frequencies than would be imputed from your immediate dataset. (An extreme case of this is when you are running –check-sex/–impute-sex separately for a bunch of single-sample filesets.) To patch in the more-accurate allele frequencies, reformat them as a PLINK –freq report if necessary, and then use –read-freq:

```
plink --bfile ldpruned_data \
    --read-freq more_accurate.frq \
    --check-sex \
    --out f_distribution
```

3.9 Reporting Linkage Disequilibrium Statistics

While the last several operations worked with a LD-pruned subset of the data, sometimes you will want to handle linkage disequilibrium in a more sophisticated manner, or study it directly. PLINK's –r2 command can be used for this purpose.

```
plink --bfile hwe_filtered_data \
    --r2 dprime \
    --ld-window 999999 \
    --ld-window-kb 1000 \
    --ld-window-r2 0.1 \
    --out r2_report
```

The command line above reports squared-allele-count-correlations and an estimate of Lewontin's D' (thanks to the additional "dprime" modifier) for each pair of variants within 1000 kilobases of each other, whenever $r^2 \geq 0.1$. The "–ld-window 999999" flag is needed because PLINK defaults to only considering variant pairs which are at most 9 lines apart from each other in the .bim file.

–r2 has a bunch of other options, including matrix output (when you want to look at literally every single pair of variants) and ways to focus on just a few SNPs; see the documentation at
https://www.cog-genomics.org/plink/1.9/ld#r

3.10 Data Export

While the PLINK 1 binary fileset format is widely supported, occasionally you will need to convert to a different format like VCF or PLINK text in order to use another tool. This can be accomplished with the PLINK –export command:

```
plink --bfile hwe_filtered_data \
    --export ped \
    --out exported_plink_text

plink2 --bfile hwe_filtered_data \
    --ref-from-fa reference.fa \
    --export vcf \
    --out exported_vcf
```

The first command generates {exported_plink_text.ped, exported_plink_text.map}, while the second one generates exported_vcf.vcf, determining the correct REF alleles from reference.fa in case they were scrambled by PLINK 1.9. Many other formats are supported; see
https://www.cog-genomics.org/plink/1.9/data#recode and
https://www.cog-genomics.org/plink/2.0/data#export.

---

4   Notes

1. SNPs with three or four alleles exist, but are rare enough such that most software developers ignore them. The usual practice as of this writing is to "split" a triallelic SNP into two biallelic records; for example, if A, C, and G nucleotides are all observed

at a single position and A is the most common, the dataset will contain one record describing an A/C SNP, and another record describing an A/G SNP. This is obviously imperfect, but it is good enough for most genomic analyses.

2. Unfortunately, there is another type of ".bed" file widely used in genomics: UCSC Browser Extensible Data, a text format for positional intervals. However, you can usually distinguish between the two simply by checking for the presence of .bim and .fam files with the same filename prefix; these are practically required to make sense of a PLINK .bed file. Or, in a pinch, you can check whether the first three bytes of the .bed file are consistent with the specification at https://www.cog-geno mics.org/plink/1.9/formats#bed.

3. Assuming your species adheres to the XY sex-determination system. For ZW species, you may want to deliberately reverse the sexes, and encode Z→X and W→Y.

4. This does not mean that both samples in each related pair are thrown out. Instead, –king-cutoff tries to keep as much data as possible, and as a consequence it usually keeps one sample out of each pair.

5. What p-value threshold does "large" correspond to, you may ask? Well, this depends on the size of your dataset and some other characteristics of your data. But a good rule of thumb is to use "ridiculous" thresholds like 1e-25 or 1e-50 for quality control when you have at least a thousand samples (or even 1e-200 if you have many more), while only using "normal" thresholds like 1e-4 or 1e-7 when you are preparing data for an analysis which actually assumes all your SNPs are in Hardy–Weinberg equilibrium.

6. The –indep-pairwise command also works in PLINK 1.9, but it may run a lot more slowly since it is not automatically parallelized.

References

1. Purcell S, Neale B, Todd-Brown K, Thomas L, Ferreira M, Bender D, Maller J, Sklar P, de Bakker P, Daly M, Sham P (2007) PLINK: a tool set for whole-genome association and population-based linkage analyses. Am J Hum Genet 81:559–575

2. Chang C, Chow C, Tellier L, Vattikuti S, Purcell S, Lee J (2015) Second-generation plink: rising to the challenge of larger and richer datasets. GigaScience 4:7

3. The 1000 Genomes Project Consortium (2012) An integrated map of genetic variation from 1,092 human genomes. Nature 491:56–65

4. Danecek P, Auton A, Abecasis G, Albers C, Banks E, DePristo M, Handsaker R, Lunter G, Marth G, Sherry S, McVean G, Durbin R (2011) The variant call format and vcftools. Bioinformatics 27:2156–2158

5. Manichaikul A, Mychaleckyj J, Rich S, Daly K, Sale M, Chen W (2010) Robust relationship

inference in genome-wide association studies. Bioinformatics 26:2867–2873

6. Wigginton J, Cutler D, Abecasis G (2005) A note on exact tests of Hardy-Weinberg equilibrium. Am J Hum Genet 76:887–893

7. Patterson N, Price A, Reich D (2006) Population structure and eigenanalysis. PLoS Genet 2

8. Price A, Patterson N, Plenge R, Weinblatt M, Shadick N, Reich D (2006) Principal components analysis corrects for stratification in genome-wide association studies. Nat Genet 38:904–909

Chapter 4

Exploring Population Structure with Admixture Models and Principal Component Analysis

Chi-Chun Liu, Suyash Shringarpure, Kenneth Lange, and John Novembre

Abstract

Population structure is a commonplace feature of genetic variation data, and it has importance in numerous application areas, including evolutionary genetics, conservation genetics, and human genetics. Understanding the structure in a sample is necessary before more sophisticated analyses are undertaken. Here we provide a protocol for running principal component analysis (PCA) and admixture proportion inference—two of the most commonly used approaches in describing population structure. Along with hands-on examples with CEPH-Human Genome Diversity Panel and pragmatic caveats, readers will learn to analyze and visualize population structure on their own data.

Key words Population structure, Admixture, Principal component analysis, Population stratification

1 Introduction

Population structure is a commonplace feature of genetic variation data, and it has importance in numerous application areas, including evolutionary genetics, conservation genetics, and human genetics. At a broad level, population structure is the existence of differing levels of genetic relatedness among some subgroups within a sample. This may arise for a variety of reasons, but a common cause is that samples have been drawn from geographically isolated groups or different locales across a geographic continuum. Regardless of the cause, understanding the structure in a sample is necessary before more sophisticated analyses are undertaken. For example, to infer divergence times between two populations requires knowing two populations even exist and which individuals belong to each.

The original version of this chapter was revised. The correction to this chapter is available at https://doi.org/10.1007/978-1-0716-0199-0_20

Electronic supplementary material: The online version of this chapter (https://doi.org/10.1007/978-1-0716-0199-0_4) contains supplementary material, which is available to authorized users.

Julien Y. Dutheil (ed.), *Statistical Population Genomics*, Methods in Molecular Biology, vol. 2090, https://doi.org/10.1007/978-1-0716-0199-0_4, © The Author(s) 2020, Corrected Publication 2021

Two of the most commonly used approaches to describe population structure in a sample are principal component analysis [5, 16, 23, 25] and admixture proportion inference [19, 26]. In brief, principal component analysis reduces a multi-dimensional dataset to a much smaller number of dimensions that allows for visual exploration and compact quantitative summaries. In its application to genetic data, the numerous genotypes observed per individual are reduced to a few summary coordinates. With admixture proportion inference, individuals in a sample are modeled as having a proportion of their genome derived from each of several source populations. The goal is to infer the proportions of ancestry from each source population, and these proportions can be used to produce compact visual summaries that reveal the existence of population structure in a sample.

The history and basic behaviors of both these approaches have been written about extensively, including by some of us, and so we refer readers to several previous publications to learn the basic background and interpretative nuances of these approaches and their derivatives [1, 2, 9, 10, 12, 17, 18, 20, 21, 23, 25–27, 29]. Here, in the spirit of this volume, we provide a protocol for running these analyses and share some pragmatic caveats that do not always arise in more abstract discussions regarding these methods.

2   Materials

The protocol we present is based on two pieces of software: (1) the ADMIXTURE software that our team developed [2] for efficiently estimating admixture proportions in the "Pritchard-Stephens-Donnelly" model of admixture [19, 26]. (2) The smartpca software developed by Nick Patterson and colleagues for carrying out PCA [25]. Both of these pieces of software are used widely. We also pair them with downstream tools for visualization, in particular pong [3], for visualizing output of admixture proportion inferences, and PCAviz [31], a novel R package for plotting PCA outputs. We also use PLINK [6, 24] as a tool to perform some basic manipulations of the data (*see* Chapter 3 for more background on PLINK).

The example data we use is derived from publicly available single-nucleotide polymorphism (SNP) genotype data from the CEPH-Human Genome Diversity Panel [4]. Specifically, we will look at Illumina 650Y genotyping array data as first described by Li et al. [15]. This sample is a global-scale sampling of human diversity with 52 populations in total, and the raw files are available from the following link: http://hagsc.org/hgdp/files.html. These data have been used in numerous subsequent publications and are an important reference set.

A few technical details are that the genotypes were filtered with a cutoff of 0.25 for the Illumina GenCall score [13] (a quality score generated by the basic genotype calling software). Further, individuals with a genotype call rate <98.5% were removed, with the logic being that if a sample has many missing genotypes it may be due to poor quality of the source DNA, and so none of the genotypes from that individual should be trusted. Beyond this, to prepare the data, we have filtered down the individuals to a set of 938 unrelated individuals. We exclude related individuals as we are not interested in population structure that is due to family relationships and methods such as PCA and ADMIXTURE can inadvertently mistake family structure for population structure. The starting data are available as plink-formatted files H938.bed H938.fam, H938.bim, and an accompanying set of population identifiers H938.clst.txt in the raw_input sub-directory of the companion data.

As a pragmatic side note, it is common (and recommended) when carrying out analyses of population structure to merge one's data with other datasets that contain populations which may be representative sources of admixing individuals. For example, in analyzing a dataset with African American individuals, it can be helpful to include datasets containing African and European individuals in the analysis. These datasets can be merged with your dataset using software such as plink. However, when merging several datasets, one should be aware of potential biases that can be introduced due to strand flips (i.e., one dataset reports genotypes on the "+" strand of the reference human genome, and another on the "−" strand). One precautionary step to detect strand flips is to group individuals by what dataset they derive from and then produce a scatterplot of allele frequencies for pairs of groups at a time. If strand flips are not being controlled correctly, one will observe numerous variants on the $y = 1 - x$ line, where x is the frequency in one dataset and y is the frequency in a second dataset. (Note: This rule of thumb assumes levels of differentiation are low between datasets, as is the case in human datasets in general, but one should still keep this in mind interpreting results.)

3 Methods

In this section we walk you through an example analysis using ADMIXTURE and smartpca. We assume the raw data files are in a directory raw_input that is below our working directory and that a second directory out exists in which outputs can be placed. If following along in an R console, you should use the setwd( ) command to set the working directory correctly.

3.1 Subsetting Data

For running some simple examples below, we will first create a subset of the HGDP sample that is restricted to only European populations. The European populations in the HGDP have the labels "Adygei," "Basque," "French," "Italian," "Orcadian," "Russian," "Sardinian" and "Tuscan," so we create a list of individuals matching these labels using an awk command, and then use plink's --keep option to make a new dataset with output prefix "H938_Euro."

```
awk '$3=="Adygei"||$3=="French_Basque"||$3=="French"|| \
$3=="North_Italian"||$3=="Orcadian"||$3=="Russian"|| \
$3=="Sardinian"||$3=="Tuscan" {print $0}' \
raw_input/H938.clst.txt > out/Euro.clst.txt

plink --bfile raw_input/H938 --keep out/Euro.clst.txt \
--make-bed --out out/H938_Euro
```

3.2 Filter Out SNPs to Remove Linkage Disequilibrium (LD)

SNPs in high LD with each other contain redundant information. More worrisome is the potential for some regions of the genome to have a disproportionate influence on the results and thus distort the representation of genome-wide structure. A nice empirical example of the problem is in figure 5 of Tian et al. [30], where PC2 of the genome-wide data is shown to be reflecting the variation in a 3.8 Mb region of chromosome 8 that is known to harbor an inversion. A standard approach to address this issue is to filter out SNPs based on pairwise LD to produce a reduced set of more independent markers. Here we use plink's commands to produce a new LD-pruned dataset with output prefix H938_Euro. LDprune. The approach considers a chromosomal window of 50 SNPs at a time, and for any pair whose genotypes have an association r^2 value greater than 0.1, it removes a SNP from the pair. Then the window is shifted by 10 SNPs and the procedure is repeated:

```
plink --bfile out/H938_Euro --indep-pairwise 50 10 0.1
plink --bfile out/H938_Euro --extract plink.prune.in --make-bed
--out out/H938_Euro.LDprune
```

(Advanced note: For particularly sensitive results, we recommend additional rounds of SNP filtering based on observed principal component loadings and/or population differentiation statistics. For example, a robust approach is to filter out large windows around any SNP with a high PCA loading, *see* ref. 22.)

3.3 Running ADMIXTURE

3.3.1 An Example Run with Visualization

The ADMIXTURE software (v 1.3.0 here) comes as a pre-compiled binary executable file for either Linux or Mac operating systems. To install, simply download the package and move the executable into your standard execution path (e.g. "/usr/local/bin" on many Linux systems). Once installed, it is straightforward to run ADMIXTURE with a fixed number of source populations, commonly denoted by K. For example, to get started let's run ADMIXTURE with $K=6$:

```
admixture out/H938_Euro.LDprune.bed 6
```

ADMIXTURE is a maximum-likelihood based method, so as the method runs, you will see updates to the log-likelihood as it converges on a solution for the ancestry proportions and allele frequencies that maximize the likelihood function. The algorithm will stop when the difference between successive iterations is small (the "delta" value takes a small value). A final output is an estimated F_{ST} value [11] between each of the source populations, based on the inferred allele frequencies. These estimates reflect how differentiated the source populations are, which is important for understanding whether the population structure observed in a sample is substantial or not (values closer to 0 reflect less population differentiation).

After running, ADMIXTURE produces two major output files. The file with suffix .P contains an $L \times K$ table of the allele frequencies inferred for each SNP in each population. The file with suffix .Q contains an $N \times K$ table of inferred individual ancestry proportions from the K ancestral populations, with one row per individual.

For our example dataset with $K=6$, this will be a file called H938.LDprune.6.Q. This file can be used to generate a plot showing individual ancestry (*see* Fig. 1). In R, this can be done using the following commands:

```
library(RColorBrewer)
tbl <- read.table("out/H938_Euro.LDprune.6.Q")
par(mar = c(1.5, 4, 2.5, 2),cex.lab=0.75,cex.axis=0.6)
barplot(t(as.matrix(tbl)),
   col = brewer.pal(6, "Set1"), ylab = "Anc. Proportions",
   border = NA, space = 0
)
```

Each thin vertical line in the barplot represents one individual and each color represents one inferred ancestral population. The length of each color in a vertical bar represents the proportion of that individual's ancestry that is derived from the inferred ancestral population corresponding to that color. The above image suggests there are some genetic clusters in the data, but it's not a well-organized data display.

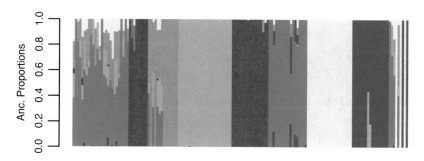

Fig. 1 Initial rough plot of the ADMIXTURE results for $K=6$ using R base graphics

To improve the visualization, one can use a package dedicated to plotting ancestry proportions [3, 14, 28]. Here we use a post-processing tool, `pong` [3], which visualizes individual ancestry with similarity between individuals within clusters. You will most likely want to install `pong` on a local machine as it initializes a local web server to display the results.

To run `pong` requires setting up a few files: (1) an ind2pop file that maps individuals to populations; (2) a `Qfilemap` file that points `pong` towards which ".Q" files to display; these are easy to build up from the command-line using the `Euro.clst.txt` file we built above, and an `awk` command to output tab-separated text to a file with the `Qfilemap` suffix added to whatever file prefix we're using to organize our runs:

```
cut -d' ' -f3 out/Euro.clst.txt > out/H938_Euro.ind2pop
FILEPREFIX=H938_Euro.LDPrune
K=6
awk -v K=$K -v file=$FILEPREFIX 'BEGIN{ \
printf("ExampleRun\t%d\t%s.%d.Q\n",K,file,K)
}' > out/$FILEPREFIX.Qfilemap
```

Note when building the `.Qfilemap` one needs to use tabs to separate the columns for `pong` to read the file correctly.

Then to run `pong`, we use the following command:

```
pong -m out/H938_Euro.LDprune.Qfilemap -i out/H938_Euro.ind2pop
```

We open a web browser to http://localhost:4000/ to view the results. Figure 2 shows an example of what you should see. From this visualization, we can see the admixture model fits most individuals of the Adygei, Sardinian, Russian, French Basque samples as being derived each from a single source population (represented by purple, red, green, and yellow, respectively). The French, Tuscan, and North Italian samples are generally estimated to have a majority component of ancestry from a single source population (blue) though with admixture with other sources. A first conclusion is that the population labels do not capture the complexity of the

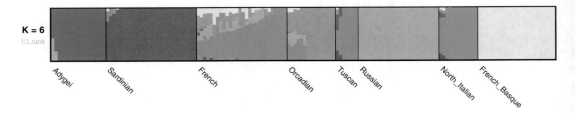

Fig. 2 Plot of the ADMIXTURE results for $K = 6$ using PONG

population structure. There is apparent cryptic structure within some samples (e.g., Orcadian) and minimal differentiation between other samples (North Italian and Tuscan samples, for instance).

Because ADMIXTURE is a "greedy," "hill-climbing" optimization algorithm it is good practice to do multiple runs from different initial random starting points. We can do this by using the -s flag to specify the random seed for each ADMIXTURE run.

```
K=6
prefix=H938_Euro.LDprune
# run admixture multiple times
for r in {1..10}
do
admixture -s ${RANDOM} out/H938_Euro.LDprune.bed $K
mv out/${prefix}.${K}.Q out/${prefix}.K${K}r${r}.Q
done
```

Pong has nice functionality for summarizing the output of the multiple ADMIXTURE runs. It can collect similar solutions into "modes" and display them in ranked order of the number of runs supporting each. In the interactive version, you use the check to highlight multimodality checkbox and whiten populations with ancestry matrices agreeing with the major mode. One can also click on and visualize only one cluster. Here we set up the PONG input files and show an example output.

```
K=6
prefix=H938_Euro.LDprune
# create a pong parameter file
for r in {1..10}
do
awk -v K=$K -v r=$r -v file=${prefix}.K${K}r${r} 'BEGIN{ \
printf("K%dr%d\t%d\t%s.Q\n",K,r,K,file)
}' >> out/${prefix}.k6multiplerun.Qfilemap
done
# run pong
pong -m out/${prefix}.k6multiplerun.Qfilemap --greedy -s .95 \
-i out/H938_Euro.ind2pop
```

The resulting figure (Fig. 3) shows that six out of ten runs converged to the same mode, which appears equivalent to our initial run above. We observe the appearance of structure within Sardinia in the second and third modes. The original run had North Italian and Tuscan samples as a mostly unadmixed, while all three minor modes model the two population as highly admixed. The fourth mode (supported by just one run) inferred sub-structure within the French Basque sample. This instability in the solution is a

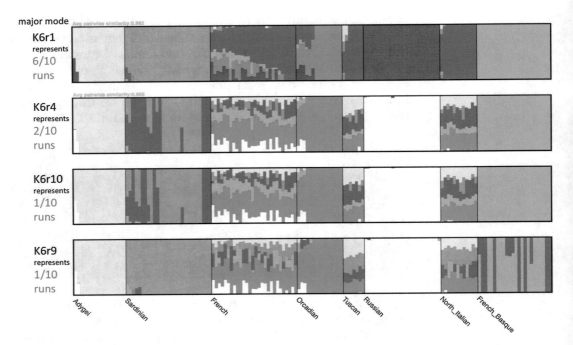

Fig. 3 PONG plot summarizing multiple ADMIXTURE runs with different random starting points. The top row shows the major mode (supported by 6 out of 10 runs as indicated in the blue text). The next three rows show three other solutions found by ADMIXTURE in 2, 1, and 1 runs, respectively

hint that the ADMIXTURE model with $K = 6$ is not a perfect fit to this data.

3.3.2 Considering Different Values of K

In a typical analysis, one wants to explore the sensitivity of the results to the choice of K. One approach is to run ADMIXTURE with various plausible values of K and compare the performance of results visually and using cross-validation error rates. Here is a piece of bash command-line code that will run ADMIXTURE for values of K from 2 to 12, and that will build a file with a table of cross-validation error rates per value of K.

```
# Run for different values of K
prefix=H938_Euro.LDprune
Klow=1
Khigh=12
for ((K=$Klow;K<=$Khigh;K++)); \
do
    admixture --cv out/$prefix.bed $K | tee log.$prefix.${K}.out
done
```

Then let's compile results on cross-validation error across values of K:

```
prefix=H938_Euro.LDprune
Klow=1
Khigh=12
echo '# CV results' > $prefix.CV.txt
for ((K=$Klow;K<=$Khigh;K++)); do
  awk -v K=$K '$1=="CV"{print K,$4}' out/log.$prefix.$K.out \
  >> out/$prefix.CV.txt;
done
```

Now let's inspect the outputs. First let's make a plot of the cross-validation error as a function of K (Fig. 4):

```
tbl <- read.table("out/H938_Euro.LDprune.CV.txt")
par(mar = c(4, 4, 2, 2),cex=0.7)
plot(tbl$V1, tbl$V2, xlab = "K", ylab = "Cross-validation error"
     pch = 16, type = "l")
```

The cross-validation error suggests a single source population can model the data adequately and larger values of K lead to over-fitting.

To inspect further, we can use the `pong` software to visualize the ancestry components inferred at different K across several runs. We need to set up some of the results and input files first.

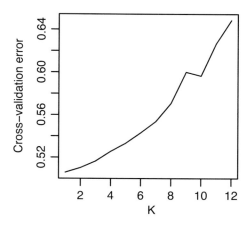

Fig. 4 Cross-validation error as a function of K for the example dataset

```
# Run for different values of K, each with 10 runs
prefix=H938_Euro.LDprune
for r in {1..10}; do for K in {2..12};
do
    admixture -s ${RANDOM} out/${prefix}.bed $K
    mv out/${prefix}.${K}.Q out/${prefix}.K${K}r${r}.Q
done; done

# create Qmap file for pong
createQmap(){
local r=$1
local K=$2
awk -v K=$K -v r=$r -v file=${prefix}.K${K}r${r} 'BEGIN{ \
printf("K%dr%d\t%d\t%s.Q\n",K,r,K,file)
}' >> out/${prefix}.multiplerun.Qfilemap
}
export -f createQmap
for K in {2..12}; do for r in {1..10}; do createQmap $r $K; \
done; done

#run pong
pong -m out/H938_Euro.LDprune.multiplerun.Qfilemap --greedy \
-s.95 -i out/H938_Euro.ind2pop
```

Here we find how as K increases through to $K=6$, the Sardinian, Basque, Adygei, and Russian samples are typically modeled as descended from unique sources, and at K of 7, 8, 9 we find structure within the Sardinian, Russian, Orcadian, and Basque samples is revealed, though for each, the substructure is not very stable (Fig. 5). The values of $K=10$ and above make increasingly finer scale divisions that are difficult to interpret, and the major modes for $K=7$ and up only consist of one to three runs, suggesting a very multi-modal likelihood surface and a poor resolution of the population structure.

Overall, it is interesting to note that the visual inspection of the results suggests several "real" clusters in the data, supported by an alignment of the clustering with known population labels, even though the cross-validation supports a value of $K=1$. This highlights a long-standing known issue with admixture modeling: the selection of K is a difficult problem to automate in a way that is robust.

3.3.3 Some Advanced Options

Running ADMIXTURE with the -B option provides estimates of standard errors on the ancestry proportion inferences. The -l flag runs ADMIXTURE with a penalized likelihood that favors more sparse solutions (i.e., ancestry proportions that are closer to zero). This is useful in settings where small, possibly erroneous ancestry proportions may be overinterpreted. By using the -P option, the population allele frequencies inferred from one dataset

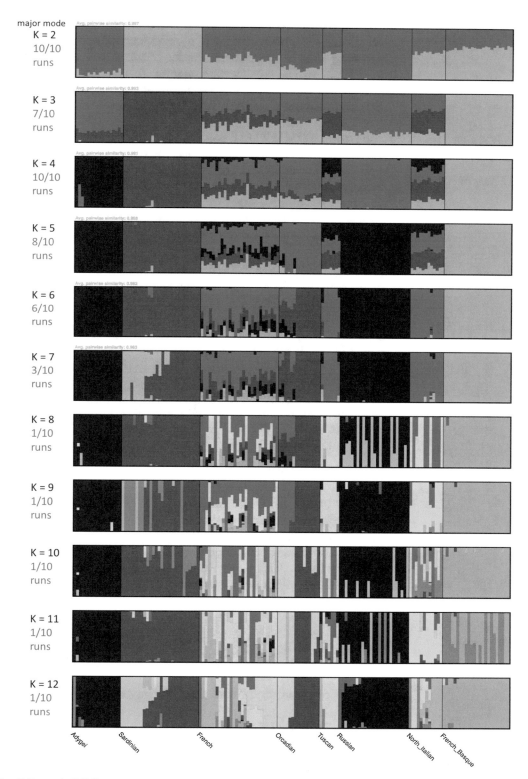

Fig. 5 Example PONG output showing results from across a range of *K* values (with ten ADMIXTURE runs per *K* value)

can be provided as input for inference of admixture proportions in a second dataset. This is useful when individuals of unknown ancestry are being analyzed against the background of a reference sample set. Please see the ADMIXTURE manual for a complete listing of options and more detail, and we encourage testing these options in test datasets such as the one provided here.

3.4 PCA with SMARTPCA

3.4.1 Running PCA

Comparing ADMIXTURE and PCA results often helps give insight and confirmation regarding population structure in a sample. To run PCA, a standard package that is well-suited for SNP data is the smartpca package maintained by Nick Patterson and Alkes Price (at http://data.broadinstitute.org/alkesgroup/EIGENSOFT/). To run it, we first set up a basic smartpca parameter file from the command-line of a bash shell:

```
PREFIX=H938_Euro.LDprune
echo genotypename: out/$PREFIX.bed > out/$PREFIX.par
echo snpname: out/$PREFIX.bim >> out/$PREFIX.par
echo indivname: out/$PREFIX.PCA.fam >> out/$PREFIX.par
echo snpweightoutname: out/$PREFIX.snpeigs >> out/$PREFIX.par
echo evecoutname: out/$PREFIX.eigs >> out/$PREFIX.par
echo evaloutname: out/$PREFIX.eval >> out/$PREFIX.par
echo phylipoutname: out/$PREFIX.fst >> out/$PREFIX.par
echo numoutevec: 20 >> out/$PREFIX.par
echo numoutlieriter: 0 >> out/$PREFIX.par
echo outlieroutname: out/$PREFIX >> out/$PREFIX.par
echo altnormstyle: NO >> out/$PREFIX.par
echo missingmode: NO >> out/$PREFIX.par
echo nsnpldregress: 0 >> out/$PREFIX.par
echo noxdata: YES >> out/$PREFIX.par
echo nomalexhet: YES >> out/$PREFIX.par
```

This input parameter file runs smartpca in its most basic mode (i.e., no automatic outlier removal or adjustments for LD—features which you might want to explore later).

As a minor issue, smartpca ignores individuals in the .fam file if they are marked as missing in the phenotypes column. This awk command provides a new .fam file that will automatically include all individuals.

```
awk '{print $1,$2,$3,$4,$5,1}' out/H938_Euro.LDprune.fam \
> out/H938_Euro.LDprune.PCA.fam
```

Now run smartpca with the following command.

```
smartpca -p ./out/H938_Euro.LDprune.par
```

You will find the output files in the `out` sub-directory as specified in the parameter file.

The `PCAviz` package can be found at https://github.com/ NovembreLab/PCAviz. It provides a simple interface for quickly creating plots from PCA results. It encodes several of our favored best practices for plotting PCA (such as using abbreviations for point characters and plotting median positions of each labelled group). To install the package use:

```
install.packages("devtools")
devtools::install_github("NovembreLab/PCAviz",
                         build_vignettes = TRUE)
```

The following command in R generates plots showing each individual sample's position in the PCA space and the median position of each labelled group in PCA space:

```
library(PCAviz)
library(cowplot)
prefix <- "out/H938_Euro.LDprune"
nPCs <- 20

# Read in individual coordinates on PCs and eignvalues
PCA <- read.table(paste(prefix, ".eigs", sep = ""))
names(PCA) <- c("ID", paste("PC", (1:nPCs), sep = ""),
                "case.control")
PCA <- PCA[, 1:(nPCs + 1)] # Remove case/control column
eig.val <- sqrt(unlist(read.table(
  paste(prefix, ".eval", sep = "")))[1:nPCs])
sum.eig <- sum(unlist(read.table(
  paste(prefix, ".eval", sep = ""))))

# Read in snp weightings matrix
snpeigs <- read.table(paste(prefix, ".snpeigs", sep = ""))
names(snpeigs) <- c("ID", "chr", "pos",
                    paste("PC", (1:nPCs), sep = ""))
snpeigs$chr <- factor(snpeigs$chr)
rownames(snpeigs) <- snpeigs$ID
snpeigs <- snpeigs[, -1]

# Note smartpca pushes the plink family and individual
# ids together so we need to extract out the ids afresh
tmp <- unlist(sapply(as.character(PCA$ID), strsplit, ":"))
ids <- tmp[seq(2, length(tmp), by = 2)]
PCA$ID <- ids

# Read in the group/cluster labels
clst <- read.table("out/Euro.clst.txt")
# Order them to match the ids of PCA object
clst_unord <- clst$V3[match(ids, clst$V2)]
PCA <- as.data.frame(PCA)
PCA <- cbind(PCA, clst_unord)
names(PCA)[ncol(PCA)] <- "sample"
```

```
# Build the PCAviz object
hgdp <- pcaviz(dat = PCA, sdev = eig.val,
               var = sum.eig, rotation = snpeigs)
hgdp <- pcaviz_abbreviate_var(hgdp, "sample")

# Make PCA plots
geom.point.summary.params <- list(
  shape = 16, stroke = 1, size = 5,
  alpha = 1, show.legend = F
)
plot1 <- plot(hgdp,
  coords = paste0("PC", c(1, 2)), color = "sample",
  geom.point.summary.params = geom.point.summary.params,
  scale.pc.axes = 0.6
)
plot2 <- plot(hgdp,
  coords = paste0("PC", c(2, 3)), color = "sample",
  geom.point.summary.params = geom.point.summary.params,
  scale.pc.axes = 0.6
)
plot3 <- plot(hgdp,
  coords = paste0("PC", c(4, 5)), color = "sample",
  geom.point.summary.params = geom.point.summary.params,
  scale.pc.axes = 0.6
)
plot4 <- plot(hgdp,
  coords = paste0("PC", c(5, 6)), color = "sample",
  geom.point.summary.params = geom.point.summary.params,
  scale.pc.axes = 0.6
)

plot_grid(plot1, plot2, plot3, plot4)
```

First one may notice several populations are separated with PC1 and PC2, with the more isolated populations being those that were most distinguished from the others by ADMIXTURE (Fig. 6). PC4 distinguishes a subset of Orcadian individuals and PC5 distinguishes two Adygei individuals. PC6 corresponds to the cryptic structure observed within Sardinians in the ADMIXTURE analysis.

As an alternative visualization, it can be helpful to see the distribution of PC coordinates per population for each labeled group in the data (*see* Fig. 7):

```
pcaviz_violin(hgdp, pc.dims = paste0("PC", c(1:3)),
              plot.grid.params = list(nrow = 3))
```

As mentioned above in the section on LD, it is useful to inspect the PC loadings to ensure that they broadly represent variation across the genome, rather than one or a small number of genomic

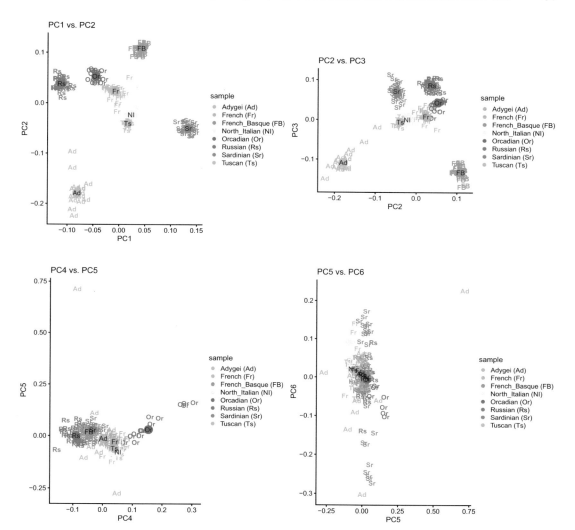

Fig. 6 Pairwise plots of PC scores generated using the PCAviz package

regions [7] (*see* Fig. 8). SNPs that are selected in the same direction as genome-wide structure can show high loadings, but what is particularly pathological is if the only SNPs that show high loadings are all concentrated in a single region of the genome, as might occur if the PCA is explaining local genomic structure (such as an inversion) rather than population structure.

```
for (i in 1:5) {
  plotname <- paste("plot", i, sep = "")
  plot <- pcaviz_loadingsplot(hgdp,
    pc.dim = paste0("PC", i),
    min.rank = 0.8, gap = 200, color = "chr",
    geom.point.params = list(show.legend = FALSE)
  ) +
    xlab("SNPs") + ylab(paste0("PC", i, " loading"))
```

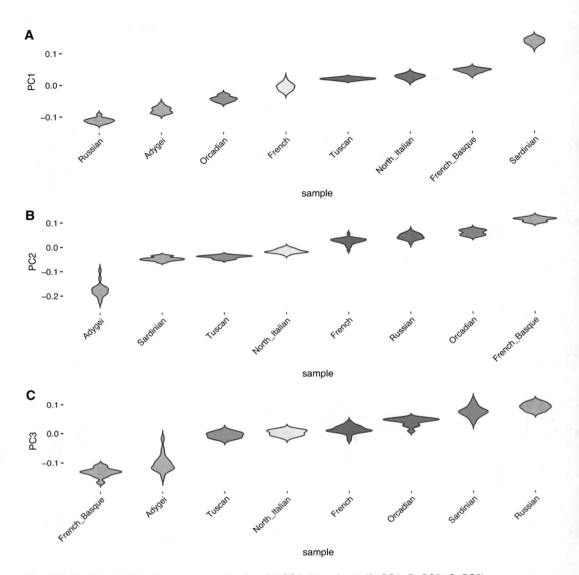

Fig. 7 Violin plots of PC values generated using the PCAviz package (A: PC1, B: PC2, C: PC3)

```
   assign(plotname, plot)
}
# grep common legend
plot <- pcaviz_loadingsplot(hgdp,
  pc.dim = paste0("PC", 1),
  min.rank = 0.8, gap = 200, color = "chr"
) +
  guides(color = guide_legend(nrow = 2, byrow = TRUE)) +
  theme(legend.position = "bottom",
        legend.justification = "center")
plot_legend <- get_legend(plot)
# plot loadings
prow <- plot_grid(plot1, plot2, plot3, plot4, plot5,
                  nrow = 5, align = "vh")
plot_grid(prow, plot_legend, ncol = 1, rel_heights = c(1, .2))
```

Fig. 8 PC loading plots generated using the PCAviz package

The proportion of total variance explained by each PC is a useful metric for understanding structure in a sample and for evaluating how many PCs one might want to include in downstream analyses (*see* Fig. 9). This can be computed as $\lambda_i/\Sigma_k\lambda_k$, with λ_i being eigenvalues in decreasing order, and is plotted below:

```
screeplot(hgdp, type = "pve") +
  ylim(0, 0.018) +
  ylab('Proportion of Variance Explained') +
  theme(axis.text.x = element_text(angle = 90, hjust = 1)) +
  theme(axis.line = element_line(size = 1, linetype = "solid"))
```

The results show that the top PCs only explain a small fraction of the variance ($<1.5\%$) and that after about $K=6$ the variance explained per PC becomes relatively constant; roughly in line with the visual inspection of the `admixture` results that revealed $K=6$ may be reasonable for this dataset.

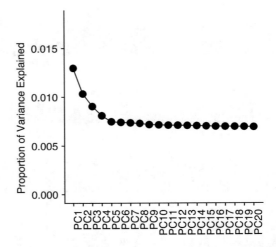

Fig. 9 Proportion of variance explained by each PC. Plot generated using the PCAviz package

4 Discussion

Our protocol above is relatively straightforward and presents the most basic implementation of these analyses. Each analysis software (ADMIXTURE and smartpca) and each visualization package (pong and PCAviz) contain numerous other options that may be suitable for specific analyses and we encourage the readers to spend time in the manuals of each. Nonetheless, what we have presented is a useful start and a standard pipeline that we use in our research.

Two broad perspectives we find helpful to keep in mind are: (1) How the admixture model and PCA framework are related to each other indirectly as different forms of sparse factor analysis [8]; (2) How the PCA framework in particular can be considered as a form of efficient data compression. Both of these perspectives can be helpful in interpreting the outputs of the methods and for appreciating how these approaches best serve as helpful visual exploratory tools for analyzing structure in genetic data. These methods are ultimately relatively simple statistical tools being used to summarize complex realities. They are part of the toolkit for analysis, and often are extremely useful for framing specific models of population structure that can be further investigated using more detailed and explicit approaches (such as those based on coalescent or diffusion theory, Chapters 7 on MSMC and 8 on CoalHMM).

References

1. Alexander DH, Lange K (2011) Enhancements to the admixture algorithm for individual ancestry estimation. BMC Bioinformatics 12:246. https://doi.org/10.1186/1471-2105-12-246

2. Alexander DH, Novembre J, Lange K (2009) Fast model-based estimation of ancestry in unrelated individuals. Genome Res 19 (9):1655–1664. https://doi.org/10.1101/gr.094052.109

3. Behr AA, Liu KZ, Liu-Fang G, Nakka P, Ramachandran S (2016) Pong: fast analysis and visualization of latent clusters in population genetic data. Bioinformatics 32(18):2817–2823. https://doi.org/10.1093/bioinformatics/btw327

4. Cann HM, de Toma C, Cazes L, Legrand MF, Morel V, Piouffre L, Bodmer J, Bodmer WF, Bonne-Tamir B, Cambon-Thomsen A, Chen Z, Chu J, Carcassi C, Contu L, Du R, Excoffier L, Ferrara GB, Friedlaender JS, Groot H, Gurwitz D, Jenkins T, Herrera RJ, Huang X, Kidd J, Kidd KK, Langaney A, Lin AA, Mehdi SQ, Parham P, Piazza A, Pistillo MP, Qian Y, Shu Q, Xu J, Zhu S, Weber JL, Greely HT, Feldman MW, Thomas G, Dausset J, Cavalli-Sforza LL (2002) A human genome diversity cell line panel. Science 296 (5566):261–262. https://doi.org/10.1126/science.296.5566.261b

5. Cavalli-Sforza LL, Menozzi P, Piazza A (1994) The history and geography of human genes. Princeton University Press, Princeton. https://doi.org/10.2307/2058750

6. Chang CC, Chow CC, Tellier LC, Vattikuti S, Purcell SM, Lee JJ (2015) Second-generation plink: rising to the challenge of larger and richer datasets. GigaScience 4(1): s13742–015–0047–8. https://doi.org/10.1186/s13742-015-0047-8

7. Duforet-Frebourg N, Luu K, Laval G, Bazin E, Blum MG (2016) Detecting genomic signatures of natural selection with principal component analysis: application to the 1000 genomes data. Mol Biol Evol 33(4):1082–1093. https://doi.org/10.1093/molbev/msv334

8. Engelhardt BE, Stephens M (2010) Analysis of population structure: a unifying framework and novel methods based on sparse factor analysis. PLOS Genet 6(9):1–12. https://doi.org/10.1371/journal.pgen.1001117

9. Falush D, Stephens M, Pritchard JK (2003) Inference of population structure using multi-locus genotype data: linked loci and correlated allele frequencies. Genetics 164(4):471–492

10. Falush D, van Dorp L, Lawson D (2016) A tutorial on how (not) to over-interpret structure/admixture bar plots. Nat Commun 9:3258. https://doi.org/10.1101/066431

11. Holsinger K, Weir B (2009) Genetics in geographically structured populations: defining, estimating and interpreting FST. Nat Rev Genet 10:639–650

12. Hubisz MJ, Falush D, Stephens M, Pritchard JK (2009) Inferring weak population structure with the assistance of sample group information. Mol Ecol Resour 9(5):1322–1332. https://doi.org/10.1111/j.1755-0998.2009.02591.x

13. Kermani BG (2006) Artificial intelligence and global normalization methods for genotyping. U.S. Patent No. 7,035,740. Washington, DC: U.S. Patent and Trademark Office

14. Kopelman NM, Mayzel J, Jakobsson M, Rosenberg NA, Mayrose I (2015) Clumpak: a program for identifying clustering modes and packaging population structure inferences across K. Mol Ecol Resour 15(5):1179–1191. https://doi.org/10.1111/1755-0998.12387

15. Li JZ, Absher DM, Tang H, Southwick AM, Casto AM, Ramachandran, S, Cann HM, Barsh GS, Feldman M, Cavalli-Sforza LL, Myers RM (2008) Worldwide human relationships inferred from genome-wide patterns of variation. Science 319(5866):1100–1104. https://doi.org/10.1126/science.1153717

16. Menozzi P, Piazza A, Cavalli-Sforza LL (1978) Synthetic maps of human gene frequencies in Europeans. Science 201(4358):786–792

17. McVean G (2009) A genealogical interpretation of principal components analysis. PLoS Genet 5(10):e1000686. https://doi.org/10.1371/journal.pgen.1000686

18. Novembre J (2014) Variations on a common structure: new algorithms for a valuable model. Genetics 197(3), 809–811. https://doi.org/10.1534/genetics.114.166264

19. Novembre J (2016) Pritchard, Stephens, and Donnelly on population structure. Genetics 204(2):391–393. https://doi.org/10.1534/genetics.116.195164

20. Novembre J, Peter BM (2016) Recent advances in the study of fine-scale population structure in humans. Curr Opin Genet Dev 41:98–105. https://doi.org/10.1016/j.gde.2016.08.007

21. Novembre J, Stephens M (2008) Interpreting principal component analyses of spatial population genetic variation. Nat Genet 40

(5):646–649. https://doi.org/10.1038/ng. 139

22. Novembre J, Johnson T, Bryc K, Kutalik Z, Boyko AR, Auton A, Indap A, King K, Bergmann S, Nelson M, Stephens M, Bustamante C (2008) Genes mirror geography within Europe. Nature 456:274

23. Patterson NJ, Price AL, Reich D (2006) Population structure and eigenanalysis. PLoS Genet 2(12):2074–2093. https://doi.org/10.1371/journal.pgen.0020190

24. Purcell S, Neale B, Todd-Brown K, Thomas L, Ferreira, ARM, Bender D, Maller J, Sklar P, de Bakker IWP, Daly M, Sham CP (2007) PLINK: a tool set for whole-genome association and population-based linkage analyses. Am J Hum Genet 81, 559–575

25. Price AL, Patterson NJ, Plenge RM, Weinblatt ME, Shadick NA, Reich D (2006) Principal components analysis corrects for stratification in genome-wide association studies. Nat Genet 38(8):904–909. https://doi.org/10.1038/ng1847

26. Pritchard JK, Stephens M, Donnelly P (2000) Inference of population structure using multilocus genotype data. Genetics 155(2):945–959

27. Raj A, Stephens M, Pritchard JK (2014) fastSTRUCTURE: variational inference of population structure in large SNP data sets. Genetics 197(2):573–589. https://doi.org/10.1534/genetics.114.164350

28. Rosenberg NA (2004) Distruct: a program for the graphical display of population structure. Mol Ecol Notes 4(1):137–138. https://doi.org/10.1046/j.1471-8286.2003.00566.x

29. Rosenberg NA, Mahajan S, Ramachandran S, Zhao C, Pritchard JK, Feldman MW (2005) Clines, clusters, and the effect of study design on the inference of human population structure. PLoS Genet 1(6):e70. https://doi.org/10.1371/journal.pgen.0010070

30. Tian C, Plenge RM, Ransom M, Lee A, Villoslada P, Selmi C, Klareskog L, Pulver AE, Qi L, Gregersen PK, Seldin MF (2008) Analysis and application of european genetic substructure using 300 K SNP information. PLoS Genet 4(1):e4. https://doi.org/10.1371/journal.pgen.0040004

31. Williams R, Pourreza H, Wang Y, Carbonetto P, Novembre J (2017) PCAviz: visualizing principal components analysis. http://github.com/NovembreLab/PCAviz

Chapter 5

Detecting Positive Selection in Populations Using Genetic Data

Angelos Koropoulis, Nikolaos Alachiotis, and Pavlos Pavlidis

Abstract

High-throughput genomic sequencing allows to disentangle the evolutionary forces acting in populations. Among evolutionary forces, positive selection has received a lot of attention because it is related to the adaptation of populations in their environments, both biotic and abiotic. Positive selection, also known as Darwinian selection, occurs when an allele is favored by natural selection. The frequency of the favored allele increases in the population and, due to genetic hitchhiking, neighboring linked variation diminishes, creating so-called selective sweeps. Such a process leaves traces in genomes that can be detected in a future time point. Detecting traces of positive selection in genomes is achieved by searching for signatures introduced by selective sweeps, such as regions of reduced variation, a specific shift of the site frequency spectrum, and particular linkage disequilibrium (LD) patterns in the region. A variety of approaches can be used for detecting selective sweeps, ranging from simple implementations that compute summary statistics to more advanced statistical approaches, e.g., Bayesian approaches, maximum-likelihood-based methods, and machine learning methods. In this chapter, we discuss selective sweep detection methodologies on the basis of their capacity to analyze whole genomes or just subgenomic regions, and on the specific polymorphism patterns they exploit as selective sweep signatures. We also summarize the results of comparisons among five open-source software releases (SweeD, SweepFinder, SweepFinder2, OmegaPlus, and RAiSD) regarding sensitivity, specificity, and execution times. Furthermore, we test and discuss machine learning methods and present a thorough performance analysis. In equilibrium neutral models or mild bottlenecks, most methods are able to detect selective sweeps accurately. Methods and tools that rely on linkage disequilibrium (LD) rather than single SNPs exhibit higher true positive rates than the site frequency spectrum (SFS)-based methods under the model of a single sweep or recurrent hitchhiking. However, their false positive rate is elevated when a misspecified demographic model is used to build the distribution of the statistic under the null hypothesis. Both LD and SFS-based approaches suffer from decreased accuracy on localizing the true target of selection in bottleneck scenarios. Furthermore, we present an extensive analysis of the effects of gene flow on selective sweep detection, a problem that has been understudied in selective sweep literature.

Key words Positive selection, Selective sweep, Software tools, Summary statistics, Machine learning

1 The Selective Sweep Theory

When a strongly beneficial mutation occurs and spreads in a population, the frequency of linked neutral (or weakly negatively selected) variants will increase. In a seminal paper, Smith and

Julien Y. Dutheil (ed.), *Statistical Population Genomics*, Methods in Molecular Biology, vol. 2090, https://doi.org/10.1007/978-1-0716-0199-0_5, © The Author(s) 2020

Haigh [86] described this process, for which they coined the term genetic hitchhiking. They showed that in large populations, where random genetic drift is negligible, hitchhiking can drastically reduce genetic variation near the site/locus favored by natural selection. Due to the local reduction of genetic diversity, which is swept by natural selection, the process is called "selective sweep."

The selective sweep model predicts that in recombining chromosomal regions diversity vanishes at the site of selection immediately after the fixation of the beneficial allele. Due to recombination, genetic diversity is predicted to increase as a function of the distance to the selected site (scaled by the selection coefficient and the recombination rate). As a result, the genetic diversity is maintained due to recombination in genomic regions that are in the proximity of a selective sweep: SNPs are not generated by novel mutations, but they are old mutations that escaped selection because of recombination. This result is also roughly correct in finite populations [49, 90]. Further signatures of the hitchhiking effect include (1) shifts in the SFS of polymorphisms such as an excess of low- and high-frequency derived alleles [22, 37], and (2) an elevated level of LD in the early phase of the fixation process of a beneficial mutation [51, 91]. It is important to note that the aforementioned signatures of a selective sweep are predicted when (1) fixation of the beneficial mutation has just been completed; (2) recombination rate is positive, i.e., the chromosome is recombining; (3) the population size is approximately constant over time; (4) the population is isolated; (5) no gene conversion has occurred in the proximity of the beneficial mutation. Despite the relatively strict assumptions of the selective sweep model, several tests have been developed that exploit the properties of the hitchhiking effect to map recent, strong, positive directional selection along recombining chromosomes of several species.

Searching for strong positive selection in the genomes of individuals of a natural population has been the focus of a multitude of studies over the past years [3, 8, 41, 52, 67, 78, 97, 100, 102]. The goals of these studies have been (1) to provide evidence of positive selection, (2) estimate the strength of selection, and (3) localize the targets of selection. Thus, these studies aim to provide insights into the genetical mechanisms of adaptation either in wild populations or during domestication. A long-term goal is that the genes that experienced recent and strong positive selection could be identified and the associated functions and phenotypes characterized.

Early studies of selective sweep localization followed a two-tier approach: at first, levels of DNA polymorphism were measured for a very large number of loci on a genome-wide scale within populations. The goal of this initial step was to identify loci with reduced diversity compared to divergence with another species. The diversity–divergence contrast highlights regions with reduced intra-population diversity compared to what is expected from the divergence data. Thus, divergence is treated as a proxy for the mutation rate. Some studies employed microsatellite markers to measure polymorphism and searched for regions of depleted variability as an indicator of a selective sweep due to genetic hitchhiking in the region. In the second step, a thorough sequencing of the candidate regions was performed and a selective sweep detection pipeline was executed. A statistical problem related to this procedure springs from the fact that regions analyzed for the occurrence of a selective sweep do not represent a random fraction of the genome. Instead, they are outliers since they are characterized by decreased amounts of diversity. A proper statistical testing for the hypothesis of a selective sweep requires the null distribution of the statistic to be built from neutral regions with the same properties (e.g., outliers for diversity levels) [93]. With the advent of next generation sequencing, the candidate gene approach is replaced by full genome screenings for positive selection, thus the statistical problem of testing outlier genes for positive selection is diminished at least for the model organisms. For non-model organisms, where a reference genome is still missing a candidate gene approach could provide insights into their adaptation processes.

2   Methods to Detect Selective Sweeps in Genome-Wide Data

2.1   Detecting Sweeps Based on Diversity Reduction

The most striking and persistent effect of genetic hitchhiking is the reduction of diversity. Smith and Haigh [86] predicted the reduction of heterozygosity immediately after the fixation of the beneficial mutation. Especially in genomic regions with reduced recombination rate per physical distance, the reduction of diversity is expected to be evident. Subsequent studies [1, 2, 15, 53, 62, 89, 90] confirmed this prediction for *D. melanogaster*, *D. simulans*, and *D. ananassae* species. Charlesworth et al. [27], however, showed that a similar prediction holds for background selection as well: if neutral variants are linked to a strongly deleterious mutation, the level of polymorphism diminishes while the deleterious mutation is gradually removed from the population. The amount of polymorphism reduction depends on the selection coefficient of the deleterious mutation [35]. For example, for lethal mutations there is no polymorphism reduction effect since it is being directly removed from the population. Innan and Stephan [47] demonstrated that in a hitchhiking model, the estimated level of diversity, $\hat{\theta}$, is negatively correlated with $\hat{\theta}/\rho$, where ρ is the recombination rate. In contrast, in a background selection model, the estimated level of diversity is positively correlated with the same quantity (see also ref. 88 for a review).

2.2 The SFS Signature of a Selective Sweep

The studies by Braverman et al. [22] and Fay and Wu [37] showed that a selective sweep shifts the SFS toward high- and low-frequency derived variants. Neutral variants that are initially linked to the beneficial variant increase in frequency, whereas variants that are initially not linked to the beneficial variant decrease in frequency during the fixation of the beneficial mutation.

A breakthrough on detecting selective sweeps was proposed by Kim and Stephan [52], known as the Kim and Stephan test. They developed a composite-likelihood-ratio (CLR) test to compare the probability of the observed polymorphism data under the standard neutral model with the probability of observing the data under a model of selective sweep. The Kim and Stephan test is a maximum-likelihood-based test that reports the value of $a = 4N_e s$, where s is the selection coefficient that maximizes the CLR. The Kim and Stephan test was the first to implement a CLR test on sweep detection. Due to its inefficient implementation, however, it has been used to detect selection only in candidate loci [16, 80]. Furthermore, it adopts several oversimplified assumptions. First, the neutral model was derived by an equilibrium neutral population, i.e., a population with constant population size. Second, the selection model was derived by Fay and Wu's model [37], where only the low- and the high-frequency derived classes are assumed.

2.3 The LD Signature of a Selective Sweep

The third signature of a selective sweep refers to a specific pattern of LD that emerges in the neighborhood of the beneficial mutation. Upon fixation of the beneficial mutation, elevated levels of LD emerge on each side of the selected site, whereas a decreased LD level is observed between polymorphisms found on different sides of the selected site. The high LD levels on the different sides of the selected locus are due to the fact that *a single* recombination event allows multiple polymorphisms *on the same side of the sweep* to escape the sweep. Between those SNPs the level of LD will be high. On the other hand, polymorphisms that reside on different sides of the selected locus need a minimum of two recombination events, thus LD is decreased. Figure 1 shows an example of the LD patterns emerging after a sweep.

The LD-based signature of a selective sweep was proposed and thoroughly investigated by Kim and Nielsen [51]. In this study, Kim and Nielsen introduced a simple statistic, named *ω-statistic*, that facilitates the detection of the specific LD patterns that emerge after a sweep. For a window of W SNPs that is split into two non-overlapping subregions L and R, with l and $W - l$ SNPs, respectively, the *ω-statistic* is computed as follows:

$$\omega = \frac{\left(\binom{l}{2} + \binom{W-l}{2}\right)^{-1}\left(\sum_{i,j\in L} r_{ij}^2 + \sum_{i,j\in R} r_{ij}^2\right)}{(l(W-l))^{-1}\sum_{i\in L, j\in R} r_{ij}^2}. \tag{1}$$

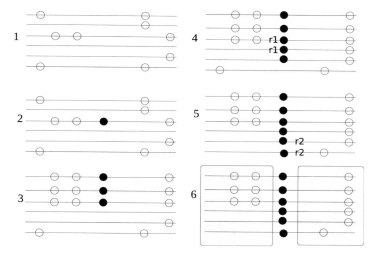

Fig. 1 The LD signature of a complete hard selective sweep. Assume a population with neutral segregating variation (1). A beneficial mutation occurs (shown as a black allele) in subfigure (2). Since the mutation is beneficial, its frequency will increase in the population. Neutral variants that are linked to the beneficial mutation will hitchhike with it (3). Due to recombination, mutations from a neutral background will get linked with the beneficial mutation (4, 5). The recombination events are depicted on the locations of the involved chromosomes by r1 and r2, respectively. Finally, the selective sweep completes (6). The LD pattern that emerges from such a process is the elevated LD on each side of the beneficial mutation and the decreased LD for SNPs that are on different sides of the beneficial mutation. The figure is adapted from [69]

2.4 Detecting Sweeps Using Machine Learning Methods

The process of detecting genomic regions that have been affected by positive selection can be treated as a classification problem for which each genomic region is classified as either neutral or selected. If the parameters of the selective sweep and the demographic model are known, then disentangling a selective sweep from demography can be treated as a typical binary classification problem. In computer science and mathematics, theoretical and algorithmic advancements have been developed the last decades that perform classification of datasets. These advancements can be grouped as machine learning methods, because first they train computers to understand patterns from the data, and then use this knowledge to classify an unknown sample. Their application in population genetics still remains limited, even though the last years a few methods have been developed [57, 71, 82]. The first application of machine learning in population genetics to our knowledge was developed by Pavlidis et al. [71], who used a support vector machine approach to perform the classification. Pavlidis et al. [71] used as features results from the CLR test (SFS-based) and the $\omega - statistic$ as well as the difference between the locations that each of the aforementioned tests pinpoint. Lin et al. [57] also developed a machine learning approach based on the "boosting" algorithm, a statistical method that combines simple classification rules using summary statistics to maximize their joint predictive performance. More recently, Schrider and Kern [82] proposed an extremely randomized trees classifier to identify soft selective sweeps, hard selective sweeps, their linked regions, and neutral regions. Their software is called "S/HIC." A new version of "S/HIC" (called diploS/HIC) was proposed by [50] that can also use unphased genotypes in contrast to "S/HIC." The application of machine learning tools in population genetics has been reviewed in [83].

Typically, in a supervised learning problem, the goal is to accurately predict previously unseen data based on a set of already seen data (training data). The problem can be formulated as training the computer to recognize the combinations of feature-values that are associated with either of the classes. Here, the class of each data point is encoded as "neutrality/selection." In contrast to other disciplines in which machine learning methods are applied, the number of well-annotated examples that the algorithm requires for its training is limited. In fact, all "known" targets of selection do not represent any established truth but are predictions of algorithms that are mostly based on simplistic models. Even though there is a general agreement about the validity of positive selection detection in loci such as the LCT [17], the historical truth, i.e., whether a locus was indeed selected by natural selection remains unknown. Even if we did know the definite true targets of selection, it would still be challenging to build an accurate predictor based on them. The reason is that those training examples would be obtained from heterogeneous populations that have experienced and would incorporate a multitude of other evolutionary forces besides positive selection. A remedy for the aforementioned problems is to use simulated results for the training of the machine learning algorithms. On one hand, simulated data ensure the control of heterogeneity of the training samples as well as the correctness of the assigned class. On the other hand, the simulation process does not capture the whole set of stochastic processes that affect the data. Thus, even though training and evaluation processes perform well on simulated data, they might perform poorly on real data. In this study, we present an extensive testing of machine learning methodologies in Subheading 6.

3   The Problem of Demography

Demography poses severe challenges on the selection detection process due to the fact that it may generate SNP patterns that resemble the signatures of genetic hitchhiking. In recombining chromosomes, selective sweep detection becomes feasible mainly due to two factors: (1) the fixation of the beneficial mutation, and (2) the fact that coalescent events occur at a higher rate in the presence of a sweep than they do in its absence. It is these two factors, along with *recombination events*, that generate the specific signatures of a selective sweep, enabling us to detect traces of positive selection in genomes. However, additional factors can also trigger a high rate of coalescent events, leading to the generation of similar (to a selective sweep) signatures in the genome, and thus misleading current selective sweep detection approaches. For instance, assume a bottleneck event that is characterized by three phases: (1) a recent phase of large effective population size, (2) a second phase, prior to the first one, of small population size (the bottleneck phase), and (3) an ancestral period of large population size. It is due to the decrease of the effective population size in the bottleneck phase that a high rate of coalescent events occur in a relatively short period of time. Furthermore, lineages can escape the bottleneck, passing to the ancestral phase of large effective population size, and therefore requiring more time to coalesce. In a recombining chromosome, genomic regions that are characterized by short coalescent trees due to massive coalescent events may alternate with genomic regions with lineages that have escaped the bottleneck phase (*see* Fig. 2). Such alternations can generate SNP

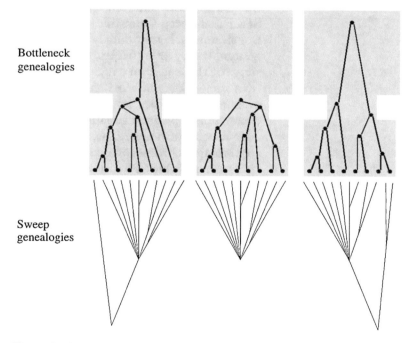

Fig. 2 Bottleneck demographic scenarios (top panel) may result in similar genealogies to a selective sweep (bottom panel). Both models may produce very short coalescent trees. As we move from the selection site, selective sweeps produce genealogies with long internal branches. Similarly, bottlenecks may produce genealogies with very long internal branches if the ancestral population size is large

patterns that are highly similar to those generated by a selective sweep, yielding the detection process very challenging, if not infeasible [70].

Besides demographic bottlenecks, other demographic scenarios may also generate SNP patterns that resemble those of a selective sweep. Recently, Alachiotis and Pavlidis [5] demonstrated that gene flow (migration) between populations poses severe challenges to existing sweep detection methods, suggesting that appropriate sweep signatures for migration models are yet to be found (figure 2 in [5]). Similarly, De and Durrett [31] demonstrated that both the LD and the SFS are affected if a stepping stone spatial structure characterizes the population; specifically, the LD decay becomes slower and the SFS is shifted toward high-frequency derived variants for migration rates that are intermediate ($4Nm = 3$, where N is the effective population size and m the probability of migration per individual and per generation; figure 5 in [31]). Similar results are obtained from island models.

It is generally believed that, unlike the localized effect of a selective sweep, neutral demographic changes generate genome-wide patterns. This idea of "local sweep effects" *vs.* "global demographic effects" in the genome has been extensively used to control the demography-induced false positive rates [56, 65, 73]. In SFS-based sweep scans, this idea translates to a two-step computational approach that entails the initial estimation of an average, genome-wide SFS (background SFS) followed by a detection step, for those genomic regions that fit the selection model better than the background SFS. An issue with such an approach, however, is that it does not take into account the fact that SFS is characterized by great variation along the genome. In bottlenecks, or in models with gene flow, which generate great variance along a recombining chromosome [13, 26, 70, 93], the usage of the average, genome-wide SFS may be problematic. Therefore, under certain bottleneck demographic scenarios, there can be neutral-like genomic regions, as well as sweep-resembling ones, regardless of the actual existence of a selective sweep. Since both recombination and the alternation of genealogies along a recombining chromosome are stochastic, it is highly challenging to determine which genealogies are shaped by neutrality and which genealogies are shaped by positive selection. Current approaches are not able to completely overcome the confounding effect of bottlenecks on positive selection in recombining chromosomes, therefore users should be careful when interpreting results of selective sweep scans. It should be noted however, that several tools, such as SweepFinder, SweepFinder2, SweeD, OmegaPlus, and RAiSD and/or the deployment of the demographic model as the null model, contribute to alleviating the problem generated by the confounding effects of demography.

Demography not only affects the false positive rate (FPR) of the detection methods, but it also affects the true positive rate (TPR). This derives from the fact that the SNP patterns that emerge from the combined action of *demography and selection* are unknown. For instance, the SFS-based tools SweepFinder and SweeD (presented in a following section) assume that if a lineage escapes the selective sweep due to a recombination event, then, prior to the sweep, its frequency is given by the neutral (or background) SFS. This is valid if the selective sweep has occurred in a constant-size population. If, however, the population has experienced population size changes (or other demographic events such as migrations), this assumption does not necessarily hold.

Given the challenges that demographic changes pose to the accurate detection of positive selection, it is unfortunate (even though expected) that most natural populations have experienced various demographic scenarios during their evolutionary history. For example, the European population of D. *melanogaster* experienced a severe bottleneck about 15,800 years ago, when the European population diverged from the African population [56]. The duration of the bottleneck was about 340 years and the effective population size during the bottleneck was only 2200 individuals [56], thus the effective population size of the European population was decreased by a factor of 500, approximately. Regarding the demography of human populations, the proposed models suggest several bottleneck (founder) events and migrations between subpopulations [36]. Domesticated animals have also experienced a series of bottleneck events during the domestication process. Using only mtDNA and the approximate Bayesian computation methodology, Gerbault et al. [40] report that goats have experienced severe bottleneck events during their domestication. Approximate Bayesian computation was also used to provide insights into the demographic history of silkworm [99]. Using 17 loci in the domesticated silkworm, they reported that the most plausible scenario explaining the demographic history of silkworm comprises both bottleneck and gene flow events [99].

4   A Guideline on Selection Detection Tools

4.1   Summary Statistics

Summary statistics are computationally inexpensive data calculations. On whole-genome data, typically they are applied following a sliding window approach. Simpler statistics such as Tajima's D or the SNP count do not require phased data, but only SNP calling, whereas LD-based ones require phased data. Several summary statistics serve as neutrality tests because their distributions are affected by the presence of positive selection (for example, Tajima's D obtains negative values in the proximity of a strongly beneficial allele).

Relying on Tajima's D, Braverman et al. [22] were able to detect genomic regions affected by recent and strong positive selection in simulated datasets, as well as to demonstrate that in regions of low genetic diversity and low recombination rate (e.g., around centromeres or at telomeres) a simple hitchhiking model is not a sufficient explanation for the observed DNA polymorphisms. Since then, Tajima's D has been deployed in numerous studies as a neutrality test to detect selection [12, 19, 20, 58, 67, 79, 94]. This summary statistic captures the difference between two estimates of the diversity level $\theta = 4 N_e \mu$, where μ is the mutation rate. The first estimate, π, is based on the number of pairwise differences between sequences, while the second one, Watterson's θ (θ_W), is based on the number of polymorphic sites. Tajima's D obtains negative values in the proximity of a selective sweep, since π decreases with both high- and low-frequency derived variants, while θ_W remains unaffected.

In 2000, Fay and Wu [37] proposed a new statistic, H, which obtains low values in regions where high-frequency derived variants are overrepresented. To distinguish between high- and low-frequency derived variants, Fay and Wu's H relies on an out-group sequence. Additionally, Fay and Wu [37] invented a new unbiased estimator for θ, named θ_H, which assumes high values in regions with overrepresented high-frequency derived variants. The H statistic is defined as the difference between π and θ_H, and as such it becomes significantly negative in the proximity of a beneficial mutation. Since a back-mutation will result in the incorrect inference of the derived polymorphic state, Fay and Wu's H requires the probability of mis-inference to be incorporated in the construction of the null distribution of the statistic. In 2006, Zeng et al. [101] improved the H statistic by adding the variance of the statistic in the denominator, thus scaling H by the variance of the statistic.

Depaulis and Veuille [34] introduced two neutrality tests relying on haplotypes. The first summary statistic, K, is simply the number of distinct haplotypes in the sample. In the presence of a selective sweep K takes low values. The second test measures haplotype diversity, denoted by H (or DVH, Depaulis and Veuille H, to be distinguished from Fay and Wu's H). DVH is calculated as $DVH = 1 - \sum_{i=1}^{K} p_i^2$, where p_i is the frequency of the ith haplotype. Both the DVH and the K summary statistics are conditioned on the number of polymorphic sites, s, which yields the construction of the null (neutral) distribution of the statistic rather problematic. Depaulis and Veuille simulated data using a fixed number of polymorphic sites s, and without conditioning on the coalescent trees. This approach is suboptimal because the number of polymorphic sites is a random variable that follows a Poisson distribution, and it is determined by the total length of the (local) coalescent tree and the mutation rate. Thus, to construct the null distribution of the statistic, a two-step approach is required: first, a coalescent tree is generated according to the demographic model and mutations are placed randomly on its branches (this step can be achieved using Hudson's ms [45]), and second, a rejection process is applied in order to condition on the number of polymorphic sites s, during which only the simulations that produced s segregating sites are kept while the rest are discarded. Thus, only a subset of coalescent trees will be accepted: the trees that given the mutation rate result in the specified number of segregating sites s.

Typically, summary statistics are applied on whole-genome data following a sliding-window approach. This allows efficient computations on large datasets for those statistics used as neutrality tests, introducing, however, two main problems. The fixed size of the window length creates the first problem since small changes (even by only a few bases) of the window length may shift the results from statistically non-significant to significant [72], regardless of whether the window size is measured in number of base pairs or number of SNPs. The second problem, which is common for most neutrality tests, is that they are not robust to demographic changes of the population. For instance, Tajima's D can assume negative values in a population expansion scenario as well as locally in genomic regions under a bottleneck scenario. It also becomes negative in genomic regions that have experienced purifying selection and in regions affected by positive selection. Fay and Wu's H can become negative in demographic models that increase the high-frequency derived variants. Such demographic models include gene flow [31] or sampling from one deme that is part of a metapopulation [87].

4.2 Detecting Sweeps in Whole Genomes

The advent of next generation sequencing (NGS) allowed the analysis of whole genomes at different geographic locations and environmental conditions, and revealed a need for more efficient processing solutions in order to handle the increased computational and/or memory requirements generated by large-scale NGS data. While typical summary statistics are generally suitable for NGS data, they are applied on fixed-size windows, and as a result they do not provide any insight on the extent of a selective sweep. More advanced methods that rely on the CLR test (e.g., SweepFinder [65], SweepFinder2 [33], and SweeD [73]) or on patterns of LD (e.g., OmegaPlus [6, 7]) perform an optimization on the size of the window and, therefore, they provide information on the genomic region affected by a selective sweep at the cost of increased execution times. The aforementioned methods have been widely used to detect recent and strong positive selection in a variety of eukaryotic or prokaryotic organisms, such as human [18, 65, 75], *D. melanogaster* [11, 25, 95, 98], lizards [54], rice [24], butterflies [59], and bacteria [63].

4.2.1 SweepFinder

In 2005, Nielsen et al. [65] released SweepFinder, an advanced method to detect selective sweeps that relies on information directly derived from the SFS, either folded or unfolded. Sweep-Finder implements a composite likelihood ratio (CLR) test. The numerator of SweepFinder represents the likelihood of a sweep at a given location in the genome, given its selection intensity α. The denominator accounts for the neutral model. An important feature of SweepFinder is that neutrality is modeled based on the empirical SFS of the entire dataset. All SNPs are considered independent, therefore allowing the likelihood score per region for the sweep model to be computed as the product of per-SNP likelihood scores over all SNPs in a region. SweepFinder was among the first software releases with the capacity to analyze whole genomes via a complete and standalone implementation. SweepFinder can process small and moderate sample sizes efficiently. However, the source code does not handle floating-point exceptions that occur when a large number of sequences are analyzed, yielding analyses with more than 1027 sequences impossible.

4.2.2 SweeD

Pavlidis et al. [73] released SweeD (**Swee**p **D**etector), a stable, parallel, and optimized implementation of the same CLR test as SweepFinder. SweeD can parse various input file formats (e.g., Hudson's ms, FASTA, and the Variant Call Format) and provides the option to employ a user-specified demographic model for the theoretical calculation of the expected neutral SFS. Also, it allows the user to provide her/his own points of interest where the CLR will be assessed (via the gridfile option). Pavlidis et al. [73] showed that sweep detection accuracy increases with an increasing sample size, and altered the mathematical operations for the CLR test implementation in SweeD to avoid numerical instability (floating-point underflows), allowing the analysis of datasets with thousands of sequences. The time-efficient analysis of large-scale datasets in SweeD is mainly due to two factors: (a) parallel processing using POSIX threads, and (b) temporary storage of frequently used values in lookup tables. Additionally, SweeD relies on a third-party library for checkpointing (Ansel et al. [10]) to allow resuming long-running analyses that have been abruptly interrupted by external factors, such as a power outage or a job queue timeout.

4.2.3 SweepFinder2

More recently, DeGiorgio et al. [33] released SweepFinder2. SweepFinder2 uses the statistical framework of SweepFinder, and additionally it takes into account local reductions in diversity caused by the action of negative selection. Therefore, it provides the opportunity to distinguish between background selection and the effect of selective sweeps. Thus, it exhibits increased sensitivity and robustness to background selection and mutation rate variations. Besides the ability to account for reductions in the diversity caused by background selection, the implementation of SweepFinder2 is

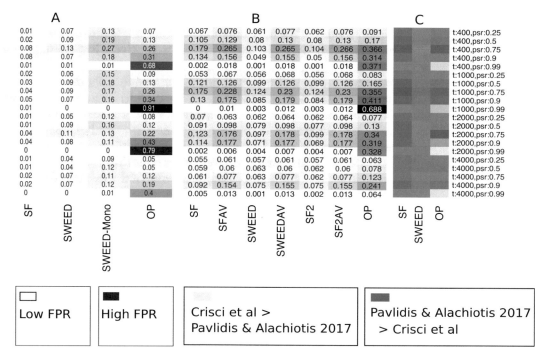

Fig. 3 False positive rates for the selective sweep detection process under various algorithms and demographic models. Demographic models consist of bottlenecks and are characterized by two parameters: t is the time in generations since the recovery of the populations, and psr the relative population size reduction during bottleneck. Prior to the bottleneck, the population size equals to the present-day population size. We show the results from the study of Crisci et al. [30] (**a**), our analysis in the current study (**b**) and the difference between **a** and **b** (**c**). Note that Crisci et al. studied SweepFinder (SF), SweeD (SWEED), SweeD with monomorphic (SWEED-Mono), and OmegaPlus (OP). In the current work, we studied SweepFinder (SF), SweepFinder with average SFS (SWEEDAV), SweeD (SWEED), SweeD with average SFS (SWEEDAV), SweepFinder2 (SF2), SweepFinder2 with average SFS (SF2AV), and OmegaPlus. Thus, in (**c**) we show only results from the common tools (SF, SWEED, OP). In (**a**) and (**b**), the darker a cell, the lower the false positive rate. In (**c**), yellow denotes that Crisci et al. report higher false positive rate than [69] while blue denotes that the reported false positive rate by Crisci et al. is lower. The figure is adapted from [69]

very similar to SweepFinder. However, there exist code modifications that increase the stability of SweepFinder2 on the calculation of likelihood values. Using simulated data with constant mutation rate and in the absence of negative selection, SweepFinder2 scores are closer to those obtained by SweeD rather than the initial SweepFinder implementation (*see* Fig. 3).

4.2.4 OmegaPlus

In 2012, Alachiotis et al. [7] released a high-performance implementation of the ω-*statistic* [51] for the detection of selective sweeps by searching for a specific pattern of LD that emerges in the neighborhood of a recently fixed beneficial mutation. The ω-*statistic* assumes a high value at a specific location in the genome, which can be indicative of a potential selective sweep in the region, if extended contiguous genomic regions of high LD are detected on both sides of the location under evaluation, while the level of LD between the high LD regions remains relatively low. OmegaPlus evaluates multiple locations along a dataset following an exhaustive per-region evaluation algorithm, which was initially introduced by Pavlidis et al. [71]. The algorithm by Pavlidis et al. [71] required large memory space for the analysis of many-SNP regions and exhibited increased complexity, yielding the analysis of regions with thousands of SNPs computationally unfeasible. OmegaPlus introduced a dynamic programming algorithm to reduce the computational and memory requirements of the exhaustive evaluation algorithm, enabling the efficient analysis of whole-genome datasets with millions of SNPs. OmegaPlus exhibits a series of four different parallelization alternatives [4, 6] for the distribution of computations to multiple cores to overcome the load balancing problem in selective sweep detection due to the difference in SNP density between regions in genomes.

4.2.5 MFDM Test

In 2011, Li et al. [55] presented a neutrality test that detects selective sweep regions using the maximum frequency of derived mutations (MFDM), which is a paramount signature of a selective sweep. According to [55], the MFDM test is robust to processes that occur in a single and isolated population. This is because there is no demographic scenario in single and isolated populations that generates a non-monotonic SFS and increases the amount of high-frequency derived variants. Thus, at least in theory, the test is robust to demographic models, such as bottlenecks, when they occur in isolated populations. However, four severe problems arise regarding the robustness of the test, which broadly apply to other tests of neutrality as well: (1) although bottlenecks generate monotonic *average* SFSs, certain genomic regions may locally exhibit increased amounts of high-frequency derived variants, even in the absence of positive selection, (2) high-frequency derived variants are a signature of selective sweeps in constant populations but it is not known whether and how they will be affected by the combined action of selection and demography, (3) in populations that exchange migrants with other demes (non-isolated), the frequency of high-frequency derived variants may increase (e.g., [31]), and (4) back-mutations (in general, the violation of the infinite-site model) may also increase the amount of high-frequency derived variants.

4.3 RAiSD

In 2018, Alachiotis and Pavlidis [5] introduced the μ statistic and released RAiSD (Raised Accuracy in Sweep Detection). The μ statistic is a composite evaluation test that scores genomic locations by relying on the enumeration of SNP vector patterns (entire alignment columns) to quantify changes in the SFS, the levels of LD, and the amount of genetic variation. RAiSD implements a SNP-driven, sliding-window algorithm that reuses calculated data between overlapping windows to considerably reduce execution times. It exhibits increased detection accuracy and sensitivity due to the fact that consecutive SNP windows with variable size in terms of base pairs are placed along a dataset with a step of 1 SNP. This achieves increased granularity in SNP-dense regions and avoids redundant operations in SNP-sparse ones, consequently improving processing speed without deteriorating the quality of the results. Furthermore, RAiSD couples the sliding window algorithm with an out-of-core approach that allocates a negligible amount of memory (typically few MBs) irrespectively of the dataset size, thus maintaining overall low memory requirements. Details on RAiSD software, its command line options, as well as working examples are available on the github repository of RAiSD (https://github.com/alachins/raisd).

5   Evaluation

The aforementioned software tools (SweepFinder, SweepFinder2, SweeD, and OmegaPlus, and RAiSD *see* Table 1) have been independently evaluated by three studies: Crisci et al. [30] studied the effect of demographic model misspecification on selective sweep detection, while Alachiotis and Pavlidis [4] conducted a performance comparison in terms of execution time for various dataset sizes and number of processing cores. Alachiotis and Pavlidis [5] evaluated all tools in terms of detection accuracy, sensitivity, and execution time, with the aim to assess RAiSD. We summarize these results in the following subsections and partially reproduce the FPR evaluation analysis by Crisci et al. [30], including SweepFinder2.

Table 1
List of software tools for selective sweep detection

	Method	Implementation	Availability (source code , web service)
SweepFinder (2005)	SFS	Sequential	http://people.binf.ku.dk/rasmus/webpage/sf.html , –
OmegaPlus (2012)	LD	Parallel	https://github.com/alachins/omegaplus , http://pop-gen.eu
SweeD (2013)	SFS	Parallel	https://github.com/alachins/sweed , http://pop-gen.eu
SweepFinder2 (2016)	SFS	Sequential	http://www.personal.psu.edu/mxd60/sf2.html , –
RAiSD (2018)	Mixed	Sequential	https://github.com/alachins/raisd , –

The table is adapted from [69]

5.1   Detection Accuracy

Crisci et al. [30] calculate the FPR for the neutrality tests using the following pipeline: (1) simulations from equilibrium models using Hudson's ms [45] and constant number of SNPs. This set of simulations is used only for the determination of the thresholds for the tools; (2) simulations using sfscode [44] (constant or bottlenecked population). These data are called empirical datasets, and are used for the estimation of the FPR; (3) execution of the neutrality tests on the empirical datasets. The FPR is estimated by assigning each empirical dataset to a threshold value from an equilibrium model with similar number of SNPs. Note that, such an approach differs from the approach that has been followed by other studies (e.g., [38, 60]), where the null model is specified by the inferred neutral demographic model. Specifying the null model by the inferred neutral demographic model controls efficiently for the FPR. Thus, Crisci et al. effectively studied how demographic model misspecification affects the FPR. Another major difference between the approach followed by Crisci et al. and other studies is that, for the SFS-based methods (SweepFinder, SweeD), Crisci et al. calculate the neutral (or *prior-to-sweep*) SFS using the candidate region itself (here 50 kb), instead of the average SFS on a chromosome-wide scale. Even though the first approach might have a lower FPR, the later is more powerful to detect selective sweeps: when the neutral SFS is calculated by a small genetic region that potentially includes a sweep, the affected (by the sweep) SFS is assumed to represent neutrality. Thus, the CLR test will assume lower values. For neutral equilibrium models, i.e., constant population size, they find that the FPR for SweepFinder ranges from 0.01 to 0.18, depending on the mutation and recombination rate: the lower the mutation and recombination rates, the higher the FPR of SweepFinder. The FPR for SweeD ranges between 0.04 and 0.07. For OmegaPlus, the FPR ranges between 0.05 and 0.07. In general, the FPR for all tools is low when the demographic model is at equilibrium.

When the assumption of an equilibrium population is violated and the empirical datasets are derived from bottlenecked populations, the FPR increases. Such an increase of the FPR is more striking when the average SFS of the empirical dataset is used to represent the SFS of the null model. The reason for such an increase is that bottlenecked datasets show great variance of the SFS from a region to another. Thus, even though, on average, a bottlenecked population will have a monotonically decreasing SFS [104], there might be regions that show an excess of high-frequency and low-frequency derived variants, and thus they mimic the SFS of a selective sweep.

Interestingly, Crisci et al. report low FPR for SweepFinder and SweeD. For OmegaPlus, they report high FPR for the very severe bottleneck scenario, where the population size has been reduced by 99%. For SweepFinder and SweeD, the FPR ranges between 0 and 0.08, and 0 and 0.13, respectively. For OmegaPlus, they report

FPR between 0.05 and 0.91. We repeated the analysis of Crisci et al. for SweeD, SweepFinder, and OmegaPlus, including SweepFinder2. Furthermore, we have included execution results of SweepFinder, SweeD, and SweepFinder2 using the average SFS instead of the regional SFS. We used Hudson's ms for *all* simulations, whereas Crisci et al. had used sfs_code for the empirical simulated data. In general our results are comparable to Crisci et al., but we report higher FPR than Crisci et al. A notable exception is the case of OmegaPlus in the severe bottleneck case, where our FPR is considerably lower. Perhaps this is due to the simulation software, as we used Hudson's ms (coalescent) simulator, while Crisci et al. used sfs_code (forward). FPR results are shown in Fig. 3.

Since FPR is considerably increasing when a false model (e.g., equilibrium) is used to construct the null hypothesis, we repeated the aforementioned analysis using a bottleneck demographic model. Using a bottleneck demographic model for the construction of the null hypothesis reduces the FPR to very low values (Fig. 4). Here, we have used the bottleneck model characterized

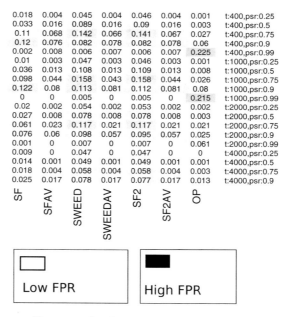

SF	SFAV	SWEED	SWEEDAV	SF2	SF2AV	OP	
0.018	0.004	0.045	0.004	0.046	0.004	0.001	t:400,psr:0.25
0.033	0.016	0.089	0.016	0.09	0.016	0.003	t:400,psr:0.5
0.11	0.068	0.142	0.066	0.141	0.067	0.027	t:400,psr:0.75
0.12	0.076	0.082	0.078	0.082	0.078	0.06	t:400,psr:0.9
0.002	0.008	0.006	0.007	0.006	0.007	0.225	t:400,psr:0.99
0.01	0.003	0.047	0.003	0.046	0.003	0.001	t:1000,psr:0.25
0.036	0.013	0.108	0.013	0.109	0.013	0.008	t:1000,psr:0.5
0.098	0.044	0.158	0.043	0.158	0.044	0.026	t:1000,psr:0.75
0.122	0.08	0.113	0.081	0.112	0.081	0.08	t:1000,psr:0.9
0	0	0.005	0	0.005	0	0.215	t:1000,psr:0.99
0.02	0.002	0.054	0.002	0.053	0.002	0.002	t:2000,psr:0.25
0.027	0.008	0.078	0.008	0.078	0.008	0.003	t:2000,psr:0.5
0.061	0.023	0.117	0.021	0.117	0.021	0.021	t:2000,psr:0.75
0.076	0.06	0.098	0.057	0.095	0.057	0.025	t:2000,psr:0.9
0.001	0	0.007	0	0.007	0	0.061	t:2000,psr:0.99
0.009	0	0.047	0	0.047	0	0	t:4000,psr:0.25
0.014	0.001	0.049	0.001	0.049	0.001	0.001	t:4000,psr:0.5
0.018	0.004	0.058	0.004	0.058	0.004	0.003	t:4000,psr:0.75
0.025	0.017	0.078	0.017	0.077	0.017	0.013	t:4000,psr:0.9

Low FPR   High FPR

Fig. 4 False positive rates for the selective sweep detection process under various algorithms and demographic models when the demographic model used for the construction of the threshold value is a bottleneck model instead of an equilibrium model. *t*: time since the population size recovery (generations). psr: relative population size reduction during bottleneck. To compute *all* threshold values, we have used the bottleneck model characterized by a population recovery at time *t* = 1000 generations, and bottleneck population size reduction by 0.90. The duration of the bottleneck was 4000 generations. FPR values have been reduced considerably compared to the case that the equilibrium model was used for the calculation of the threshold values (Fig. 3). The figure is adapted from [69]

by a population size reduction of 0.99, a recovery time of 1000 generations, and bottleneck duration of 4000 generations, even though empirical datasets were composed by additional models. The ancestral population size was equal to the present-day population size.

Regarding the true positive rate (TPR), Crisci et al. report that under strong selection in an equilibrium population ($2N_e s = 1000$, where s is the selection coefficient), TPR for SweepFinder and SweeD is moderate and ranges between 0.32 and 0.34. For OmegaPlus, TPR is higher and equals to 0.46. For weaker selection ($2N_e s = 100$), OmegaPlus is still the most powerful tool to detect selective sweeps. For selective sweep models in bottlenecked populations, OmegaPlus outperforms SFS-based methods and it is the only test studied by Crisci et al. able to detect selective sweeps. Finally, regarding recurrent hitchhiking event (RHH), OmegaPlus reports higher values of TPR.

5.2 Execution Time

The performance comparisons conducted by Alachiotis and Pavlidis [4] aimed at evaluating the effect of the number of sequences and SNPs on execution time, as well as the capacity of each code to employ multiple cores effectively to achieve faster execution. Table 2 shows execution times on a single processing core for different dataset sizes, ranging from 100 sequences to 1000 sequences, and from 10,000 SNPs up to 100,000 SNPs. Additionally, the table provides (in parentheses) how many times faster are SweeD and OmegaPlus than SweepFinder.

The comparison between SweepFinder and SweeD is the most meaningful one since both tools implement the same floating-point-intensive CLR test based on the SFS, thus requiring the same type and amount of arithmetic operations. The significantly faster execution of OmegaPlus on the other hand, which relies on LD, is attributed to the fact that a limited number of computationally intensive floating-point operations are required, with the majority of operations being performed on integers, such as the enumeration of ancestral and derived alleles.

The execution times in Table 2 refer to sequential execution. Multiple cores can be employed by SweeD and OmegaPlus,

Table 2
Comparison of execution times (in seconds) for different dataset sizes (format: D—number of sequences—number of SNPs) on a single processing core [4]

	$D-10^2-10^4$	$D-10^2-10^5$	$D-10^3-10^4$	$D-10^3-10^4$
SweepFinder	540 (1×)	4138 (1×)	132,938 (1×)	135,996 (1×)
SweeD	125 (4.3×)	1169 (3.5×)	283 (469×)	1345 (101×)
OmegaPlus	6 (90×)	652 (6.4×)	7 (18,991×)	753 (180×)

achieving speedups that vary depending on the number of sequences and SNPs. The parallel efficiency of SweeD decreases with an increasing sample size, whereas the respective parallel efficiency of OmegaPlus increases. As the number of SNPs increases, both SweeD and OmegaPlus exhibit poorer parallel efficiency, which is attributed to load balancing issues that arise with an increasing variance in the SNP density along the datasets.

6 Machine Learning for Population Genetics

6.1 Machine Learning Background

One of the main problems of model-based methods, such as SweeD [73], SweepFinder [65], and OmegaPlus [7], is their inability to provide accurate results when their assumptions are violated. Since, however, in natural populations several of the assumptions of model-based methods (e.g., constant population size) are violated, there is a need for more flexible methodologies. Machine learning was introduced in population genetics as an alternative methodology to detect genomic regions that evolve under selection by treating the problem of detecting selection as a classification problem [83].

The inspiration behind the field of machine learning (ML) was the concept of artificial intelligence (AI). In AI, the main goal was to successfully recognize patterns previously unseen by the algorithm. For this purpose, the process of learning began via observing examples to search for patterns in data and attempt to improve decisions in the future based on the provided examples. The aim is for computers to learn, or rather be trained, by these examples without human assistance, similarly to how humans, and many other living organisms learn from experience. ML enables the analysis of massive quantities of data. The data used in ML tasks can be split into two categories: training data and test data. Training data are used for learning, whereas test data are used to test/evaluate performance, or, in other words, how well the algorithm learned to work for the given task.

6.2 Categories of Machine Learning

The field of ML can be split into three different categories in terms of the learning approach. The first category is supervised learning, which is concerned with predicting the value of a response variable or label (either a categorical or continuous value) on the basis of the input variables/features. Supervised learning accomplishes this feat through the use of a training set of labeled data examples whose true response values are known. The second category is unsupervised machine learning, where, contrary to supervised learning, these learning algorithms are used when the information in the training set is neither classified nor labeled. Unsupervised learning studies how systems can infer a function to describe a hidden structure from unlabeled data. The system does not infer the classes of the data, but it explores the data and infers hidden structures. The third category is reinforcement learning, a field strongly linked to artificial intelligence and game theory. Reinforcement is a learning method that interacts with its environment by producing actions and discovering errors or rewards. Trial-and-error search and delayed reward are the most relevant characteristics of reinforcement learning. This method allows machines and software agents to automatically determine the ideal behavior within a specific context to maximize its performance. Simple reward feedback is required for the agent to learn which action is the best. A further categorization can be made between classification and regression tasks. Classification deals with identifying a group membership where the output variable takes class labels. Regression involves predicting a response when the output variable takes continuous values.

For an in-depth description of machine learning, Alpaydin's introduction to machine learning [66], Michel's machine learning [61], and Bishop's pattern recognition and machine learning [64] are highly recommended.

6.3 Algorithms in Machine Learning

There are various approaches to train machines, ranging from basic decision trees to multilayer artificial neural networks (which evolved to deep learning), depending on what task should be accomplished and the type and amount of available data. Here, we investigate the performance of various well-known and widely used ML algorithms in the classification problem of selection versus neutrality. Our goal is to examine whether machine learning algorithms, used in population genetics analyses, can accurately infer selection. Classification algorithms can be either generative or discriminative. A generative algorithm models how the data was generated in order to categorize them in different classes. Thus, its aim is to find the category that is most likely to generate the observed result. A discriminative algorithm does not care about how the data was generated, it simply categorizes the given set of features. A general concern of the ML-related problems is overfitting. Overfitting [43] is the phenomenon when results of training cannot reliably capture previously unseen data due to being tailored on just the given training data. In other words, if a model performs significantly better on the training data than on unseen/test data, then the model probably suffers from overfitting.

In this study, the ML classifiers that we will evaluate are logistic regression (LR), random forests (RF), k-nearest neighbors (kNN), and support vector machines (SVM). The ML framework was implemented in Python using the *sklearn* [74] package.

Naive Bayes A generative classifier that uses the Bayes rule to describe the joint probability of data and classes is called naive Bayes (NB). An important drawback of NB is that it assumes conditional independence of features given the class label. However, in population genetics, conditional independence does not hold for most of the features neither under neutrality nor under selection. Thus, naive Bayes may not be appropriate for population genetics data, and we do not evaluate it in this study.

Logistic Regression It is a classifier that assumes a parametric form for the distribution $P(Y|X)$ *and directly estimates its parameters from the training data. The central premise of LR is the assumption that the input space can be separated into two regions, one for each class, by a linear boundary. Unlike NB, LR does not assume that the features are conditionally independent.*

k-Nearest Neighbor kNN is not strictly a learning classifier but rather a memory-based classifier. It classifies each of the test data by its position based on its k *closest/nearest neighbors for which the class is known. To the best of our knowledge, kNN has not been examined as a selection/neutrality classification algorithm yet.*

Random Forests RF is a classifier that works well for classification problems as it is able to exploit both high- and low-"informative" features and to deal with the problem of overfitting. The original classification algorithm that inspired RF was the decision trees method. Based on the values each of the features may take, "decision" nodes are created resulting in a tree structure. Upon reaching a leaf of this tree, a decision is achieved for the label of the input data. The features with lower entropy (most informative) appear closer to the root of the tree. However, a single tree might be heavily biased and as a result the algorithm may overfit. The solution to the overfitting problem is RF, a classifier that consists of several different decision trees whose outcomes are combined, usually by averaging the results, to predict the class of the input.

Support Vector Machines It is a machine learning algorithm proposed by Cortes et al. [28]. SVMs attempt to split the dataset into two classes via using a hyperplane that separates those classes. The goal is to find the ideal hyperplane that best separates those classes. It uses specific data points of each class to determine the position of the hyperplane. These points are called the support vectors. A distance between the hyperplane and the closest support vector from each class is kept, namely the margin. SVMs attempt to maximize this margin to maximize the probability of correctly classifying new data. Due to the ability of SVMs to reach higher dimensions, they do not suffer from the "curse of dimensionality," making them a suitable algorithm for classifying between selection and neutrality.

7   Methods

7.1   Data Generation

There exist various models that produce single nucleotide polymorphisms (SNP) from demographic models. To generate our data, we used the *ms* tool, a Monte Carlo computer program written in C, that generates samples drawn from a population evolving according to a Wright–Fisher neutral model [42]. The program assumes an infinite-sites model of mutation, and allows recombination, gene conversion, symmetric migration among subpopulations, and a variety of demographic histories. For each sample, the program generates a random genealogical history of a segment of a chromosome. Conditional on the genealogy of a sample, mutations are randomly placed on the genealogy according to the usual assumption that the number of mutations on a branch is Poisson distributed with mean given by the product of the mutation rate and the branch length. The times between nodes in the genealogy are approximated by continuous (exponential) distributions.

We simulated neutral datasets and datasets with selection for 60 demographic models that include a variety of bottleneck scenarios (from mild to severe). For the selection data, we used an extension of *ms*, called *mssel*, kindly provided by R.R. Hudson. Each bottleneck model is characterized by a reduction in population size at some point in time and a recovery to the original population size (backwards in time). For each demographic model we generated 1000 datasets to incorporate the genealogical uncertainty in the training process. The mutation parameter of the model was set to $4N\mu = 2000$. In our simulations, we used a constant value for $4N\mu$. We could also sample this parameter from a distribution (e.g., Gaussian). Even though, results of the neutrality tests could be affected, at least partially, we expect that this effect will be minor because there is no direct involvement of the number of SNPs in the tests' results.

7.2   Computing Summary Statistics

The raw data, generated from *ms*, cannot be used directly for the classification task. Thus, from each polymorphic dataset, we compute a vector of summary statistics that will serve as data features. We used the software *CoMuStats* [68] to calculate a multitude of summary statistics from the *ms* simulations, such as Tajima's D [92], Wall's B and Q statistics [96], FST values [46], the site frequency spectrum [42], and others (Table 3).

Table 3
Description of a subset of the summary statistics generated by *CoMuStats*

Summary statistic	Definition
θ_W	Watterson's estimator of θ using the number of segregating sites and the sample size
Tajima's D	Computed as the difference between two measures of genetic diversity: the mean number of pairwise differences and the number of segregating sites, each scaled so that they are expected to be the same in a neutrally evolving population of constant size
B and Q	The number of pairs of adjacent segregating sites that are congruent
F_{ST}	A measure of population differentiation due to genetic structure. It is frequently estimated from genetic polymorphism data, such as single-nucleotide polymorphisms (SNP) or microsatellites
SFS	The number of segregating sites where the derived allele occurs i times out of n samples

7.3 Application of Classification Algorithms

7.4 Dataset Manipulation

When we obtain a collection of datasets, each belonging to an a priori known class, we follow the next steps for optimal and unbiased results. First, (1) we split the data in two parts. A part of the data is used for training, whereas the remaining is used as the test set. We used 20% of the generated dataset as a test set leaving 80% for training each model. Each classifier has parameters that need to be set before training begins. Thus, an important step of classification is to find the optimal parameter values. This process is called tuning. Thus, in step (2) we tune the classifier. Finally, in step (3) we evaluate the performance of the tuned classifier based solely on the (unseen) test set.

The simplest form of tuning is to use a part of the data for training and the remaining part for test. Tuning the parameters takes place by repeatedly evaluating the performance of the algorithm for different parameter values on the test set. This process, however, leads to overfitting. Another part of the dataset, which is named as validation set, is held in order to tackle the problem of overfitting. Using this approach, training proceeds only on the training set, while tuning the parameters of the classifier is performed on the validation set. When tuning is complete, a final evaluation can be done on the test set. This method is called cross validation (CV), and it remedies overfitting by ensuring that the parameters estimation of the classifier is not strictly associated with the data we used to estimate them. However, this simple approach results in tuning the classifier parameters based on a small part of the data, thus results may be suboptimal. A better strategy is the so-called k-fold CV, in which the training set is split into k folds. We use all but one of the folds for training and the resulting model is validated on the remaining part of the data. This is repeated k times with a different validation set. The parameters of the classifier that result in the greatest accuracy, on average, are stored. As a final step, we train the classifier using the optimal parameters from the tuning step in the whole training set (training and validation). The accuracy is measured solely using the test set.

By using a single test set, the evaluation of the classification performance may be biased depending on the specific test set. Thus, an approach called nested k-fold CV can be followed. Nested k-fold CV effectively uses a series of train/validation/test set splits. In the inner loop (k-fold CV), the accuracy is approximately maximized by fitting a model to each training set, and then inferring the optimal parameter values using the validation set. In the outer loop (nested), the generalization error is estimated by averaging test set scores over several dataset splits.

Another popular method is the stratified nested k-fold cross validation, which ensures that representation of classes in each fold is according to their frequency in the original dataset. However, since our data are simulated, both classes are balanced (equally represented) by design and, therefore, there was no need to use stratified CV [77].

7.5 Feature Selection

To further increase the performance of our classifiers, we can use for training only those features (variables) that mostly enable classification between the two classes. In other words, by removing those features that do not contribute enough to the classification, the performance of the classifier will be increased. This method, which is widely used in machine learning, is called feature selection [39].

There are two problems related to feature selection. The first is how much does each feature contribute to solving the classification problem. Here, we use the mutual information [29] of each feature with respect to the others. The second problem is related to the number of features that will be used. This is performed via the SelectKBest package from python's sklearn [74]. In detail, we rank our features from the most informative one to the least informative one. We use the top m features ($2 \leq m \leq 40$) successively, and evaluate their performance.

8 Results

8.1 Reducing the Feature Space

We first perform the feature selection step. The number of features we kept was decided solely on the SVM classifier as described in Subheading 7.5. Each pair of datasets, one for selection and its neutral counterpart, was studied separately. We calculated the

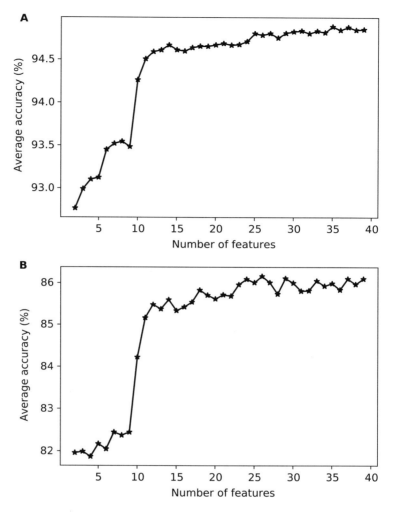

Fig. 5 Average accuracy of the feature selection procedure among the 60 datasets used in the study. (**a**) Average across all 60 datasets. (**b**) Average across the five datasets with the most severe bottleneck

average accuracy for each number of features across all 60 pairs. The results showed that 36 features produced the optimal results (94.865% average accuracy), as seen in Fig. 5.

The 60 datasets implement bottleneck demographic scenarios of various severities. Among them, some scenarios are mild, whereas others are severe. In mild bottlenecks, selection detection is a rather simple task. However, in severe bottlenecks disentangling selection from neutrality is challenging and often the accuracy of the algorithms is diminished [71]. To test the performance of the feature selection process in challenging scenarios, we chose to evaluate feature selection only on the five most severe bottleneck models. As seen in Fig. 5, the number of best performing features was reduced to 26, achieving an average accuracy of 86.16%. Average accuracy is, as expected, lower overall since the best performing pairs were excluded.

8.2   Evaluation

Since the dataset is created from simulated data, we chose to have balanced classes by generating the same number of samples for both selection and neutrality. Thus, the trivial (random) classifier would achieve an average accuracy of 50%. All the classifiers were tested using the 36 best features.

8.2.1   Logistic Regression

Logistic regression works by separating the two classes using a linear boundary. It starts by setting the line according to the features. Then LR modifies the initial guess by changes its position or its slope to try and improve the accuracy of the classifier. A parameter to tune is the maximum number of attempts to optimize the accuracy. We set the parameter of maximum number of attempts to the values 100, 150, 200, and 250. To prevent the model from overfitting or underfitting, logistic regression uses a regularization penalty. The goal of that penalty is to not allow extreme values to influence the classifier. Two options for regularization are Ridge (L2 regularization) [21] and Lasso (L1 regularization) [21]. Ridge adds penalty equivalent to square of the magnitude of coefficients. Lasso adds a penalty equivalent to the absolute value of the magnitude of the coefficients. Both were considered during tuning, each with its own hyperparameters. Ridge uses sag [48] and lbfgs [9]. Lasso uses saga [32] and liblinear [74].

The highest accuracy (94.92%) was achieved by using Ridge regression with saga for at most 150 attempts/iterations of the algorithm attempting to converge, whereas the performance dropped for more than 150 iterations (Fig. 6).

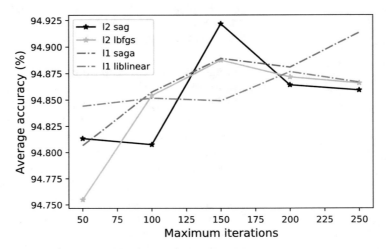

Fig. 6 Accuracy of logistic regression classifier for Lasso(l1) and Ridge (l2) regularization, while increasing maximum iteration allowed in order to converge

8.2.2 Random Forests

For the random forests classifier, the tuning parameters are the maximum depth the tree was allowed to reach, the maximum number of features to consider for each split, and the number of decision trees generated. Forests consisting of 50, 100, 150, and 200 trees were examined. For these trees, a maximum depth of 10, 20, 30, and 36 splits was allowed. For each split, either the square root (F_{SQRT}) or the logarithm (F_{LOG}) of the 36 features was the maximum number considered. We also used bagging, a method designed to improve the stability and accuracy of machine learning algorithms. According to [23], bagging is defined as:

Given a standard training set $\mathbf{D}$ of size n, bagging generates m new training sets $\mathbf{D\{i\}}$ each of size n', by sampling from D uniformly and with replacement. By sampling with replacement, some observations may be repeated in each $\mathbf{D\{i\}}$. If $\mathbf{n'=n}$, then for large n the set $\mathbf{D\{i\}}$ is expected to have the fraction $(1 - 1/e)$ ($\approx 63.2\%$) of the unique examples of D, the rest being duplicates. This kind of sample is known as a bootstrap sample. The m models are fitted using the above m bootstrap samples and combined by averaging the output (for regression) or voting (for classification).

As Fig. 7a, b shows, increasing the maximum depth of the decision trees, RFs achieve better accuracy up to a depth of 30 features. Further increasing the number of features results in a lower accuracy. Also, setting the maximum features considered to F_{SQRT} in each split performed better than F_{LOG}. A forest consisting of 150 trees performed optimally for both F_{SQRT} and F_{LOG}, and by comparing the two we can deduce that F_{SQRT} is the better performing method, as seen in Fig. 8.

8.2.3 K Nearest Neighbors

The two parameters to be tuned for the kNN classifier are the number of neighbors to consider and the distance metric used to calculate the distance between two neighbors. The two distance metrics under consideration are the Euclidean and the Chebyshev distance.

Euclidean is a better distance metric than Chebyshev for all neighbors considered (Fig. 9). By increasing the number of neighbors, the accuracy was increasing, reaching a maximum performance of 94.25% for 36 neighbors. For more than 36 neighbors, the accuracy declines.

8.2.4 Support Vector Machines

Support vector machines map the data to a predetermined high-dimensional space via a kernel function that enables classification of non-linearly separable data. The kernel function is used as a measure of similarity [81]. In particular, the kernel function $k(x, \cdot)$ defines the distribution of similarities of points around a given point x. $k(x, y)$ denotes the similarity of point x with another given point y. The polynomial kernel [81] and the random Bayesian forests (rbf) [81] are the kernels considered here. For the polynomial kernel, the maximum degree/dimension of the kernel

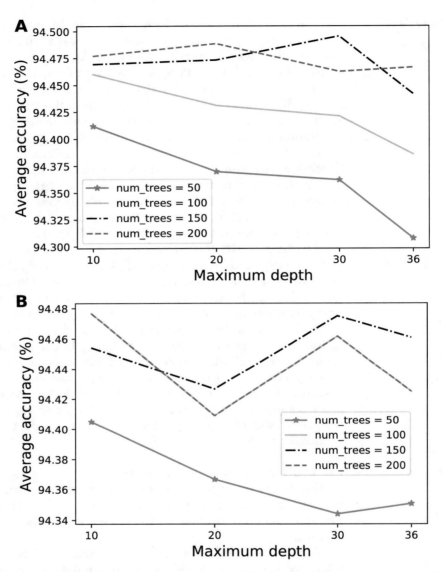

Fig. 7 Accuracy of random forest classifier for half, 75% and unlimited max depth allowed. Each line represents a different number of trees spawned (num_trees). (**a**) Log2 maximum features considered for each split. (**b**) Square root of maximum features considered for each split

function assumes the values 1 (the equivalent of a linear kernel), 2, 3, 4, and 5. For the rbf, the gamma hyperparameter ranged from −8 to 4. In our setup, we use the Soft-Margin SVMs. Soft-Margin SVMs permit some errors while trying to find the optimal classification surface, thus the model is more robust to overfitting. Soft-Margin SVMs require a cost parameter that determines the number of errors we allow. The cost parameter ranges from 1 to 10 and its optimal value is 1. Based on Fig. 10, we can deduce that polynomial

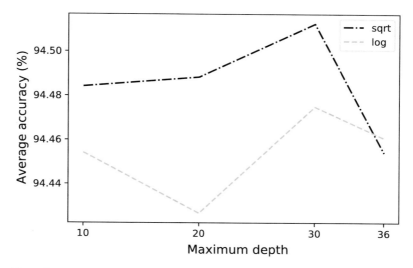

Fig. 8 Comparison of best performing cases (150 trees in the random forest) for log and sqrt maximum features

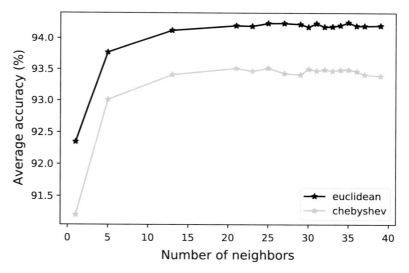

Fig. 9 Accuracy of kNN classifier for Euclidean and Chebyshev distances, for 1, 5, 21, 25, 27, 29, 31, 35, 36 neighbors

is the best performing kernel. It reaches the highest accuracy, 0.9484%, with a degree equal to 1.

The classifier that achieved the best overall performance is the support vector machines peaking at 0.9484%. Figure 11 compares the tuned versions of each classifier on each pair of datasets.

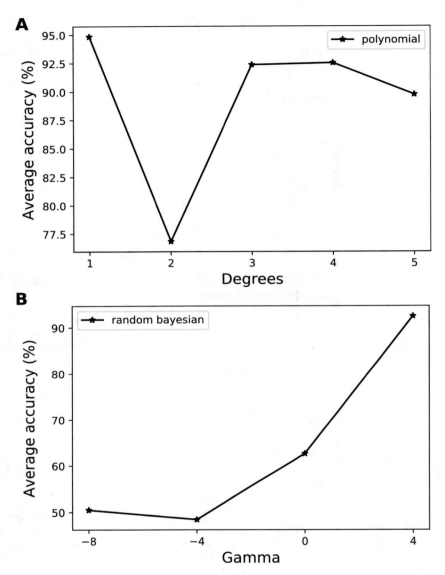

Fig. 10 Accuracy of support vector machines classifier for polynomial (**a**) and random Bayesian (**b**) forests kernel

9   Discussion

We demonstrate the use of machine learning algorithms in disentangling selection from neutrality. This task is treated as a supervised learning classification. We evaluated logistic regression, k-nearest neighbors, random forests, and support vector machines. All classifiers outperformed the trivial classifier and showed high accuracy, which, however, depends on the bottleneck severity.

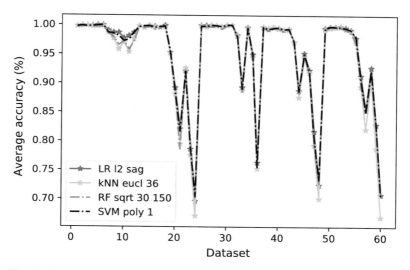

Fig. 11 Comparison of tuned classifier across all datasets

Among the classifiers tested in this survey, the kNN classifier had the worst performance among the examined algorithms, as seen in Fig. 11. Logistic regression had the best performance in the datasets with a mild bottleneck, implying that selection can be separated from neutrality linearly in mild bottlenecks. On the contrary, it showed the second worst performance, only better that kNN, in severe bottlenecks. Random forests classifier showed better performance in severe bottleneck models compared to kNN and LR. An additional advantage of RF is the ability to handle missing data, which real-world scenarios will likely include. This makes random forests a suitable classifier for selection inference. Finally, support vector machines achieved the best average performance. SVM was slightly outperformed by LR in mild bottlenecks, but achieved the best accuracy in severe bottlenecks. As a result, we suggest SVMs as the most robust classifier out of those examined in this survey.

Choosing the parameters that maximize nested k-fold CV often yields an optimistic accuracy [77]. In addition, since we use simulated data, the accuracies calculated in this report may be slightly optimistic. Still, results clearly highlight the potential of machine learning in population genetics.

In this work, we focused on a small part of the genome. Because of the advancements in sequencing technologies, whole-genome data are constantly produced, allowing to infer selection forces acting on genomes. Applying the algorithms on the genome as a whole will presumably fail to detect selection. This is due to the fact that recent selection has operated only on small parts of the whole genome, leaving the rest of it effectively neutral. Thus, if a classifier has to take a single decision for the whole genome, this will favor neutrality. A better approach is to split the whole genome into smaller regions (sliding windows) and infer selection in each one separately. The split of the dataset into regions is performed by a sliding window algorithm that requires two parameters, the size of each window (in base pairs) and an offset that defines the starting position of the next window relative to the previous one. The pseudocode shown in Algorithm 1 describes this process.

Algorithm 1: Whole-genome selection inference in sliding windows:

> **Data**: whole genome
> **Result**: selection inference in sliding windows
> read dataset;
> offset = 0;
> **while** *start < size(dataset) - size(window)* **do**
> region = dataset(start:start+offset) # take an offset-sized
> region from the genome
> outcome = prediction(region) # infer selection
> store(outcome) # store results
> start = start + offset # move to the next region
> **end**

Despite the success of machine learning, it still faces challenges. First, machine learning algorithms are expensive in terms of both time and computational resources. This is a problem that will be mitigated as computer hardware and software technology advances. A general issue of the machine learning field is the dependence on quality training data. Even if an ideal algorithm would exist, it would fail to produce valuable results if the quality of the training dataset was poor. In complex problems, the need for appropriate training data that, on one hand, are labeled accurately and, on the other hand, represent correctly real scenarios is of utmost importance. Especially in population genetics, training examples are obtained via simulations because it is not possible to obtain real training example data with accurate class labeling. However, simulated data only capture a part of the evolutionary processes that may have shaped real data. Using simulated data guarantees the data quality, but it comes with the drawback of obtaining optimistic results during testing. A further approach to improve results is feature selection, which improves the quality of our data by removing noisy features. In the matter of selection inference, feature selection can improve results. If datasets contain missing or corrupted parts, then preprocessing methods exist [14].

For a most thorough study of methods related to learning, other approaches different than machine learning need to be examined as well, for example, artificial neural networks and deep learning algorithms. Currently, there are only a few studies related to deep learning in population genetics [76, 84, 103], but the potential of the field is already apparent. Recently, a breakthrough algorithm was implemented that outcompeted both real and AI players in the strategy game Go by just knowing the rules of the game [85]. The idea of learning without human knowledge in the field of population genetics is currently far from being formulated as a proper scientific approach.

References

1. Aguadé M, Langley CH (1994) Polymorphism and divergence in regions of low recombination in Drosophila. In: Non-neutral evolution. Springer, Boston, pp 67–76

2. Aguade M, Miyashita N, Langley CH (1989) Reduced variation in the yellow-achaete-scute region in natural populations of Drosophila melanogaster. Genetics 122(3):607–615

3. Akey JM, Eberle MA, Rieder MJ, Carlson CS, Shriver MD, Nickerson DA, Kruglyak L (2004) Population history and natural selection shape patterns of genetic variation in 132 genes. PLoS Biol 2(10):e286

4. Alachiotis N, Pavlidis P (2016) Scalable linkage-disequilibrium-based selective sweep detection: a performance guide. GigaScience 5(1):7. https://doi.org/10.1186/s13742-016-0114-9

5. Alachiotis N, Pavlidis P (2018) Raised detects positive selection based on multiple signatures of a selective sweep and SNP vectors. Commun Biol 1(1):79

6. Alachiotis N, Pavlidis P, Stamatakis A (2012) Exploiting multi-grain parallelism for efficient selective sweep detection. In: International conference on algorithms and architectures for parallel processing. Springer, Berlin, pp 56–68

7. Alachiotis N, Stamatakis A, Pavlidis P (2012) Omegaplus: a scalable tool for rapid detection of selective sweeps in whole-genome datasets. Bioinformatics 28(17):2274–2275

8. Andersen EC, Gerke JP, Shapiro JA, Crissman JR, Ghosh R, Bloom JS, Félix MA, Kruglyak L (2012) Chromosome-scale selective sweeps shape Caenorhabditis elegans genomic diversity. Nat Genet 44(3):285

9. Andrew G, Gao J (2007) Scalable training of l 1-regularized log-linear models. In: Proceedings of the 24th international conference on machine learning. ACM, New York, pp 33–40

10. Ansel J, Arya K, Cooperman G (2009) DMTCP: transparent checkpointing for cluster computations and the desktop. In: IEEE international symposium on parallel & distributed processing, IPDPS 2009. IEEE, Piscataway, pp 1–12

11. Arguello JR, Cardoso-Moreira M, Grenier JK, Gottipati S, Clark AG, Benton R (2016) Extensive local adaptation within the chemosensory system following Drosophila melanogaster/'s global expansion. Nat Commun 7:11855 (2016)

12. Bachtrog D (2004) Evidence that positive selection drives Y-chromosome degeneration in Drosophila miranda. Nat Genet 36(5):518–522 (2004). https://doi.org/10.1038/ng1347

13. Barton NH (1998) The effect of hitch-hiking on neutral genealogies. Genet Res 72(2):123–133

14. Batista GE, Monard MC (2003) An analysis of four missing data treatment methods for supervised learning. Appl Artif Intell 17(5–6):519–533

15. Begun DJ, Aquadro CF (1991) Molecular population genetics of the distal portion of the x chromosome in Drosophila: evidence for genetic hitchhiking of the yellow-achaete region. Genetics 129(4):1147–1158

16. Beisswanger S, Stephan W (2008) Evidence that strong positive selection drives neofunctionalization in the tandemly duplicated polyhomeotic genes in Drosophila. Proc Nat Acad Sci 105(14):5447–5452

17. Bersaglieri T, Sabeti PC, Patterson N, Vanderploeg T, Schaffner SF, Drake JA, Rhodes M, Reich DE, Hirschhorn JN (2004) Genetic signatures of strong recent positive selection at the lactase gene. Am J Hum Genet 74(6):1111–1120

18. Besnier F, Kent M, Skern-Mauritzen R, Lien S, Malde K, Edvardsen RB, Taylor S, Ljungfeldt LE, Nilsen F, Glover KA (2014) Human-induced evolution caught in action: SNP-array reveals rapid amphi-atlantic spread of pesticide resistance in the salmon ectoparasite Lepeophtheirus salmonis. BMC Genom 15(1):1

19. Bigham AW, Mao X, Mei R, Brutsaert T, Wilson MJ, Julian CG, Parra EJ, Akey JM, Moore

LG, Shriver MD (2009) Identifying positive selection candidate loci for high-altitude adaptation in Andean populations. Hum Genom 4(2):1

20. Bigham A, Bauchet M, Pinto D, Mao X, Akey JM, Mei R, Scherer SW, Julian CG, Wilson MJ, López Herráez D, Brutsaert T, Parra EJ, Moore LG, Shriver MD (2010) Identifying signatures of natural selection in Tibetan and Andean populations using dense genome scan data. PLoS Genet 6(9):e1001116. https://doi.org/10.1371/journal.pgen.1001116

21. Boyd S, Parikh N, Chu E, Peleato B, Eckstein J, et al (2011) Distributed optimization and statistical learning via the alternating direction method of multipliers. Found Trends Mach Learn 3(1):1–122

22. Braverman JM, Hudson RR, Kaplan NL, Langley CH, Stephan W (1995) The hitchhiking effect on the site frequency spectrum of DNA polymorphisms. Genetics 140 (2):783–796

23. Breiman L (2001) Random forests. Mach Learn 45(1):5–32

24. Caicedo AL, Williamson SH, Hernandez RD, Boyko A, Fledel-Alon A, York TL, Polato NR, Olsen KM, Nielsen R, McCouch SR, et al (2007) Genome-wide patterns of nucleotide polymorphism in domesticated rice. PLoS Genet 3(9):e163

25. Catalán A, Glaser-Schmitt A, Argyridou E, Duchen P, Parsch J (2016) An indel polymorphism in the MtnA 3′untranslated region is associated with gene expression variation and local adaptation in Drosophila melanogaster. PLoS Genet 12(4), e1005987

26. Celine Becquet (2003) Signatures of a population bottleneck can be localised along a recombining chromosome. Tech. rep. http://przeworski.uchicago.edu/cbecquet/MasterThesis.pdf

27. Charlesworth B, Morgan M, Charlesworth D (1993) The effect of deleterious mutations on neutral molecular variation. Genetics 134 (4):1289–1303

28. Cortes C, Vapnik V (1995) Machine learning. Supp Vector Netw 20:273–297

29. Cover TM, Thomas JA (2012) Elements of information theory. Wiley, Hoboken

30. Crisci JL, Poh YP, Mahajan S, Jensen JD (2013) The impact of equilibrium assumptions on tests of selection. Front Genet 4:235

31. De A, Durrett R (2007) Stepping-stone spatial structure causes slow decay of linkage disequilibrium and shifts the site frequency spectrum. Genetics 176(2):969–981.

https://doi.org/10.1534/genetics.107.071464

32. Defazio A, Bach F, Lacoste-Julien S (2014) Saga: a fast incremental gradient method with support for non-strongly convex composite objectives. In: Advances in neural information processing systems, pp 1646–1654

33. DeGiorgio M, Huber CD, Hubisz MJ, Hellmann I, Nielsen R (2016) Sweepfinder2: increased sensitivity, robustness and flexibility. Bioinformatics 32(12):1895–1897

34. Depaulis F, Veuille M (1998) Neutrality tests based on the distribution of haplotypes under an infinite-site model. Mol Biol Evol 15 (12):1788–1790

35. Ewing GB, Jensen JD (2016) The consequences of not accounting for background selection in demographic inference. Mol Ecol 25(1):135–141

36. Excoffier L, Dupanloup I, Huerta-Sánchez E, Sousa VC, Foll M (2013) Robust demographic inference from genomic and SNP data. PLoS Genet 9(10):e1003905

37. Fay JC, Wu CI (2000) Hitchhiking under positive Darwinian selection. Genetics 155 (3):1405–1413

38. Frantz LA, Schraiber JG, Madsen O, Megens HJ, Cagan A, Bosse M, Paudel Y, Crooijmans RP, Larson G, Groenen MA (2015) Evidence of long-term gene flow and selection during domestication from analyses of Eurasian wild and domestic pig genomes. Nat Genet 47 (10):1141–1148

39. Friedman J, Hastie T, Tibshirani R (2001) The elements of statistical learning. Springer series in statistics New York, vol 1. Springer, Berlin

40. Gerbault P, Powell A, Thomas MG (2012) Evaluating demographic models for goat domestication using mtDNA sequences. Anthropozoologica 47(2):64–76)

41. Glinka S, Ometto L, Mousset S, Stephan W, De Lorenzo D (2003) Demography and natural selection have shaped genetic variation in Drosophila melanogaster: a multi-locus approach. Genetics 165(3):1269–1278

42. Hartl DL, Clark AG, Clark AG (1997) Principles of population genetics, vol 116. Sinauer Associates, Sunderland

43. Hawkins DM (2004) The problem of overfitting. J Chem Inf Comput Sci 44(1):1–12

44. Hernandez RD (2008) A flexible forward simulator for populations subject to selection and demography. Bioinformatics 24 (23):2786–2787

45. Hudson RR (2002) Generating samples under a Wright-Fisher neutral model of genetic variation. Bioinformatics 18 (2):337–338

46. Hudson RR, Slatkin M, Maddison W (1992) Estimation of levels of gene flow from DNA sequence data. Genetics 132(2), 583–589

47. Innan H, Stephan W (2003) Distinguishing the hitchhiking and background selection models. Genetics 165(4):2307–2312 (2003)

48. Johnson R, Zhang T (2013) Accelerating stochastic gradient descent using predictive variance reduction. In: Advances in neural information processing systems, pp 315–323

49. Kaplan NL, Hudson R, Langley C (1989) The "hitchhiking effect" revisited. Genetics 123(4):887–899

50. Kern AD, Schrider DR (2018) diploS/HIC: an updated approach to classifying selective sweeps. G3 Genes Genom Genet 8:1959–1970

51. Kim Y, Nielsen R (2004) Linkage disequilibrium as a signature of selective sweeps. Genetics 167(3):1513–1524. https://doi.org/10. 1534/genetics.103.025387

52. Kim Y, Stephan W (2002) Detecting a local signature of genetic hitchhiking along a recombining chromosome. Genetics 160 (2):765–777

53. Langley CH, MacDonald J, Miyashita N, Aguade M (1993) Lack of correlation between interspecific divergence and intraspecific polymorphism at the suppressor of forked region in Drosophila melanogaster and Drosophila simulans. Proc. Natl. Acad. Sci. 90(5):1800–1803

54. Laurent S, Pfeifer SP, Settles ML, Hunter SS, Hardwick KM, Ormond L, Sousa VC, Jensen JD, Rosenblum EB (2016) The population genomics of rapid adaptation: disentangling signatures of selection and demography in white sands lizards. Mol Ecol 25(1):306–323

55. Li H (2011) A new test for detecting recent positive selection that is free from the confounding impacts of demography. Mol Biol Evol 28(1):365–375. https://doi.org/10. 1093/molbev/msq211

56. Li H, Stephan W (2006) Inferring the demographic history and rate of adaptive substitution in Drosophila. PLoS Genet 2(10):e166

57. Lin K, Li H, Schlötterer C, Futschik A (2011) Distinguishing positive selection from neutral evolution: boosting the performance of summary statistics. Genetics 187(1):229–44. https://doi.org/10.1534/genetics.110. 122614

58. Luo Q, Ahmad K, Fu HY, Wang JD, Chen RK, Gao SJ (2016) Genetic diversity and population structure of Sorghum mosaic virus infecting Saccharum spp. hybrids. Ann. Appl. Biol. 169(3):398–407

59. Martin SH, Möst M, Palmer WJ, Salazar C, McMillan WO, Jiggins FM, Jiggins CD (2016) Natural selection and genetic diversity in the butterfly Heliconius melpomene. Genetics 203(1):525–541

60. McManus KF, Kelley JL, Song S, Veeramah KR, Woerner AE, Stevison LS, Ryder OA, Kidd JM, Wall JD, Bustamante CD, et al (2015) Inference of gorilla demographic and selective history from whole-genome sequence data. Mol Biol Evol 32(3):600–612

61. Michalski RS, Carbonell JG, Mitchell TM (2013) Machine learning: an artificial intelligence approach. Springer, New York

62. Miyashita NT (1990) Molecular and phenotypic variation of the Zw locus region in Drosophila melanogaster. Genetics 125 (2):407–419

63. Montano V, Didelot X, Foll M, Linz B, Reinhardt R, Suerbaum S, Moodley Y, Jensen JD (2015) Worldwide population structure, long-term demography, and local adaptation of helicobacter pylori. Genetics 200 (3):947–963

64. Nasrabadi NM (2007) Pattern recognition and machine learning. J Electron Imag 16 (4):049901

65. Nielsen R, Williamson S, Kim Y, Hubisz MJ, Clark AG, Bustamante C (2005) Genomic scans for selective sweeps using SNP data. Genom Res 15(11):1566–1575

66. Alpaydin E (2009) Introduction to machine learning. MIT press (Cambridge)

67. Orengo DJ, Aguadé M (2004) Detecting the footprint of positive selection in a European population of Drosophila melanogaster: multilocus pattern of variation and distance to coding regions. Genetics 167 (4):1759–1766. https://doi.org/10.1534/ genetics.104.028969

68. Papadantonakis S, Poirazi P, Pavlidis P (2016) CoMuS: simulating coalescent histories and polymorphic data from multiple species. Mol Ecol Resour 16(6):1435–1448

69. Pavlidis P, Alachiotis N (2017) A survey of methods and tools to detect recent and strong positive selection. J Biol Res-Thessaloniki 24 (1):7

70. Pavlidis P, Hutter S, Stephan W (2008) A population genomic approach to map recent positive selection in model species. Mol Ecol 17(16):3585–3598

71. Pavlidis P, Jensen JD, Stephan W (2010) Searching for footprints of positive selection in whole-genome SNP data from nonequilibrium populations. Genetics 185(3):907–922

72. Pavlidis P, Jensen JD, Stephan W, Stamatakis A (2012) A critical assessment of storytelling: gene ontology categories and the importance of validating genomic scans. Mol Biol Evol 29 (10):3237–3248. https://doi.org/10.1093/molbev/mss136

73. Pavlidis P, Živković D, Stamatakis A, Alachiotis N (2013) SweeD: likelihood-based detection of selective sweeps in thousands of genomes. Mol Biol Evol 30(9):2224–2234

74. Pedregosa F, Varoquaux G, Gramfort A, Michel V, Thirion B, Grisel O, Blondel M, Prettenhofer P, Weiss R, Dubourg V, *et al* (2011) Scikit-learn: machine learning in python. J Mach Learn Res 12:2825–2830

75. Pickrell JK, Coop G, Novembre J, Kudaravalli S, Li JZ, Absher D, Srinivasan BS, Barsh GS, Myers RM, Feldman MW, *et al* (2009) Signals of recent positive selection in a worldwide sample of human populations. Genom Res 19(5):826–837

76. Quang D, Xie X (2016) DanQ: a hybrid convolutional and recurrent deep neural network for quantifying the function of DNA sequences. Nucleic Acids Res 44(11), e107–e107

77. Refaeilzadeh P, Tang L, Liu H (2009) Cross-validation. In: Encyclopedia of database systems. Springer, Berlin, pp 532–538

78. Rubin CJ, Zody MC, Eriksson J, Meadows JR, Sherwood E, Webster MT, Jiang L, Ingman M, Sharpe T, Ka S, *et al* (2010) Whole-genome resequencing reveals loci under selection during chicken domestication. Nature 464(7288):587

79. Sabeti PC, Schaffner SF, Fry B, Lohmueller J, Varilly P, Shamovsky O, Palma A, Mikkelsen T, Altshuler D, Lander E (2006) Positive natural selection in the human lineage. Science 312(5780):1614–1620

80. Schlenke TA, Begun DJ (2004) Strong selective sweep associated with a transposon insertion in Drosophila simulans. Proc Natl Acad Sci USA 101(6):1626–1631

81. Scholkopf B, Smola AJ (2001) Learning with kernels: support vector machines, regularization, optimization, and beyond. MIT Press, Cambridge (2001)

82. Schrider DR, Kern AD (2016) S/HIC: robust identification of soft and hard sweeps using machine learning. PLoS Genet 12(3): e1005928

83. Schrider DR, Kern AD (2018) Supervised machine learning for population genetics: a new paradigm. Trends Genet 34(4):301–312

84. Sheehan S, Song, YS (2016) Deep learning for population genetic inference. PLoS Comput Biol 12(3):e1004845

85. Silver D, Schrittwieser J, Simonyan K, Antonoglou I, Huang A, Guez A, Hubert T, Baker L, Lai M, Bolton A, *et al* (2017) Mastering the game of go without human knowledge. Nature 550(7676):354

86. Smith JM, Haigh J (1974) The hitch-hiking effect of a favourable gene. Genet Res 23 (1):23–35

87. Städler T, Haubold B, Merino C, Stephan W, Pfaffelhuber P (2009) The impact of sampling schemes on the site frequency spectrum in nonequilibrium subdivided populations. Genetics 182(1):205–216

88. Stephan W (2010) Genetic hitchhiking versus background selection: the controversy and its implications. Philos Trans R Soc Lond Ser B Biol Sci 365(1544):1245–1253. https://doi.org/10.1098/rstb.2009.0278

89. Stephan W, Langley CH (1989) Molecular genetic variation in the centromeric region of the x chromosome in three Drosophila ananassae populations. i. contrasts between the vermilion and forked loci. Genetics 121 (1):89–99 (1989)

90. Stephan W, Wiehe THE, Lenz MW (1992) The effect of strongly selected substitutions on neutral polymorphism: analytical results based on diffusion theory. Theor. Popul Biol 41(2):237–254. https://doi.org/10.1016/0040-5809(92)90045-U

91. Stephan W, Song YS, Langley CH (2006) The hitchhiking effect on linkage disequilibrium between linked neutral sites. Genetics 172 (4):2647–2663

92. Tajima, F (1989) Statistical method for testing the neutral mutation hypothesis by DNA polymorphism. Genetics 123(3):585–595

93. Thornton KR, Jensen JD (2007) Controlling the false positive rate in multilocus genome scans for selection. Genetics 175(2):737–750

94. Trujillo JT, Beilstein MA, Mosher RA (2016) The Argonaute-binding platform of NRPE1 evolves through modulation of intrinsically disordered repeats. New Phytol 212 (4):1094–1105. https://doi.org/10.1111/nph.14089

95. Voigt S, Laurent S, Litovchenko M, Stephan W (2015) Positive selection at the polyhomeotic locus led to decreased thermosensitivity of gene expression in temperate

Drosophila melanogaster. Genetics 200 (2):591–599

96. Wall JD (1999) Recombination and the power of statistical tests of neutrality. Genet Res 74(1):65–79

97. Wang MS, Zhang RW, Su LY, Li Y, Peng MS, Liu HQ, Zeng L, Irwin DM, Du JL, Yao YG, et al (2016) Positive selection rather than relaxation of functional constraint drives the evolution of vision during chicken domestication. Cell Res 26(5):556

98. Wilches R, Voigt S, Duchen P, Laurent S, Stephan W (2014) Fine-mapping and selective sweep analysis of QTL for cold tolerance in Drosophila melanogaster. G3 Genes Genom Genet 4(9):1635–1645

99. Yang SY, Han MJ, Kang LF, Li ZW, Shen YH, Zhang Z (2014) Demographic history and gene flow during silkworm domestication. BMC Evol Biol 14(1):185

100. Yuan Y, Zhang Q, Zeng S, Gu L, Si W, Zhang X, Tian D, Yang S, Wang L (2017) Selective sweep with significant positive selection serves as the driving force for the differentiation of japonica and indica rice cultivars. BMC Genom 18(1):307

101. Zeng K, Fu YX, Shi S, Wu CI (2006) Statistical tests for detecting positive selection by utilizing high-frequency variants. Genetics 174(3):1431–1439

102. Zhang Z, Jia Y, Almeida P, Mank JE, van Tuinen M, Wang Q, Jiang Z, Chen Y, Zhan K, Hou S, et al (2018) Whole-genome resequencing reveals signatures of selection and timing of duck domestication. GigaScience 7(4):giy027

103. Zhou J, Troyanskaya OG (2015) Predicting effects of noncoding variants with deep learning–based sequence model. Nat Methods 12 (10):931

104. Živković D, Stephan W (2011) Analytical results on the neutral non-equilibrium allele frequency spectrum based on diffusion theory. Theor Popul Biol 79(4):184–191

Chapter 6

polyDFE: Inferring the Distribution of Fitness Effects and Properties of Beneficial Mutations from Polymorphism Data

Paula Tataru and Thomas Bataillon

Abstract

The possible evolutionary trajectories a population can follow is determined by the fitness effects of new mutations. Their relative frequencies are best specified through a distribution of fitness effects (DFE) that spans deleterious, neutral, and beneficial mutations. As such, the DFE is key to several aspects of the evolution of a population, and particularly the rate of adaptive molecular evolution (α). Inference of DFE from patterns of polymorphism and divergence has been a longstanding goal of evolutionary genetics.

polyDFE provides a flexible statistical framework to estimate the DFE and α from site frequency spectrum (SFS) data. Several probability distributions can be fitted to the data to model the DFE. The method also jointly estimates a series of nuisance parameters that model the effect of unknown demography as well data imperfections, in particular possible errors in polarizing SNPs. This chapter is organized as a tutorial for polyDFE. We start by briefly reviewing the concept of DFE, α, and the principles underlying the method, and then provide an example using central chimpanzees data (Tataru et al., Genetics 207 (3):1103–1119, 2017; Bataillon et al., Genome Biol Evol 7(4):1122–1132, 2015) to guide the user through the different steps of an analysis: formatting the data as input to polyDFE, fitting different models, obtaining estimates of parameters uncertainty and performing statistical tests, as well as model averaging procedures to obtain robust estimates of model parameters.

Key words Distribution of fitness effects, Rate of adaptive molecular evolution, Beneficial mutations, Polymorphism and divergence data

1 Introduction

The following tutorial requires the successful installation of polyDFE-v1.1 (see manual for details on installation), and basic skills in using the command line and R. The latest version of

---

The original version of this chapter was revised. The correction to this chapter is available at https://doi.org/10.1007/978-1-0716-0199-0_20

Electronic supplementary material: The online version of this chapter (https://doi.org/10.1007/978-1-0716-0199-0_6) contains supplementary material, which is available to authorized users.

`polyDFE`, its manual as well as an R script `postprocessing.R` that contains functions which facilitate post-processing of `polyDFE` output files can be found on https://github.com/paula-tataru/polyDFE.

1.1 Modelling the Properties of Mutations on Fitness

Genome and exome sequencing studies open the possibility to survey systematically nucleotide variation in genomes. Several model-based methods have been developed to infer the properties of mutations from these surveys. In a nutshell, population genetics models introduced in Chapter 1 can formalize a fundamental intuition: the fitness effect of a new mutation will influence the frequency at which it segregates in a population. More formally, assuming a set of independent SNPs, mathematical expectations can be obtained that relate mutation rates and the fitness effect of mutations to observable quantities such as the number of SNPs that are found at a given frequency (i.e., the counts of the site frequency spectrum, SFS) in a sample of individuals that were re-sequenced or genotyped [1].

The effects of new mutations on fitness are expected to vary depending on the region where the mutation happens and what types of changes are incurred by the mutation. We model the variation in effects of mutations by making a number of assumptions:

- We can make an a priori distinction between sites where only neutrally evolving mutations are segregating and sites that harbor mutations potentially under selection. We refer to these as neutral/selected sites, respectively.

- The number of mutations in a region of known length of nucleotides arises randomly as a Poisson process with a certain intensity that depends on the length of the region and the mutation rate per nucleotide.

- Mutations that happen at neutral sites are lost or drift to fixation solely due to genetic drift.

- Mutations happening at selected sites are ascribed a fitness effect through a scaled selection coefficient. Each mutation at a selected site is treated as exchangeable: no sites are identified a priori as yielding mutations that are intrinsically good or bad for fitness. The scaled selection coefficient $4N_e s$ of a mutation is drawn at random from an underlying distribution, also called the distribution of fitness effects (DFE).

`polyDFE` performs maximum likelihood (ML) inference of DFE parameters from polymorphism data. Various probability distributions have been used to model DFEs [2]. Currently, `polyDFE` uses four types of distributions, referenced as models A through D and described in detail in the `polyDFE` manual. In this chapter, we focus solely on examples where models A and C are used.

Under model A, the DFE is given by a reflected and displaced Γ distribution. This distribution is parameterized through a mean scaled selection coefficient $\overline{S}$, a shape b, and a maximum scaled selection coefficient S_{max}. This continuous distribution is theoretically motivated as the approximation for the DFE expected under an explicit fitness landscape where fitness is determined by k traits under stabilizing selection and where each mutation will pleiotropically affect every trait [2]. In the general case where S_{max} is not restricted to be 0, the DFE will comprise some beneficial mutations, otherwise the DFE will only comprise deleterious mutations.

Under model C, the DFE is given by a mixture of two distributions. A proportion p_b of mutations are favorable and their scaled selection coefficient is drawn from an exponential distribution with mean S_b, and the remaining $1 - p_b$ are deleterious mutations with scaled selection coefficient drawn from a reflected Γ distribution with mean S_d and shape b. If the proportion p_b is restricted to be 0, the DFE will only comprise deleterious mutations. Note that the DFEs containing only deleterious mutations obtained from either model A ($S_{max}=0$) or model C ($p_b=0$) are equivalent and are given by a reflected Γ distribution.

When inferring DFEs from SFS data, a DFE with only deleterious mutations (henceforth a deleterious DFE) is typically assumed. In order to obtain information about the selection coefficients of beneficial mutations, available methods rely on the amount of divergence data between the species of interest (ingroup) and an outgroup. polyDFE departs from this approach as it allows the user to obtain estimates of a DFE also containing beneficial mutations (henceforth a full DFE) solely from SFS data. In doing so, polyDFE has the advantage of not assuming that the DFE is a constant in both the ingroup and outgroup. The price to pay for relaxing this assumption is that by only using the SFS, the ML estimates have more sampling variance, reflecting the uncertainty due to reduced amounts of data.

1.2 Calculating the Rate of Adaptive Evolution, α

Obtaining estimates of the DFE allows one to learn more about factors governing the rate of adaptive molecular evolution, commonly defined as the proportion of fixed adaptive mutations, α. Besides ML estimates of DFE, polyDFE can be used to obtain ML estimates of α. Once the DFE is estimated, α can be obtained using the divergence data as:

$$\alpha = \frac{\text{expected number of beneficial substitutions}}{\text{observed divergence selected counts}}$$

$$= \frac{\text{observed divergence selected counts} - \begin{array}{c}\text{expected number of neutral}\\\text{and deleterious substitutions}\end{array}}{\text{observed divergence selected counts}},$$

where the expected number of neutral and deleterious substitutions is obtained from the DFE [1].

Alternatively, α can be estimated without using the divergence data by replacing the observed divergence selected counts with expectations derived from the DFE [1]. The two different estimations of α are referred to α_{div} and α_{dfe} [1], to reflect the type of data/information used.

Note that alternative statistics exist for measuring molecular adaptation, such as ω_A, the rate of adaptive evolution relative to the mutation rate, or K_{a^+}, the rate of adaptive amino acid substitutions [3, 4]. These have different properties and might be better suited for studying various aspects of adaptation [4]. Currently, polyDFE only calculates α, but these statistics can also be obtained once the DFE is estimated.

2   Pre-processing of the Data

2.1   The Type of Information Required by polyDFE

polyDFE requires as input the derived SFS at both neutral and selected sites. The input file can optionally also contain counts of divergence.

When preparing the data, there are three elements that require some careful attention:

- what is the length of the region that was called for the potential occurrence of SNPs;
- how are SNPs polarized into an ancestral and a derived allele;
- how missing data is removed.

Software that enables SNP calling will also report the length of the region, calculated as the number of nucleotide positions where SNP calling could be performed. This length has to be then (correctly) divided between the data containing neutral sites and a priori selected sites. For an example, see the end of Subheading 2.2.

polyDFE assumes that the SFS is derived (polarized) and the given counts (see Subheading 2.2) are for derived SNP alleles. Various methods are available to orient SNPs, including parsimony and more rigorous probabilistic methods [5–7]. All methods require access to at least one outgroup.

polyDFE cannot deal with missing data. If the SFS contains missing data, several strategies can be used. If local linkage disequilibrium is known, SNP imputation can be used to estimate the missing genotypes. Alternatively, projection methods can be used to down-sample the SNP data to build a complete SFS with a reduced number of samples [8, 9].

2.2   Example of a polyDFE Input File

This tutorial uses central chimpanzee data [1, 10] to exemplify the different steps of an analysis. The central chimpanzee data and all additional files used here can be found on https://github.com/StatisticalPopulationGenomics. Here is a snippet of the chimpanzee data found in the input file central_chimp_sfs:

```
# lines starting with # are comments and can be present anywhere in the file
# Autosomes

1 1 24
14492   6138   [...]   845    4292115      44048    4290192
12645   4573   [...]   469    16146528     26481    16139295
```

The first non-empty non-comment line specifies sequentially that there is only one (1) neutral and one (1) selected region and that 24 haplotypes were re-sequenced to obtain the SFS data.

Note that polyDFE can, in principle, analyze jointly multiple regions with different mutation rates that share the DFE parameters [1]. For the remainder of this chapter, we analyze data pooled into a single SFS for neutral and selected sites. For details on variability of mutation rates, see the polyDFE manual.

For each region (here two in total), the neutral followed by the selected ones, there is one line of input that gives sequentially the entries of the SFS, i.e. how many SNPs had the derived allele in 1 copy (14492 for the neutral region and 12645 for the selected region), 2 copies (6138 for the neutral region and 4573 for the selected region), up to 23 copies (845 for the neutral region and 469 for the selected region), followed by the length of the region (4292115 for the neutral region and 16146528 for the selected region), the divergence counts for the same region (44048 for the neutral region and 26481 for the selected region) and the length of the region where divergence data was obtained (4290192 for the neutral region and 16139295 for the selected region). The presence of divergence data in the input file is optional. For further details on the data format, see the polyDFE manual.

The central chimpanzee data was obtained from exome sequencing. To divide it into a neutrally evolving region and one potentially containing sites under selection, SNPs have been classified into synonymous and non-synonymous, with the first class assumed to be neutral [10]. This is a general practice when working with exome data. The length of the neutral (here, synonymous) and selected (here, non-synonymous) regions is typically calculated from the total length by using a proportionality principle where a proportion of the sites are deemed, respectively, synonymous and non-synonymous. Roughly we expect 3/4 of the sites in exome regions to be non-synonymous, but there are more rigorous and precise ways to calculate this quantity [11].

2.3 Note on SFS Data

A priori, one expects an L or U shaped SFS, where a lot of derived alleles are present in low frequencies and possibly a few more have high frequencies, especially when beneficial mutations contribute substantially to the SFS. This is borne out of population genetics theory and the fact that we expect that at selected sites a substantial fraction of the variation is deleterious to some degree. Large counts in intermediate frequencies are only caused by a significant amount of balancing selection or by cryptic and highly pronounced genetic differentiation in the sample of sequenced individuals.

3 Model Fitting with `polyDFE`

3.1 Specifying a DFE Model to Fit Using `polyDFE`

We use a series of examples to illustrate how to run `polyDFE`, and how to specify two key things:

- the input file containing the data to be analyzed;
- the DFE model used for the analysis.

The aim of the analysis is to fit a series of models differing by:

- the type of DFE assumed;
- the presence or not of beneficial mutations in the DFE;
- the inclusion of two types of nuisance parameters that apply to both neutral and selected sites:
  - a polarization error ϵ_{an};
  - a series of nuisance parameters r_i, one for each class of frequency in the SFS.

The polarization error ϵ_{an} accounts for the fact that methods for orienting the SNPs are not perfect and still leave errors in the data, where the inferred derived allele is, in fact, the ancestral allele. When ϵ_{an} is set and fixed to 0, it is assumed that the data contains no error. To the best of our knowledge, `polyDFE` is the only available method that explicitly incorporates polarization error.

The nuisance parameters r_i describe how the SFS can be distorted by sampling, demography, and/or linkage, relative to what is expected in a stable Wright-Fisher population at mutation-selection-drift equilibrium. When these parameters are fixed to 1, it is assumed that the data does not depart from standard expectations. For some datasets, errors in the SNP orientation can be efficiently captured by the distortion parameters r_i [1, 12], but this is not always the case [1]. We always recommend inferring a full model where all parameters are estimated and use hypothesis testing to decide whether such additional parameters are needed or not (*see* Subheading 5.2). To simply run `polyDFE` on the chimpanzee data using default settings, the following command line can be used:

```
polyDFE -d central_chimp_sfs
```

where

- `-d central_chimp_sfs` specifies that `polyDFE` runs on the input file `central_chimp_sfs`.

However, it is more useful to customize the behavior of polyDFE through the command line arguments by specifying, for example, which DFE model is used or which parameters should be estimated or not:

```
polyDFE -d central_chimp_sfs -m A -i init_A.txt 1 -e
        -b params_basinhop.txt 1 > central_chimp_A
```

where

- -m A specifies that model A will be used to infer the DFE.
- -i init_A.txt 1 specifies that the parameter configuration with ID 1 from the initialization file init_A.txt should be used. The initialization file is used to control which parameters should polyDFE estimate and to provide their initial values used during the estimation process. For example, polyDFE can be forced to estimate a deleterious DFE. The init_A.txt file provides examples on how parameters are set to be fixed or estimated.
- -e specifies that the parameters' initial values should be estimated automatically, using a combination of approximate analytical results and a grid search. Using -e is highly recommended when running an initial analysis.
- -b params_basinhop.txt 1 specifies polyDFE to run a more involved likelihood maximization, *see* Subheading 3.2 for details.
- >central_chimp_A (the redirection command) specifies that the output of polyDFE should be stored in the file central_chimp_A.

To run a full analysis on a dataset, models with increasing complexity (i.e. deleterious or full DFE, and allowing or not for nuisance parameters) should be used, as specified using -i. The example file run_polyDFE.txt contains all the command lines that allow a full analysis of the example dataset using both estimation under models A and C. As the true shape of the DFE is not known (i.e., is the DFE in the form of an A or C model?), it is recommended to run poly-DFE with multiple models and finally use hypothesis testing to find the best fitting model (*see* Subheading 5.2).

If the input data also contains divergence counts, polyDFE uses this information for estimation by default. To use exclusively the SFS data during the estimation process, the command line argument -w is used.

3.2 Note on Likelihood Maximization

One of the key issues in obtaining reliable ML estimates is to ensure that the likelihood function is properly maximized over the space of parameters. This is not trivial, and besides the limitations of the method itself, this is what in practice will cause polyDFE to return poor estimates.

polyDFE implements multiple steps to ensure, as much as possible, that good estimates are found.

1. The maximization algorithm requires initial values for the parameters, and these values can have a big effect on the estimates found. They can be provided by the user in the initialization file given through the command line argument -i, but for an initial run of polyDFE, it is strongly recommended to use the command line argument -e. This ensures that the maximization is started with sensible parameter values.

2. Standard maximization algorithms allow parameters to take any value over the whole real line. Many of the parameters polyDFE estimates are constrained within a specific range. For example, the shape b for the DFE distribution is constrained to be positive. To allow for such constraints, polyDFE transforms the parameters from a given range to the whole real line. The range of each parameter can be controlled through the command line argument -r. polyDFE uses large ranges by default, but providing a range that is tighter around the potential maximum could improve the maximization process. For more details, see the polyDFE manual.

3. Standard maximization algorithms, including the ones used in polyDFE, only ensure that a local maximum likelihood is found. However, a better solution can possibly exist. To avoid that polyDFE is stuck in a local solution, the basin hopping algorithm [13] can be used. This is a stochastic algorithm that runs the standard maximization algorithm multiple times, using different initial values for the parameters. polyDFE runs basin hopping when the command line argument -b is provided, as in the previous example. The basin hopping algorithm can be customized through a parameterization file. In the previous example, we used -b params_basinhop.txt 1, which specified that polyDFE should use the parameterization found in params_basinhop.txt with ID 1. The params_basinhop.txt file provides examples on how the basin hopping algorithm can be customized. For more details, see the polyDFE manual.

polyDFE uses maximization algorithms that rely on the gradient of the likelihood. If a set of parameters truly has a locally maximum likelihood, then the corresponding gradient is 0. If the gradient of the best solution found is far away from 0, this is warning sign that a good solution was not found. For different ways to change the run of polyDFE to aim for a better gradient, see the polyDFE manual.

4   Post-Processing of the polyDFE Output

polyDFE is accompanied by an R script postprocessing.R contains a series of functions written that parse the output of polyDFE and calculate other quantities of interest. We provide examples of polyDFE output files, as well as how to use the R functions to parse the output and perform hypothesis testing/model averaging on DFE and α. The analysis found below is also provided as an example file analysis.R.

4.1 Example of a polyDFE Output File

The output starts with a summary of how polyDFE was run, the progression of the maximization procedure, the best estimates found, various expectations under the best estimates (such as, the expected SFS) and, finally, estimates of α.

Here is a snippet of the output file central_chimp_A created in the previous example:

```
---- Running command
---- ./polyDFE -d central_chimp_sfs -b params_basinhop.txt 1 -m A -i init_A.txt
---- Performing inference on  central_chimp_sfs using model A

---- Warning: mutation variability is not used
     when only one neutral and one selected fragment is available.
---- No mutation variability. Using Poisson likelihood.

---- Starting a maximum of 10 basin hopping iterations

---- Calculating initial values for neutral parameters
---- Calculating initial values for selection parameters

[...]

---- Basin hopping performed 10 iterations
---- Basin hopping reached maximum number of iterations allowed

---- Best joint likelihood found -213.134021489085512 with gradient 0.01906
--  Model: A
--     eps_an     lambda   theta_bar           a
-- 0.00200487 0.00996581 0.00337419          -1
--      S_bar          b       S_max
-- -18585.547 0.19723338 0.62130240
--        r 2        r 3        r 4        r 5         [...]
-- 0.85918273 0.66327008 0.55830194 0.46158509         [...]


---- Expected P_neut(i), 0 < i < n (neutral SFS per site)
E[P_neut(1)] = 0.0033677960
[...]

---- Expected P_sel(i), 0 < i < n (selected SFS per site)
E[P_sel(1)] = 0.0007854124
[...]

---- Expected D_neut and D_sel (neutral and selected divergence per site)
E[D_neut] = 0.0102841198
E[D_sel] = 0.0016362652

---- Expected neutral and selected misattributed polymorphism per site
E[mis_neut] = 0.0003183091
E[mis_sel] = 0.0000503808

---- alpha_div = 0.835613
---- alpha_dfe = 0.834864
```

For more details on the misattributed polymorphism, *see* Sub-heading 4.4.

4.2 Merging and Parsing Output Files

The R functions from postprocessing.R can parse files containing multiple outputs from polyDFE, which enables an easier analysis. For this, multiple output files can be merged, for example, as follows:

```
cat central_chimp_A_nor_noeps  >  central_chimp_A.txt
cat central_chimp_A_nor        >> central_chimp_A.txt
cat central_chimp_A_noeps      >> central_chimp_A.txt
cat central_chimp_A            >> central_chimp_A.txt
```

The file run_polyDFE.txt contains the necessary polyDFE commands to obtain all of the above output files, which contain results for model A where:

- central_chimp_A_nor_noeps: r_i parameters are fixed to 1 and ϵ_{an} is fixed to 0;
- central_chimp_A_nor: r_i parameters are fixed to 1, but ϵ_{an} is estimated;
- central_chimp_A_noeps: r_i parameters are estimated, but ϵ_{an} is fixed to 0;
- central_chimp_A: both r_i and ϵ_{an} are estimated.

Corresponding files, as given in run_polyDFE.txt, can be obtained for model C and for a deleterious DFE model. Restricting either model A or model C to only deleterious mutations yields the same type of DFE: a reflected Γ distribution.

The output file central_chimp_A.txt from above can be parsed using the parseOutputR function (line 2) into a list where each entry corresponds to a different run of polyDFE found in the parsed file (line 3):

```
1   > source("postprocessing.R")
2   > est = parseOutput("central_chimp_A.txt")
3   > length(est)
4   > names(est[[1]])
5   > est[[1]]$input
6   > est[[1]]$model
7   > est[[1]][c("lk", "grad")]
8   > est[[1]]$values
9   > est[[1]]$estimated
10  > est[[1]]$expec
11  > est[[1]]$alpha
```

One entry in the list contains:

- the name of the input file polyDFE was run on (line 5),
- the DFE model used (line 6),
- the best likelihood found and corresponding gradient (line 7),
- the values of all parameters, including those that were not estimated (line 8),

- which parameters have been estimated (line 9),
- expectations for SFS and divergence (line 10),
- estimates for α (line 11).

We can also get an overview of the gradients obtained during the optimization process:

```
1  > est = c(parseOutput("central_chimp_A.txt"), parseOutput("central_chimp_C.tx
2  +         parseOutput("central_chimp_Del.txt"))
3  > grad = sapply(est, function(e) e$grad)
4  > names(grad) = sapply(est, getModelName)
5  > grad[which(grad > 0.001)]
6  A + r - eps    A + r + eps    C + r + eps
7       0.01503        0.01906        0.00536
```

The function getModelName (line 4) returns a string briefly describing the estimated model: which DFE model was used and whether the r_i and ϵ_{an} parameters were estimated (+) or not (-). The runs using model A where ϵ_{an} was not estimated (A + r - eps), and models A and C where all parameters have been estimated (A + r + eps, C + r + eps), have gradients that are larger than 0.001. For these, running additional iterations of basin-hopping might lead to an improved solution (*see* Subheading 3.2).

4.3 Summarizing the DFE Estimated by polyDFE

The estimated DFEs under the different models can be discretized using the getDiscretizedDFER function (lines 3–5), and then plotted for visual comparison (lines 6–13). This is exemplified below, where the DFE is binned in six classes of $4N_e s$ values, as shown in Fig. 1:

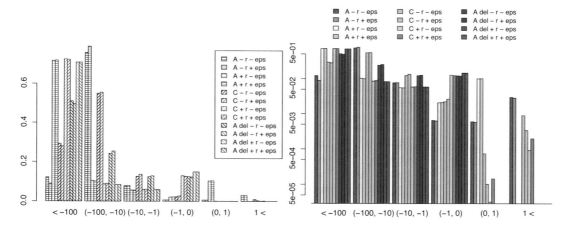

Fig. 1 Estimated discretized DFEs for models A, A del (which contains only deleterious mutations) and C, where the r_i parameters were estimated (+ r) or not (− r) and ϵ_{an} was estimated (+ eps) or not (− eps). The plot to the right has the *y*-axis on log scale

```
1  > est = c(parseOutput("central_chimp_A.txt"), parseOutput("central_chimp_C.tx
2  +        parseOutput("central_chimp_Del.txt"))
3  > getDiscretizedDFE(est[[1]])
4  > getDiscretizedDFE(est[[1]], c(-100, -10, -1, 0, 1))
5  > dfe = t(sapply(est, getDiscretizedDFE, c(-100, -10, -1, 0, 1)))
6  > rownames(dfe) = sapply(est, getModelName)
7  > par(mfrow = c(1, 2))
8  > barplot(dfe, beside = TRUE, legend.text = TRUE,
9  +        angle = rep(c(0, 45, -45), each = 4), density = 20, col = 1:4)
10 > dfe[which(dfe == 0)] = NA
11 > col = rainbow(n = nrow(dfe), end = 0.9)
12 > barplot(dfe, beside = TRUE, log = 'y', col = col)
13 > legend("topright", legend = rownames(dfe), fill = col, bty = "n", ncol =
```

The majority of the mutations are typically deleterious and only few mutations have beneficial effects (Fig. 1). To more easily compare visually the discretized DFE, it can be useful to use a log-scale on the y-axis for plotting (lines 10–13, Fig. 1).

To obtain a discretized DFE, the continuous DFE has to be integrated. This is prone to numerical issues and sometimes the resulting discretized DFE does not sum exactly to 1. When this is the case, getDiscretizedDFE issues a warning.

4.4 Estimating α

polyDFE automatically calculates and outputs α. If divergence data was used when polyDFE was run, then both α_{div} and α_{dfe} are estimated, otherwise only the latter is calculated:

```
1  > est = parseOutput("central_chimp_C.txt")
2  > sapply(est, function(e) e$alpha)
```

When a strictly deleterious DFE model is estimated, as by construction the DFE does not contain any beneficial mutations, the estimated α_{dfe} is 0:

```
1  > est = parseOutput("central_chimp_Del.txt")
2  > sapply(est, function(e) e$alpha["alpha_dfe"])
```

α can also be calculated in R from the estimated DFE using the function estimateAlpha (lines 5, 7). By default, α_{dfe} is calculated (line 5), but divergence data can be parsed in R using the function parseDivergenceData (line 3) and provided to estimateAlpha (line 7), which then returns α_{div}. We can compare the values that polyDFE outputted with the estimates obtained in R (lines 4–13). As polyDFE is implemented in C, the values for α returned by polyDFE are typically slightly different than the ones calculated in R (lines 8–13), though the difference should be minor:

```
1   > est = c(parseOutput("central_chimp_A.txt"), parseOutput("central_chimp_C.tx
2   +          parseOutput("central_chimp_Del.txt"))
3   > div = parseDivergenceData("central_chimp_sfs")
4   > alpha = sapply(est, function(e) c(e$alpha["alpha_dfe"],
5   +                                    "R alpha_dfe" = estimateAlpha(e),
6   +                                    e$alpha["alpha_div"],
7   +                                    "R alpha_div" = estimateAlpha(e, div =
8   > summary(alpha["alpha_dfe", ] - alpha["R alpha_dfe", ])
9         Min.    1st Qu.    Median        Mean    3rd Qu.          Max.
10  -2.865e-03 -1.005e-03 -4.989e-05 -6.723e-04  0.000e+00   0.000e+00
11  > summary(alpha["alpha_div", ] - alpha["R alpha_div", ])
12        Min.    1st Qu.    Median        Mean    3rd Qu.          Max.
13  -4.066e-03 -2.254e-05 -3.168e-06 -6.729e-04  1.562e-06   1.148e-04
```

Divergence data can contain misattributed polymorphism: SNPs that were misidentified as substitutions as they were fixed within the sample [1]. When calculating α_{div}, polyDFE automatically corrects for this. Using R, this correction can be turned off (line 4), but typically the difference in the estimated α_{div} is not very large (lines 3–7):

```
1   > est = c(parseOutput("central_chimp_A.txt"), parseOutput("central_chimp_C.tx
2   +          parseOutput("central_chimp_Del.txt"))
3   > alpha = sapply(est, function(e) c("+ corr" = estimateAlpha(e, div = div),
4   +                                    "- corr" = estimateAlpha(e, div = div,
5   > summary(alpha["+ corr", ] - alpha["- corr", ])
6         Min.    1st Qu.    Median        Mean  3rd Qu.        Max.
7   5.047e-05  5.905e-05  4.373e-04  1.762e-03  1.105e-03  8.540e-03
```

When calculating α, the expected number of substitutions that are either non-beneficial (deleterious or neutral) or beneficial is calculated. polyDFE and the R function estimateAlpha assume that a mutation that has a positive selection coefficient is beneficial. However, one could argue that a mutation with a very low positive selection coefficient is effectively neutral and only mutations that have a selection coefficient above a threshold S_{sup} should be considered as beneficial when calculating α [1, 12]. To obtain such estimates of α, the supLimit can be changed in R (lines 2–4). Setting a higher supLimit will mechanically decrease α, as fewer substitutions will be considered beneficial (lines 2–10):

```
1   > est = c(parseOutput("central_chimp_A.txt"), parseOutput("central_chimp_C.tx
2   > alpha = sapply(est, function(e) c("supLimit = 0" = estimateAlpha(e),
3   +                                    "supLimit = 5" = estimateAlpha(e, supLimit
4   +                                    "supLimit = 10" = estimateAlpha(e, supLimit
5   > summary(alpha["supLimit = 0", ] - alpha["supLimit = 5", ])
6      Min. 1st Qu.  Median    Mean 3rd Qu.     Max.
7   0.00572 0.01328 0.22622 0.32156 0.52148 0.83711
8   > summary(alpha["supLimit = 5", ] - alpha["supLimit = 10", ])
9         Min.    1st Qu.    Median       Mean  3rd Qu.      Max.
10  0.000000 0.004442 0.020051 0.133093 0.186817 0.468287
```

5   Hypothesis Testing and Model Averaging

The estimation of DFE and α entails substantial statistical uncertainty. We discuss how to obtain the sampling variance of parameter estimates and approximate confidence intervals using a bootstrap approach. We also outline how to perform hypothesis testing and how to use model averaging as an alternative to the hard thresholding inherent to hypothesis testing. The flexibility of polyDFE allows performing model averaging with the advantage of generating parameter estimates where model uncertainty is also incorporated.

5.1 Bootstrap-Based Confidence Intervals

In principle, likelihood profiles can be obtained for one or more parameters by fixing these parameters to a set of values and maximizing the likelihood for all other parameters. Using polyDFE, this can be achieved by using the command line argument -i (*see* Subheading 3.1). The profile likelihood can then be used to obtain approximate confidence intervals for the parameters of interest.

In practice, we recommend to use a bootstrap approach:

1. Generate 100–500 bootstrap datasets.

2. Run polyDFE on these datasets.

3. Calculate the sampling distribution of the ML estimates returned by polyDFE.

Although a crude likelihood profile can be obtained for a single parameter with as few as 20–30 runs of polyDFE, the bootstrap approach has the advantage of yielding sampling distributions for all parameters of interest in one go, as well as capturing the patterns of covariation between parameters.

Bootstraps are generated by re-sampling the data at the site level. More specifically, bootstrap datasets are obtained by parametric bootstrapping and assuming that all counts in the SFS and, possibly, divergence data are independent variables following a Poisson distribution, with means given by the observed data. This is in line with the modelling assumption that the number of mutations in each SFS entry and, possibly, divergence data follows a Poisson process. Using R, 500 bootstrap datasets can, for instance, be obtained using the commands:

```
1  > bootstrapData("central_chimp_sfs", rep = 500)
2  > bootstrapData("central_chimp_sfs", outputfile = "boostrap_central", rep
```

which generate 500 datasets each stored in files cental_-chimp_sfs_j (line 1) and bootstrap_central_j (line 2), with $1 \leq j \leq 500$, respectively. The name of the output files can be optionally specified through outputfile (line 2).

To speed up the analysis when running `polyDFE` on bootstrap data, the command line argument `-i` (*see* Subheading 3.1) can be used to initialize the parameters to the values that were found when running `polyDFE` on the full dataset. Using the R function `createInitLines` (see the `polyDFE` manual for details), these values from a `polyDFE` output file can be written automatically to an initialization file that is then given to `-i`.

Once `polyDFE` is run on bootstrap data, confidence intervals for parameters, expected SFS, and discretized DFE can be obtained using the quantiles of the bootstrap distributions of the quantities of interest [14].

5.2 Hypothesis Testing

`polyDFE` returns ML estimates and therefore, likelihood ratio tests (LRT) and the Akaike information criteria (AIC) can be used to compare models.

The likelihood ratio test entails fitting two nested models using `polyDFE`, where one reduced model is a special case of a more general model. A p-value can be obtained by assuming that the log of the ratio of the maximum likelihoods of the two models follows a χ^2 distribution parameterized by the difference in the number of degrees of freedom (i.e., number of estimated parameters) between the two models. A small *p*-value means that the reduced model is rejected in favor of the more parameter-rich model.

For instance, one can formally test for the occurrence of beneficial mutations by fitting two models A that differ by the maximum allowed scaled selection coefficient S_{max}, where in the general model S_{max} is freely estimated, while in the reduced model S_{max} is fixed to 0 and thus a deleterious DFE is estimated. Similarly, this can be done under model C by fixing the amount of beneficial mutations p_b to 0. Note that the two reduced models under models A and C yield the same type of deleterious DFE, therefore testing for the occurrence of beneficial mutations under model C can be obtained by comparing the reduced model A with the general model C.

Using LRT requires that models are nested, which does not allow to test whether the full DFE model A or C fits the data better. For this, the Akaike information criteria can be used

$$\text{AIC} = 2m - 2\log(L)$$

where m is the number of estimated parameters (or degrees of freedom) and L is the maximum likelihood. Then the preferred model is the one with the minimum AIC value.

To test for the occurrence of beneficial mutations under model A, the p-values from the LRTs and AIC values can be obtained using the R function `compareModels` as follows:

```
1  > compareModels("central_chimp_A.txt", "central_chimp_Del.txt")
```

The function sequentially compares model number j found in central_chimp_A.txt with model number j found in central_chimp_Del.txt (in the order of appearance in the files), which vary in the estimation of r_i and ϵ_{an}, as detailed in Subheading 4.2 and run_polyDFE.txt.

The models found within central_chimp_A.txt and the other output files are also nested. For example, for models number 3 and 4, the r_i parameters were estimated, but ϵ_{an} was either estimated (model number 4) or fixed to 0 (model number 3). The LRT for these two models can be obtained as follows:

```
> est = parseOutput("central_chimp_A.txt")
> compareModels(est[3], est[4])$LRT
```

The compareModels function automatically detects nestedness when the same DFE model was used. Recall that the deleterious DFEs obtained from either model A or model C are equivalent, but as the deleterious model was obtained from model A, to test for the occurrence of beneficial mutations under model C, nestedness has to be enforced by setting nested = TRUE:

```
> compareModels("central_chimp_C.txt", "central_chimp_Del.txt", nested = TRUE)
```

As noted in Subheading 3.1, sometimes the r_i parameters can also account for polarization errors. This is the case here, as we can observe that, when the r_i parameters are not estimated and fixed to 1 (- r), the LRT and AIC indicate that inferring ϵ_{an} (+ eps) leads to a better fit of the data. However, when the r_i parameters are estimated (+ r), ϵ_{an} is not needed for fitting the data (- eps). This is also supported by the estimated value of ϵ_{an}, which is very small when the r_i parameters are estimated (+ r), but much larger otherwise (Table 1):

Table 1
Model testing for $\epsilon_{an} \neq 0$

	r_i fixed to 1				r_i estimated			
	AIC - r - eps	AIC - r + eps	*p*-Value	ϵ_{an}	AIC + r - eps	AIC + r + eps	*p*-Value	ϵ_{an}
Model A	10473	10283	1.07e–43	0.011	481	484	2.68e–01	0.002
Model C	10114	9758	4.91e–80	0.016	475	478	5.23e–01	0.00012
Model del	13709	13562	2.51e–34	0.010	633	635	8.31e–01	0.00002

Note: - r: r_i parameters are not estimated and fixed to 1, + r: r_i parameters estimated, - eps: ϵ_{an} is not estimated and fixed to 0, + eps: ϵ_{an} is estimated

```
1  > aic = list()
2  > for (f in c("central_chimp_A.txt", "central_chimp_C.txt", "central_chimp_Del.txt"))
3  + {
4  +    est = parseOutput(f)
5  +    without_r = compareModels(est[1], est[2])
6  +    with_r = compareModels(est[3], est[4])
7  +    aic[[f]] = rbind("- r" = c(without_r$AIC[, "AIC model 1"], without_r$AIC[,
8  +                              without_r$LRT[, "p-value"], est[[2]]$values["eps_an"]),
9  +                     "+ r" = c(with_r$AIC[, "AIC model 1"], with_r$AIC[, "AIC
10 +                              with_r$LRT[, "p-value"], est[[4]]$values["eps_an"]))
11 + }
```

5.3 Model Averaging with `polyDFE`

Using model averaging provides a way to obtain the most honest estimates that account for model uncertainty. In particular, we can have a series of estimates that can differ by the DFE model assumed (A, B, C or D), whether or not distortions of the SFS are accounted for and possibly whether errors happen when polarizing the SNPs. One can choose the model most supported by the data using LRT or AIC as described in Subheading 5.2, but model averaging has the advantage of avoiding a strict thresholding where we decide that a given model is the best and exclude all other competing models. This might be necessary, for example, when the data contain only limited information about the DFE [15, 16] and therefore different models cannot be differentiated, due to AIC values that are very close (see also Subheading 5.4).

In practice, the parameter of interest x (such as any DFE parameter, α, entries in the discretized DFE or any other quantities) is estimated as x_j under each model j that has an AIC value of AIC_j. These values are combined to yield a model-averaged estimate where the contribution of each model is averaged using Akaike weights [17]:

$$x_{avg} = \frac{\sum_j x_j e^{-1/2\Delta AIC_j}}{\sum_j e^{-1/2\Delta AIC_j}} \tag{1}$$

where ΔAIC_j is obtained as

$$\Delta AIC_j = AIC_j - \min_j (AIC_j).$$

When doing model averaging, models that are most supported by the data have a ΔAIC_j that is close to 0 and consequently a weight that is close to 1, while models fitting badly have high ΔAIC_j values and, accordingly, weights that shrink towards 0.

The AIC weights can be calculated using the R function `getAICweights` (line 3). Then the average value of, for example, α_{div} (lines 4, 5) and α_{dfe} (lines 6, 7) can be calculated as follows:

```
1  > est = c(parseOutput("central_chimp_A.txt"), parseOutput("central_chimp_C.tx
2  +           parseOutput("central_chimp_Del.txt"))
3  > aic = getAICweights(est)
4  > alpha_div = sum(sapply(1:length(est),
5  +                        function(i) aic[i, "weight"] * est[[i]]$alpha["alpha_div"]))
6  > alpha_dfe = sum(sapply(1:length(est),
7  +                        function(i) aic[i, "weight"] * est[[i]]$alpha["alpha_dfe"]))
```

Note that `getAICweights` returns weights that are already rescaled so that they sum to 1, so the average value of α does not have to be normalized by the sum of the weights as in Eq. 1.

5.4 Note on Divergence Data

One of the advantages of `polyDFE` is that it does not require the use of divergence data for inferring the DFE and α. This is free of the assumption that the ingroup and outgroup share the DFE, which is needed when divergence data is used. Violating this assumption can lead to biases in the estimates [1]. However, divergence data is always available, as it is needed to orient the SNPs to calculate the derived SFS (*see* Subheading 2.1). So then the question arises on whether divergence data should be used or not in the inference.

The impact of using divergence data can be observed when investigating the AIC values (Table 2). When divergence data was not used, they are much closer to each other: the best 6 models are within an AIC difference of approximately 4, while when divergence data is used, only the first two models have an AIC difference of approximately 2, while the rest of the models have a much poorer fit. This is because using less data for the inference makes it more difficult to differentiate between the models. Using or not the divergence data also has a big impact on the estimates of α (Table 2):

```
1  > est = list(c(parseOutput("central_chimp_A.txt"),
2  +               parseOutput("central_chimp_C.txt"),
3  +               parseOutput("central_chimp_Del.txt")),
4  +             c(parseOutput("central_chimp_A_no_div.txt"),
5  +               parseOutput("central_chimp_C_no_div.txt"),
6  +               parseOutput("central_chimp_Del_no_div.txt")))
7  > for (i in 1:2)
8  + {
9  +     aic = getAICweights(est[[i]])
10 +     rownames(aic) = sapply(est[[i]], getModelName)
11 +     aic = cbind(aic, t(sapply(est[[i]], function(e) e$alpha)))
12 +     aic = aic[order(aic[, "delta AIC"]), ]
13 +     print(aic)
14 +     print(sum(apply(aic, 1, function(a) a["weight"] * a["alpha_dfe"])))
15 + }
```

The model-averaged α_{dfe} (line 14) is estimated to be 0.388 when divergence data is not used, but only 0.225 when divergence data is used.

Table 2
Model comparison and estimates of α

	df	log lk	Δ AIC	α_{div}	α_{dfe}
Using divergence data					
C + r − eps	29	−208.94	0.00	0.179	0.179
C + r + eps	30	−209.14	2.41	0.197	0.197
A + r − eps	28	−212.52	5.17	0.838	0.837
A + r + eps	29	−213.13	8.39	0.836	0.835
A del + r − eps	27	−289.91	157.95	0.066	0.000
A del + r + eps	28	−289.94	159.99	0.066	0.000
C − r + eps	7	−4872.04	9282.21	0.767	0.767
C − r − eps	6	−5051.49	9639.10	0.798	0.800
A − r + eps	6	−5135.66	9807.44	0.929	0.928
A − r − eps	5	−5231.75	9997.62	0.927	0.927
A del − r + eps	5	−6776.33	13086.79	0.187	0.000
A del − r − eps	4	−6850.96	13234.05	0.178	0.000
Not using divergence data					
A + r − eps	26	−196.05	0.00	0.770	0.741
A del + r − eps	25	−197.30	0.51	0.159	0.000
A del + r + eps	26	−196.90	1.71	0.167	0.000
A + r + eps	27	−196.07	2.04	0.770	0.741
C + r − eps	27	−196.13	2.16	0.385	0.308
C + r + eps	28	−196.33	4.56	0.580	0.514
C − r + eps	6	−4844.06	9256.02	0.787	0.945
C − r − eps	5	−4878.07	9322.04	0.814	0.970
A del − r + eps	4	−4881.24	9326.37	0.731	0.000
A − r + eps	5	−4881.24	9328.37	0.731	0.000
A − r − eps	4	−5199.53	9962.96	0.909	0.873
A del − r − eps	3	−5328.36	10218.61	0.652	0.000

Note: Results for models A, A del (which contains only deleterious mutations) and C, where the r_i parameters were estimated (+ r) or not (− r) and ϵ_{an} was estimated (+ eps) or not (− eps)

Deciding on whether or not divergence data should be used is not necessarily straightforward. The previous approaches of LRT and AIC are not applicable here, as we want to compare two models that are fitted on different datasets. One way is to compare the observed and expected SFS obtained under the estimated models. The best models found in both cases contain estimated r_i parameters, but an ϵ_{an} that was fixed to 0. When divergence data was used, the best model is model C, while when divergence data is not used, the model of choice is model A (Table 2).

The observed SFS (line 1) is given as counts for the total length of the region, while the expected one (lines 2–4) is given per site, so before the comparison, the expected SFS has to be normalized by the length of the region (lines 5, 6). Using a log scale on the y-axis (lines 10, 11) and coloring the background in alternating colors (lines 12–16) makes it easier to visually compare the SFS for $1 \leq i < n$, where n is the sample size (line 7). Then the expected SFS counts scan be plotted (lines 19, 20) next to the observed SFS counts (line 21):

```
1  > sfs = parseSFSData("central_chimp_sfs")
2  > est = parseOutput("central_chimp_C.txt")[[3]]
3  > sfs_div = est$expec
4  > sfs_no_div = parseOutput("central_chimp_A_no_div.txt")[[3]]$expec
5  > sfs_div = sfs_div * sfs[, "length_sfs"]
6  > sfs_no_div = sfs_no_div * sfs[, "length_sfs"]
7  > n = max(grep("E[P", colnames(sfs_div), fixed = TRUE))
8  > m = min(sfs[, 1:n], sfs_div[, 1:n], sfs_no_div[, 1:n])
9  > M = max(sfs[, 1:n], sfs_div[, 1:n], sfs_no_div[, 1:n])
10 > plot(1, 1, type = "n", ylim = c(m, M), xlim = c(1, n),
11 +      xlab = "i", ylab = "SFS", log = "y")
12 > xleft = seq(from = 1, to = n, by = 2)
13 > rect(xleft = xleft - 0.5, ybottom = m * 0.001,
14 +      xright = xleft + 0.5, ytop = M * 100,
15 +      col = gray(0.9), border = NA)
16 > box()
17 > for (i in 1:2)
18 + {
19 +   points(1:n - 0.15, sfs_div[i, 1:n], col = "red", pch = 19 + i)
20 +   points(1:n + 0.15, sfs_no_div[i, 1:n], col = "blue", pch = 19 + i)
21 +   segments(x0 = 1:n - 0.35, x1 = 1:n + 0.35, y0 = sfs[i, 1:n], col = "black")
22 + }
23 > legend("topright", bty = "n", pch = c(NA, 21, 21, 20, 21), lwd = c(1, NA,
24 +        col = c("black", "red", "blue", "black", "black"),
25 +        legend = c("obs", "expec with div", "expec without div", "neutral",
```

Visualizing the SFS (Fig. 2) can give insights into how well the models fit the data, but it does not give any statistical measure on how well the expected SFS matches the observed. To test that, we can use a χ^2 goodness-of-fit test (lines 4–6). This indicates that both using the divergence data or not gives a good fit to the SFS, and that, in general, the selected SFS is more difficult to fit well (lines 8–10). The χ^2 statistic could be used as a way to decide if divergence data should be used or not (lines 11–13) which shows that, overall, not using divergence data seems to give a closer fit to the data:

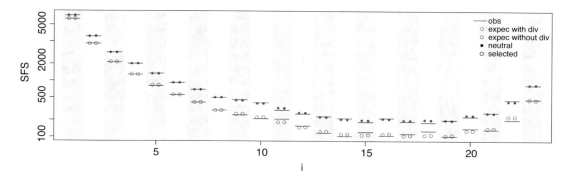

Fig. 2 Observed and expected SFS for best fitted models when divergence data was used or not. The *y*-axis is on log scale

```
 1  > sfs_both = rbind(sfs_div, sfs_no_div)
 2  > fit = sapply(1:nrow(sfs_both),
 3  +                function(i)
 4  +                    unlist(chisq.test(x = sfs_both[i, 1:n],
 5  +                                       p = sfs[(i + 1) %% 2 + 1, 1:n],
 6  +                                       rescale.p = TRUE)[c("statistic", "p.value")]))
 7  > colnames(fit) = paste(rep(c("div", "no div"), each = 2), rownames(sfs_div))
 8  > fit["p.value", ]
 9     div neut      div sel   no div neut    no div sel
10    0.9693439    0.6584354    0.9773883    0.6852477
11  > c(div = sum(fit["statistic.X-squared", 1:2]), no_div = sum(fit["statistic.X-squared"
12        div   no_div
13    30.13997 29.15413
```

The above check can also be done using a model-averaged expected SFS, as detailed in `analysis.R`. Doing so does not change the above conclusion.

6   Conclusion

`polyDFE` provides a flexible likelihood framework to infer the DFE and properties of beneficial mutations from SFS and possibly divergence data. The estimation procedure accounts for uncertainty of the models and data imperfection. `polyDFE` is continually developed further with updates posted on https://github.com/paula-tataru/polyDFE and is currently being extended to test for heterogeneous DFE among species or gene categories within a single species.

References

1. Tataru P, Mollion M, Glémin S, Bataillon T (2017) Inference of distribution of fitness effects and proportion of adaptive substitutions from polymorphism data. Genetics 207 (3):1103–1119

2. Bataillon T, Bailey SF (2014) Effects of new mutations on fitness: insights from models and data. Ann New York Acad Sci 1320 (1):76–92

3. Gossmann TI, Song BH, Windsor AJ, Mitchell-Olds T, Dixon CJ, Kapralov MV, Filatov DA, Eyre-Walker A (2010) Genome wide analyses reveal little evidence for adaptive evolution in many plant species. Mol Biol Evol 27:1822–1832

4. Castellano D, Coronado-Zamora M, Campos JL, Barbadilla A, Eyre-Walker A (2016) Adaptive evolution is substantially impeded by Hill-Robertson interference in *Drosophila*. Mol Biol Evol 33:442–455

5. Hernandez RD, Williamson SH, Bustamante CD (2007) Context dependence, ancestral misidentification, and spurious signatures of natural selection. Mol Biol Evol 24(8):1792–1800

6. Keightley PD, Campos JL, Booker TR, Charlesworth B (2016) Inferring the frequency spectrum of derived variants to quantify adaptive molecular evolution in protein-coding genes of *Drosophila melanogaster*. Genetics 203(2):975–984

7. Keightley PD, Jackson BC (2018) Inferring the probability of the derived versus the ancestral allelic state at a polymorphic site. Genetics 209 (3):897–906

8. Nielsen R, Bustamante C, Clark AG, Glanowski S, Sackton TB, Hubisz MJ, Fledel-Alon A, Tanenbaum DM, Civello D, White TJ, Sninsky JJ, Adams MD, Cargill M (2005) A scan for positively selected genes in the genomes of humans and chimpanzees. PLoS Biol 3(6):e170

9. James JE, Piganeau G, Eyre-Walker A (2016) The rate of adaptive evolution in animal mitochondria. Mol Ecol 25(1):67–78

10. Bataillon T, Duan J, Hvilsom C, Jin X, Li Y, Skov L, Glemin S, Munch K, Jiang T, Qian Y, Hobolth A (2015) Inference of purifying and positive selection in three subspecies of chimpanzees (*Pan troglodytes*) from exome sequencing. Genome Biol Evol 7 (4):1122–1132

11. Bierne N, Eyre-Walker A (2003) The problem of counting sites in the estimation of the synonymous and nonsynonymous substitution rates: implications for the correlation between the synonymous substitution rate and codon usage bias. Genetics 165 (3):1587–1597

12. Galtier N (2016) Adaptive protein evolution in animals and the effective population size hypothesis. PLoS Genet 12(1):e1005774

13. Wales DJ, Doye JP (1997) Global optimization by basin-hopping and the lowest energy structures of Lennard-Jones clusters containing up to 110 atoms. J Phys Chem A 101 (28):5111–5116

14. Efron B, Tibshirani RJ (1993) An introduction to the bootstrap: monographs on statistics and applied probability, vol 57. Chapman and Hall/CRC, New York/London

15. Boyko AR, Williamson SH, Indap AR, Degenhardt JD, Hernandez RD, Lohmueller KE, Adams MD, Schmidt S, Sninsky JJ, Sunyaev SR, White TJ (2008) Assessing the evolutionary impact of amino acid mutations in the human genome. PLoS Genet 4(5):e1000083

16. Wilson DJ, Hernandez RD, Andolfatto P, Przeworski M (2011) A population genetics-phylogenetics approach to inferring natural selection in coding sequences. PLoS Genet 7 (12):e1002395

17. Posada D, Buckley TR (2004) Model selection and model averaging in phylogenetics: advantages of Akaike information criterion and Bayesian approaches over likelihood ratio tests. Syst Biol 53(5):793–808

Chapter 7

MSMC and MSMC2: The Multiple Sequentially Markovian Coalescent

Stephan Schiffels and Ke Wang

Abstract

The Multiple Sequentially Markovian Coalescent (MSMC) is a population genetic method and software for inferring demographic history and population structure through time from genome sequences. Here we describe the main program MSMC and its successor MSMC2. We go through all the necessary steps of processing genomic data from BAM files all the way to generating plots of inferred population size and separation histories. Some background on the methodology itself is provided, as well as bash scripts and python source code to run the necessary programs. The reader is also referred to community resources such as a mailing list and github repositories for further advice.

Key words Demographic inference, Complete genome sequencing, Phasing, Population structure, Coalescent modelling

1 Introduction

1.1 MSMC

MSMC [1] is an algorithm and program for analyzing genome sequence data to answer two basic questions: How did the effective population size of a population change through time? When and how did two populations separate from each other in the past? As input data, MSMC analyzes multiple phased genome sequences simultaneously (separated into haplotypes, i.e. maternal and paternal haploid chromosomes) to fit a demographic model to the data.

MSMC models an approximate version of the coalescent under recombination across the input sequences. Specifically, the coalescent under recombination is approximated by a Markov model along multiple sequences [2, 3], which describes how local genealogical trees change due to ancestral recombinations (Fig. 1).

The original version of this chapter was revised. The correction to this chapter is available at https://doi.org/10.1007/978-1-0716-0199-0_20

Electronic supplementary material: The online version of this chapter (https://doi.org/10.1007/978-1-0716-0199-0_7) contains supplementary material, which is available to authorized users.

Julien Y. Dutheil (ed.), *Statistical Population Genomics*, Methods in Molecular Biology, vol. 2090,
https://doi.org/10.1007/978-1-0716-0199-0_7, © The Author(s) 2020, Corrected Publication 2021

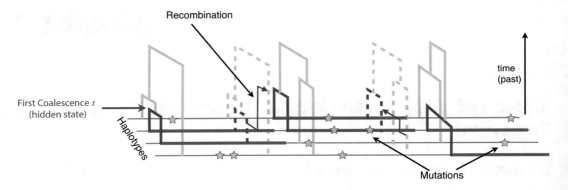

Fig. 1 Schematic description of MSMC as a hidden Markov model along multiple sequences. The sequences are related by local genealogical trees that change due to ancestral recombination events. The trees and recombination events are hidden states of the model and can be probabilistically inferred from the patterns of mutations

These local genealogies as well as the recombination events are of course invisible and therefore act as latent variables that are to be integrated out of the joint probability distribution. Since it is infeasible to do this integration across the entire space of possible trees, MSMC focuses only on one particular aspect of those trees: the first coalescence event. This variable (dark blue in Fig. 1) acts as a hidden state in the Hidden Markov Model (HMM). Using standard HMM algorithms, the hidden state (trees and recombination events) can be integrated out efficiently using dynamic programming. We can thus efficiently compute the likelihood of the data given a demographic model, and iteratively find a demographic model that maximizes this likelihood.

The demographic model itself is—in the simplest case of just one population—parameterized by a sequence of piecewise constant coalescence rates, i.e. inverse effective population sizes. The time segments are chosen such that they cover the distribution of times to first coalescence. Therefore, the more sequences are analyzed, the more recent the window of analysis will be (Fig. 2).

If the input individuals come from two populations, the demographic model is parameterized by three coalescent rates through time: A coalescence rate between lineages sampled within the first population, a coalescence rate between lineages sampled within the second population, and a coalescence rate between lineages sampled across the two populations (Fig. 3a). As introduced in Schiffels and Durbin [1], to simplify interpretation of the three inferred rates, we can plot a simple summary by taking the ratio of the across-rate and the mean within-rate, which is termed the relative cross coalescence rate (rCCR) (Fig. 3b). This summary variable ranges between 0 and 1, and indicates when and how the two populations diverged. Values close to 1 indicate that the two populations were really one population at that time. At the time when the rCCR drops to zero, the two populations likely separated into two isolated populations. Heuristically, the mid-point of that

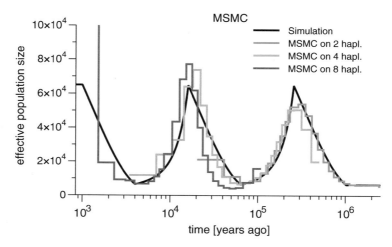

Fig. 2 Population size inference with MSMC from simulated data. Time segments are chosen to cover the distribution of first coalescence times. They cover younger time segments if more sequences are analyzed

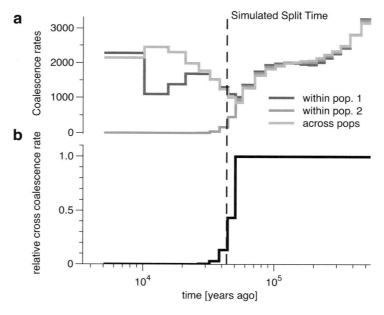

Fig. 3 Studying divergence processes through the cross-coalescence rate. When sequences are sampled from two different populations, MSMC estimates not one but three coalescence rates (two within and one across populations) over time (**a**), here for a simple scenario of two populations with a clean split about 43,500 years ago (1500 generations with generation time of 29 years). The split time is indicated by the dashed black line. The relative cross coalescence rate (**b**) is the cross coalescence rate divided by the mean within-rate. As it drops from 1 to zero (forward in time), it indicates when the two populations split. The drop agrees well with the simulated split time

decline (i.e., the time when the rCCR hits 0.5) is often taken to be an estimate for the split time between the two populations.

MSMC has been widely applied to human data (for example [4–8]) and non-human organisms (for example [9–13]).

1.2 MSMC2

MSMC2 is a newer algorithm, and the tool is still actively being developed. A first version was used in Malaspinas et al. [6] for analyzing Australian genomes. At the time of this writing, a manuscript that presents the new algorithm in more detail is in preparation. MSMC2 was developed to overcome some problems that we saw with MSMC. In particular, MSMC is computationally intensive, and for all practical purposes limited to analyzing eight haplotypes at most. But even within this scope, we see that coalescence rate estimates for more than four haplotypes are sometimes biased (see, for example, Fig. 2, red curve), with some systematic over- and underestimations of the true coalescence rates. These biases are in part caused by approximations in the emission rate of the HMM, which requires knowledge of the local lengths of leaf branches of trees. This variable is estimated by a separate HMM that is heuristic and cannot easily be improved, and which apparently performs poorly for larger trees. This means that even if we improved the computational aspects, we could not scale up this algorithm easily to more haplotypes.

MSMC2 takes a step back from these complications and approaches the problem of modelling multiple samples in a much simpler way: Instead of analyzing all input haplotypes simultaneously, it uses a much simpler pairwise HMM (very similar to PSMC) on all pairs of haplotypes. The likelihood of the data is then simply multiplied across all pairs as a composite likelihood. This has two interesting consequences: First, the pairwise model is—in contrast to the MSMC—an exact model under the Sequentially Markovian Coalescent, and does not suffer from biases with increasing number of genomes. Second, the pairwise model describes the entire distribution of pairwise coalescence times, not just the time to first coalescence. MSMC2 can therefore estimate coalescent rates across the entire distribution of pairwise coalescence times, with increasing resolution in more recent times, and importantly without biased estimates (Fig. 4). In contrast, MSMC loses power in ancient times with increasing numbers of input genomes (*see* Fig. 2).

MSMC2 can also analyze population separations via the relative cross coalescence rate, and gives similar results as MSMC, but with computational improvements, as we will point out further below.

We caution that at the time of writing, MSMC2 is still in beta and some aspects of the interface and algorithm may still change. Nevertheless, we will cover its use throughout this chapter alongside MSMC.

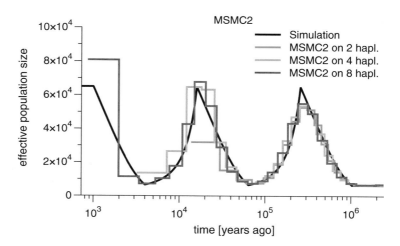

Fig. 4 MSMC2 population size estimates. Time segments are covering more recent times with increasing number of input haplotypes, without losing power in ancient times (compare with Fig. 2)

2   Software Overview

MSMC has been implemented in three open source software packages, summarized in the following. A mailing list for discussions around all three packages exists under https://groups.google.com/forum/#!forum/msmc-popgen

The main program is written in the D programming language (www.dlang.org).

A tutorial can be found at https://github.com/stschiff/msmc-tools/blob/master/msmc-tutorial/guide.md and general documentation can be found within each package.

2.1   MSMC

The main program used in the original publication [1] is accessible at http://www.github.com/stschiff/msmc. Pre-compiled packages for Mac and Linux can be found under the Releases tab. For compilation from source code, a D language compiler is needed (see www.dlang.org for details).

2.2   MSMC2

MSMC2 (*see* Subheading 1) can be accessed at http://www.github.com/stschiff/msmc2. MSMC2 is still under development, but has been used in a key publication [6], which can be used to cite this program. A publication describing the novel aspects and comparison to other state-of-the-art methods is in preparation at the time of this writing.

2.3   MSMC-Tools

Utilities for preparing input files for MSMC, as well as some other tasks, can be found in a separate repository at http://www.github.com/stschiff/msmc-tools and mainly contains python scripts that help with generating the input data and with processing the output data.

2.4 Data Requirements

MSMC normally operates on diploid, phased, complete, high coverage genomes. Here we discuss these conditions one by one.

2.4.1 Diploid Data

Technically, it is not a strict condition that input sequences be diploid. However, most populations/organisms that are not diploid do not follow a coalescent under recombination. For example, bacteria and viruses are asexual without recombination, which breaks several key assumptions that the MSMC model makes.

In some diploid model organisms, inbred lines are available and sequenced (for example, in Drosophila). Such inbred lines are effectively haploid, but originate from a diploid outbred population. In this case we think MSMC should work OK, by using each homozygous haploid input genome as a single "haplotype," although we lack explicit experience and overview of potential caveats in this case.

2.4.2 Phasing

When sequencing diploid genomes, modern sequencing platforms generate unphased data, which randomly permutes the association of heterozygous alleles to the paternal and maternal haplotypes. For MSMC, knowledge of the paternal vs. maternal allele is important when more than two haplotypes are analyzed. Note that for a single diploid genome as input (i.e., two haplotypes), no phasing is necessary.

Phasing can be a laborious preprocessing step, which requires external tools, such as shapeit (https://jmarchini.org/shapeit3/) or beagle (https://faculty.washington.edu/browning/beagle/beagle.html). As a general rule, what helps phasing quality a lot are:

- availability of a reference panel of phased populations
- presence of related individuals (e.g., parent–child duos or father–mother–child trios)
- long sequencing reads
- long-insert libraries in combination with paired-end sequencing.

Note that MSMC and MSMC2 can in principle handle unphased data within the input data format (see below), but for some analyses we recommend to exclude those sites from the analysis, which can be done within MSMC. Note also that MSMC2 now can optionally run on unphased genomes for population size analysis, but not for population separation analysis. As described below, this is achieved by running the MSMC2-HMM only within each diploid genome, but not across pairs of genomes. This will give lower resolution than with phased data, but may be a good compromise if phasing is not possible and only population sizes need to be estimated.

| 2.4.3 | Complete Genomes |

MSMC and MSMC2 cannot run on Array data, with selected SNPs, but require contiguous sequence segments. For many organisms, genomes are shorter than in humans, and from our experience, MSMC still works fine for much smaller genomes, but we recommend in these cases to run simulations with shorter genome length and specified heterozygosity to test performance of the program on shorter genomes.

For many non-model organisms, reference genomes are only available via assembly scaffolds, which are sometimes as short as a few hundred thousand basepairs (compared to hundred million basepairs for a human chromosome). In our experience, MSMC works still fine in many such cases, as long as scaffolds are not too short. Although the exact threshold depends on an organisms mean heterozygosity, in my experience scaffolds on the order of 500 kb and longer often work OK. We again recommend simulations of short chromosomes to assess the power in those cases.

| 2.4.4 | High Coverage Data |

MSMC requires good resolution of heterozygous vs. homozygous genotypes across the genome, which is only available with high coverage sequencing data. In our experience, 20-fold coverage and higher is sufficient. MSMC may work on lower coverage data as well, but detailed analyses of the effects of false negative/positives in genotype calling need to be assessed in these cases, ideally again through simulated data, into which sequencing errors are randomly introduced to test their effect on the estimates.

3   Input Data Format

MSMC/MSMC2 take several files as input, one for each chromosome, each with a list of segregating sites, including a column to denote how many sites have been called since the last segregating site. Note that here we use the term "chromosomes" to refer to coordinate blocks in a reference genome (which could also be an assembly scaffolds). We use the term "haplotypes," when we refer to the phased input sequences from multiple individuals. Here is an example part of an input file for chromosome 1 for four haplotypes (two diploid individuals):

```
1   58432    63      TCCC
1   58448    16      GAAA
1   68306    15      CTTT
1   68316    10      TCCC
1   69552    8       GCCC
1   69569    17      TCCC
1   801848   9730    CCCA
1   809876   1430    AAAG
1   825207   1971    CCCT,CCTC
1   833223   923     TCCC
```

The four (tab-separated) columns are:

1. The chromosome (can be an arbitrary string, but has to be the same for all rows in a file).

2. The position on the chromosome.

3. The number of called homozygous sites since the last segregating site, which includes the given location. This number must always be greater than zero and cannot be larger than the difference between the current position and the previous position.

4. The ordered and phased alleles of the multiple haplotypes. If the phasing is unknown or only partially known, multiple phasings can be given, separated by a comma to indicate the different possibilities (see the second-last line in the example). Unknown alleles can be indicated by "?", but they can also simply be left out and expressed by a reduced number of called sites in the line of the next heterozygous site.

The third column is needed to indicate where missing data is. For simulated data, without any missing data, this column should simply contain the distance in bp from the previous segregating site, indicating that all sites between segregating sites are called homozygous reference, without missing data. To the extent that this number is lower than the distance from the previous site do the input data contain missing data. Information about missing vs. homozygous reference calls is crucial for MSMC: If, for example, missing data is not correctly annotated, long distances between segregating sites may falsely be seen as long homozygous blocks, indicating a very recent time to the common ancestor between the lineages, thereby skewing model estimates.

The generation of such an input file follows three steps:

1. Generating VCF and mask files from individual BAM files.

2. Phasing the input.

3. Combining multiple phased individuals.

In the following, we describe these steps in order

3.1 Generating VCF and Mask Files from Individual BAM Files

Starting with a BAM file, `bamCaller.py` (included in the MSMC-Tools package) can be used for generating a sample-specific VCF file and a mask file. This script reads samtools mpileup data from stdin, so it has to be used in a pipe in which a reference file in fasta format is also required. Here is an example bash script using samtools 1.0 or higher for generating chromosome-specific VCF files (`sample1.chr*.vcf.gz`) and mask files (`sample1.mask.chr*.bed.gz`) from a human BAM file:

Listing 1.1. bash script to call genotypes and masks from a single BAM file

```
#!/bin/bash
BAM=sample1.bam

# estimating the average sequencing depth using all
    sites on chromosome 20
DEPTH=$(samtools depth -r 20 <in.bam> | awk '{sum +=
    $3} END {print sum / NR}')$

for CHR in {1..22}; do
    samtools mpileup -B -q 20 -Q 20 -C 50 -g -r $CHR
        -f <ref.fa> <in.bam> | bcftools call -c -V
        indels | ./bamCaller.py $DEPTH <out.mask.
        chr$CHR.bed.gz> | gzip -c > <out.chr$CHR.vcf.
        gz>
done
```

Further options of `bamCaller.py` are:

`--minMapQ`	to set the minimum mapping quality, which defaults to 20.0
`--minConsQ`	to set the minimum consensus quality, which defaults to 20.0
`--legend_file`	If you aim to phase your data against a reference panel, e.g. from 1000 Genomes (*see* Subheading 3.2), you need your VCF to not only contain the variant sites of the sample, but also the genotypes at additional sites at which the panel is genotyped. This option takes a gzipped file of a format that is used in the IMPUTE and SHAPEIT reference panels. It is a simple tab-separated tabular file format with one header line which gets ignored. The only important columns for this purpose are: (1) the chromosome; (2) the position; (3) the reference allele; (4) the alternative allele; (5) the type of the variant, only sites of type SNP are considered here.

3.2 Phasing

If your samples are unrelated and you want to run MSMC on more than two haplotypes at a time, you would need to statistically phase the VCFs with a tool like shapeit. There are two different phasing strategies using shapeit, either with a reference panel or without a reference panel. If a good reference panel is available for your samples, shapeit phasing with a reference panel is recommended.

Here, as an example, we describe phasing a single human diploid sample against the 1000 Genomes Phase 3 reference panel. In the following, we assume that shapeit2 is installed, the 1000 Genomes (phase 3) reference panel is available locally (can be downloaded from https://mathgen.stats.ox.ac.uk/impute/

1000GP_Phase3.html), and that the unphased VCF file contains all variable positions in the sample, plus all variable positions in the 1000 Genomes reference panel. This can be achieved using the `--legend_file` option in `bamCaller.py`.

The script first removes multi-allelic sites in your VCF, generating .noMultiAllelicSites.vcf.gz with bcftools. Then it makes a list of sites to be excluded in the main run for phasing by `shapeit -check`, because shapeit can only phase SNPs that are in both the sample and the reference panel with the same allele type. Apart from the main log file per chromosome `sample1.chr$CHR.alignments.log`, the two following files will be generated from `shapeit -check`:

1. sample1.chr$CHR.alignments.strand: this file describes all sites in detail that either have incompatible allele types in the sample and the reference panel or found in the sample but not in the reference panel.

2. sample1.chr$CHR.alignments.strand.exclude: this file gives a simple list of physical positions of sites to be excluded from phasing.

Then the script runs shapeit with `--exclude-snp` and `-no-mcmc`, generating two output files including phased sites only `sample1.chr$CHR.phased.haps.gz` and `sample1.chr$CHR.phased.samples`. These two files can be converted into VCF format by `shapeit -convert`. Afterwards, we merge the phased VCF `sample1.chr$CHR.onlyPhased.vcf.gz` and the unphased (original) VCF `sample1.chr$CHR.fixedformat.vcf.gz`, keeping all unphased sites from the original VCF, but replacing the phased ones.

```bash
#!/bin/bash
for CHR in {1..22}; do
    UNPHASED_VCF=sample1.chr$CHR.vcf.gz
    UNPHASED_VCF_NOMAS=sample1.chr$CHR.
        noMultiAllelicSites.vcf.gz
    GEN_MAP=genetic_map_chr${CHR}_combined_b37.txt
    REF_HAPS=1000GP_Phase3_chr$CHR.hap.gz
    REF_LEGEND=1000GP_Phase3_chr$CHR.legend.gz
    REF_SAMPLE=1000GP_Phase3.sample
    LOG_ALIGN=sample1.chr$CHR.alignments
    EXCLUDE_LIST=$LOG_ALIGN.strand.exclude
    LOG_MAIN=sample1.chr$CHR.main

    PHASED_HAPS=sample1.chr$CHR.phased.haps.gz
    PHASED_SAMPLE=sample1.chr$CHR.phased.samples
    PHASED_VCF=sample1.chr$CHR.onlyPhased.vcf

    LOG_CONVERT=sample1.chr$CHR.convert
    FINAL_VCF=sample1.chr$CHR.phased.vcf.gz
```

```
#Preparation
bcftools view -M 2 -O z $UNPHASED_VCF >
    $UNPHASED_VCF_NOMAS
shapeit -check -V $UNPHASED_VCF_NOMAS -M
    $GEN_MAP --input-ref $REF_HAPS
    $REF_LEGEND $REF_SAMPLE --output-log
    $LOG_ALIGN

#Main run
shapeit -V $UNPHASED_VCF_NOMAS -M $GEN_MAP
    --input-ref $REF_HAPS $REF_LEGEND
    $REF_SAMPLE -O $PHASED_HAPS
    $PHASED_SAMPLE --exclude-snp
    $EXCLUDE_LIST --no-mcmc --output-log
    $LOG_MAIN

shapeit -convert --input-haps $PHASED_HAPS
    $PHASED_SAMPLE --output-vcf $PHASED_VCF --
    output-log $LOG_CONVERT

#Zipping and indexing
bcftools view -O z $PHASED_VCF > $PHASED_VCF
    .gz
bcftools index -f $PHASED_VCF

#Merging phased and unphased vcfs, keeping
    all unphased sites from the original vcf
    , but replacing the phased ones.
bcftools merge --force-samples $UNPHASED_VCF
    $PHASED_VCF | awk 'BEGIN {ofs"=\"t}
$0 ~ /^#CHROM/ {print $1, $2, $3, $4, $5, $6
    , $7, $8, $9, $10}
$0 !~ /^#/ {
    if(substr($11, 1, 3) != "./.")
        $10 = $11
    print $1, $2, $3, $4, $5, $6, $7, $8, $9
        , $10
}' | bcftools view -O z > $FINAL_VCF
done
```

Note that this script can also be found in the git repository accompanying this book chapter (https://github.com/StatisticalPopulationGenomics/MSMCandMSMC2).

3.3 Combining Multiple Individuals into One Input File

At this point, we assume that you have a phased VCF for each individual per chromosome (potentially containing some unphased sites not in the reference panel), and one mask file for each individual per chromosome. In addition, you will need one mappability mask file per chromosome, which is universal per chromosome and does not depend on the input individuals. Mappability masks ensure that only regions in the genome are included, which have sufficiently high mappability, i.e. no repeat regions and other features that are hard to map with next-generation sequencing data. Mappability masks can be generated using the SNPable pipeline described at http://lh3lh3.users.sourceforge.net/snpable.shtml. For the human reference genome hs37d5, they can be downloaded here   https://oc.gnz.mpg.de/owncloud/index.php/s/ RNQAkHcNiXZz2fd.

For generating the input files for MSMC for one chromosome, the script `generate_multihetsep.py` from MSMC-tools is required, which merges VCF and mask files together, and also performs simple trio-phasing in case the data contains trios. Here is an example of generating multihetsep files for two (previously phased) diploid individuals on chromosome 1.

```
#!/bin/bash
generate_multihetsep.py --chr 1 --mask sample1.mask.
    chr1.bed.gz --mask sample2.mask.chr1.bed.gz --
    mask mappability_mask.chr1.bed sample1.chr1.
    phased.vcf.gz sample2.chr1.phased.vcf.gz >
    sample1_sample2.chr1.multihetsep.txt
```

Another useful option in generate_multihetsep.py is `--trio <child>,<father>,<mother>`, allowing the three members of a trio. All three fields must be integers specifying the index of the child/father/mother within the VCFs you gave as input, in order. So for example, if you had given three VCF files in the order of father, mother, child, you need to give –trio 2,0,1. This option will automatically apply a constraint for phasing and also strip the child genotypes from the result.

4   Running MSMC and MSMC2

4.1   Resource Requirements

Resource usage for MSMC and MSMC2 depend on the size of the dataset, the number of haplotypes analyzed, the number of time segments and on the number of CPUs used. The following numbers are example use cases and need to be somewhat extrapolated to other use cases. As a general rule of thumb, run time and number of CPUs are inversely proportional, and memory and number of CPUs are linearly proportional. Also, the number of haplotypes and the number of time segments affect both memory and run time quadratically.

Use cases for MSMC, assuming 22 human chromosomes and 11 CPUs, default time patterning:

- A single diploid genome: 30 min, 17Gb of RAM.
- Two diploid genomes, same population: 90 min, 32 Gb of RAM.

- Two diploid genomes, two populations: 11 h, 35 Gb of RAM.
- Four diploid genomes, two populations: 21 h, 170 Gb of RAM.

Use cases for MSMC2, assuming 22 human chromosome and 11 CPUs, default time patterning:

- A single diploid genome: 18 min, 7 Gb of RAM.
- Two diploid genomes, same population: 2 h, 36 Gb of RAM.
- Two diploid genomes, two populations: 90 min, 21 Gb of RAM.
- Four diploid genomes, two populations: 8 h, 100 Gb of RAM.

4.2 Test Data

We provide input files for MSMC and MSMC2 for four diploid human individuals, two Yoruba and two French individuals. The test input data consists of 22 text files for 22 autosomes in the MSMC input format described above. The test data can be accessed at https://github.com/StatisticalPopulationGenomics/MSMCandMSMC2.

4.3 Running MSMC

A typical command line to run MSMC on the test data is

```
msmc -t 11 -R -o out_prefix Yoruba_French.double.chr
    *.multihetsep.txt
```

which runs the program on 11 CPUs (option -t), keeps the recombination rate fixed at the initial value (option -R), and uses as output-prefix the file prefix `out_prefix`. The parallelization, here specified by the number of CPUs (-t 11), goes across input files. So when given 22 input chromosomes as in the test data, which is typical for human data, running on 11 CPUs means that the first 11 chromosomes can be run in parallel, and then the second 11. Using more CPUs will help a bit to make things even faster, but only to the extent that the number of chromosomes exceeds or equals the number of CPUs. The -R option is recommended for MSMC except when running on two haplotypes only. Additional options can be viewed by running `msmc -h`.

In order to run MSMC to obtain estimates of cross-population divergences, you need to prepare your input files to contain individuals from multiple populations. For example, in order to run MSMC on one Yoruba and one French individual from the test data, you run (here for chr1 only):

```
msmc -t 11 -R -s -I 0,1,4,5 -P 0,0,1,1 -o
    crosspop_out_prefix Yoruba_French.double.chr1.
    multihetsep.txt
```

There are two changes here with respect to the first run. First, we use the options -I 0,1,4,5 -P 0,0,1,1, which specifies that only the first two haplotypes in each subpopulation should be used (indices 0,1 are the first Yoruba individual, indices 4,5 the first French), and that those selected four haplotypes belong to two subpopulations. Second, we set -s, which instructs MSMC to skip ambiguously phased sites. This is important if you have phased your samples against a reference panel and have private variants unphased. Empirically, we have found that MSMC is quite robust to unphased sites when analyzing population size changes in a single population, but that results on cross-population divergence are affected by unphased sites, and results are less biased if those sites are removed [1].

Upon running either of the two commands above, MSMC produces several output files. First, a file containing log output, called prefix.log. Second, a file containing the parameter estimates at each iteration step, called prefix.loop.txt. And third, a file containing the final results, called prefix.final.txt. This last file looks like this:

time_index	left_time_boundary	right_time_boundary	lambda_00
0	-0	2.79218e-06	2605.47
1	2.79218e-06	5.68236e-06	6451.92
2	5.68236e-06	8.67766e-06	3152.31
3	8.67766e-06	1.1786e-05	2526.36

...

Each row of this output file lists one time segment, with scaled start and end time indicated by second and third column. The fourth column contains the scaled coalescent rate in each time segment. In case of cross-population analysis (using the -P flag), the output will contain two more columns, titled lambda_01 and lambda_11, giving the coalescence rate estimates between populations and within the second population, respectively.

Times and rates are scaled. In order to convert to real values, you need a mutation rate μ per site per generation. All times can then be converted to generations by dividing the scaled time by μ. In order to convert generations into years, a generation time is needed (for humans we typically take 29 years). Population size estimates are obtained by first taking the inverse of the scaled coalescence rate, and then dividing that inverse rate by 2μ.

To get the relative cross coalescence rate (rCCR, see Fig. 3), you need to compute $2\lambda_{01}/(\lambda_{00} + \lambda_{11})$, without any additional scaling. It can then be informative to compute the time point at which the relative CCR hits 0.5, to reflect an estimate of the split time between two populations (provided that a clean-split scenario is appropriate).

4.4 Running MSMC2

Running MSMC2 is very similar to running MSMC if samples come from a single population. In that case, a typical command line may look like this:

```
msmc2 -t 11 -o out_prefix msmc_input_chr*.
    multihetsep.txt
```

Note that here we have omitted the option -R, since MSMC2 can robustly infer recombination rates simultaneously with population sizes, so there is no need to keep the recombination rate fixed. The output of the program is the same as in MSMC.

To analyze individuals from multiple populations, as in the provided test data the procedure is different from MSMC. In that case, MSMC2 needs to be run three times independently: Once each for estimating coalescence rates within population 1, within population 2, and across populations. This has two advantages: First, since runs can be parallelized, the combined running should be faster on computer clusters. Second, if many pairs of populations are analyzed, estimates of coalescence rates within populations need to be run only once and not co-estimated with each cross-coalescence rate estimates.

So taking the test data as an example, we have four diploid individuals from two populations in a single input file, and we can run on only one individual from each population like this:

```
msmc2 -t 11 -s -I 0,1 -o pop1_out_prefix Yoruba_French.double.
    chr*.multihetsep.txt
msmc2 -t 11 -s -I 4,5 -o pop2_out_prefix Yoruba_French.double.
    chr*.multihetsep.txt
msmc2 -t 11 -s -I 0-4,0-5,1-4,1-5 -o crosspop4hap_out_prefix
    Yoruba_French.double.chr*.multihetsep.txt
```

or if we want to run on all individuals:

```
msmc2 -t 11 -s -I 0,1,2,3 -o pop1_4hap_out_prefix
    Yoruba_French.double.chr*.multihetsep.txt
msmc2 -t 11 -s -I 4,5,6,7 -o pop2_4hap_out_prefix
    Yoruba_French.double.chr*.multihetsep.txt
msmc2 -t 11 -s -I
    0-4,0-5,0-6,0-7,1-4,1-5,1-6,1-7,2-4,2-5,2-6,2-7,3-4,3-5,3-6,3-7
    -o crosspop_8hap_out_prefix Yoruba_French.double.chr*.
    multihetsep.txt
```

Here, we have again used the option -s to remove unphased sites. A key difference to MSMC is how haplotype pairs in MSMC2 are specified using the -I option. In MSMC2, haplotype configurations passed via -I can be given in two flavors. First, you can enter a single comma-separated list, like this -I 0,1,4,5. In this case, MSMC2 will run over all pairs of haplotypes within this set of indices. This is useful for running on multiple phased diploid genomes sampled from one population. In the second flavor, you can give a list of pairs, like above: -I 0-4,0-5,1-4,1-5. In this case, MSMC2 will run only those specified pairs, which are all pairs between the first Yoruba and first French individual in this case. Note that if you do not use this parameter altogether, MSMC2 will run on all pairs of input haplotypes and assume that they all belong to one population.

As a special feature in MSMC2, the option -I can be used also to run MSMC2 to get population size estimates from entirely unphased genomes, using the composite likelihood approach to run on all pairs of unphased diploids, but not across them. For example, if your input file contains four diploid unphased samples, you could use -I 0-1,2-3,4-5,6-7 to instruct MSMC2 to estimate coalescence rates only within each diploid genome.

In order to simplify plotting and analysis of the relative cross coalescence rate from MSMC2, we provide a tool in the MSMC-tools repository called combineCrossCoal.py. This tool takes as input three result files from MSMC2, obtained by running within each population and across. It will then use interpolation to create a single joint output file with all three rates that can then be plotted exactly as in the MSMC case above. To use the script on the three estimates obtained with the three MSMC2 runs above, simply run

```
python3 combineCrossCoal.py crosspop_out_prefix.
    final.txt pop1_out_prefix.final.txt
    pop2_out_prefix.final.txt > combined_pop1_pop2.
    final.txt
```

and then use the combined file to proceed with plotting.

4.5 Plotting Results

Here is an example of plotting population sizes and relative CCR in python, as well as computing the midpoint of the rCCR curve, using the numpy, pandas, and matplotlib libraries. To try this out, we provide result files for MSMC2 within the book chapter repository (https://github.com/StatisticalPopulationGenomics/MSMCandMSMC2), and those result files are used in this script, which is also included in the same repository:

```
#!/usr/bin/env python3

import pandas as pd
import numpy as np
import matplotlib.pyplot as plt
mu = 1.25e-8
gen = 29
dir_ = "MSMC2_OUTPUT"
msmc_out=pd.read_csv("{}/Yoruba_French.8haps.combined.msmc2.
    final.txt".format(dir_), sep='\t', header=0)
t_years=gen * ((msmc_out.left_time_boundary + msmc_out.
    right_time_boundary)/2) / mu
```

```
plt.figure(figsize=(8, 10))
plt.subplot(211)
plt.semilogx(t_years, (1/msmc_out.lambda_00)/(2*mu), drawstyle
    ='steps',color='red', label='Yoruba')
plt.semilogx(t_years, (1/msmc_out.lambda_11)/(2*mu), drawstyle
    ='steps',color='blue', label='French')
plt.xlabel("years ago")
plt.ylabel("population Sizes")
plt.legend()
plt.subplot(212)
relativeCCR=2.0 * msmc_out.lambda_01 / (msmc_out.lambda_00 +
    msmc_out.lambda_11)
plt.semilogx(t_years,relativeCCR, drawstyle='steps')
plt.xlabel("years ago")
plt.ylabel("Relative CCR")
plt.savefig("MSMC_plot.pdf")

def getCCRintersect(df, val):
    xVec = gen * ((df.left_time_boundary + df.
        right_time_boundary)/2) / mu
    yVec = 2.0 * df.lambda_01 / (df.lambda_00 + df.lambda_11)
    i = 0
    while yVec[i] < val:
        i += 1
    assert i > 0 and i <= len(yVec), "CCR intersection index
        out of bounds: {}".format(i)
    assert yVec[i - 1] < val and yVec[i] >= val, "this should
        never happen"
    intersectDistance = (val - yVec[i - 1]) / (yVec[i] - yVec[
        i - 1])
    return xVec[i - 1] + intersectDistance * (xVec[i] - xVec[i
        - 1])
print(getCCRintersect(msmc_out, 0.5)) #Print out the time when
    relativeCCR=0.5
```

This script produces the plot shown in Fig. 5 and prints out the midpoint of the cross-coalescence rate, which is `69405.8165002096` for the test data, i.e. around 70,000 years ago for a rough estimate of the split time between French and Yoruba.

5 Tips and Tricks

5.1 Bootstrapping

It is often important to obtain confidence intervals around coalescence rate estimates (either for population size estimates or for rCCR estimates). This can be done using block-bootstrapping. We provide a script called `multihetsep_bootstrap.py` in the MSMC-tools repository. You can run `python3 multihetsep_bootstrap.py -h` to get some inline help. The program generates artificial "bootstrapped" datasets from an input dataset consisting of MSMC input files, by chopping up the input data into blocks (5 Mb long by default) and randomly sampling with replacement to create artificial 3 Gb long genomes out of these blocks. By default, 20 datasets are generated. You can run the tool via

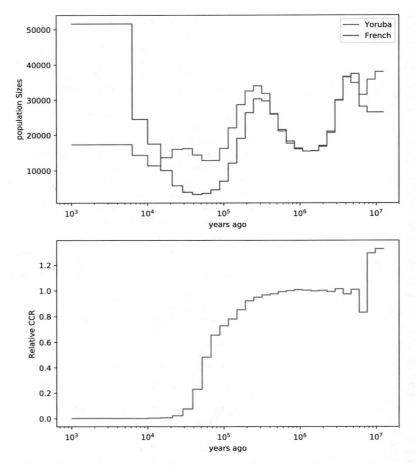

Fig. 5 The figure produced by the plotting script using the test data results

```
python3 multihetsep_bootstrap.py bootstrap_dir
    msmc_input_chr*.multihetsep.txt
```

which creates 20 subdirectories, here beginning with boot-strap_dir, each containing 30 multihetsep input files created with the block-sampling strategy described above. You should then run MSMC or MSMC2 on each of these datasets separately and plot all results together with the original estimates to visualize confidence intervals.

5.2 Controlling Time Patterning

Often, MSMC creates extremely large estimates in the most recent or the most ancient time intervals. This is a sign of overfitting, and can be mitigated by using fewer free parameters. By default, MSMC uses 40 time segments, with 25 free parameters (some neighboring time segments are forced to have the same coalescence rate). MSMC2 by default uses 32 time segments with 28 free parameters.

You can use the -p flag to control the time patterning in detail. For example, to change the patterning of MSMC2 from 32 to 20 time segments with 18 free parameters, you could try -p 1*2+16*1 +1*2, which would use 20 time segments, and merge together the first two and last two to have just one free coalescence rate parameter, respectively. We recommend to experiment with these settings, in particular when non-human data is analyzed, where sometimes the default settings in MSMC and MSMC2 are not appropriate because the genomes are substantially shorter and hence fewer parameters should be estimated.

References

1. Schiffels S, Durbin R (2014) Inferring human population size and separation history from multiple genome sequences. Nat Genet 46(8), 919–925

2. McVean GAT, Cardin NJ (2005) Approximating the coalescent with recombination. Philos Trans R Soc Lond B Biol Sci 360(1459), 1387–1393

3. Marjoram P, Wall JD (2006) Fast "coalescent" simulation. BMC Genet 7, 16

4. Pagani L, Schiffels S, Gurdasani D, Danecek P, Scally A, Chen Y, Xue Y, Haber M, Ekong R, Oljira T, Mekonnen E, Luiselli D, Bradman N, Bekele E, Zalloua P, Durbin R, Kivisild T, Tyler-Smith C (2015), Tracing the route of modern humans out of Africa by using 225 human genome sequences from Ethiopians and Egyptians. Am J Hum Genet 96(6), 986–991

5. Raghavan M, Steinrücken M, Harris K, Schiffels S, Rasmussen S, DeGiorgio M, Albrechtsen A, Valdiosera C, Ávila-Arcos MC, Malaspinas AS, Eriksson A, Moltke I, Metspalu M, Homburger JR, Wall J, Cornejo OE, Moreno-Mayar JV, Korneliussen TS, Pierre T, Rasmussen M, Campos PF, Damgaard PdB, Allentoft ME, Lindo J, Metspalu E, Rodríguez-Varela R, Mansilla J, Henrickson C, Seguin-Orlando A, Malmström H, Stafford T, Shringarpure SS, Moreno-Estrada A, Karmin M, Tambets K, Bergström A, Xue Y, Warmuth V, Friend AD, Singarayer J, Valdes P, Balloux F, Leboreiro I, Vera JL, Rangel-Villalobos H, Pettener D, Luiselli D, Davis LG, Heyer E, Zollikofer CPE, Ponce de León MS, Smith CI, Grimes V, Pike KA, Deal M, Fuller BT, Arriaza B, Standen V, Luz MF, Ricaut F, Guidon N, Osipova L, Voevoda MI, Posukh OL, Balanovsky O, Lavryashina M, Bogunov Y, Khusnutdinova E, Gubina M, Balanovska E, Fedorova S, Litvinov S, Malyarchuk B, Derenko M, Mosher MJ, Archer D, Cybulski J, Petzelt B, Mitchell J,

Worl R, Norman PJ, Parham P, Kemp BM, Kivisild T, Tyler-Smith C, Sandhu MS, Crawford M, Villems R, Smith DG, Waters MR, Goebel T, Johnson JR, Malhi RS, Jakobsson M, Meltzer DJ, Manica A, Durbin R, Bustamante CD, Song YS, Nielsen R, Willerslev E (2015), Population Genetics. Genomic evidence for the pleistocene and recent population history of native Americans. Science 349(6250), aab3884–aab3884

6. Malaspinas AS, Westaway MC, Muller C, Sousa VC, Lao O, Alves I, Bergström A, Athanasiadis G, Cheng JY, Crawford JE, Heupink TH, Macholdt E, Peischl S, Rasmussen S, Schiffels S, Subramanian S, Wright JL, Albrechtsen A, Barbieri C, Dupanloup I, Eriksson A, Margaryan A, Moltke I, Pugach I, Korneliussen TS, Levkivskyi IP, Moreno-Mayar JV, Ni S, Racimo F, Sikora M, Xue Y, Aghakhanian FA, Brucato N, Brunak S, Campos PF, Clark W, Ellingvåg S, Fourmile G, Gerbault P, Injie D, Koki G, Leavesley M, Logan B, Lynch A, Matisoo-Smith EA, McAllister PJ, Mentzer AJ, Metspalu M, Migliano AB, Murgha L, Phipps ME, Pomat W, Reynolds D, Ricaut FX, Siba P, Thomas MG, Wales T, Wall CM, Oppenheimer SJ, Tyler-Smith C, Durbin R, Dortch J, Manica A, Schierup MH, Foley RA, Lahr MM, Bowern C, Wall JD, Mailund T, Stoneking M, Nielsen R, Sandhu MS, Excoffier L, Lambert DM, Willerslev E (2016) A genomic history of aboriginal Australia. Nature 538, 207–214

7. Mallick S, Li H, Lipson M, Mathieson I, Gymrek M, Racimo F, Zhao M, Chennagiri N, Nordenfelt S, Tandon A, Skoglund P, Lazaridis I, Sankararaman S, Fu Q, Rohland N, Renaud G, Erlich Y, Willems T, Gallo C, Spence JP, Song YS, Poletti G, Balloux F, van Driem G, de Knijff P, Romero IG, Jha AR, Behar DM, Bravi CM, Capelli C, Hervig T, Moreno-Estrada A, Posukh OL, Balanovska E,

Balanovsky O, Karachanak-Yankova S, Sahakyan H, Toncheva D, Yepiskoposyan L, Tyler-Smith C, Xue Y, Abdullah MS, Ruiz-Linares A, Beall CM, Di Rienzo A, Jeong C, Starikovskaya EB, Metspalu E, Parik J, Villems R, Henn BM, Hodoglugil U, Mahley R, Sajantila A, Stamatoyannopoulos G, Wee JTS, Khusainova R, Khusnutdinova E, Litvinov S, Ayodo G, Comas D, Hammer MF, Kivisild T, Klitz W, Winkler CA, Labuda D, Bamshad M, Jorde LB, Tishkoff SA, Watkins WS, Metspalu M, Dryomov S, Sukernik R, Singh L, Thangaraj K, Pääbo S, Kelso J, Patterson N, Reich DE (2016) The simons genome diversity project: 300 genomes from 142 diverse populations. Nature 538(7624), 201–206

8. Pagani L, Lawson DJ, Jagoda E, Mörseburg A, Eriksson A, Mitt M, Clemente F, Hudjashov G, DeGiorgio M, Saag L, Wall JD, Cardona A, Mägi R, Sayres MAW, Kaewert S, Inchley C, Scheib CL, Järve M, Karmin M, Jacobs GS, Antao T, Iliescu FM, Kushniarevich A, Ayub Q, Tyler-Smith C, Xue Y, Yunusbayev B, Tambets K, Mallick CB, Saag L, Pocheshkhova E, Andriadze G, Muller C, Westaway MC, Lambert DM, Zoraqi G, Turdikulova S, Dalimova D, Sabitov Z, Sultana GNN, Lachance J, Tishkoff S, Momynaliev K, Isakova J, Damba LD, Gubina M, Nymadawa P, Evseeva I, Atramentova L, Utevska O, Ricaut FX, Brucato N, Sudoyo H, Letellier T, Cox MP, Barashkov NA, Skaro V, Mulahasanovic L, Primorac D, Sahakyan H, Mormina M, Eichstaedt CA, Lichman DV, Abdullah S, Chaubey G, Wee JTS, Mihailov E, Karunas A, Litvinov S, Khusainova R, Ekomasova N, Akhmetova V, Khidiyatova I, Marjanović D, Yepiskoposyan L, Behar DM, Balanovska E, Metspalu A, Derenko M, Malyarchuk B, Voevoda M, Fedorova SA, Osipova LP, Lahr MM, Gerbault P, Leavesley M, Migliano AB, Petraglia M, Balanovsky O, Khusnutdinova EK, Metspalu E, Thomas MG, Manica A, Nielsen R, Villems R, Willerslev E, Kivisild T, Metspalu M (2016) Genomic analyses inform on migration events during the peopling of Eurasia. Nature 538(7624), 238–242

9. Malinsky M, Challis RJ, Tyers AM, Schiffels S, Terai Y, Ngatunga BP, Miska EA, Durbin R, Genner MJ, Turner GF (2015) Genomic islands of speciation separate cichlid ecomorphs in an East African crater lake. Science 350 (6267), 1493–1498

10 1001 Genomes Consortium (2016), 1,135 genomes reveal the global pattern of polymorphism in Arabidopsis thaliana. Cell 166(2), 481–491

11. Frantz LAF, Mullin VE, Pionnier-Capitan M, Lebrasseur O, Ollivier M, Perri A, Linderholm A, Mattiangeli V, Teasdale MD, Dimopoulos EA, Tresset A, Duffraisse M, McCormick F, Bartosiewicz L, Gál E, Nyerges ÉA, Sablin MV, Bréhard S, Mashkour M, Bălăşescu A, Gillet B, Hughes S, Chassaing O, Hitte C, Vigne JD, Dobney K, Hänni C, Bradley DG, Larson G (2016) Genomic and archaeological evidence suggest a dual origin of domestic dogs. Science 352(6290), 1228–1231

12. Beissinger TM, Wang L, Crosby K, Durvasula A, Hufford MB, Ross-Ibarra J (2016) Recent demography drives changes in linked selection across the maize genome. Nat Plants 2, 16084

13. Hung CM, Shaner PJL, Zink RM, Liu WC, Chu TC, Huang WS, Li SH (2014) Drastic population fluctuations explain the rapid extinction of the passenger pigeon. Proc Natl Acad Sci USA 111(29), 10636–10641

Chapter 8

Ancestral Population Genomics with Jocx, a Coalescent Hidden Markov Model

Jade Yu Cheng and Thomas Mailund

Abstract

Coalescence theory lets us probe the past demographics of present-day genetic samples and much information about the past can be gleaned from variation in rates of coalescence event as we trace genetic lineages
back in time. Fewer and fewer lineages will remain, however, so there is a limit to how far back we can
explore. Without recombination, we would not be able to explore ancient speciation events because of
this—any meaningful species concept would require that individuals of one species are closer related than
they are to individuals of another species, once speciation is complete. Recombination, however, opens a
window to the deeper past. By scanning along a genomic alignment, we get a sequential variant of the
coalescence process as it looked at the time of the speciation. This pattern of coalescence times is fixed at
speciation time and does not erode with time; although accumulated mutations and genomic rearrangements will eventually hide the signal, it enables us to glance at events in the past that would not be
observable without recombination. So-called coalescence hidden Markov models allow us to exploit this,
and in this chapter, we present the tool Jocx that uses a framework of these models to infer demographic
parameters in ancient speciation events.

Key words Genome analysis, Coalescence, Hidden Markov models, Population history inference

1 Introduction

Understanding how species form and diverge is a central topic of
biology, and by observing emerging species today, we can understand many of the genetic and environmental processes involved.
Through such observations, we can understand the underlying
forces that drive speciation, but to understand how specific speciation events occurred in the past, and understand the specifics of
how existing species formed, we must make the inference from the

The original version of this chapter was revised. The correction to this chapter is available at https://doi.org/
10.1007/978-1-0716-0199-0_20

Electronic supplementary material: The online version of this chapter (https://doi.org/10.1007/978-1-0716-
0199-0_8) contains supplementary material, which is available to authorized users.

Julien Y. Dutheil (ed.), *Statistical Population Genomics*, Methods in Molecular Biology, vol. 2090,
https://doi.org/10.1007/978-1-0716-0199-0_8, © The Author(s) 2020, Corrected Publication 2021

signals these events have left behind. The speciation processes leave genetic "fossils" in the genome of the resulting species, and through what you might call genetic paleontology we can study past events from the signals they left behind.

The main objectives of the methods we describe in this chapter are to infer demographic parameters, Θ, given genetic data, D, through the model likelihood: $L(\Theta \mid D) = \Pr(D \mid \Theta)$. Here, we assume that Θ contains information such as effective population sizes, time points where population structure changes (populations split or admix), or migration rates between populations. We can connect data and demographics through coalescence theory [8]. This theory gives us a way to assign probability densities to genealogies; densities that depend on the demographic parameters, $f(G \mid \Theta)$. Then, if we know the underlying genealogy, we can assign probabilities to observed data using standard algorithms such as Felsenstein's likelihood recursion [7] and get $\Pr(D \mid G, \Theta)$. Theoretically, we now simply need to integrate away the nuisance parameter G to get the desired likelihood

$$\mathscr{L}(\Theta \mid D) = \Pr(D \mid \Theta) = \int \Pr(D \mid G, \Theta) f(G \mid \Theta) \, \mathrm{d}\, G. \tag{1}$$

In practice, however, the space of all possible genealogies prevents this beyond a small sample size of sequences and for any sizeable length of genetic material. Approximations are needed, and *the sequential Markov coalescent* (*see* Chapter 1) and *coalescent hidden Markov models* approximate the likelihood in two steps: they assume that sites are independent given the genealogy, i.e.,

$$\Pr(D \mid G, \Theta) \approx \prod_{i=1}^{L} \Pr(D_i \mid G_i, \Theta) \tag{2}$$

where L is the length of the sequence and D_i is the data and G_i the genealogy at site i, and assume that the dependency between genealogies is Markovian:

$$f(G \mid \Theta) \approx f(G_1 \mid \Theta) \prod_{i=2}^{L} f(G_i \mid G_{i-1}, \Theta). \tag{3}$$

Both assumptions are known to be invalid, but simulation studies indicate that this model captures most important summary statistics from the coalescent [17, 18] and that it can be used to accurately infer parameters in various demographic models [2, 14, 16]. Because of the form the likelihood now has,

$$f(D, G \mid \Theta) = f(G_1 \mid \Theta) \prod_{i=2}^{L} f(G_i \mid G_{i-1}, \Theta) \prod_{i=1}^{L} \Pr(D_i \mid G_i, \Theta),$$

$$\tag{4}$$

which is the form of a *hidden Markov model,* we can compute the likelihood efficiently using the so-called *Forward* algorithm (see Chapter 3 in Durbin et al. [3]).

This efficiency has permitted us and others (*see* Chapters 7 and 10) to apply this approximation to the coalescence to infer demographic parameters on whole genome data [1, 9, 11–13, 19, 24, 25, 27] in addition to inferring recombination patterns [20, 21] and scanning for signs of selection [4, 22].

2   Software

We have created a theoretical framework for constructing coalescent hidden Markov models from demographic specifications [2, 14–16] and used it to implement various models in the software package Jocx, available at

```
https://github.com/jade-cheng/Jocx.git
```

Jocx handles the state space explosion problem of dealing with many sequences by creating hidden Markov models for all *pairs* of sequences and then combining these into a composite likelihood when estimating parameters. In brief, a full analysis looks something like the following. In the remainder of this chapter, we describe in detail how to apply Jocx to sequence data and how to interpret the results.

```
Jocx.py init . iso a.fasta b.fasta
```

It is very important that the verbatim (typewriter font) sections are left exactly as in the input. They contain ascii art that is output from our program.

```
Jocx.py run . iso nm 0.0001 1000 0.1
```

Jocx executes CoalHMMs by specifying a model and an optimizer. It uses sequence alignments in the format of "ziphmm" directories, which is also prepared by Jocx. The program prints to standard output the progression of the estimated parameters and the corresponding log likelihood. The source package contains a set of Python files, and it requires no installation.

2.1   Preparing Data

Jocx takes two or more aligned sequences as input; the number of sequence pairs depends on the CoalHMM model specified for a particular execution. We will discuss CoalHMM model specification later. For example, for inference in a two-population isolation scenario [14], we need a minimum of one pair of aligned sequences, with one sequence from each of the two populations. The input should be FASTA files with names matching the names of the sequences we will use in the analysis, and since the sequences will be interpreted as aligned, they should all have the same lengths. The preprocessing will skip indels and handle all symbols except A, C, G, and T as the wildcard N.

In the following example, sequence a and sequence b form an alignment. Each sequence may have multiple data segments (e.g., contigs or chromosomes). In the example, we have two segments, 1 and 2. The names for these data segments need to be consistent between the two sequences. In the software we have the data-preparation step and model-inference step. In the data-preparation step, we supply Fasta sequences by providing their file names, e.g., a.fasta and b.fasta.

```
$ ls
a.fasta   b.fasta

$ cat a.fasta | wc -c
1827

$ cat b.fasta | wc -c
1827

$ head *.fasta -n 7
==> a.fasta <==
>1
aaaaaaaaaaaaaaaaaaaaaaaaaaaaaaaaaaaagaaaaaaaaaaaaa
aaaaaaaaaaaaaaaaaaaaattaaaaaaaaaaaacaaaaaaaaaaaaaa

>2
aaataaaaaaaaaaaaaaaaaaaaaaaagacaaaaaaaaaaaaaaaaaa
aaaaaaaaaaaaaaaaaaataaaaaaaaaaaaaaaaaaaaaaaaaaaaaa
==> b.fasta <==
>1
ataaaaaaaaaaaaaaaaaaacaaaaaagaaaaaaaaaaaaaaaaaaaa
aaaaaaaaaaaaaaaaaaaagaaaaaacaaaaaaaaaaaaaaaaaaaaa

>2
aaaaaaaaaacaaaaaaaaaaaaaaaaaaaaaaaaaaaaaaaaaaaaaa
aaaaaaaaaagaaaaaaaaaaaaaaaaaaaataaaaaaaaaataaaaa
```

We use the ZipHMM framework [26] to calculate likelihoods—in previous experiments we have found that ZipHMM gives us one or two orders of magnitude speedup in full genome analyses. To use ZipHMM in Jocx, we must preprocess the sequence files. The preprocessing step is customized to each demographic model and is done using the

```
$ Jocx.py init
```

command. This command takes a variable number of arguments, depending on how many sequences are needed for the demographic model we intend to use. The first two arguments are the directory in which to put the preprocessed alignment and the demographic model to use. The sequences used for the alignment, the number of which depends on the model, must be provided as the remaining arguments. In the aforementioned two-population isolation scenario, the model iso, we need to process two aligned sequences, so the init command will take four arguments in total. To create a pairwise alignment for the isolation model, we would execute the following command:

```
$ Jocx.py init . iso a.fasta b.fasta
# Creating directory: ./ziphmm_iso_a_b
# creating uncompressed sequence file
# using output directory "./ziphmm_iso_a_b"
# parsing "a.fasta"
# parsing "b.fasta"
# comparing sequence "1"
# sequence length: 900
# creating "./ziphmm_iso_a_b/1.ziphmm"
# comparing sequence "2"
# sequence length: 900
# creating "./ziphmm_iso_a_b/2.ziphmm"
#Creating5-statealignmentindirectory:./ziphmm_iso_a_b/1.ziphmm
#Creating5-statealignmentindirectory:./ziphmm_iso_a_b/2.ziphmm
```

The result of the init command is the directory ziphmm_iso_a_b that contains information about the alignment of a.fasta and b.fasta in a format that ZipHMM can use to efficiently analyze the isolation model. Each Fasta data segment forms its own ZipHMM subdirectory. In the above example, we have two data segments, named 1 and 2, so we have two ZipHMM subdirectories.

```
$ ls
a.fasta  b.fasta  ziphmm_iso_a_b

$ find ziphmm_iso_a_b/
ziphmm_iso_a_b/
ziphmm_iso_a_b/1.ziphmm
```

```
ziphmm_iso_a_b/1.ziphmm/data_structure
ziphmm_iso_a_b/1.ziphmm/nStates2seq
ziphmm_iso_a_b/1.ziphmm/nStates2seq/5.seq
ziphmm_iso_a_b/1.ziphmm/original_sequence
ziphmm_iso_a_b/2.ziphmm
ziphmm_iso_a_b/2.ziphmm/nStates2seq
ziphmm_iso_a_b/2.ziphmm/nStates2seq/5.seq
ziphmm_iso_a_b/2.ziphmm/data_structure
ziphmm_iso_a_b/2.ziphmm/original_sequence
```

The exact structure of this directory is not important to how Jocx is used, but you must preprocess input sequences to match each demographic model you will analyze.

To see the list of all supported models, use the --help option. Here iso is the two-population two-sequence isolation scenario, shown below.

```
$ Jocx.py ----help
:
ISOLATION MODEL (iso)
   *
 / \ tau
A    B

3 params -> tau, coal_rate, recomb_rate
2 seqs   -> A, B
1 group  -> AB
:
```

For each model, the tool implements, the --help command will show an ASCII image of the model, annotated with the parameters of the model and with leaves labelled by populations. Below the image, the parameters are listed in the order they will be output when optimizing the model, followed by the sequences in the order they must be provided to the init command when creating the ZipHMM file. Finally, the help lists the pairs of sequences that will be used in the composite likelihood in the list of "groups." When initializing a sequence alignment, you will get a ZipHMM directory per group.

The two-population isolation demographic model is symmetric, so the order of input Fasta sequences does not matter. This is not always the case. For example, in a three-population admixture model, shown below, the roles the populations take are different. Population C is admixed, and it is formed from ancestral siblings of the two source populations, A and B. The order of input Fasta sequences, therefore, needs to match.

In this model, we have five unknown time points and durations to be estimated, they are three-population isolation time (iso_time), two time points where the admixed population merges with each of the two source populations (buddy23_time_1a, buddy23_time_2a), and finally the duration before all populations find their common ancestry for the first population (greedy1_time_1a). The last unknown duration can be calculated: greedy1_time_2a = greedy1_time_1a + buddy23_time_1a - buddy23_time_2a.

```
$ Jocx.py ----help
:
THREE POP ADMIX 2 3 MODEL (admix23)
                    *
                   / \       greedy1_time_1a
buddy23_time_1a /\   \
               /  \_/\   buddy23_time_2a
 admix_prop   /  <-|  \  iso_time
          A       C   B


7 params -> iso_time, buddy23_time_1,
          buddy23_time_2, greedy1_time_1,
          coal_rate, recomb_rate, admix_prop
3 seqs   -> A, B, C
3 groups -> AC, BC, AB
:
```

When executing the init command, the order of the Fasta sequences should match the order of species names in the help command:

```
$ ls
a1.fasta  b1.fasta  c1.fasta

$ Jocx.py init . admix23 a1.fasta b1.fasta c1.fasta
# Creating directory: ./ziphmm_admix23_a_c
# creating uncompressed sequence file
:


$ ls
a1.fasta  b1.fasta  c1.fasta
ziphmm_admix23_a_b  ziphmm_admix23_a_c  ziphmm_admix23_b_c
```

In the two examples above, each population contributes a single sequence to the CoalHMM's construction. Jocx also has models that support two sequences per population.

```
$ Jocx.py ----help
:
THREE POP ADMIX 2 3 MODEL 6 HMM (admix23-6hmm)
                        *
                       / \      greedy1_time_1a
        buddy23_time_1a /\  \
                      /  \_/\   buddy23_time_2a
          admix_prop /  <-|  \  iso_time
                    A1   C1   B1
                    A2   C2   B2


7 params -> iso_time,          buddy23_time_1a,
            buddy23_time_2a, greedy1_time_1a,
            coal_rate, recomb_rate, admix_prop
6 seqs   -> A1, A2, B1, B2, C1, C2
6 groups -> A1C1, B1C1, A1B1, A1A2, B1B2, C1C2
:
```

In this example, we have the same admixture demographic model as before but with each population contributing two sequences to form six pairwise alignments, which are then used to construct six HMMs for the inference.

```
$ ls
a1.fasta  a2.fasta  b1.fasta  b2.fasta  c1.fasta  c2.fasta

$ Jocx.py init . admix23-6hmm a1.fasta a2.fasta \
$                      b1.fasta b2.fasta \
$                      c1.fasta c2.fasta
# Creating directory: ./ziphmm_admix23-6hmm_a1_c1
:


$ ls
a1.fasta  b1.fasta  c1.fasta
a2.fasta  b2.fasta  c2.fasta
ziphmm_admix23-6hmm_a1_a2  ziphmm_admix23-6hmm_a1_c1  ziphmm_admix23-6hmm_b1_c1
ziphmm_admix23-6hmm_a1_b1  ziphmm_admix23-6hmm_b1_b2  ziphmm_admix23-6hmm_c1_c2
```

In the two-population isolation model, we have one demographic transition for a pair of samples. That is from a two-population isolation scenario (Fig. 1a) to a single ancestral population scenario (Fig. 1b). In the three-population admix model, we have three kinds of demographic transitions for a pair of samples. They are from a two-population duration (Fig. 1a) to a three-population duration (Fig. 2a), then to another two-population duration (Fig. 2b), finally to a single ancestral population (Fig. 1b). In the three-population duration, only two

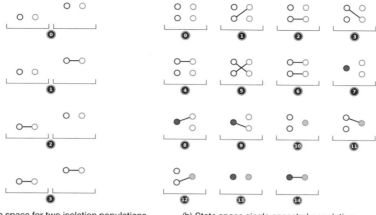

(a) State space for two isolation populations. (b) State space single ancestral population.

Fig. 1 Demographic transition in the two-population isolation model for a pair of samples. Backwards in time, the state space transits from a two-population isolation scenario (**a**) to a single ancestral population scenario (**b**)

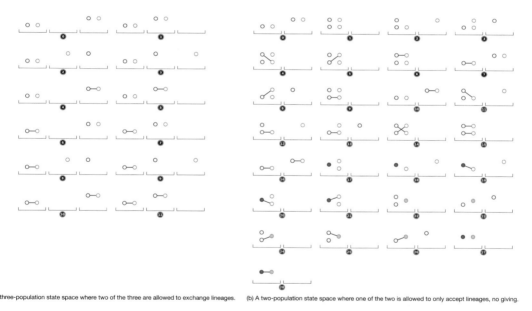

(a) A three-population state space where two of the three are allowed to exchange lineages. (b) A two-population state space where one of the two is allowed to only accept lineages, no giving.

Fig. 2 Demographic transitions in the three-population admix model for a pair of samples. Backwards in time, the state space transits from a two-population isolation scenario (Fig. 1a) to a three-population scenario (**a**), then to another two-population scenario (**b**), and finally to a single ancestral population scenario (Fig. 1b)

populations are allowed to exchange lineages, shown in Fig. 2a as the second and third populations; hence we call this duration `buddy23`. In the second two-population duration, one of the two populations only accepts lineages because it is not involved in the admixture event at the previous state space transition. Since we have one population that never gives lineages during this time, we call this duration `greedy1`.

**2.2 Inferring
Parameters**

To infer parameters, we maximize the model likelihood. Jocx implements three optimization subroutines, Nelder–Mead (NM), genetic algorithms (GA), and particle swarm optimization (PSO). After preparing the ZipHMM directories, user can run the CoalHMM to maximize the likelihood using one of these three algorithms using the `run` command.

```
$ Jocx.py run . iso nm 0.0001 1000 0.1
```

The first argument of this command, like for the `init` command, is the directory where the ZipHMM preprocessed data is found. The next argument is the demographic model. If we preprocessed the ZipHMM data with the `iso` model, we can use `iso` here to fit that model. The third argument is the optimization algorithm, one of `nm`, `ga`, and `pso`.

Following the optimizer option are the initialization values for the optimization. These arguments should match the number and order of parameters given by the `--help` command. In the `iso` model, for example, the parameters are these:

```
$ Jocx.py ----help
:
ISOLATION MODEL (iso)
   *
 / \ tau
A   B


3 params -> tau, coal_rate, recomb_rate
2 seqs   -> A, B
1 group  -> AB
:
```

In this model, we infer three parameters: the population split time, `tau`, the coalescent rate, `coal_rate`, and the recombination rate, `recomb_rate`. In this model, populations are assumed to have the same coalescent rate, which is why there is only one parameter for this.

2.2.1 NM

NM was introduced by John Nelder and Roger Mead in 1965 [23] as a technique to minimize a function in a many-dimensional space. This method uses several algorithm coefficients to determine the amount of effect of possible actions.

```
$ Jocx.py run . iso nm 0.0001 1000 0.1
# algorithm          = _NMOptimiser
# timeout            = None
# max_executions     = 1
#
```

```
# 2017-10-11 11:29:08.069462
:
# execution state score param0 param1 param2
0 init -38.2023478685 0.000376954454165 7480.36836670 0.337649514816
1 fmin-in -40.5337262711 0.000385595244114 661.208520686 0.920281817958
1 fmin-cb -40.3804021200 0.000385595244114 694.268946721 0.920281817958
:
1 fmin-cb -37.8927822292 0.000695082517418 200504630.601 32081.6528250
Optimization terminated successfully.
      Current function value: 37.892782
      Iterations: 262
      Function evaluations: 533
1 fmin-out -37.8927822292 0.000695082517418 200504630.601 32081.652825
```

In the output of NM's execution, we have a final report of whether or not the execution was successful together with the optimal solution. It is possible for the optimizer to fail for various reasons, the number of parameters being a major cause of this. If the parameter space is too large, the Nelder–Mead optimizer often fail and one of the other optimizers will do better.

2.2.2 GA

GA was introduced by John Holland in the 1970s [10]. The idea is to encode each solution as a chromosome-like data structure and operate on them through actions analogous to genetic alterations, which usually involves selection, recombination, and mutation. For each type of alteration, various authors have developed different techniques.

```
$ Jocx.py run . iso ga 0.0001 1000 0.1
# algorithm            = _GAOptimiser
# timeout              = None
# elite_count          = 1
# population_size      = 50
# initialization       = UniformInitialisation
# selection            = TournamentSelection
# tournament_ratio     = 0.1
# selection_ratio      = 0.75
# mutation             = GaussianMutation
# point_mutation_ratio = 0.15
# mu                   = 0.0
# sigma                = 0.01
#
# 2017-10-23 10:31:32.821761
#
# param0 = (1.0000000000000016e-05, 0.001)
# param1 = (99.99999999999996, 10000.0)
# param2 = (0.009999999999999995, 1.0)
```

```
#
#
# POPULATION FOR GENERATION 1
# average_fitness = -5.32373335161
# min_fitness    = -10.7962322739
# max_fitness    = -0.613544122419
#
# gen idv    fitness       param0          param1       param2
   1   1   -0.61354412  0.00002825   6305.95175380  0.04139445
   1   2   -1.38710619  0.00004282   2182.61708962  0.03027973
   1   3   -4.45085424  0.00001133    254.73764392  0.01081756
   1   4   -9.37092993  0.00067074    116.84983427  0.13757425
   1   5  -10.79623227  0.00071728    142.34535478  0.81564586
   :

#
# POPULATION FOR GENERATION 2
# average_fitness = -5.83495296756
# min_fitness    = -10.5697879572
# max_fitness    = -0.613544122419
#
# gen idv    fitness       param0          param1       param2
   2   1   -0.61354412  0.00002825   6305.95175380  0.04139445
   2   2   -0.61382451  0.00002825   6305.95175380  0.13757425
   2   3   -6.89850999  0.00002825    116.84983427  0.14110664
   2   4  -10.47909826  0.00067074    145.01523656  0.81564586
   2   5  -10.56978796  0.00067074    142.34535478  0.81564586

   :

:
```

In the output of GA's execution, we have multiple generations of solutions, and multiple solutions per generation. Solutions in each generation are ordered by the fitness, i.e., best solution is at the top. The final solution is, therefore, the first solution in the last generation.

2.2.3 PSO

PSO was introduced by Eberhart and Kennedy in 1995 [5] as an optimization technique relying on stochastic processes, similar to GA. As its name implies, each individual solution mimics a particle in a swarm. Each particle holds a velocity and keeps track of the best positions it has experienced and best position the swarm has experienced. The former encapsulates the social influence, i.e., a force pulling towards the swarm's best. The latter encapsulates the cognitive influence, i.e., a force pulling towards the particle's best. Both forces act on the velocity and drive the particle through a hyperparameter space.

```
$ Jocx.py run . iso pso 0.0001 1000 0.1
# algorithm            = _PSOptimiser
# timeout              = None
# max_iterations       = 50
# particle_count       = 50
# max_initial_velocity = 0.02
# omega                = 0.9
# phi_particle         = 0.3
# phi_swarm            = 0.1
#
# 2017-10-23 10:32:29.123305
#
# param0 = (1.0000000000000016e-05, 0.001)
# param1 = (99.99999999999996, 10000.0)
# param2 = (0.009999999999999995, 1.0)
#
#
# PARTICLES FOR ITERATION 1
# swarm_fitness           = -0.832535308472
# best_average_fitness    = -4.40169918533
# best_minimum_fitness    = -9.77654933959
# best_maximum_fitness    = -0.832535308472
# current_average_fitness = -4.40169918533
# current_minimum_fitness = -9.77654933959
# current_maximum_fitness = -0.832535308472
#
#                                       best-   best-     best-     best-
# gen idv  fitness  param0   param1  param2 fitness param0   param1    param2
   1   0   -0.83  0.000044  4619.31  0.20   -0.83  0.000044  4619.31   0.20
   1   1   -0.86  0.000048  4502.80  0.26   -0.86  0.000048  4502.80   0.26
   1   2   -0.89  0.000061  4669.48  0.58   -0.89  0.000061  4669.48   0.58
   1   3   -1.10  0.000035  2970.77  0.31   -1.10  0.000035  2970.77   0.31
   1   4   -1.46  0.000057  2148.93  0.15   -1.46  0.000057  2148.93   0.15
   :
#
# PARTICLES FOR ITERATION 2
# swarm_fitness           = -0.810479293858
# best_average_fitness    = -4.02436023707
# best_minimum_fitness    = -9.12434788412
# best_maximum_fitness    = -0.810479293858
# current_average_fitness = -4.02984771812
# current_minimum_fitness = -9.12434788412
# current_maximum_fitness = -0.810479293858
#
#                                       best-   best-     best-     best-
# gen idv  fitness  param0   param1  param2 fitness param0   param1    param2
   2   0   -0.81  0.000045  4854.87  0.25   -0.81  0.000045  4854.87   0.25
```

```
2   1   -0.82   0.000040   4622.38   0.21   -0.82   0.000040   4622.38   0.21
2   2   -0.91   0.000064   4599.97   0.59   -0.89   0.000061   4669.48   0.58
2   3   -1.12   0.000038   2917.40   0.29   -1.10   0.000035   2970.77   0.31
2   4   -1.39   0.000058   2308.29   0.14   -1.39   0.000058   2308.29   0.14
:
:
```

In the output of the PSO's execution, we have multiple generations and multiple particles (solutions) per generation. Each particle contains two sets of solutions, the current solution and the best solution that this particle has encountered throughout the PSO's execution. The latter is never worse than the former. Similar to GA, each generation is ordered by the particles' fitness. The final solution is, therefore, the second solution of the first particle in the last generation.

3   Simulation, Execution, and Result Summarization

In this section, we will use a simulation experiment to show how to perform a full analysis and extract the final solution. We will use the software fastSIMCOAL2 [6] to simulate sequences under given demographic parameters, and we will use the two-population isolation model. All scripts and input files used here can be found in the Companion Material of this book.

We execute the following command to generate variable sites of a two-sequence alignment.

```
$ ./fsc251 -i input.par -n 1
```

The first argument points to a file containing the demographic parameters, shown below. The second argument specified the number of simulations to perform. We need only one pairwise alignment.

```
$ cat input.par
//Number of population samples (demes)
2
//Population effective sizes (number of genes)
12000
12000
//Sample sizes
1
1
//Growth rates: negative growth implies population expansion
0
0
//Number of migration matrices : 0 implies no migration between demes
0
```

```
//historical event: time, source, sink, migrants, ...
1 historical event
10000 0 1 1 2 0 0
//Number of independent loci [chromosome]
1 0
//Per chromosome: Number of linkage blocks
1
//per Block: data type, num loci, rec. rate ...
    DNA 8000000 0.00000001 0.00000002 0.33
```

This simulation input file corresponds to the isolation model demography and model parameters. Our goal is to recover these parameters through CoalHMM model-based inference. The `historical event` line contains seven parameters. They are the time of the event (in generations), source population id, destination population id, the proportion of a population that migrated in this event, the new population size of the source population, the new growth rate, and the new migration matrix to use after this event. The last line contains five parameters. They are the type of data, the size of simulated sequence, the recombination rate, and the migration rate.

```
ISOLATION MODEL
      *
    / \ Tau
   A   B

Tau
    = Sim_Time * Sim_Mutation_rate
    = 10000   * 0.00000002
    = 0.0002
Coal_rate
    = 1 / (2 * Sim_Population_size * Sim_Mutation_rate)
    = 1 / (2 * 12000 * 0.00000002)
    = 2083
Recombination_rate
    = Sim_Recombination_rate / Sim_Mutation_rate
    = 0.00000001 / 0.00000002
    = 0.5
```

The direct output from the simulation program is a directory of the same name as the input file, and in this case this directory contains three files:

```
$ ls input
input_1.arb  input_1.simparam  input_1_1.arp
```

The fist file `input_1.arb` lists the file paths and names of the generated alignments. The second file `input_1.simparam` records simulation conditions and serves as a log. The last file `input_1_1.arp` contains the variable sites of a sequence alignment. The content of this file is shown below.

```
$ less -S ./input/input_1_1.arp

#Arlequin input file written by the simulation program fas-
tsimcoal2

[Profile]
        Title="A series of simulated samples"
        NbSamples=2

        GenotypicData=0
        GameticPhase=0
        RecessiveData=0
        DataType=DNA
        LocusSeparator=NONE
        MissingData='?'

    [Data]
        [[Samples]]

#Number of independent chromosomes: 1
#Total number of polymorphic sites: 10960
# 10960 polymorphic positions on chromosome 1
#414, 1380, 2815, 3855, 4036, 5364, 5772, 5816, ...

#Total number of recombination events: 5381
#Positions of recombination events:
# Chromosome 1
#      3350, 8236, 9270, 10691, 11097, 12316, ...

                SampleName="Sample 1"
                SampleSize=1
                SampleData= {
1_1     1       CCTCGGTTGTTGTCAAGGACAGTAACTATG...
}
                SampleName="Sample 2"
                SampleSize=1
                SampleData= {
2_1     1       GAATAAAAAAAACGTGAATGCAAGTACGAA...
}

    [[Structure]]
```

```
        StructureName="Simulated data"
        NbGroups=1
        Group={
            "Sample 1"
            "Sample 2"
        }
```

We use the script `arlequin2fasta.py` to convert the Arlequin alignment into Fasta files. Since the Arlequin file contains only the variable sites, we need to specify the total length of the simulated sequence, which should match the simulation parameter in the input file `intput.par`, e.g., 8,000,000 in this example.

```
$ ./arlequin2fasta.py input/input_1_1.arp 8000000
```

This creates two Fasta sequences for the pairwise alignment, and they are ready for Jocx's analysis.

```
$ ls
input  input.par

$ ./arlequin2fasta.py ./input/input_1_1.arp 8000000

$ ls
input  input.par
input_1_1-sample_1-1_1.fasta  input_1_1-sample_2-2_1.fasta
```

Analysis using Jocx follows a two-step procedure as described earlier. We first prepare the ZipHMM data directory using the `init` command and then infer parameters using the `run` command. The following commands conduct a full analysis, and it tests all three optimization options using ten independent executions per optimizer.

```
Jocx.py init . iso \
  ./input_1_1-sample_1-1_1.fasta \
  ./input_1_1-sample_2-2_1.fasta
Jocx.py run . iso pso 0.0001 1000 0.1 > pso-0.stdout
Jocx.py run . iso pso 0.0001 1000 0.1 > pso-1.stdout
:
Jocx.py run . iso pso 0.0001 1000 0.1 > pso-9.stdout
Jocx.py run . iso ga  0.0001 1000 0.1 > ga-0.stdout
Jocx.py run . iso ga  0.0001 1000 0.1 > ga-1.stdout
:
Jocx.py run . iso ga  0.0001 1000 0.1 > ga-9.stdout
Jocx.py run . iso nm  0.0001 1000 0.1 > nm-0.stdout
Jocx.py run . iso nm  0.0001 1000 0.1 > nm-1.stdout
:
Jocx.py run . iso nm  0.0001 1000 0.1 > nm-9.stdout
```

Upon completion, we receive ten sets of parameter estimates per optimization method. The format of the stand output, which contains the inference results, is different for each optimization method. We can use the following commands to summarize and plot the outcome. This plotting script is also provided in the Companion Material.

```
tail nm*.stdout -n 1 -q > nm-summary.txt
grep '500    1' ga-*.stdout > ga-summary.txt
grep '500    1' pso-*.stdout > pso-summary.txt
./box-plot-simple.py nm-summary.txt 3 nm-summary.png
./box-plot-simple.py ga-summary.txt 3 ga-summary.png
./box-plot-simple.py pso-summary.txt 3 pso-summary.png
```

The results are shown in Fig. 3. The first command collects the inference results from the NM optimizer. The last line in a NM execution's standard output contains the final estimates. The second two commands collect the inference results from the GA and PSO optimizers. The first solution/particle in the last generation/iteration, which is 500 in this experiment, contains the estimates.

```
$ head *summary.txt
==> ga-summary.txt <==
ga-0.stdout: 500 1 -81395.70680891   0.00011837   1815.42025279   0.42354064
ga-1.stdout: 500 1 -81470.10001761   0.00019243   1938.38996498   0.12963492
ga-2.stdout: 500 1 -81424.59984134   0.00021634   1846.60957248   0.19741876
ga-3.stdout: 500 1 -81430.96932585   0.00021685   1886.66976041   0.18309926
ga-4.stdout: 500 1 -81386.45366757   0.00019324   1916.03941578   0.32995308
ga-5.stdout: 500 1 -81463.45628041   0.00004345   1915.25301917   0.23921500
ga-6.stdout: 500 1 -81373.58453032   0.00018669   1968.26116983   0.52133035
ga-7.stdout: 500 1 -81504.94579193   0.00021242   1500.28846236   0.10292456
ga-8.stdout: 500 1 -81374.56618397   0.00019414   2046.25788612   0.52203350
ga-9.stdout: 500 1 -81433.41521075   0.00022051   1876.14477389   0.17886387


==> nm-summary.txt <==
1 fmin-out  -81373.5832257   0.000186088241216   1966.58533828   0.52229387809
1 fmin-out  -81373.5832257   0.000186088706436   1966.58675497   0.52229303544
1 fmin-out  -81373.5832257   0.00018608870264    1966.58640056   0.522294017033
1 fmin-out  -81373.5832257   0.000186088642201   1966.5864041    0.522295006576
1 fmin-out  -81373.5832257   0.000186088168201   1966.58599993   0.522295026609
1 fmin-out  -81373.5832257   0.00018608835674    1966.58624347   0.522297122163
1 fmin-out  -81373.5832257   0.000186088509117   1966.58601275   0.52229560587
1 fmin-out  -81373.5832257   0.000186088949749   1966.58644739   0.522293271654
1 fmin-out  -81373.5832257   0.000186088354698   1966.58755713   0.522294573711
1 fmin-out  -81373.5832257   0.000186088870812   1966.5853934    0.522294569147
```

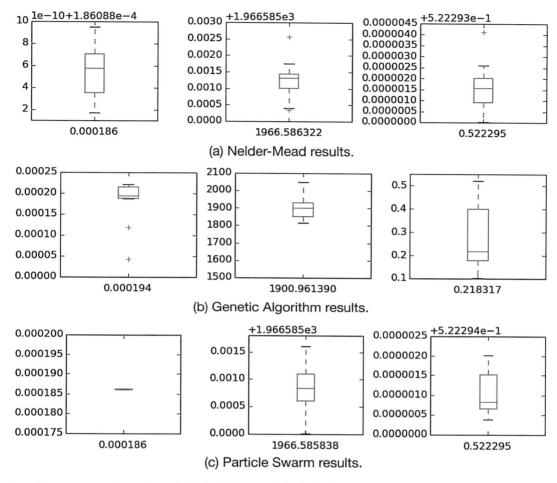

(a) Nelder-Mead results.

(b) Genetic Algorithm results.

(c) Particle Swarm results.

Fig. 3 Summary of ten independent simulations and CoalHMM executions on the two-population isolation model using the three optimisation methods. The three columns show parameters speciation time, coalescence rate, and recombination rate, respectively. The simulated values of these parameters are 0.0002, 2083, and 0.5. The number written below each box-plot is the median value of the estimates shown on the y-axis. This median can be used as a point estimate for the parameters

```
==> pso-summary.txt <==
pso-0.stdout: 500 1 -81373.583 0.000186 1966.585 0.522 -81373.583 0.000186
1966.585 0.522
pso-1.stdout: 500 1 -81373.583 0.000186 1966.586 0.522 -81373.583 0.000186
1966.585 0.522
pso-2.stdout: 500 1 -81373.583 0.000186 1966.586 0.522 -81373.583 0.000186
1966.586 0.522
pso-3.stdout: 500 1 -81373.583 0.000186 1966.586 0.522 -81373.583 0.000186
1966.586 0.522
pso-4.stdout: 500 1 -81373.583 0.000186 1966.585 0.522 -81373.583 0.000186
1966.585 0.522
pso-5.stdout: 500 1 -81373.583 0.000186 1966.585 0.522 -81373.583 0.000186
```

```
1966.585 0.522
pso-6.stdout: 500 1 -81373.583 0.000186 1966.586 0.522 -81373.583 0.000186
1966.585 0.522
pso-7.stdout: 500 1 -81373.583 0.000186 1966.586 0.522 -81373.583 0.000186
1966.586 0.522
pso-8.stdout: 500 1 -81373.583 0.000186 1966.585 0.522 -81373.583 0.000186
1966.585 0.522
pso-9.stdout: 500 1 -81373.583 0.000186 1966.586 0.522 -81373.583 0.000186
1966.586 0.522
```

The plotting script simply places these estimates in box plots. The first parameter indicates the summary file to plot. The second parameter indicates the number of parameters that this model has. For the two-population isolation model, we have three parameters. Each particle in PSO contains two sets of results, the local best and swarm best. The second set, swarm's best, should be used. The last parameter specifies the output file's name. At the bottom of each box plot we print the median value of the estimates.

The demographic parameters we use in this experiment are 0.0002, 2083, and 0.5. They are the split time of the two isolated populations, the coalescent rate, and the recombination rate, respectively. These values are roughly recovered by CoalHMM for all the optimizers.

In summary, the following commands conduct a full simulation and estimation data analysis, and it summarizes the final results by creating box plots and printing the median estimate for each parameter.

```
$ ./fsc251 -i input.par -n 1
$ ./arlequin2fasta.py input/input_1_1.arp 8000000
$ ./Jocx.py init . iso \
  ./input_1_1-sample_1-1_1.fasta \
  ./input_1_1-sample_2-2_1.fasta
$ ./Jocx.py run . iso pso 0.0001 1000 0.1 > pso-0.stdout
:
$ grep '500    1' pso-*.stdout > pso-summary.txt
$ ./box-plot-simple.py pso-summary.txt 3 pso-summary.png
```

The first command simulates a pairwise sequence alignment using the fastSIMCOAL2 program. The second command uses a custom script to convert the simulated alignment from the Arlequin format to the Fasta format. The third command prepares the ZipHMM directories using the Fasta sequences. The fourth command executes CoalHMM's model inference and dumps the output to a file. Potentially, multiple independent runs are dispatched and a HPC cluster is involved in this step. The fifth command obtains the inference results from the output file. The number 500 here is the maximum iteration count for this experiment, and the number 1 indicates the first particle in the last iteration. Finally, the sixth command plots the parameters and presents the median estimates as the final results.

4   Conclusions

We have presented the Jocx tool for estimating parameters in ancestral population genomics. The tool uses a framework of pairwise coalescent hidden Markov models combined in a composite likelihood to implement various demographic scenarios. A full list of available demography models are available through the tool's help command. Using a simple isolation model, we described an analysis pipeline based on simulating data and then analyzing it using the three different optimizers implement in Jocx. This pipeline is available in the Companion Material associated with this chapter, and serves as a good starting point for getting familiar with Jocx before moving to more involved models.

References

1. Abascal F, Corvelo A, Cruz F, Villanueva-Cañas JL, Vlasova A, Marcet-Houben M, Martínez-Cruz B, Cheng JY, Prieto P, Quesada V, Quilez J, Li G, García F, Rubio-Camarillo M, Frias L, Ribeca P, Capella-Gutiérrez S, Rodríguez JM, Câmara F, Lowy E, Cozzuto L, Erb I, Tress ML, Rodriguez-Ales JL, Ruiz-Orera J, Reverter F, Casas-Marce M, Soriano L, Arango JR, Derdak S, Galán B, Blanc J, Gut M, Lorente-Galdos B, Andrés-Nieto M, López-Otín C, Valencia A, Gut I, García JL, Guigó R, Murphy WJ, Ruiz-Herrera A, Marquès-Bonet T, Roma G, Notredame C, Mailund T, Albà MM, Gabaldón T, Alioto T, Godoy JA (2016) Extreme genomic erosion after recurrent demographic bottlenecks in the highly endangered Iberian lynx. Genome Biol 17(1):251

2. Cheng JY, Mailund T (2015) Ancestral population genomics using coalescence hidden Markov models and heuristic optimisation algorithms. Comput Biol Chem 57:80–92

3. Durbin R, Eddy SR, Krogh A, Mitchison G (1998) Biological sequence analysis: probabilistic models of proteins and nucleic acids. Cambridge University Press, Cambridge. http://www.amazon.com/Biological-Sequence-Analysis-Probabilistic-Proteins/dp/0521629713

4. Dutheil JY, Munch K, Nam K, Mailund T, Schierup MH (2015) Strong selective sweeps on the X chromosome in the human-chimpanzee ancestor explain its low divergence. PLoS Genet 11(8):e1005451

5. Eberhart R, Kennedy J (1995) A new optimizer using particle swarm theory. In: Proceedings of the sixth international symposium on micro machine and human science, 1995. MHS'95. IEEE, Piscataway, pp 39–43

6. Excoffier L, Dupanloup I, Huerta-Sánchez E, Sousa VC, Foll M (2013) Robust demographic inference from genomic and SNP data. PLoS Genet 9(10):e1003905

7. Felsenstein J (1981) Evolutionary trees from DNA sequences: a maximum likelihood approach. J Mol Evol. https://doi.org/10.1007/BF01734359

8. Hein J, Schierup M, Wiuf C (2004) Gene genealogies, variation and evolution: a primer in coalescent theory. Oxford University Press, New York

9. Hobolth A, Dutheil JY, Hawks J, Schierup MH, Mailund T (2011) Incomplete lineage sorting patterns among human, chimpanzee, and orangutan suggest recent orangutan speciation and widespread selection. Genome Res 21(3):349–356

10. Holland JH (1992) Genetic algorithms. Sci Am 267(1):66–73

11. Jónsson H, Schubert M, Seguin-Orlando A, Ginolhac A, Petersen L, Fumagalli M, Albrechtsen A, Petersen B, Korneliussen TS, Vilstrup JT, Lear T, Myka JL, Lundquist J, Miller DC, Alfarhan AH, Alquraishi SA, Al-Rasheed KAS, Stagegaard J, Strauss G, Bertelsen MF, Sicheritz-Ponten T, Antczak DF, Bailey E, Nielsen R, Willerslev E, Orlando L

(2014) Speciation with gene flow in equids despite extensive chromosomal plasticity. PNAS 111(52):18655–18660

12. Li H, Durbin R (2011) Inference of human population history from individual whole-genome sequences. Nature 475 (7357):493–496

13. Locke DP, Hillier LW, Warren WC, Worley KC, Nazareth LV, Muzny DM, Yang SP, Wang Z, Chinwalla AT, Minx P, Mitreva M, Cook L, Delehaunty KD, Fronick C, Schmidt H, Fulton LA, Fulton RS, Nelson JO, Magrini V, Pohl C, Graves TA, Markovic C, Cree A, Dinh HH, Hume J, Kovar CL, Fowler GR, Lunter G, Meader S, Heger A, Ponting CP, Marquès-Bonet T, Alkan C, Chen L, Cheng Z, Kidd JM, Eichler EE, White S, Searle S, Vilella AJ, Chen Y, Flicek P, Ma J, Raney B, Suh B, Burhans R, Herrero J, Haussler D, Faria R, Fernando O, Darré F, Farré D, Gazave E, Oliva M, Navarro A, Roberto R, Capozzi O, Archidiacono N, Della Valle G, Purgato S, Rocchi M, Konkel MK, Walker JA, Ullmer B, Batzer MA, Smit AFA, Hubley R, Casola C, Schrider DR, Hahn MW, Quesada V, Puente XS, Ordoñez GR, López-Otín C, Vinar T, Brejova B, Ratan A, Harris RS, Miller W, Kosiol C, Lawson HA, Taliwal V, Martins AL, Siepel A, Roychoudhury A, Ma X, Degenhardt J, Bustamante CD, Gutenkunst RN, Mailund T, Dutheil JY, Hobolth A, Schierup MH, Ryder OA, Yoshinaga Y, de Jong PJ, Weinstock GM, Rogers J, Mardis ER, Gibbs RA, Wilson RK (2011) Comparative and demographic analysis of orang-utan genomes. Nature 469(7331):529–533

14. Mailund T, Dutheil JY, Hobolth A, Lunter G, Schierup MH (2011) Estimating divergence time and ancestral effective population size of Bornean and Sumatran orangutan subspecies using a coalescent hidden Markov model. PLoS Genet 7(3):e1001319

15. Mailund T, Halager AE, Westergaard M (2012) Using colored petri nets to construct coalescent hidden Markov models: automatic translation from demographic specifications to efficient inference methods. Springer, Berlin, pp 32–50

16. Mailund T, Halager AE, Westergaard M, Dutheil JY, Munch K, Andersen LN, Lunter G, Prüfer K, Scally A, Hobolth A, Schierup MH (2012) A new isolation with migration model along complete genomes infers very different divergence processes among closely related great ape species. PLoS Genet 8(12):e1003125

17. Marjoram P, Wall JD (2006) Fast "coalescent" simulation. BMC Genet 7(1):16

18. McVean GAT, Cardin NJ (2005) Approximating the coalescent with recombination. Philos Trans R Soc Lond B Biol Sci 360 (1459):1387–1393

19. Miller W, Schuster SC, Welch AJ, Ratan A, Bedoya-Reina OC, Zhao F, Kim HL, Burhans RC, Drautz DI, Wittekindt NE, Tomsho LP, Ibarra-Laclette E, Herrera-Estrella L, Peacock E, Farley S, Sage GK, Rode K, Obbard M, Montiel R, Bachmann L, Ingólfsson O, Aars J, Mailund T, Wiig O, Talbot SL, Lindqvist C (2012) Polar and brown bear genomes reveal ancient admixture and demographic footprints of past climate change. Proc Natl Acad Sci U S A 109(36): E2382–E2390

20. Munch K, Mailund T, Dutheil JY, Schierup MH (2014) A fine-scale recombination map of the human-chimpanzee ancestor reveals faster change in humans than in chimpanzees and a strong impact of GC-biased gene conversion. Genome Res 24(3):467–474

21. Munch K, Schierup MH, Mailund T (2014) Unraveling recombination rate evolution using ancestral recombination maps. BioEssays 36(9):892–900

22. Munch K, Nam K, Schierup MH, Mailund T (2016) Selective sweeps across twenty millions years of primate evolution. Mol Biol Evol 33 (12):3065–3074

23. Nelder JA, Mead R (1965) A simplex method for function minimization. Comput J 7 (4):308–313

24. Prado-Martinez J, Sudmant PH, Kidd JM, Li H, Kelley JL, Lorente-Galdos B, Veeramah KR, Woerner AE, O'Connor TD, Santpere G, Cagan A, Theunert C, Casals F, Laayouni H, Munch K, Hobolth A, Halager AE, Malig M, Hernandez-Rodriguez J, Hernando-Herraez I, Prüfer K, Pybus M, Johnstone L, Lachmann M, Alkan C, Twigg D, Petit N, Baker C, Hormozdiari F, Fernández-Callejo M, Dabad M, Wilson ML, Stevison L, Camprubí C, Carvalho T, Ruiz-Herrera A, Vives L, Mele M, Abello T, Kondova I, Bontrop RE, Pusey A, Lankester F, Kiyang JA, Bergl RA, Lonsdorf E, Myers S, Ventura M, Gagneux P, Comas D, Siegismund H, Blanc J, Agueda-Calpena L, Gut M, Fulton L, Tishkoff SA, Mullikin JC, Wilson RK, Gut IG, Gonder MK, Ryder OA, Hahn BH, Navarro A, Akey JM, Bertranpetit J, Reich D, Mailund T, Schierup MH, Hvilsom C, Andrés AM, Wall JD, Bustamante CD, Hammer MF, Eichler EE, Marquès-Bonet T (2013) Great ape genetic

diversity and population history. Nature 499 (7459):471–475

25. Prüfer K, Munch K, Hellmann I, Akagi K, Miller JR, Walenz B, Koren S, Sutton G, Kodira C, Winer R, Knight JR, Mullikin JC, Meader SJ, Ponting CP, Lunter G, Higashino S, Hobolth A, Dutheil J, Karakoç E, Alkan C, Sajjadian S, Catacchio CR, Ventura M, Marquès-Bonet T, Eichler EE, André C, Atencia R, Mugisha L, Junhold J, Patterson N, Siebauer M, Good JM, Fischer A, Ptak SE, Lachmann M, Symer DE, Mailund T, Schierup MH, Andrés AM, Kelso J, Pääbo S (2012) The bonobo genome compared with the chimpanzee and human genomes. Nature 486(7404):527–531

26. Sand A, Kristiansen M, Pedersen CNS, Mailund T (2013) zipHMMlib: a highly optimised HMM library exploiting repetitions in the input to speed up the forward algorithm. BMC Bioinf 14(1):339

27. Scally A, Dutheil JY, Hillier LW, Jordan GE, Goodhead I, Herrero J, Hobolth A, Lappalainen T, Mailund T, Marquès-Bonet T, McCarthy S, Montgomery SH, Schwalie PC, Tang YA, Ward MC, Xue Y, Yngvadottir B, Alkan C, Andersen LN, Ayub Q, Ball EV, Beal K, Bradley BJ, Chen Y, Clee CM, Fitzgerald S, Graves TA, Gu Y, Heath P, Heger A, Karakoç E, Kolb-Kokocinski A, Laird GK, Lunter G, Meader S, Mort M, Mullikin JC, Munch K, O'Connor TD, Phillips AD, Prado-Martinez J, Rogers AS, Sajjadian S, Schmidt D, Shaw K, Simpson JT, Stenson PD, Turner DJ, Vigilant L, Vilella AJ, Whitener W, Zhu B, Cooper DN, de Jong P, Dermitzakis ET, Eichler EE, Flicek P, Goldman N, Mundy NI, Ning Z, Odom DT, Ponting CP, Quail MA, Ryder OA, Searle SM, Warren WC, Wilson RK, Schierup MH, Rogers J, Tyler-Smith C, Durbin R (2012) Insights into hominid evolution from the gorilla genome sequence. Nature 483 (7388):169–175

Chapter 9

Coalescent Simulation with msprime

Jerome Kelleher and Konrad Lohse

Abstract

Coalescent simulation is a fundamental tool in modern population genetics. The msprime library provides unprecedented scalability in terms of both the simulations that can be performed and the efficiency with which the results can be processed. We show how coalescent models for population structure and demography can be constructed using a simple Python API, as well as how we can process the results of such simulations to efficiently calculate statistics of interest. We illustrate msprime's flexibility by implementing a simple (but functional) approximate Bayesian computation inference method in just a few tens of lines of code.

Key words Population genetics, Coalescent theory, Simulation, Python

1 Introduction

Thanks to the rapid advances in sequencing technology, generating genome-wide sequence datasets for many species has become routine and there is great interest in learning about the history of populations from sequence variation. The coalescent [15, 25, 40] gives an elegant mathematical description of the ancestry of a sample of sequences from a more or less idealized population and, given its focus on samples, has become the backbone of modern population genetics [16, 43]. However, despite the flood of sequence data and the plethora of coalescent-based inference tools now available, many analyses of genome wide variation remain superficial or entirely descriptive. Progress on developing efficient inference methods has been hindered in two ways. First, analytic results for models of population structure and/or history are often restricted to average coalescence times and small (often pairwise) samples. Even when it is possible to derive the full distribution of

The original version of this chapter was revised. The correction to this chapter is available at https://doi.org/10.1007/978-1-0716-0199-0_20

Electronic supplementary material: The online version of this chapter (https://doi.org/10.1007/978-1-0716-0199-0_9) contains supplementary material, which is available to authorized users.

Julien Y. Dutheil (ed.), *Statistical Population Genomics*, Methods in Molecular Biology, vol. 2090, https://doi.org/10.1007/978-1-0716-0199-0_9, © The Author(s) 2020, Corrected Publication 2021

191

genealogies for realistic models and sample sizes, the results are cumbersome and generally rely on automation using symbolic mathematics software [28]. Second, and perhaps more fundamentally, dealing with recombination has proven extremely challenging and we still lack analytic results for basic population genetic quantities for a linear sequence with recombination even under the simplest null models of genetic drift. Thus, inference methods that incorporate linkage information [12, 26] generally rely on substantial simplifying assumptions about recombination [31].

Because analytic approaches relating sequence variation to mechanistic models of population structure and history are severely limited, simulations—in particular, coalescent simulations—have become an integral part of inference in a number of ways. First, comparisons between analytic results and simulations serve as an important sanity check for both. Second, while it is often possible to use analytic approaches to obtain unbiased point estimates of demographic parameters by ignoring linkage [10], quantifying the uncertainty and potential biases in such estimates requires parametric bootstrapping on data simulated with linkage. Finally, a range of inference methods directly rely on coalescent simulations to approximate the likelihood (or in a Bayesian setting, the posterior) of parameters under arbitrarily complex models of demography. Inference based on approximate Bayesian computation (ABC) [2, 6] or approximate likelihoods can be based either on single nucleotide polymorphisms (SNPs) [9] or multilocus data [3, 4].

This chapter is a tutorial for running and analyzing coalescent simulations using msprime [23]. As the name implies, msprime is heavily indebted to the classical ms program [17], and largely follows the simulation model that it popularized. The methods for representing genealogies that underlie msprime are based on earlier work on simulating coalescent processes in a spatial continuum [21, 22]. There are many other coalescent simulators available—see refs. 1, 5, 14, 27, 46 for reviews—but msprime has some distinct advantages. Firstly, msprime is capable of simulating sample sizes far larger than any other simulator, and is generally extremely efficient. The ability to simulate hundreds of thousands of realistic human genomes has already enabled simulation studies that were hitherto impossible [29]. Secondly, msprime can simulate realistic models of recombination over whole chromosomes without resorting to approximations. The Sequentially Markov Coalescent (SMC) approximation [31] was largely motivated by the need to efficiently simulate chromosome-length sequences under the effects of recombination, which was unfeasible with simulators such as ms [17]. However, for large sample sizes, msprime is significantly faster than the most efficient SMC simulator [39], rendering this approximation unnecessary for simulation purposes [23]. (The SMC is an important analytic approximation, however, and has led to many important advances in inference; see, e.g., [12, 26, 36, 37]. See also Chapter 1 in this volume for formal definitions of the SMC approximation, and Chapters 7, 8, and 10 for further applications.) Thirdly, the data structure that `msprime` uses to represent the results of simulations is extremely concise and can be efficiently processed. This data structure is known as a *succinct tree sequence* (or tree sequence for brevity), and its applications to other areas of population genomics is an active research topic [24]. The tree sequence data structure reduces the amount of space required to store simulations and removes the significant overhead of loading and parsing large volumes of text in order to analyze simulation data. As we see in Section 3, it also leads to powerful algorithms for analyzing variation data. Finally, `msprime`'s primary interface is through a simple but powerful Python API, providing many advantages over command-line or GUI based alternatives. One of the advantages of this approach is the ease with which we can integrate with state-of-the-art analysis tools from the Python ecosystem such as NumPy [42], SciPy [20], Matplotlib [18], Pandas [30], Seaborn [44], and Jupyter Notebooks [35]. Part of the goal of this tutorial is to provide idiomatic examples for interacting with these toolkits.

We assume a minimal working knowledge of Python, although it should be possible to follow and replicate the examples given here with no prior knowledge. All of the examples given here can be found in the accompanying Jupyter notebook (see the Online Resources section at the end of this chapter for details.) For those beginning with Python, we recommend the tutorial that is part of the official documentation. We also assume a basic knowledge of coalescent theory; [43] is an excellent introduction.

The chapter is organized as follows. Section 2 provides an overview of how to run coalescent simulations in `msprime`, including some of the most important extensions to the basic model. Section 3 illustrates by way of simple examples how we can efficiently process the results of such simulations, with particular emphasis on the methods required to work with large sample sizes. We then provide some examples of how to compare simulations with analytic predictions in Section 4, emphasizing idiomatic ways of interacting with toolkits such as Pandas and Seaborn. In Section 5, we show how `msprime` can be used to set up a simple ABC inference. Inference tools are generally implemented with a command line or graphical user interface and designed for a more or less narrow set of inference problems. Thus the aim of Section 5 is to illustrate how `msprime`'s flexible Python API can be used to build inference tools for arbitrary demographic histories from first principles. Finally, we outline some future plans for `msprime` in Section 6.

2   Running Simulations

In the following subsections we examine some basic examples of running simulations with `msprime`, starting with the simplest possible models and adding the various complexities required to model biological populations. We use a notebook-like approach throughout, where we intersperse code chunks and their results freely within the text.

2.1 Trees and Replication

At the simplest level, coalescent simulation is about generating trees (or genealogies). These trees (which are always rooted) represent the simulated history of a sample of individuals drawn from an idealized population (in later sections we show how to vary the properties of this idealized population). The function `msprime.simulate` runs these simulations and the parameters that we provide define the simulation that is run. It returns a `TreeSequence` object, which represents the full coalescent history of the sample. In later sections we discuss the effects of recombination, when this `TreeSequence` contains a sequence of correlated trees. For now, we focus on non-recombining sequences and use the method `first()` to obtain the tree object from this tree sequence. (In general, we can use the `trees()` iterator to get all trees; see Section 2.7.) For example, here we simulate a history for a sample of three chromosomes:

```
1  import msprime
2  ts = msprime.simulate(3)
3  tree = ts.first()
4  SVG(tree.draw())
```

This code chunk illustrates the basic approach required to draw a tree in a Jupyter notebook. We first generate a tree sequence object (`ts`), and we then obtain the tree object representing the first (and only) tree in this sequence. Finally, we draw a representation of this tree using the IPython SVG function on the output of the `tree.draw()` method. By default, `tree.draw()` returns a depiction of the tree in SVG format, but also supports plain text rendering. For example, `print( tree.draw( format=unicode) )` prints the tree to the console using Unicode box-drawing characters. This is a very useful debugging tool. We have omitted the `import` statements required for the SVG function here as it is rather specific to the Jupyter notebook environment. All code chunks in this chapter are included in the accompanying Jupyter notebook and are fully runnable.

The output of one random realization of this process is shown in Fig. 1. The resulting tree has five nodes: nodes 0, 1, and 2 are *leaves*, and represent our samples. Node 3 is an *internal* node, and is the parent of 0 and 2. Node 4 is also an internal node, and is the root of the tree. In `msprime`, we always refer to nodes by their integer IDs and obtain information about these nodes by calling methods on the tree object. For example, the code `tree.children(4)` will return the tuple `(1, 3)` here, as these are the node IDs of the children of the root node. Similarly, `tree.parent(0)` will return 3.

Fig. 1 Coalescent tree with mutations using the `tree.draw()` method

The height of a tree node is determined by the *time* at which the corresponding ancestor was born. So, contemporary samples always have a node time of zero, and time values increase as we go upwards in the tree (i.e., further back in time). Times in `msprime` are always measured in *generations*.

When we run a single simulation, the resulting tree is a single random sample drawn from the probability distribution of coalescent trees. Since a single random draw from any distribution is usually uninformative, we nearly always need to run many different *replicate* simulations to obtain useful information. The simplest way to do this in `msprime` is to use the `num_replicates` argument.

```
1  import msprime
2  N = 1000
3  mean_T_mrca = 0
4  for ts in msprime.simulate(10, num_replicates=N):
5      tree = ts.first()
6      mean_T_mrca += tree.time(tree.root)
7  mean_T_mrca = mean_T_mrca / N
8  print(mean_T_mrca)
9
10 >>> 3.6717548653768133
```

In this example we run 1000 independent replicates of the coalescent for a sample of 10 chromosomes, and compute the mean time to the MRCA of the entire sample, i.e., the root of the tree. The value of 3.7 generations in the past we obtain is of course highly unrealistic. However, by default, time is measured in units of $4N_e$ generations (see the next section for details on how to specify population models and interpret times). It is important to note here that although time is measured in units of generations, this is of course an approximation and we may have fractional values. Internally, during a simulation time is scaled into coalescent units using the Ne parameter and once the simulation is complete, times are scaled back into units of generations before being presented to the user. This removes the burden of such tedious time scaling calculations from the user. We discuss these time scaling issues in more detail in the next section.

The simulate function behaves slightly differently when it is called with the num_replicates argument: rather than returning a single tree sequence, we return an *iterator* over the individual replicates. This means that we can use the convenient **for** loop construction to consider each simulation in turn, but without actually storing all these simulations. As a result, we can run millions of replicates using this method without using any extra storage.

When simulating coalescent trees, we are often interested in more than just the mean of the distribution of some statistic. Rather than compute the various summaries by hand (as we have done for the mean in the last example), it is convenient to store the result for each replicate in a NumPy array and analyze the data after the simulations have completed. For example:

```
1  import msprime
2  import numpy as np
3  N = 1000
4  T_mrca = np.zeros(N)
5  for j, ts in enumerate(msprime.simulate(10, num_replicates=N)):
6      tree = ts.first()
7      T_mrca[j] = tree.time(tree.root)
8  print([np.mean(T_mrca), np.var(T_mrca)])
9
10 >>> [3.6690718290544053, 4.8541533617765706]
```

Here we simulate 1000 replicates, storing the time to the MRCA for each replicate in the array T_mrca. We use the Python enumerate function to simplify the process of efficiently inserting values into this array, which simply ensures that j is 0 for the first replicate, 1 for the second, and so on. Thus, by the time we finish the loop, the array has been filled with T_{MRCA} values generated under the coalescent. We then use the NumPy library (which has an extensive suite of statistical functions) to compute the mean and variance of this array. This example is idiomatic, and we will use this type of approach throughout. In the interest of brevity, we will omit all further import statements from code chunks.

It is usually more convenient to use the num_replicates argument to perform replication, but there are situations in which it is desirable to specify random seeds manually. For example, if simulations require a long time to run, we may wish to use multiple processes to run these simulations. To ensure that the seeds used in these different processes are unique, it is best to manually specify them. For example,

```
1  def run_simulation(seed):
2      ts = msprime.simulate(10, random_seed=seed)
3      tree = ts.first()
4      return tree.time(tree.root)
5
6  N = 1000
7  seeds = np.random.randint(1, 2**32 - 1, N)
8  with multiprocessing.Pool(4) as pool:
9      T_mrca = np.array(pool.map(run_simulation, seeds))
10 print(np.mean(T_mrca))
11
12 >>> 3.6459775450221832
```

In this example we create a list of 1000 seeds between 1 and $2^{32} - 1$ (the range accepted by msprime) randomly. We then use the multiprocessing module to create a worker pool of four processes, and run our different replicates in these subprocesses. The results are then collected together in an array so that we can easily process them. This approach is a straightforward way to utilize modern multi-core processors.

Specifying the same random seed for two different simulations (with the same parameters) ensures that we get precisely the same results from both simulations (at least, on the same computer and with the same software versions). This is very useful when we wish to examine the properties of a specific simulation (for example, when debugging), or if we wish to illustrate a particular example. We will often set the random seed in the examples in this tutorial for this reason.

2.2 Population Models

In the previous section the only parameters we supplied to simulate were the sample_size and num_replicates parameters. This allows us to randomly sample trees with a given number of nodes, but, as it leaves the population unspecified, has little connection with biological reality. The most fundamental population parameter is the *effective population size*, or N_e. This parameter simply rescales time; larger effective population sizes correspond to older coalescence times:

```
1  def pairwise_T_mrca(Ne):
2      N = 10000
3      T_mrca = np.zeros(N)
4      for j, ts in enumerate(
5              msprime.simulate(2, Ne=Ne, num_replicates=N)):
6          tree = ts.first()
7          T_mrca[j] = tree.time(tree.root)
8      return np.mean(T_mrca)
9
10 print(
11     pairwise_T_mrca(0.5), pairwise_T_mrca(10),
12     pairwise_T_mrca(100))
13
14 >>> (0.99569690432656333, 19.816809844176138, 196.42125227336615)
```

Thus, when we specify $N_e = 10$ we get a mean pairwise coalescence time of about 20 generations, and with $N_e = 100$, the mean coalescence time is about 200 generations. *See* ref. 43 for details on the biological interpretation of effective population size.

By default, $N_e = 1$ in msprime, which is equivalent to measuring time in units of N_e generations. It is very important to note that N_e in msprime is the *diploid* effective population size, which means that all times are scaled by $2N_e$ (rather than N_e for a haploid coalescent). Thus, if we wish to compare the results that are given in the literature for a haploid coalescent, then we must set N_e to 1/2 to compensate. For example, we know that the expected coalescence time for a sample of size 2 is 1, and this is the value we obtain from the pairwise_T_mrca function when we have $N_e = 0.5$. We will usually assume that we are working in haploid coalescent time units from here on, and so set $N_e = 0.5$ in most examples. However, when running simulations of a specific organism and/or population, it is substantially more convenient to use an appropriate estimated value for N_e so that times are directly interpretable.

2.2.1 Exponentially Growing/Shrinking Populations

When we provide an N_e parameter, this specifies a fixed effective population size. We can also model populations that are exponentially growing or contracting at some rate over time. Given a population size at the present s and a growth rate α, the size of the population t generations in the past is $se^{-\alpha t}$. (Note again that time and rates are measured in units of *generations*, not coalescent units.)

In msprime, the initial size and growth rate for a particular population are specified using the PopulationConfiguration object. A list of these objects (describing the different populations; see Section 2.4) are then provided to the simulate function. When providing a list of PopulationConfiguration objects, the Ne parameter to simulate is not required, as the initial_size of the population configurations performs the same task. For example,

```
 1  def pairwise_T_mrca(growth_rate):
 2      N = 10000
 3      T_mrca = np.zeros(N)
 4      replicates = msprime.simulate(
 5          population_configurations=[
 6              msprime.PopulationConfiguration(
 7                  sample_size=2, initial_size=0.5,
 8                  growth_rate=growth_rate)],
 9          num_replicates=N, random_seed=100)
10      for j, ts in enumerate(replicates):
11          tree = ts.first()
12          T_mrca[j] = tree.time(tree.root)
13      return np.mean(T_mrca)
14
15  print(
16      pairwise_T_mrca(0.05), pairwise_T_mrca(0),
17      pairwise_T_mrca(-0.05))
18  >>> (0.96598072124289924, 1.0124999939843193, 1.0694803236032397)
```

Here we simulate the pairwise T_{MRCA} for positive, zero, and negative growth rates. When we have a growth rate of zero, we see that we recover the usual result of 1.0 (as our initial size, and hence N_e, is set to 1/2). When the growth rate is positive, we see that the mean coalescence time is reduced, since the population size is getting smaller as we go backwards in time, resulting in an increased rate of coalescence. Conversely, when we have a negative growth rate, the population is getting larger as we go backwards in time, resulting in a slower coalescence rate. (Care must be taken with negative growth rates, however, as it is possible to specify models in which the MRCA is never reached. In some cases this will lead to an error being raised, but it is also possible that the simulator will keep generating events indefinitely. This is particularly important in simulation based approaches to inference from real data.)

2.3 Mutations

We cannot directly observe gene genealogies; rather, we observe mutations in a sample of sequences which ultimately have occurred on genealogical branches. We are therefore very often interested not just in the genealogies generated by the coalescent process, but also in the results of mutational processes imposed on these trees. msprime currently supports simulating mutations under the infinitely many sites model (arbitrarily complex mutations are supported by the underlying data model, however). This is accessed by the mutation_rate parameter to the simulate function. As usual, this rate is the per-generation rate.

```
1  ts = msprime.simulate(3, mutation_rate=1, random_seed=7)
2  tree = ts.first()
3  SVG(tree.draw())
```

The tree produced by this code chunk is shown in Fig. 2. Here we have two mutations, shown by the red squares. Mutations occur above a given node in the tree, and all samples beneath this node will inherit the mutation. The infinite sites mutations used here are

Fig. 2 Coalescent tree with mutations

simple binary mutations, that is, the ancestral state is 0 and the derived state is 1. One convenient way to access the resulting sample genotypes is to use the `genotype_matrix( )` method, which returns an $m \times n$ NumPy array, if we have m variable sites and n samples. Thus, if G is the genotype matrix, $G[j, k]$ is the state of the kth sample at the jth site. In our example above, the site 0 has a mutation over node 3, and site 1 has a mutation over node 1, and so we get the following matrix:

```
print(ts.genotype_matrix())

>>> array([[1, 0, 1],
           [0, 1, 0]], dtype=uint8)
```

The genotype matrix gives a convenient way of accessing genotype information, but will consume a great deal of memory for larger simulations. See Section 3.4 for more information on how to access genotype data efficiently.

When comparing simulations to analytic results, it is very important to be aware of the way in which the mutation rates are defined in coalescent theory. For historical reasons, the scaled mutation rate θ is defined as $2N_e\mu$, where μ is the per-generation mutation rate. Since all times and rates are specified in units of generations in `msprime`, we must divide by a factor of two if we are to compare with analytic predictions. For example, the mean number of segregating sites for a sample of two is θ; to run this in `msprime` we do the following:

```
N = 10000
theta = 5
S = np.zeros(N)
replicates = msprime.simulate(
    2, Ne=0.5, mutation_rate=theta / 2, num_replicates=N)
for j, ts in enumerate(replicates):
    S[j] = ts.num_sites   # Number of segregrating sites.
print(np.mean(S))

>>> 4.8276000000000003
```

Note that here we set the mutation rate to $\theta/2$ (to cancel out the factor of 2 in the definition of θ) and $N_e = 1/2$ (so that time is measured in haploid coalescent time units). Such factor-of-two gymnastics are unfortunately unavoidable in coalescent theory.

2.4 Population Structure

Following `ms` [17], `msprime` supports a discrete-deme model of population structure in which d panmictic populations exchange migrants according to the rates defined in an $d \times d$ matrix. This approach is very flexible, allowing us to simulate island models (in which all populations exchange migrants at a fixed rate), one-

and two-dimensional stepping stone models (where migrants only move to adjacent demes) and other more complex migration patterns. This population structure is declared in `msprime` via the `population_configurations` and `migration_matrix` parameters in the `simulate` function. The list of population configurations defines the populations; each element of this list must be a `PopulationConfiguration` instance (each population has independent initial population size and growth rate parameters). The migration matrix is a NumPy array (or list of lists) of per-generation migration rates; $m[j, k]$ defines the fraction of population j that consists of migrants from population k in each generation. (Note that when running simulations on the coalescence scale, i.e. setting $N_e = 1/2$, this is equivalent to the number of migrants per deme and generation $M[j, k] = 2 N_e m[j, k]$.)

```
pop_configs = [
    msprime.PopulationConfiguration(sample_size=2),
    msprime.PopulationConfiguration(sample_size=2)]
M = np.array([
    [0, 0.1],
    [0, 0]])
ts = msprime.simulate(
    population_configurations=pop_configs, migration_matrix=M,
    random_seed=2)
tree = ts.first()
colour_map = {0:"red", 1:"blue"}
node_colours = {
    u: colour_map[tree.population(u)] for u in tree.nodes()}
SVG(tree.draw(node_colours=node_colours))
```

We create our model by first making a list of two `PopulationConfiguration` objects. For convenience here, we use the `sample_size` argument to these objects to state that we wish to have two samples from each population. This results in samples being allocated sequentially to the populations when `simulate` is called: 0 and 1 are placed in population 0, and samples 2 and 3 are placed in population 1. We then declare our migration matrix, which is asymmetric in this example. Because $M[0, 1] = 0.1$ and $M[1, 0] = 0$, forwards in time, individuals can migrate from population 1 to population 0 but not vice versa. This is illustrated in Fig. 3a which shows the tree produced by this simulation. Each node has been colored by its population (red is population 0 and blue population 1). Thus, the leaf nodes 0 and 1 are both from population 0, and 2 and 3 are both from population 2 (as explained above). As we go up the tree, the first event that occurs is 2 and 3 coalescing in population 1, creating node 4. After this, 4 coalesces with node 0, which has at some point before this migrated into deme 1, creating node 5. Node 1 also migrates into deme 1, where it coalesces with 5. Because migration is asymmetric here, the MRCA of the four samples *must* occur within deme 1.

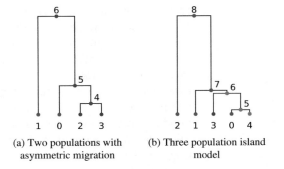

(a) Two populations with   (b) Three population island
    asymmetric migration   model

Fig. 3 Example trees produced in models with multiple populations and migration. Nodes are colored by population. **(a)** Two populations with asymmetric migration. **(b)** Three-population island model

The exact history of migration events is available if we use the `record_migrations` option. In the next example, we set up a symmetric island model and track every migration event:

```
1   pop_configs = [
2       msprime.PopulationConfiguration(sample_size=3),
3       msprime.PopulationConfiguration(sample_size=1),
4       msprime.PopulationConfiguration(sample_size=1)]
5   M = [
6       [0, 1, 1],
7       [1, 0, 1],
8       [1, 1, 0]]
9   ts  = msprime.simulate(
10      population_configurations=pop_configs, migration_matrix=M,
11      record_migrations=True, random_seed=101)
12  tree = ts.first()
13  colour_map = {0:"red", 1:"blue", 2: "green"}
14  node_colours = {
15      u: colour_map[tree.population(u)] for u in tree.nodes()}
16  SVG(tree.draw(node_colours=node_colours))
```

Figure 3b shows the tree produced by this code chunk. Here we sample three nodes from population 0, but because there is a lot of migration, the locations of coalescences are quite random. For example, the first coalescence occurs in deme 2 (green), after node 0 has migrated in. To see the details of these migration events, we can examine the "migration records" that are stored by `msprime`. (These are not stored by default, as they may consume a substantial amount of memory. The `record_migrations` parameter must be supplied to `simulate` to turn on this feature.) Migration records store complete information about the time, source, and destination demes and the genomic interval in question. Here we are interested in the total number of migration events experienced by each node:

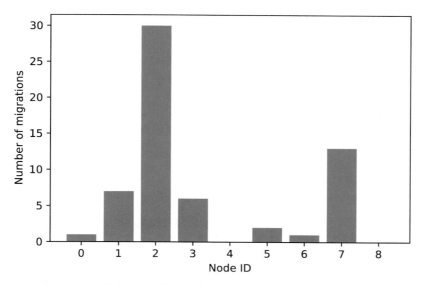

Fig. 4 Number of migration events for each tree node in a simulation with migration

```
1  node_count = np.zeros(ts.num_nodes)
2  for migration in ts.migrations():
3      node_count[migration.node] += 1
4  plt.bar(np.arange(ts.num_nodes), node_count)
5  plt.xlabel("Node_ID")
6  plt.ylabel("Number_of_migrations");
```

This code produces the plot in Fig. 4. We can see that node 0 experienced very few migration events before it ended up in deme 2, where it coalesced with 4 (which never migrated). Node 2, on the other hand, migrated 30 times before it finally coalesced with 7 in deme 0. Note that there are many more migration events than nodes here, implying that most migration events are not identifiable from a genealogy in real data [38].

Other forms of migration are also possible between specific demes at specific times. These different demographic events are dealt with in the next section.

2.5 Demographic Events

Demographic events allow us to model more complex histories involving changes to the population structure over time, and are specified using the demographic_events parameter to simulate. Each demographic event occurs at a specific time, and the list of events must be supplied in the order they occur (backwards in time). There are a number of different types of demographic event, which we examine in turn.

2.5.1 Migration Rate Change

Migration rate change events allow us to update the migration rate matrix at some point in time. We can either update a single cell in the matrix or all (non-diagonal) entries at the same time.

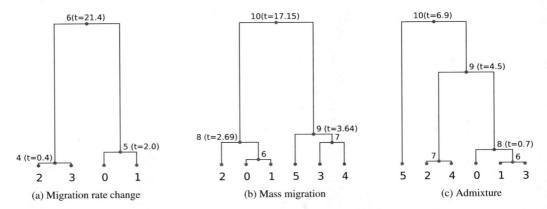

Fig. 5 Example trees produced in models with demographic events. Nodes are colored by population. **(a)** Migration rate change. **(b)** Mass migration. **(c)** Admixture

```
1  ts   = msprime.simulate(
2      population_configurations=[
3          msprime.PopulationConfiguration(sample_size=2),
4          msprime.PopulationConfiguration(sample_size=2)],
5      demographic_events=[
6          msprime.MigrationRateChange(20, rate=1.0, matrix_index=(0, 1))],
7      random_seed=2)
8  tree = ts.first()
```

The tree produced by this code chunk is shown in Fig. 5a (in this example and those following we have omitted the code required to draw the tree). The samples 0 and 1, and 2 and 3 coalesce quickly within their own populations. However, because the migration rate between the populations is zero these lineages are isolated and would never coalesce without some change in demography. The migration rate change event happens at time 20, resulting in node 5 migrating to deme 1 soon afterwards. The lineages then coalesce at time 21.4.

2.5.2 Mass Migration

This class of event allows us to move some proportion of the lineages in one deme to another at a particular time. This allows us to model population splits and admixture events. Population splits occur when (backwards in time) all the lineages in one population migrate to another.

```
1  ts   = msprime.simulate(
2      population_configurations=[
3          msprime.PopulationConfiguration(sample_size=3),
4          msprime.PopulationConfiguration(sample_size=3)],
5      demographic_events=[
6          msprime.MassMigration(15, source=1, dest=0, proportion=1)],
7      random_seed=20)
8  tree = ts.first()
```

The tree produced by this code chunk is shown in Fig. 5b. In this case we also have two isolated populations which coalesce down to a single lineage. The population split at time 15 (which, forwards in time produced all the individuals in population 1) results in this lineage migrating back to population 0, where it coalesces with the ancestor of the samples 0, 1, and 2.

Admixture events (i.e., where some fraction of the lineages move to a different deme) are specified in the same way:

```
1  ts   = msprime.simulate(
2      population_configurations=[
3          msprime.PopulationConfiguration(sample_size=6),
4          msprime.PopulationConfiguration(sample_size=0)],
5      demographic_events=[
6          msprime.MassMigration(0.5, source=0, dest=1, proportion=0.5),
7          msprime.MigrationRateChange(1.1, rate=0.1),
8      ],
9      random_seed=26)
10 tree = ts.first()
```

The tree produced by this code chunk is shown in Fig. 5c. We begin in this example with six lineages sampled in population 0, zero samples in population 1, and with no migration between these populations. At time 0.5, we specify an admixture event where each of the four extant lineages (5, 7, 0, and 6) has a probability of $1/2$ of moving to deme 1. Linages 0 and 6 migrate, and subsequently coalesce into node 8. Further back in time, at $t = 1.1$, another demographic event occurs, changing the migration rate between the demes to 0.1, thereby allowing lineages to move between them. Eventually, all lineages end up in deme 1, where they coalesce into the MRCA at time $t = 6.9$.

2.5.3 Population Parameter Change

This class of event represents a change in the growth rate or size of a particular population. Since each population has its own individual size and growth rates, we can change these arbitrarily as we go backwards in time. Keeping track of the actual sizes of different populations can be a little challenging, and for this reason msprime provides a DemographyDebugger class.

To illustrate this, we consider a very simple example in which we have a single population experiencing a phase of exponential growth from 750 to 100 generations ago. The size of the population 750 generations ago was 2000, and it grew to 20,000 over the next 650 generations. The size of the population has been stable at this value for the past 100 generations. We encode this model as follows:

```
1  N1 = 20000   # Population size at present
2  N2 = 2000    # Population size at start (forwards in time) of exponential growth.
3  T1 = 100     # End of exponential growth period (forwards in time)
4  T2 = 750     # Start of exponential growth period (forwards in time)
5  # Calculate growth rate; solve N2 = N1 * exp(-alpha * (T2 - T1))
6  growth_rate = -np.log(N2 / N1) / (T2 - T1)
7  population_configurations = [
8      msprime.PopulationConfiguration(initial_size=N1)
9  ]
10 demographic_events = [
11     msprime.PopulationParametersChange(time=T1, growth_rate=growth_rate),
12     msprime.PopulationParametersChange(time=T2, growth_rate=0),
13 ]
14 dd = msprime.DemographyDebugger(
15     population_configurations=population_configurations,
16     demographic_events=demographic_events)
17 dd.print_history()
```

It gives the following output:

```
==============================
Epoch: 0 -- 100.0 generations
==============================
        start      end     growth_rate |     0
       -------- --------    -------- | --------
  0 |  2e+04    2e+04              0 |     0


Events @ generation 100.0
     - Population parameter change for -1: growth_rate -> 0.0035


=================================
Epoch: 100.0 -- 750.0 generations
=================================
        start      end     growth_rate |     0
       -------- --------    -------- | --------
  0 |  2e+04    2e+03        0.00354 |     0


Events @ generation 750.0
     - Population parameter change for -1: growth_rate -> 0


===============================
Epoch: 750.0 -- inf generations
===============================
        start      end     growth_rate |     0
       -------- --------    -------- | --------
  0 |  2e+03    2e+03              0 |     0
```

After we set up our model, we use the DemographyDebugger to check our calculations. We see that time has been split into three "epochs." From the present until 100 generations ago, the population size is constant at 20,000. Then, we have a demographic event that changes the growth rate to 0.0035, which applies over the next epoch (from 100 to 750 generations ago). Over this time, the population grows from 2000 to 20,000 (note that the "start" and "end" of each epoch is looking *backwards* in time, as we consider epochs starting from the present and moving backwards). At generation 750, another event occurs, setting the growth rate for the population to 0. Then, the population size is constant at 20,000 from generation 750 until the indefinite past.

A more complex example involving a three-population out-of-Africa human model is available in the online documentation.

2.6 Ancient Samples   Up to this point we have assumed that all samples are taken at the present time. However, msprime allows us to specify arbitrary sampling times and locations, allowing us to simulate (for example) ancient samples.

```
 1  ts = msprime.simulate(
 2      samples=[
 3          msprime.Sample(0, 0), msprime.Sample(0, 0),
 4          msprime.Sample(0, 0),
 5          msprime.Sample(1, 0.75), # Ancient sample in deme 1
 6      ],
 7      population_configurations=[
 8          msprime.PopulationConfiguration(),
 9          msprime.PopulationConfiguration()],
10      migration_matrix=[
11          [0, 1],
12          [1, 0]],
13      random_seed=22)
14  tree = ts.first()
```

The tree produced by this code chunk is shown in Fig. 6. All of the trees that we previously considered had leaf nodes at time zero. In this case, the samples 0, 1, and 2 are taken at time 0 in population 0, but node 3 is sampled at time 0.75 in population 1. Note that in this case we used the samples parameter to simulate to specify our samples. This is the most general approach to assigning

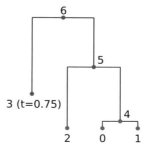

Fig. 6 Example tree produced by simulation with ancient samples

samples, and allows samples to be assigned to arbitrary populations and at arbitrary times.

2.7 Recombination

One of the key innovations of msprime is that it makes simulation of the full coalescent with recombination possible at whole-chromosome scale. Adding recombination to a simulation is simple, requiring very minor changes to the methods given above.

```
1  ts = msprime.simulate(
2      10, Ne=1e4, length=1e5, recombination_rate=1e-8, random_seed=3)
3  print(ts.num_trees)
4  >>> 82
```

In this case, we provide two extra parameters: length, which defines the length of the genomic region to be simulated, and recombination_rate, which defines the rate of recombination per unit of sequence length, per generation. It is often useful to think of both sequence lengths and recombination rates as defined in units of base-pairs. (Note, however, that these are continuous values, so this correspondence should not be taken too literally. Note also that because msprime assumes an infinite sites mutation model the length parameter is not connected to the number of mutational *sites*. Thus any number of mutations can occur on a given sequence length, depending on the mutation rate specified.) For this example, we defined a sequence length of 100 kb, and a recombination rate of 10^{-8} per base per generation. The result of this particular simulation is a *tree sequence* that contains 82 distinct trees. Other replicate simulations with different random seeds will usually result in different numbers of trees.

Up to this point we have focused on simulations that returned a single tree representing the genealogy of a sample. The inclusion of recombination, however, means that there may be more than one tree relating our samples. The TreeSequence object returned by msprime is a very concise and efficient representation of these highly correlated trees. To process the trees, we simply consider them one at a time, using the trees() iterator.

```
1  tmrca = np.zeros(ts.num_trees)
2  breakpoints = np.zeros(ts.num_trees)
3  for tree in ts.trees():
4      tmrca[tree.index] = tree.time(tree.root)
5      breakpoints[tree.index] = tree.interval[0]
6  plt.ylabel("T_mrca_(Generations)")
7  plt.xlabel("Position_(kb)")
8  plt.plot(breakpoints / 1000, tmrca, "o");
```

This code generates the plot in Fig. 7 showing the time of the MRCA of the sample for each tree across the sequence. We find the T_{MRCA} as before, and plot this against the left coordinate of the

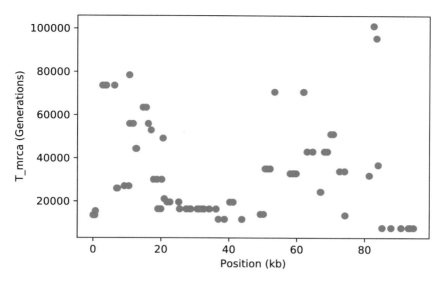

Fig. 7 Time to the MRCA of a sample across a 100 kb region

genomic interval that each tree covers. A full description of *tree sequences* and the methods for working with them is beyond the scope of this chapter (but see the online documentation for more details).

It is also possible to simulate data with recombination rates varying across the genome (for example, in recombination hotspots). To do this, we first create a RecombinationMap instance that describes the properties of the recombination landscape that we wish to simulate. We then supply this object to simulate using the recombination_map argument. In the following example, we simulate 100 samples using the human chromosome 22 recombination map from the HapMap project [19]. Figure 8 shows the recombination rate and the locations of breakpoints from the simulation, and the density of breakpoints closely follows the recombination rate, as expected.

```
1  # Read in the recombination map and run the simulation.
2  infile = "genetic_map_GRCh37_chr22.txt"
3  recomb_map = msprime.RecombinationMap.read_hapmap(infile)
4  ts = msprime.simulate(
5      sample_size=100,
6      Ne=10**4,
7      recombination_map=recomb_map,
8      random_seed=1)
```

Although coordinates are specified in floating point values, msprime uses a discrete loci model when performing simulations. By default, the number of loci is very large ($\sim 2^{32}$), and the locations of breakpoints are translated back into the coordinate system defined by the recombination map. However, the number of loci is

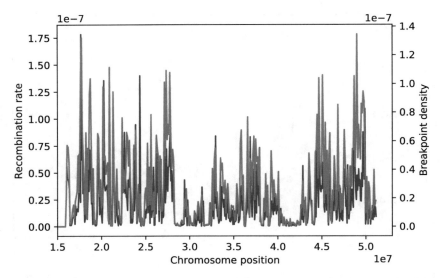

Fig. 8 The HapMap genetic map for chromosome 22 (blue) matches the density of breakpoints for a simulated chromosome (green) well

configurable and it is possible to simulate a specific number of discrete loci.

```
1  recomb_map = msprime.RecombinationMap.uniform_map(
2      length=10, rate=1, num_loci=10)
3  ts = msprime.simulate(2, recombination_map=recomb_map)
4  print(list(ts.breakpoints()))
5  >>> [0, 1.0, 2.0, 3.0, 5.0, 6.0, 7.0, 8.0, 9.0, 10.0]
```

Here we simulate the history of two samples in a system with ten loci, each of length 1 with recombination rate of 1 between adjacent loci per generation. In the output, we see that the breakpoints between trees now occur exactly at the integer boundaries between these loci. This shows that we can also simulate models of recombination with discrete loci in msprime, as well as the more standard continuous genome.

3   Processing Results

In the previous section we showed how to run simulations in msprime, and how to construct population models and demographic histories. In this section we show how to process the results of simulations. This is not a comprehensive review of the capabilities of the msprime Python API, but concentrates on some useful examples. msprime is specifically designed to enable very large simulations, and the processing methods we demonstrate below are all very efficient. To illustrate this, we consider a simulation of

200,000 samples of ten megabases from a simple two-population model with human-like parameters:

```
1  ts = msprime.simulate(
2      population_configurations=[
3          msprime.PopulationConfiguration(sample_size=10**5),
4          msprime.PopulationConfiguration(sample_size=10**5)],
5      demographic_events=[
6          msprime.MassMigration(time=50000, source=1, destination=0)],
7      Ne=10**4, recombination_rate=1e-8, mutation_rate=1e-8, length=10*10**6,
8      random_seed=3)
9  print((ts.num_trees, ts.num_sites))
10
11 >>> (93844, 102270)
```

This simulation required about 20 s to complete.

3.1 Computing MRCAs

We are often interested in finding the most recent common ancestor (MRCA) of a pair (or many pairs) of samples. For example, identity-by-descent (IBD) tracts are defined as contiguous stretches of genome in which the MRCA for a pair of samples is the same. Computing IBD segments for a pair of samples is very straightforward:

```
1  def ibd_segments(ts, a, b):
2      trees_iter = ts.trees()
3      tree = next(trees_iter)
4      last_mrca = tree.mrca(a, b)
5      last_left = 0
6      segment_lengths = []
7      for tree in trees_iter:
8          mrca = tree.mrca(a, b)
9          if mrca != last_mrca:
10             left = tree.interval[0]
11             segment_lengths.append(left - last_left)
12             last_mrca = mrca
13             last_left = left
14     segment_lengths.append(ts.sequence_length - last_left)
15     return np.array(segment_lengths) / ts.sequence_length
16
17 sns.distplot(ibd_segments(ts, 0, 1), label="Within_population")
18 sns.distplot(ibd_segments(ts, 0, 10**5), label="Between_populations")
19 plt.xlim(-0.0001, 0.003)
20 plt.legend()
21 plt.xlabel("Fraction_of_genome_length");
22 plt.ylable("Count")
```

In this example we create a function `ibd_segments` that returns a NumPy array of the lengths of IBD segments for a given pair of samples, a and b. It works simply by computing the MRCA for the samples at the left-hand side of the sequence and then, moving rightwards, records a segment each time the MRCA changes. We then plot the distribution of tract lengths for samples 0 and 1 (which are both in population 0), and also the tract lengths

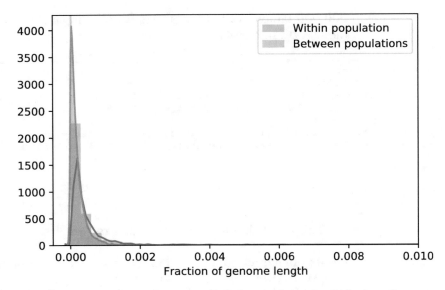

Fig. 9 The distribution of the length of IBD segments for a pair of samples taken from the same or different populations

for a pair of samples from different populations. The results are shown in Fig. 9. As we might expect, the tract lengths are shorter for the between population pair.

Of course, we would need to sample many such pairs of samples or longer sequences to get a reasonable approximation of the real distribution of block lengths. Because the main cost of this function is the iteration over all the trees in the sequence, it would be more efficient to keep track of the MRCAs for different pairs in a single iteration rather than repeatedly call the above `ibd_segments` function.

3.2 Sample Counts

The `msprime` API provides an extremely efficient way to count the number of samples that are beneath a particular node in a tree. This can be used, for example, to compute allele frequencies efficiently and is the basis for many of the fast algorithms in the API. As a simple illustration of this technique, consider the following code to compute the number of sites with derived allele frequency less than 1%:

```
1   N = ts.num_samples
2   threshold = 0.01
3   num_rare_derived = 0
4   for tree in ts.trees():
5       for site in tree.sites():
6           # Only works for infinite sites mutations.
7           assert len(site.mutations) == 1
8           mutation = site.mutations[0]
9           if tree.num_samples(mutation.node) / N < threshold:
10              num_rare_derived += 1
11  print((num_rare_derived, num_rare_derived / ts.num_sites))
12
13  >>> (65258, 0.638095238095238)
```

In this example we iterate over all the trees in the tree sequence, and then iterate over all the sites in each tree. We find the frequency of the derived allele at each site using the `num_samples` method, which returns the number of samples subtending a given node. The underlying implementation ensures that this operation requires constant time, and so it is *very* efficient. We see that such rare alleles are common. (We reiterate that `msprime` currently generates mutations under the infinitely many sites model so that each mutation occurs at a unique site. Future versions of `msprime` or other software packages may produce tree sequences with back or recurrent mutations, where this simple approach will not work. To emphasize this point and to ensure that the above code chunk is not accidentally applied in such situations we have included an `assert` statement. We use asserts in a similar way in later code chunks.)

A powerful feature of this sample-counting approach is that we can perform the same operation over an arbitrary subset of the samples. For example, suppose we wished to count the number of sites that are private to a specific population:

```
1  def num_private_sites(pop_id):
2      pop_samples = ts.samples(pop_id)
3      num_private = 0
4      for tree in ts.trees(tracked_samples=pop_samples):
5          for site in tree.sites():
6              # Only works for infinite sites mutations.
7              assert len(site.mutations) == 1
8              mutation = site.mutations[0]
9              total = tree.num_samples(mutation.node)
10             within_pop = tree.num_tracked_samples(mutation.node)
11             if total == within_pop:
12                 num_private += 1
13      return num_private
14
15 private_0 = num_private_sites(0)
16 private_1 = num_private_sites(1)
17 print((ts.num_sites, private_0 + private_1, private_0, private_1))
18
19 >>> (102270, 101607, 51295, 50312)
```

This example is very similar, except we provide an extra argument to `ts.trees`. The `tracked_samples` argument specifies a list of samples to be tracked, which can be any arbitrary subset of the samples in the simulation. Here we indicate that we are interested in tracking the set of samples within the population in question. Again, we iterate over all trees and over all sites within trees. Then, for each infinite sites mutation we compute two frequencies: the overall number of samples that inherit from the mutation's node, and the number of tracked samples *within the focal population* that inherit from this node. If the total count is equal to the within-population count, we know that this mutation is private to the population.

3.3 Obtaining Subsets

In some situations it is useful to analyze data for different subsets of the samples separately. This is possible using the `simplify` method:

```
1  samples = [1, 3, 5, 7]
2  ts_subset = ts.simplify(samples)
3  print((
4      ts_subset.num_sites, ts_subset.num_trees,
5      ts.num_sites, ts.num_trees))
6  >>> (11939, 5483, 102270, 93844)
```

Here we extract the tree sequence representing the history of a tiny subset of the original samples, with IDs 1, 3, 5, and 7. The subset tree sequence contains all the genealogical information relevant to the subsamples, but no more. Concretely, both coalescences that are not ancestral to the subsample and coalescences that predate the MRCA of the subsample are excluded. Thus, the number of distinct trees is greatly reduced. By default, we also remove any sites that have no mutations within these subtrees (i.e., those that are fixed for the ancestral state). These can be retained by using the `filter_sites=False` argument.

Node IDs in the simplified tree sequence are not the same as in the original. The `map_nodes` argument allows us to obtain the mapping from IDs in the original tree sequence to their equivalent nodes in the new tree sequence.

```
1  ts_subset, node_map = ts.simplify(samples, map_nodes=True)
2  tree = ts_subset.first()
3  node_labels = {
4      node_map[j]: "{}({})".format(node_map[j], str(j))
5      for j in range(ts.num_nodes)}
6  SVG(tree.draw(node_labels=node_labels, width=400))
```

The result of running this code chunk is shown in Fig. 10. Here we draw the first tree in the subset tree sequence, showing the new node IDs along with the IDs from the original tree sequence in parentheses. The number of nodes is greatly reduced from the original.

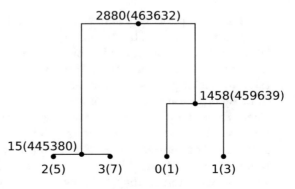

Fig. 10 Tree of a subset of the samples in a large simulation. Node IDs in the subset and full tree sequences are shown

3.4 Processing Variants

While it is nearly always more efficient to work with mutations in terms of their context within the trees, it is sometimes more convenient to work with the allelic states of the samples. This information is obtained in msprime using the `variants()` iterator, which returns a `Variant` object for each site in the tree sequence. A `Variant` consists of: (a) a reference to the `Site` in question; (b) the `alleles` at this site (the strings representing the actual states); and (c) the `genotypes` representing the observed state for each sample. The `genotypes` are encoded in a NumPy array, such that `variant.alleles[variant.genotypes[j]]` gives the allelic state for sample `j`. The values in the `genotypes` array are therefore indexes into the `alleles` list. The ancestral state at a given site is guaranteed to be the first element in the `alleles` list, but no other assumptions about ordering of the alleles list should be made.

For biallelic sites, working with genotypes is straightforward as the genotypes array can only contain 0 and 1 values, which correspond to the ancestral and derived states, respectively. The `genotypes` values are returned as a NumPy array, and so the full NumPy library is available for efficient processing. As an example, we show here how to count the number of sites at which the derived allele is at frequency less than 10%. Using the genotypes in this way is convenient, as complex patterns of back and recurrent mutations can be handled without difficulty.

```
1  %%time
2  threshold = 0.1
3  num_rare = 0
4  for variant in ts.variants():
5      # Will work for any biallelic sites; back/recurrent mutations OK
6      assert len(variant.alleles) == 2
7      if np.sum(variant.genotypes) / ts.num_samples < threshold:
8          num_rare += 1
9  print(num_rare)
10 >>> 83081
11 CPU times: user 1min 30s, sys: 4 ms, total: 1min 30s
```

This code is straightforward, as we simply iterate over all variants and count the number of one values in the genotypes array. Using the np.sum function, this operation is efficient. Generating all the genotypes for 200,000 samples at 100,000 sites, however, is an expensive operation and the overall calculation takes about 1.5 min to complete.

In the case of infinite sites mutations, we can recast this operation to use the efficient sample counting methods described in Section 3.2. This approach is far more efficient, requiring less than 2 s to compute the same value.

```
1  %%time
2  num_rare = 0
3  for tree in ts.trees():
4      for site in tree.sites():
5          # Only works for infinite sites mutations.
6          assert len(site.mutations) == 1
7          mutation = site.mutations[0]
8          if tree.num_samples(mutation.node) / ts.num_samples < threshold:
9              num_rare += 1
10 print(num_rare)
11 >>> 83081
12 CPU times: user 1.75 s, sys: 36 ms, total: 1.79 s
```

3.5 Incremental Calculations

A powerful property of the tree sequence representation is that we can efficiently find the differences between adjacent trees. This is very useful when we have some value that we wish to compute that changes in a simple way between trees. The edge_diffs iterator provides us with the information that we need to perform such incremental calculations. Here we use it to keep a running track of the total branch length of our trees, without needing to perform a full traversal each time.

```
1  def get_total_branch_length(ts):
2      current = 0
3      total_branch_length = np.zeros(ts.num_trees)
4      for j, (_, edges_out, edges_in) in enumerate(ts.edge_diffs()):
5          for e in edges_out:
6              current -= ts.node(e.parent).time - ts.node(e.child).time
7          for e in edges_in:
8              current += ts.node(e.parent).time - ts.node(e.child).time
9          total_branch_length[j] = current
10     return total_branch_length
```

This function returns the total branch length value for each tree in the sequence as a NumPy array. It works by keeping track of the total branch length as we proceed from left to right, and storing this value in the output array for each tree. The edge_diffs method returns a list of the edges that are removed for each tree transition (edges_out) and a list of edges that are inserted (edges_in). Computing the current value for the total branch length is then simply a case of subtracting the branch lengths for all outgoing edges and adding the branch lengths for all incoming edges. This is extremely efficient because, after the first tree has been constructed there is at most four incoming and outgoing edges [23]. Thus, each tree transition costs *constant time*.

```
1  %%time
2  tbl = get_total_branch_length(ts)
3
4  CPU times: user 7.67 s, sys: 64 ms, total: 7.74 s
```

In contrast, if we compute the total branch length by performing a full traversal for each tree, each tree transition is very costly when we have a large sample size. In this example, computing the array of branch lengths using the incremental approach given here took 8 s. Computing the same array using the `tree.total_branch_length` for each tree in a straightforward way still had not completed after *twenty minutes*. (This is because `msprime` currently implements this operation by a full traversal in Python; in future, this may change to using the algorithm given here.) Full tree traversals of large trees are expensive, and great gains can be made if calculations can be expressed in an incremental manner using `edge_diffs`.

3.6 Exporting Variant Data

If the `msprime` API doesn't provide methods to easily calculate the statistics you are interested in, it's straightforward to export the variant data into other libraries using the `genotype_matrix()` or `variants()` methods. We recommend the excellent `scikit-allel` [32] and `pylibseq` [https://pypi.python.org/pypi/pylibseq] libraries (`pylibseq` is a Python interface to `libsequence` [41]). If you wish to export data to external programs, VCF may be best option, which is supported using the `write_vcf` method. The `simplify` method is useful here if you wish to export data from a subset of the simulated samples.

However, it is worth noting that for large sample sizes, exporting genotype data may require a great deal of memory and take some time. One of the advantages of the `msprime` API is that we do not need to explicitly generate genotypes in order to compute many statistics of interest.

4 Validating Analytic Predictions

In this section we show some examples of validating simple analytic predictions from coalescent theory using simulations. The number of segregating sites is the total number of mutations that occurred in the history of the sample (assuming the infinite sites mutation model). Since mutations happen as a Poisson process along the branches of the tree, what we are really interested in is the distribution of the total branch length of the tree. The results in this section are well-known classical results from coalescent theory; this section is intended as a demonstration of how to proceed when comparing analytic results to simulations. We show some idiomatic examples for integrating with the state-of-the-art data analysis packages such as Pandas [30] and Seaborn [44]. All analytic predictions are taken from [43].

4.1 Total Branch Length and Segregating Sites

The first properties we are interested in are the mean and the variance of the total branch length of coalescent trees. (Note that, as before, we set $N_e = 1/2$ to convert between msprime's diploid time scaling to the haploid time scaling of these analytic results.)

```
1  ns = np.array([5, 10, 15, 20, 25, 30])
2  num_reps = 10000
3  n_col = np.zeros(ns.shape[0] * num_reps)
4  T_total_col = np.zeros(ns.shape[0] * num_reps)
5  row = 0
6  for n in ns:
7      for ts in msprime.simulate(n, Ne=0.5, num_replicates=num_reps):
8          tree = ts.first()
9          n_col[row] = n
10         T_total_col[row] = tree.total_branch_length
11         row += 1
12 df = pd.DataFrame({"n": n_col, "T_total": T_total_col})
```

We first create an array of the six different n values that we wish to simulate, and then create arrays to hold the results of the simulations. Because we are running 10,000 replicates for each sample size, we allocate arrays to hold 60,000 values. This approach of storing the data in arrays is convenient because it allows us to use Pandas dataframes in an idiomatic fashion. We then iterate over all of our sample sizes and run 10,000 replicates of each. For each simulation, we simply store the sample size value and the total branch length in a Pandas dataframe. This gives us access to many powerful data analysis tools (including the Seaborn library, which we use for visualization here).

After we have created our simulation data, we define our analytic predictions and plot the data.

```
1  def T_total_mean(n):
2      return 2 * np.sum(1 / np.arange(1, n))
3
4  def T_total_var(n):
5      return 4 * np.sum(1 / np.arange(1, n)**2)
6
7  mean_T = np.array([T_total_mean(n) for n in ns])
8  stddev_T = np.sqrt(np.array([T_total_var(n) for n in ns]))
9  ax = sns.violinplot(
10     x="n", y="T_total", data=df, color="grey", inner=None)
11 ax.plot(mean_T, "-");
12 ax.plot(mean_T - stddev_T, "--", color="black");
13 ax.plot(mean_T + stddev_T, "--", color="black");
14 group = df.groupby("n")
15 mean_sim = group.mean()
16 stddev_sim = np.sqrt(group.var())
17 x = np.arange(ns.shape[0])
18 ax.plot(x, mean_sim, "o")
19 line, = ax.plot(x, mean_sim - stddev_sim, "^")
20 ax.plot(x, mean_sim + stddev_sim, "^", color=line.get_color());
```

The plot in Fig. 11a shows the simulated distribution of the total branch length over replicate simulations (each violin is a

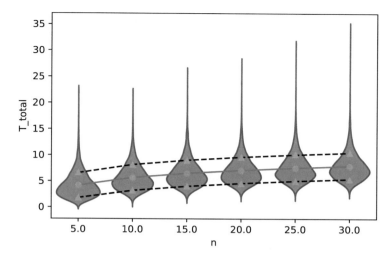

(a) The full distribution of simulated values (violin plots) along with observed and predicted mean and standard deviations for a range of sample sizes.

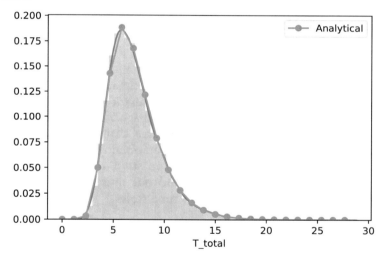

(b) The full simulated and predicted distribution of total branch length for $n = 20$.

Fig. 11 Comparisons of the distribution of simulated total branch lengths with analytic results. (a) The full distribution of simulated values (violin plots) along with observed and predicted mean and standard deviations for a range of sample sizes. (b) The full simulated and predicted distribution of total branch length for $n = 20$

distribution for a given sample size). We also show our analytic prediction for the mean and variance of each distribution (the dashed lines show ± 1 standard deviation from the mean). Also shown are the observed means and standard deviations from the simulations, as green circles and red triangles, respectively. We can see that the simulated values match our theoretical predictions for mean and variance very well. We can also see, however, that these one-dimensional summaries of the distribution capture some essential properties but lose some important aspects of the distribution.

Ideally, we wish to capture the full distribution analytically. In the following code chunk we define the analytic prediction for the total branch length distribution, and compare it with the simulated distribution for a sample of size 20. The results are shown in Fig. 11b. We can see an excellent agreement between the smoothed kernel density estimate produced by Seaborn and the theoretical prediction.

```python
def T_total_density(n, t):
    e_t2 = np.exp(-t / 2)
    return 0.5 * (n - 1) * e_t2 * (1 - e_t2)**(n - 2)

n = 20
T_total_20 = T_total_col[n_col == n]
ts = np.linspace(0, np.max(T_total_20), 25)
t_densities = np.array([T_total_density(n, t) for t in ts])
sns.distplot(T_total_20)
plt.plot(ts, t_densities, marker="o", label="Analytical")
plt.xlabel("T_total")
plt.legend();
```

Since we cannot directly observe branch lengths, we are usually more interested in mutations when working with data. The mutation process is intimately related to the distribution of branch lengths, since mutations occur randomly along tree branches. One simple summary of the mutational process is the total number of segregating sites, that is, the number of sites at which we observe variation. We can obtain this very easily from simulations simply by specifying a mutation rate parameter. (Note again that we set $N_e = 1/2$ and our mutation rate $= \theta/2$ in order to convert to msprime's time scales.)

```python
def S_dist(n, theta, k):
    S = 0
    for i in range(2, n + 1):
        S += ((-1)**i * scipy.special.binom(n - 1, i - 1)
              * (i - 1) / (theta + i - 1)
              * (theta / (theta + i - 1))**k)
    return S

n = 20
theta = 5
num_replicates = 1000
simulation = np.zeros(num_replicates)
replicates = msprime.simulate(
    n, Ne=0.5, mutation_rate=theta / 2, num_replicates=num_replicates)
for j, ts in enumerate(replicates):
    simulation[j] = ts.num_sites  # number of seg. sites
ks = np.arange(np.max(simulation))
analytical = np.array([S_dist(n, theta, k) for k in ks])
sns.distplot(simulation)
plt.plot(ks, analytical, marker='o', label="Analytical")
plt.xlabel("Segregating_sites")
plt.legend();
```

Here we take 1000 replicate simulations, store the number of infinite sites mutations for each, and plot this distribution in Fig. 12a. Also plotted is the analytic prediction, which again provides an excellent fit.

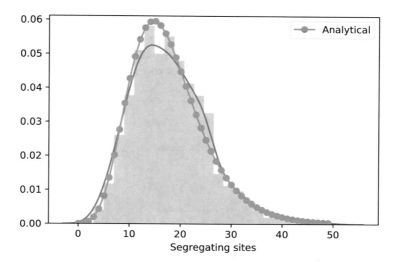

(a) The distribution of the number of segregating sites for $n = 20$, $\theta = 5$ and no recombination over 1000 simulation replicates, along with analytic prediction.

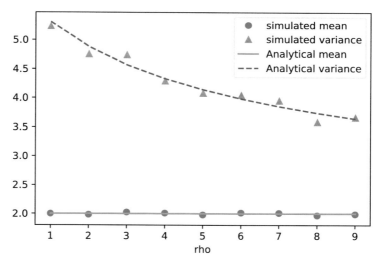

(b) The mean and variance of the number of segregating sites over 10000 simulation replicates with $n = 2$, $\theta = 2$ and varying recombination rate, along with analytic predictions.

Fig. 12 Simulations of the number of segregating sites, and comparisons with analytic predictions. (**a**) The distribution of the number of segregating sites for $n = 20$, $\theta = 5$ and no recombination over 1000 simulation replicates, along with analytic prediction. (**b**) The mean and variance of the number of segregating sites over 10000 simulation replicates with $n = 2$, $\theta = 2$ and varying recombination rate, along with analytic predictions

4.2   Recombination

In the previous section we saw how to run simulations to generate trees under the assumptions of the single-locus coalescent and compare these with analytic predictions. This assumes that our data is not affected by recombination, which is often unrealistic. Here we show how to compute empirical distributions of equivalent quantities, and compare these with classical results from the literature. Since analytic results for many quantities are generally unknown for the case of recombination along a linear sequence, we limit ourselves to the pairwise samples.

```
1  theta = 2
2  num_replicates = 10000
3  rhos = np.arange(1, 10)
4  N = rhos.shape[0] * num_replicates
5  rho_col = np.zeros(N)
6  s_col = np.zeros(N)
7  row = 0
8  for rho in rhos:
9      replicates = msprime.simulate(
10         sample_size=2, Ne=0.5, mutation_rate=theta / 2,
11         recombination_rate=rho / 2, num_replicates=num_replicates)
12     for ts in replicates:
13         rho_col[row] = rho
14         s_col[row] = ts.num_sites
15         row += 1
16 df = pd.DataFrame({"rho": rho_col, "s": s_col})
```

In this code chunk we again run 10^4 replicate simulations for a range of input parameters, and store the results in a Pandas data frame. We are interested in the effects of recombination rate in this example, and so the parameter that we vary is the scaled recombination rate ρ (noting, again, that we set $N_e = 1/2$ and `recombination_rate` $= \rho/2$ to convert to `msprime`'s time scales).

```
1  def pairwise_S_mean(theta):
2      return theta
3
4  def f2(rho):
5      return (rho + 18) / (rho**2 + 13 * rho + 18)
6
7  def pairwise_S_var(theta, rho):
8      integral = scipy.integrate.quad(lambda x: (rho - x) * f2(x), 0, rho)
9      return theta + 2 * theta**2 * integral[0] / rho**2
10
11 group = df.groupby("rho")
12 plt.plot(group.mean(), "o", label="simulated_mean")
13 plt.plot(group.var(), "^", label="simulated_variance")
14 plt.plot(
15     rhos, [pairwise_S_mean(theta) for rho in rhos], "-",
16     label="Analytical_mean")
17 plt.plot(rhos, [pairwise_S_var(theta, rho) for rho in rhos], "--",
18 label="Analytical_variance")
19 plt.xlabel("rho")
20 plt.legend();
```

After defining our analytic predictions for the mean and variance of the number of segregating sites, we then plot the observed and predicted values in Fig. 12b. Comparing the simulated results to analytic predictions we see excellent agreement. The mean number of segregating sites is not affected by recombination, but recombination does substantially reduce the variance.

5 Example Inference Scheme

The analytical challenges of deriving likelihood functions even under highly idealized models of population structure and history have led to the development of likelihood-free inference methods, in particular Approximate Bayesian Computation (ABC) [2]. ABC approximates the posterior distribution of model parameters by drawing from simulations. Because of its flexibility ABC has become a standard inference tool in statistical population genetics (see ref. 7, for a review). We will demonstrate how msprime can be used to set up an ABC inference by means of a simple toy example. We stress that this is meant as an illustration rather than an inference tool for practical use. However, given the flexibility of msprime, it should be relatively straightforward to implement more a realistic framework focused on specific inference applications.

We assume that data for 200 loci or sequence blocks (these could be RAD loci in practice) for a single diploid individual have been generated from each of two populations. We would like to infer the amount of gene flow between the two populations. For the sake of simplicity, we will assume the simplest possible model of population structure; that is, two populations, of the same effective size exchanging migrants at a constant rate of m migrants per generation.

The function run_sims simulates a dataset consisting of a specified number of loci (num_loci) given a migration rate M. We generate a single dataset of 50 loci assuming a migration rate $M = 0.3$ migrants per generation, which we will use as a (pseudo) observed dataset in the ABC implementation.

```
1  nsamp = 2
2  theta = 2
3  true_M = 0.3
4  num_loci = 200
5
6  def run_sims(m, num_loci=1,theta=0):
7      return msprime.simulate(
8          Ne=1/2,
9          population_configurations=[
10             msprime.PopulationConfiguration(sample_size=nsamp),
11             msprime.PopulationConfiguration(sample_size=nsamp)],
12         migration_matrix=[[0, m], [m, 0]],
13         num_replicates=num_loci,
14         mutation_rate=theta / 2)
15
16 def get_joint_site_frequency_spectra(reps):
17     data = np.zeros((num_loci, nsamp + 1, nsamp + 1))
18     for rep_index, ts in enumerate(reps):
19         # Track the samples from population 0.
20         for tree in ts.trees(tracked_samples=[0, 1]):
21             for site in tree.sites():
22                 # Only works for infinite sites mutations.
23                 assert len(site.mutations) == 1
24                 mutation = site.mutations[0]
25                 nleaves0 = tree.num_tracked_samples(mutation.node)
26                 nleaves1 = tree.num_samples(mutation.node) - nleaves0
27                 data[rep_index, nleaves0, nleaves1] += 1
28     return data
29
30 truth = get_joint_site_frequency_spectra(
31     run_sims(true_M, num_loci=num_loci, theta=2))
```

The run_sims function returns an iterator with the complete tree sequence and mutational information of each locus. We use the function get_joint_site_frequency_spectra to summarize the polymorphism information as the joint site frequency spectrum (jSFS) of each locus, i.e. the blockwise site frequency spectrum or bSFS [sensu 28]. Note that higher level population genetic summaries, e.g. pairwise measures of divergence and diversity such as D_{XY} [33] and F_{ST} [45] or multi-population F statistics [8, 34] which are often used in ABC inference are just further (and lossy) summaries of the jSFS.

Since msprime simulates rooted trees, the columns and rows of the unfolded jSFS correspond to the frequency of derived mutations in each population and the entries of the jSFS are simply mutation counts. For example, for the first locus we have:

```
1  print(truth[0])
2
3  >>> [[ 0.  1.  4.]
4      [ 5.  0.  7.]
5      [ 0.  0.  0.]]
```

One could base inference on the bSFS [4, 28], but we will for the sake of simplicity use a simpler (and lossy) summary of the data: the average jSFS across loci. For analyses based on SNPs, it is convenient to normalize the jSFS by the total number of mutations:

```
1  truth_mean = np.mean(truth, axis=0)
2  truth_mean /= np.sum(truth_mean)
3  print(truth_mean)
4
5  >>> [[ 0.          0.22099954  0.16139386]
6      [ 0.25630445  0.03255387  0.08482348]
7      [ 0.16093535  0.08298945  0.        ]]
```

To illustrate a simple ABC inference, we will focus on a single parameter of interest, the migration rate M. ABC measures the fit of data simulated under the prior to the observed data via a vector of summary statistics. We will use the jSFS as a summary statistic and approximate the jSFS for each M value as the mean length of genealogical branches across 100 simulation replicates (num_-reps). Below we draw 10,000 M values from the prior and use the functions run_sims and approx_jSFS to approximate the jSFS for replicate. We assume an exponential distribution, a common choice of prior [13].

```
1  num_reps = 100
2  num_prior_draws = 10000
3  prior_M = np.random.exponential(0.1, num_prior_draws)
4
5  def approx_jSFS(m):
6      reps = run_sims(m, num_loci=num_reps)
7      B = np.zeros((num_reps, nsamp + 1, nsamp + 1))
8      for rep_index, ts in enumerate(reps):
9          samp1 = ts.samples(population_id=0)
10         for tree in ts.trees(tracked_samples=samp1):
11             # Note that this will be inefficient if we have
12             # lots of trees. Should use an incremental update
13             # strategy using edge_diffs in this case.
14             for u in tree.nodes():
15                 n1 = tree.num_tracked_samples(u)
16                 n2 = tree.num_samples(u) - n1
17                 if tree.parent(u) != msprime.NULL_NODE:
18                     B[rep_index, n1, n2] += tree.branch_length(u)
19     data = np.mean(B, axis=0)
20     return data / np.sum(data)
21
22 with multiprocessing.Pool() as pool:
23     prior_jSFS = pool.map(approx_jSFS, prior_M)
```

Here we run 100 simulation replicates for each of the 10,000 m values drawn from the prior, giving a total of one million individual simulations. We use the multiprocessing module to distribute these computations over the available CPU cores. Once this has completed, we compute the Euclidean distance between the estimated jSFS for each draw from the prior (prior_jSFS) and the jSFS in the (pseudo)observed data (truth_mean):

```
1  distances = np.zeros(num_prior_draws)
2  for j in range(num_prior_draws):
3      distances[j] = np.sqrt(np.sum((prior_jSFS[j] - truth_mean)**2))
```

In its simplest form, ABC approximates the posterior by sampling from the simulated data via an acceptance threshold. Here we approximate the posterior distribution of m using the 5% of simulation replicates that most closely match the average jSFS of the observed data. Figure 13a shows that the posterior distribution (shown in green) is centered around $m = 0.25$.

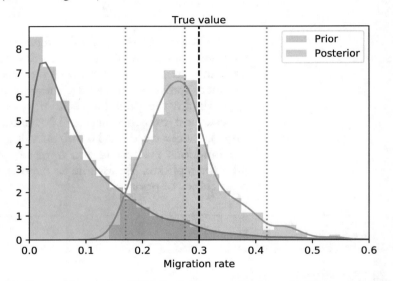

(a) Prior and posterior ABC distributions and estimated 95% approximate credible interval.

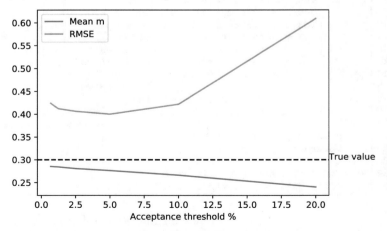

(b) Mean and root-mean-square-error of migration rate estimates computed from pseudo-observed data sets.

Fig. 13 ABC results. (**a**) Prior and posterior ABC distributions and estimated 95% approximate credible interval. (**b**) Mean and root-mean-square-error of migration rate estimates computed from pseudo-observed data sets

```
1  cutoff = np.percentile(distances, 5)
2  keep = np.where(distances < cutoff)
3  post_m = prior_m[keep]
4  mean_m = np.mean(post_m)
5  ci_m = np.percentile(post_m, 2.5), np.percentile(post_m, 95.75)
6  sns.distplot(prior_m, label="Prior")
7  sns.distplot(post_m, label="Posterior")
8  # Plotting code omitted.
```

The mean and the 95% approximate posterior credible interval for *m* are:

```
1  print([mean_m, ci_m])
2
3  >>> [0.22494687232052613, (0.14574598726102159, 0.32315656482448107)]
```

Although the true value of $m = 0.3$ is contained within the 95% credible interval, the posterior distribution is clearly downwardly biased. This bias is in fact expected given that our prior is also strongly biased towards low *m*. We can check the effect the acceptance threshold on the inference and get a sense of the expected information about *m* using a cross-validation procedure: we repeat the inference on pseudo-observed data sets (PODS) simulated under a known truth. Since we can re-use the same set of replicates simulated under the prior for inference, such cross-validation is computationally efficient.

Figure 13b shows the mean and the root mean square error (RMSE) of *m* estimates (across 100 PODS) against the acceptance threshold and confirms that both the downward bias in *m* estimates and the associated RMSE increase with larger acceptance thresholds. While this toy example illustrates the principle of ABC inference, sampling only a small fraction of simulations generated under the prior is clearly computationally inefficient and more efficient sampling strategies for ABC inference have been developed [2]. In practice, we are generally interested in fitting parameter-rich models and it would be straightforward to implement ABC inference for complex model of population structure and demography in msprime.

6   Discussion

In this chapter we have focused on the usage of msprime as a coalescent simulator, and illustrated its flexibility through concrete examples. While many examples discuss how to create and run the simulations themselves, others are concerned with how we analyze the *output* of these simulations. We have shown particularly in Section 3 that these methods can be very efficient, allowing us to easily analyze chromosome scale data for hundreds of thousands of samples. The data structures and APIs used in `msprime` are currently being developed to increase their generality and applicability. Recent work [11, 24] has shown that forward-time simulations can also benefit from these methods. By recording all genealogical information for the simulated population in the form of a succinct tree sequence, we avoid the need to generate and carry forward neutral mutations; by definition, they do not affect the genealogies, and can therefore be placed on them afterwards. Not only does this provide us with much more complete information about the forward-time simulation, it also leads to substantially faster running times (up to $50\times$ faster, in the simulations performed). Through the use of a well-documented interchange API and thoroughly specified data formats, forward-time simulators can output data that is compatible with the `msprime` API, and precisely the same techniques described here can be used to analyze the results. Thus, code written to analyze coalescent simulations can equally be applied to analyze forwards simulations.

There is currently a great deal of activity from a growing community around `msprime`. We plan to separate the tree sequence processing code from the simulator and create a library, provisionally known as `tskit`. This standalone library (C and Python interfaces are planned) will greatly facilitate integration with forwards-time simulators, allowing them to easily offload tree sequence processing to `tskit`. Algorithms for efficiently calculating statistics using the incremental techniques outlined in Section 3.5 are in development, and promise to be significantly more efficient than the state of the art. Also in development are methods to estimate the tree sequence data structure from real data, which would allow us to use these efficient algorithms on observed as well as simulated data. New features are being added to the `msprime` simulator also, with support for a discrete time Wright-Fisher model and a family of multiple-merger coalescent models in development. We hope that in the coming years a diverse ecosystem of tools and applications using these APIs and data structures will emerge.

Online Resources:

Jupyter notebook	https://github.com/StatisticalPopulationGenomics/msprime
Documentation	https://msprime.readthedocs.io/en/stable/
GitHub	https://github.com/tskit-dev/msprime
Mailing list	https://groups.google.com/forum/#!forum/msprime-users

Acknowledgements

We would like to thank Simon Aeschbacher for comments on the ABC inference example, and to thank Yan Wong, Joseph Marcus, and Julien Dutheil for detailed and insightful feedback. JK is supported by Wellcome Trust grant 100956/Z/13/Z to Gil McVean. KL is supported by an Independent Research fellowship from the Natural Environment Research Council (NE/L011522/1).

References

1. Arenas M (2012) Simulation of molecular data under diverse evolutionary scenarios. PLoS Comput Biol 8(5):e1002495
2. Beaumont MA, Zhang W, Balding DJ (2002) Approximate Bayesian computation in population genetics. Genetics 162:2025–2026
3. Becquet C, Przeworski M (2007) A new approach to estimate parameters of speciation models with application to apes. Genome Res 17(10):1505–1519
4. Beeravolu Reddy C, Hickerson MJ, Frantz LAF, Lohse K (2017) Blockwise site frequency spectra for inferring complex population histories and recombination, bioRxiv. https://doi.org/10.1101/077958
5. Carvajal-Rodríguez A (2008) Simulation of genomes: a review. Curr Genomics 9(3):155–159
6. Cornuet JM, Santos F, Beaumont MA, Robert CP, Marin JM, Balding DJ, Guillemaud T, Estoup A (2008) Inferring population history with DIY ABC: a user-friendly approach to approximate Bayesian computation. Bioinformatics 24(23):2713–2719
7. Csilléry K, Blum M, Gaggiotti OE, François O (2010) Approximate Bayesian computation (ABC) in practice. Trends Eco Evol 25(7):410–418
8. Durand EY, Patterson N, Reich D, Slatkin M (2011) Testing for ancient admixture between closely related populations. Mol Biol Evol 28(8):2239–2252
9. Excoffier L, Dupanloup I, Huerta-Sánchez E, Sousa VC, Foll M (2013) Robust demographic inference from genomic and SNP data. PLoS Genet 9(10):e1003905
10. Gutenkunst RN, Hernandez RD, Williamson SH, Bustamante CD (2009) Inferring the joint demographic history of multiple populations from multidimensional SNP frequency data. PLoS Genet 5(10):e1000695
11. Haller BC, Galloway J, Kelleher J, Messer PW, Ralph PL (2018) Tree-sequence recording in SLiM opens new horizons for forward-time simulation of whole genomes, bioRxiv. https://doi.org/10.1101/407783. https://www.biorxiv.org/content/early/2018/09/04/407783
12. Harris K, Nielsen R (2013) Inferring demographic history from a spectrum of shared haplotype lengths. PLoS Genet 9(6):e1003521
13. Hey J, Nielsen R (2004) Multilocus methods for estimating population sizes, migration rates and divergence time, with applications to the divergence of Drosophila pseudoobscura and D. persimilis. Genetics 167(2):747–760
14. Hoban S, Bertorelle G, Gaggiotti OE (2012) Computer simulations: tools for population and evolutionary genetics. Nat Rev Genet 13(2):110
15. Hudson RR (1983) Testing the constant-rate neutral allele model with protein sequence data. Evolution 37(1):203–217
16. Hudson RR (1990) Gene genealogies and the coalescent process. Oxf Surv Evol Biol 7:1–44
17. Hudson RR (2002) Generating samples under a Wright-Fisher neutral model of genetic variation. Bioinformatics 18(2):337–338
18. Hunter JD (2007) Matplotlib: a 2d graphics environment. Comput Sci Eng 9(3):90–95
19. International HapMap Consortium (2003) The international HapMap project. Nature 426(6968):789
20. Jones E, Oliphant T, Peterson P, et al (2018) SciPy: open source scientific tools for Python (2001–2018). http://www.scipy.org/ [Online; Accessed 30 Jan 2018]
21. Kelleher J, Barton NH, Etheridge AM (2013) Coalescent simulation in continuous space. Bioinformatics 29(7):955–956
22. Kelleher J, Etheridge A, Barton N (2014) Coalescent simulation in continuous space: algorithms for large neighbourhood size. Theor Popul Biol 95:13–23
23. Kelleher J, Etheridge AM, McVean G (2016) Efficient coalescent simulation and

genealogical analysis for large sample sizes. PLoS Comput Biol 12(5):e1004842

24. Kelleher J, Thornton K, Ashander J, Ralph P (2018) Efficient pedigree recording for fast population genetics simulation. PLoS Comput Biol 14(11):e1006581

25. Kingman JFC (1982) The coalescent. Stoch Processes Appl 13(3):235–248

26. Li H, Durbin R (2011) Inference of human population history from individual whole-genome sequences. Nature 475:493–496

27. Liu Y, Athanasiadis G, Weale ME (2008) A survey of genetic simulation software for population and epidemiological studies. Hum Genomics 3(1):79

28. Lohse K, Chmelik M, Martin SH, Barton NH (2016) Efficient strategies for calculating blockwise likelihoods under the coalescent. Genetics 202(2):775–786

29. Martin AR, Gignoux CR, Walters RK, Wojcik GL, Neale BM, Gravel S, Daly MJ, Bustamante CD, Kenny EE (2017) Human demographic history impacts genetic risk prediction across diverse populations. Am J Hum Genet 100 (4):635–649

30. McKinney W, et al (2010) Data structures for statistical computing in python. In: Proceedings of the 9th Python in science conference, Austin, TX, vol 445, pp 51–56

31. McVean GAT, Cardin NJ (2005) Approximating the coalescent with recombination. Philos Trans R Soc Lond B Biol Sci 360 (1459):1387–1393

32. Miles A, Harding N (2017) scikit-allel. https://doi.org/10.5281/zenodo.822784

33. Nei M (1972) Genetic distance between populations. Am Nat 106(949):283–292

34. Patterson N, Moorjani P, Luo Y, Mallick S, Rohland N, Zhan Y, Genschoreck T, Webster T, Reich D (2012) Ancient admixture in human history. Genetics 192(3): 1065–1093

35. Pérez F, Granger BE (2007) Ipython: a system for interactive scientific computing. Comput Sci Eng 9(3):21–29

36. Rasmussen MD, Hubisz MJ, Gronau I, Siepel A (2014) Genome-wide inference of ancestral recombination graphs. PLoS Genet 10(5): e1004342

37. Schiffels S, Durbin R (2014) Inferring human population size and separation history from multiple genome sequences. Nat Genet 46:919–925

38. Sousa VC, Grelaud A, Hey J (2011) On the nonidentifiability of migration time estimates in isolation with migration models. Mol Ecol 20(19):3956–3962

39. Staab PR, Zhu S, Metzler D, Lunter G (2014) scrm: efficiently simulating long sequences using the approximated coalescent with recombination. Bioinformatics 31(10):1680–1682

40. Tajima F (1983) Evolutionary relationship of DNA sequences in finite populations. Genetics 105(2):437–460

41. Thornton K (2003) Libsequence: a C++ class library for evolutionary genetic analysis. Bioinformatics (Oxf, Engl) 19(17):2325–2327

42. van der Walt S, Colbert SC, Varoquaux G (2011) The NumPy array: a structure for efficient numerical computation. Comput Sci Eng 13(2):22–30

43. Wakeley J (2008) Coalescent theory: an introduction. Roberts and Company, Englewood

44. Waskom M, Botvinnik O, O'Kane D, Hobson P, Lukauskas S, Gemperline DC, Augspurger T, Halchenko Y, Cole JB, Warmenhoven J, de Ruiter J, Pye C, Hoyer S, Vanderplas J, Villalba S, Kunter G, Quintero E, Bachant P, Martin M, Meyer K, Miles A, Ram Y, Yarkoni T, Williams ML, Evans C, Fitzgerald C, Brian, Fonnesbeck C, Lee A, Qalieh A (2017) mwaskom/seaborn: v0.8.1 (September 2017). https://doi.org/10. 5281/zenodo.883859

45. Wright S (1950) Genetical structure of populations. Nature 166:247–249

46. Yuan X, Miller DJ, Zhang J, Herrington D, Wang Y (2012) An overview of population genetic data simulation. J Comput Biol 19 (1):42–54

Chapter 10

Inference of Ancestral Recombination Graphs Using ARGweaver

Melissa Hubisz and Adam Siepel

Abstract

This chapter describes the usage of the program ARGweaver, which estimates the ancestral recombination graph for as many as about 100 genome sequences. The ancestral recombination graph is a detailed description of the coalescence and recombination events that define the relationships among the sampled sequences. This rich description is useful for a wide variety of population genetic analyses. We describe the preparation of data and major considerations for running ARGweaver, as well as the interpretation of results. We then demonstrate an analysis using the *DARC* (*Duffy*) gene as an example, and show how ARGweaver can be used to detect signatures of natural selection and Neandertal introgression, as well as to estimate the dates of mutation events. This chapter provides sufficient detail to get a new user up and running with this complex but powerful analysis tool.

Key words Ancestral recombination graph, Sequentially Markov coalescent, Markov chain Monte Carlo, Local ancestry

1 Overview

The ancestral recombination graph (ARG) can be considered the holy grail of statistical population genetics. The ARG represents the history of a collection of related genome sequences, in terms of the *coalescence* events by which segments of genomes trace to common ancestral segments and the historical *recombination* events that cause patterns of ancestry to differ from one genomic site to the next (*see* Chapter 1 for more introduction to these concepts). Provided the sequences under study are orthologous and co-linear—meaning that they trace to a common ancestral sequence without genomic duplications or rearrangements—the ARG is a

The original version of this chapter was revised. The correction to this chapter is available at https://doi.org/ 10.1007/978-1-0716-0199-0_20

Electronic supplementary material: The online version of this chapter (https://doi.org/10.1007/978-1-0716-0199-0_10) contains supplementary material, which is available to authorized users.

Julien Y. Dutheil (ed.), *Statistical Population Genomics*, Methods in Molecular Biology, vol. 2090, https://doi.org/10.1007/978-1-0716-0199-0_10, © The Author(s) 2020, Corrected Publication 2021

complete description of their evolutionary relationships. Moreover, in statistical terms, the ARG provides a highly compact and precise description of the correlation structure of such a collection of sequences. Importantly, the ARG naturally defines a set of recombination breakpoints, a set of haplotypes, and a genealogy for each non-recombining interval in the genome—all objects that are useful starting points for countless population genetic analyses.

Many questions in applied population genetics can be reframed as questions about ARG structure. For example:

- *Recombination rate estimation.* Recombination rates can be estimated by simply counting recombination events and dividing by the total branch-length of the ARG.

- *Estimation of allele ages or mutation rates.* Mutation events can easily be mapped to branches within the ARG by maximum parsimony, enabling straightforward estimation of allele ages and mutation rates.

- *Local ancestry inference.* The local ancestry structure of an admixed individual (i.e., which genomic segments derive from which distinct source populations) can be determined by tracing the individual's two diploid lineages in the ARG and identifying the source population with which each genomic segment clusters, as well as the recombination events that terminate these segments.

- *Demography inference.* More general information about demographic history (such as population sizes, migration rates, and divergence times) is also embedded in the ARG. A demographic model can fairly easily be estimated from a known ARG by making use of the counts of coalescence events within and between populations.

- *Detection of sequences under selection.* Natural selection can be detected by identifying local distortions in the ARG, for example, unusual clusters of coalescence events or extremely deep times to most recent common ancestry.

In practice, however, the true ARG is impossible to know with certainty. The "ARG space," consisting of every possible ancestral history of a set of genomes, is astronomically large, and the information in genome sequences is insufficient to choose a specific ARG above all others. But, given a model of coalescence, recombination, and nucleotide substitution, it is possible to compute the probability of an observed data set under particular ARGs, and it will generally be true that some ARGs are much more likely to have produced the data than others. The approach taken by ARGweaver is to sample from the posterior distribution of ARGs, given a collection of genome sequence data and a reasonable set of modeling assumptions. This approach is computationally expensive, and it has the drawback of producing a complex and unwieldy output—a collection of potential ARGs, none of which is exactly correct, but which, in the aggregate, reflect certain properties of the true ARG. Nevertheless, as we will show, this approach can be extremely powerful, potentially providing insights into the structure of the data and the evolutionary history of the sample that are not easily obtained using simpler methods. In this chapter we will discuss how ARGweaver works, how and when a user might want to apply it, and what can be done with sampled ARGs once they have been obtained.

1.1 What Is an ARG?   An ARG represents all ancestral relationships among a collection of genomes (*see* Fig. 1). If n is the number of (haploid) genomes under study (usually from $\frac{n}{2}$ diploid individuals), then at the present day, there are n lineages in the ARG. As we trace these lineages back in time at a particular genomic location, we will find that distinct lineages gradually *coalesce* into shared ancestral lineages, until all

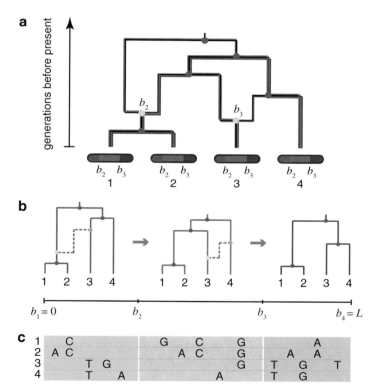

Fig. 1 (a) Schematic of an ARG with four lineages in the present, and two ancestral recombination events along a region of length *L*. Tracing the lineages upwards from present day, two lineages merge when a coalescence event is encountered, whereas a lineage splits into two when a recombination event is encountered at a particular breakpoint (b_2 or b_3). The ARG continues tracing the history backwards until all lineages have reached a common ancestor. (b) An alternative view of the ARG depicted in A, showing the local tree between each pair of recombination breakpoints. The dotted lines on the tree show the recombination event which transforms the tree on the left side of the breakpoint into the tree on the right side. (c) The data underlying this ARG, where only derived alleles at variant sites are shown. Figure adopted from [23]

n lineages have found a single most recent common ancestor. These coalescence events define a tree known as a *genealogy* that fully describes the evolutionary relationships among the present-day genomes at the locus in question.

However, *recombination* events in the history of the sample can cause the genealogy to change from one genomic location to the next. Looking backward in time, a recombination at a particular genomic location has the effect of splitting a lineage into two, with one path representing the evolutionary history to one side of the breakpoint and another path representing the history to the other side. The ARG captures these recombination events together with the coalescence events. As one follows a lineage upward in the ARG, that lineage may either merge with another lineage, representing a coalescence event, or it may split into two lineages, representing a recombination event (Fig. 1a). In the case of recombination events, the junction in the ARG is also labeled with the genomic position of the recombination (this information is not relevant for coalescence events).

Based on these labels for recombination events, one can extract a local tree for any position in the genome from the ARG. First, one identifies the lineage associated with each present-day sample. These lineages are traced backward through the ARG, and coalescences between them are noted. When a recombination event is identified, one of the two possible paths is selected based on the relationship of the position in question to the annotated recombination breakpoint. Specifically, if the position is to the left of the breakpoint, then the left path is taken; and if the position is to the right of the breakpoint, then the right path is taken. (Because recombination breakpoints by definition occur between nucleotides, one of these two cases must hold.) Thus, the paths from the present-day samples to the root will coalesce only, never splitting, and therefore must define a tree. Furthermore, the tree will be the same for all genomic positions between two recombination breakpoints, differing only between positions on opposite sites of a breakpoint.

Another way to think about the ARG is that it defines a series of operations on trees along the length of a chromosome. As one walks along a chromosome from left to right, the local tree remains fixed until a recombination breakpoint is encountered, and then that tree is altered to form a new tree, in the specific manner defined by the change in path at the corresponding recombination node in the ARG (Fig. 1b). The ARG, therefore, can be thought of as being interchangeable with a sequence of local trees and the associated recombination events that transform each tree to the next. In practice, this is the representation of the ARG assumed by the Sequentially Markov Coalescent (SMC') and used by ARGweaver, and in this chapter we will generally treat the ARG as a collection of trees and recombination events. Nevertheless, it should be noted that this representation does not strictly capture all of the information in the ARG. The full ARG also describes "trapped genetic material" that falls between two linked ancestral loci, but is not passed on to any present-day sample. Ignoring this trapped material substantially simplifies modeling and inference algorithms, with what appear to be only minor costs in accuracy [12, 14, 23].

1.2 Why Would You Want to Estimate an ARG?

As discussed above, if the ARG could be estimated accurately and easily, it would be useful for almost every question in population genetics. In practice, of course, there are limitations in the accuracy of inferred ARGs, and they require substantial time and effort to obtain. So, when does it make sense to take the trouble to run ARGweaver, instead of making use of simpler or more standard population genetic summary statistics and tools? Some reasons to consider sampling ARGs with ARGweaver include:

- *Trees/genealogies.* ARGweaver estimates explicit genealogies (with branch lengths) along the genome, considering both patterns of local mutation and local linkage disequilibrium. It may be particularly interesting to inspect trees at particular regions suspected to be under selection or to have experienced introgression.

- *Times/dates.* These trees allow the timings of various events to be estimated, including times to most recent common ancestry, other coalescence times, and the ages of derived alleles. If desired, posterior expected values of these times can be computed by averaging over the sampled trees.

- *Ancient introgression.* ARGweaver is a powerful method for detecting introgression and identifying specific introgressed haplotypes, particularly ancient introgression events that conventional methods may miss (e.g., [6]).

- *Bayesian treatment of uncertainty.* Unlike many simpler methods, ARGweaver attempts to fully account for the uncertainty in the ARG given the sequence data and an evolutionary model, by sampling from a posterior distribution of ARGs. This approach can mitigate biases from the inference method in addressing biological questions of interest.

- *Flexibility in addressing "custom" evolutionary questions.* By producing explicit ARGs, ARGweaver allows almost any evolutionary question to be addressed, including unusual ones not easily addressed with standard summary statistics (For example: at what fraction of sites do individuals *A* and *B* coalesce with one another before either coalesces with individual *C*? What is the average TMRCA for genes of functional category *X*? Are recombination events more likely to occur in introns or intergenic regions?)

- *Technical limitations of the data.* ARGweaver can accommodate unphased data, low-coverage sequences, archaic samples, and other unusual data types that may not be easy to analyze using other methods.

1.3 Practical Considerations

ARGweaver is designed to run on genome sequencing data for small to moderate numbers of individuals—anywhere from 2 to a maximum of about 100. These individuals should be unrelated but come from the same species or from recently diverged species (such as humans and chimpanzees). Phasing of diploid genome sequences is not necessary—ARGweaver can phase "on the fly," integrating over possible phasings—but the algorithm converges faster and, in some cases, performs better on phased data (depending on the rate of phasing errors). Similarly, ARGweaver can be used on low-coverage sequencing data, making use of genotype probabilities to weight the observed bases, but high-coverage sequence data is always preferable.

In gauging the feasibility of ARG inference, it is important to recognize that the processes of mutation and recombination are opposing forces in reconstructing an ARG. The more mutations there are, the more information there is to guide the inference of tree topologies (genealogies). Recombination events, however, break up the sequences into smaller blocks, effectively limiting the information for tree inference in each block. Thus, the quality of ARG inference depends on the ratio of mutation to recombination rates per nucleotide position. In human data, this ratio is close to one, but recombination events tend to be concentrated in recombination hotspots, which makes the effective ratio greater than one for most of the genome. ARGweaver appears to work quite well in this setting. Nevertheless, the method works better when this ratio is even higher, and it will break down if this ratio falls significantly below one. Another consideration is ARGweaver's assumption of at most one recombination event per site (see below), which generally appears to have little effect but could lead to biased estimates in cases of particularly high recombination rates, large sample sizes, large evolutionary distances, or large effective population sizes. Finally, because ARGweaver depends on haplotype-scale information for inference, it is generally not useful for short sequences, deriving, for example, from RAD-seq or a de novo short-read assembly.

In terms of the number of genomes analyzed, the "sweet spot" for ARGweaver is generally between a handful of individuals and a few dozen. As the number of genomes increases, more approximate models (such as the Li and Stevens model [8]) or conventional population genetic summary statistics become increasingly accurate and informative, and the relative advantage of using ARGweaver over other methods decreases. In addition, the run time and size of the ARGweaver output increase with the number of genomes, and these factors become prohibitive with more than about 100 samples. Running ARGweaver genome-wide generally requires breaking the genome into chunks of a few megabases and running ARGweaver in parallel on each chunk using a computer cluster. When running ARGweaver genome-wide is not a realistic possibility, it may still be of interest to apply ARGweaver to specific genomic regions of interest, such as candidate selective sweeps or introgressed regions. It may also be useful to run ARGweaver on subsets of the available genome sequences, for example, to shed light on genealogy structure, ancient introgression, or allele age— features ARGweaver may estimate more accurately than other methods.

Another practical consideration is that while ARGweaver's output is richly informative, it is not straightforward to interpret. The program does come with tools to compute various local summary statistics from sampled ARGs, including times to the most recent common ancestor, allele ages, and distances between samples. But many less standard analyses will require custom programs to extract the desired information from ARGs or local genealogies.

1.4 ARGweaver Algorithm Overview

ARGweaver uses a Markov chain Monte Carlo (MCMC) algorithm to sample ARGs at frequencies proportional to their probability, conditional on the observed DNA sequence data (X) and the model parameters (θ). The MCMC algorithm starts with an initial ARG, G^0, and then repeatedly removes a subset of the ARG and resamples that subset from an appropriate conditional probability distribution. This process generates a sequence of ARGs, G^0, G^1, ..., G^m, where m is the total number of iterations of the algorithm. Although G^0 may be a poor guess with low probability, by sampling each new G^i according to the appropriate distribution, the chain will eventually converge to the desired distribution—i.e., for sufficiently large i, G^i will represent a draw from the posterior distribution over ARGs given the data and the model, $P(G^i|X, \theta)$. In practice, it is customary to plot the posterior probability as a function of the iteration number, i, observe the point at which it ceases to trend upward and becomes stable, and then to discard the ARGs sampled before this point (from what is known as the "burn-in" of the MCMC algorithm).

Even once the algorithm has converged, successive samples G^i and G^{i+1}—while they both represent samples from the posterior distribution—are not *independent* samples. Rather they are strongly correlated, since only part of the ARG is resampled on each step of the algorithm. Therefore, in order to achieve a distribution of nearly independent ARGs—both to save space and processing time, and to better assess the variance of estimates derived from the samples—it is useful to "thin" the chain, recording only every jth sample (the default thinning parameter in ARGweaver is

$j = 10$). After discarding the initial "burn-in" and performing thinning, the ARGs G^i that remain can be stored and treated as a collection of samples representative of the distribution of ARGs given the data and the model, $P(G|X, \theta)$.

The technical details of the ARGweaver algorithm will not be reviewed here (*see* ref. 23), but the main idea is to remove a single haploid genome from the ARG, and then to "thread" this genome back through the ARG, by sampling both its coalescence points with the remaining sequences and the associated recombination points. There is also another, slightly more complicated, version of this threading operation, called "subtree threading," that resamples internal branches in genealogies, and is essential for ARGweaver to efficiently explore the full space of possible ARGs. In both cases, a hidden Markov model (HMM) is used to efficiently sample new coalescent points for the new lineage across the chromosome. This HMM depends on several key modeling assumptions, which are important for users to understand, and which, therefore, will be reviewed in the next section.

1.4.1 ARGweaver Model and Assumptions

The HMM underlying ARGweaver depends on the following assumptions:

- *SMC' or SMC*: The Sequentially Markov Coalescent model [14] or the closely related SMC' [12] is assumed. These models posit that the distribution over genealogies at each nucleotide position directly depends only on the genealogy at the previous position, not on the genealogies at positions further upstream—a feature known in probability theory as the *Markov property*, after the Russian mathematician Andrey Markov. More formally, the SMC and SMC' assume that the genealogy at position $i + 1$ is independent of the genealogies at positions $1, \ldots, i - 1$, given the genealogy at position i. The SMC' slightly improves on the original SMC (*see* ref. 12 for details). The differences between these models are not important here, and the choice of model seems to have only a subtle effect on the inferred ARGs. While the SMC' is technically more accurate, the SMC model may be considerably faster on data sets with large numbers of samples. ARGweaver therefore allows the user to choose either model (SMC by default, `--smc-prime` for the SMC').

- *Discrete time*: All recombination and coalescent events are assumed to occur at a predefined collection of discrete time points. The total number of time points, K, can be chosen by the user (using `--ntimes <K>`) and can be arbitrarily large, with the ARGweaver model approaching a continuous-time model as K approaches infinity. However, the computational complexity of the threading algorithm is proportional to K^2, so, in practice, K must be kept modest in size. The default value of K in ARGweaver is 20. The time points are uniformly spaced on a logarithmic scale, so that they are more closely clustered at recent time points, when there are more lineages and coalescence rates are larger. The algorithm forces all lineages to coalesce by the final time point, t_K.

- *No more than one recombination event between neighboring nucleotides.* For simplicity, the algorithm permits at most one recombination event at every "step" along the sequence, meaning between two adjacent nucleotide positions. This assumption means that adjacent genomic positions must either have identical genealogies or ones that differ by a single recombination event. In practice, this assumption is minimally restrictive, because the information about genealogies comes primarily from variable sites, which tend to be sparse along the genome. If ARGweaver should need to account for multiple recombination events between variable sites, it typically can spread those events across a series of intervening invariant sites with minimal impact on accuracy. If the data are such that multiple recombinations between neighboring sites occur frequently, then it is likely that the haplotype structure is too broken down to make use of ARGweaver.

- *Population size known*: ARGweaver assumes that the effective population size N_e (which determines the coalescence rate) is provided by the user. In the simplest case, a single global value of N_e can be provided. But ARGweaver can accommodate different values of N_e for different discrete time intervals. Values of N_e can typically be obtained from the literature or estimated from the same data using one of the many available programs for inferring demographic histories (such as SMC++ [29], PSMC [7], MSMC [27, see also Chapter 7], G-PhoCS [3], and diCal [28]). Note the user-provided values of N_e define a "prior" for coalescence rates in ARGweaver, so it is not necessary for them to be perfectly estimated; ARGweaver will consider the data together with this prior distribution in sampling coalescence events.

- *Mutation and recombination rates known.* The ARGweaver model also depends on pre-defined mutation and recombination rates. These rates can be assumed to be constant across the genome, or variable rates can be provided in a position-specific map along the genome. These values are also "priors" in the same sense as the population size (see above).

- *Jukes-Cantor model of base substitution.* ARGweaver makes use of a Jukes-Cantor model for nucleotide substitutions. This model assumes that all nucleotide substitutions are equally probable—an obvious oversimplification, but one that seems to have minimal costs at the close evolutionary distances typically considered by ARGweaver. The symmetries inherent in the Jukes-Cantor model can be exploited to optimize the likelihood calculations in ARGweaver.

2 Ancient Hominins Analysis

In the remainder of this chapter we will set up, and then walk through, an analysis of real sequence data using ARGweaver. We will use three high-quality ancient hominin genome sequences that are freely available: the Altai Neandertal [21], Vindija Neandertal [22], and Denisovan [15] genome sequences, as well as a diverse set of 14 human genomes that were sequenced to high coverage for the Altai Neandertal paper [21]. All steps of the analysis will be described in detail, with code snippets, and the complete set of commands required to replicate the analysis is provided in the companion material for this book.

The data set used in our example is ideal for several reasons. First, the Neandertal and Denisovan genome sequences provide an exciting opportunity to examine many interesting aspects of human history and adaptation. Neandertals and Denisovans are sister groups of archaic hominins that diverged from humans roughly 600 kya, and then split from each other around 400 kya [22]. Importantly, these divergence times are recent enough that modern and archaic humans share many polymorphic sites across their genomes. As we will show, this shared variation can be informative about evolutionary history. In addition, the genetic evidence strongly suggests multiple cases of interbreeding among these three groups following their initial divergence [15, 21, 24], an intriguing topic that can be examined using ARGweaver. On a more practical level, these genomes have all been sequenced to high coverage and all sequencing reads have been processed consistently. The genotypes are published in standard VCF format with genotype quality and sequencing depth information given at every genomic position with aligned reads, which, as we will discuss, simplifies the set up for ARGweaver.

Before launching into the actual analysis, which is presented in Subheading 5, we will discuss some important preliminaries relating to program installation and file formats (remainder of Subheading 2), model parameters (Subheading 3), and commonly used program options (Subheading 4).

2.1 Pre-requisites

ARGweaver is designed to run under either the Linux or Mac OSX operating system. Windows(c) users may run ARGweaver via a Linux virtual machine. The specific commands for performing the example analysis in this chapter are available as a bash script, provided in the online supplement for this book. Besides a bash shell, the script requires:

- *Python*: (http://python.org), used by several ARGweaver scripts.
- *SAMtools* [9]: (http://htslib.org), used here for the `tabix` and `bgzip` tools, which are useful for indexing and fast retrieval of VCF and BED files.
- *bedops* [18]: (http://bedops.readthedocs.io), a useful tool for computing intersections of genomic intervals.
- *PHAST* [20]: (http://compgen.cshl.edu/phast), used for computing neutral substitution rates.
- *R* (https://www.r-project.org), used for plotting results.
- *The R package "ape"* [19] (https://bioconductor.org), used for plotting trees.
- *git* (https://git-scm.com) for downloading ARGweaver.
- *g++* (https://gcc.gnu.org) or any C++ compiler for compiling ARGweaver.

2.2 Obtaining and Installing ARGweaver

The first step in our example is to download and install ARGweaver. The program is available at http://github.com/CshlSiepelLab/ARGweaver.git. It can be downloaded and compiled on Linux or Mac machines with the following commands:

```
git clone https://github.com/CshlSiepelLab/ARGweaver.git
cd ARGweaver.git
make
```

These commands create several executables in the `bin/` directory, the most important of which is called `arg-sample`. All the executables are meant to be run from the command line in a Unix shell such as Bash.

Within the ARGweaver software is also suite of R tools useful for plotting ARGweaver results. This package is optional, but was used to create many of the plots in this chapter. It can be installed from the same directory with the command:

```
R CMD INSTALL R/argweaver
```

2.3 Sequence File Format

The main data required by ARGweaver is sequence data for every individual. ARGweaver accepts Variant Call Format (VCF) files [2], provided that they are indexed (this can be done with the command `tabix -p vcf file.vcf.gz`, which creates a file `file.vcf.gz.tbi`). A single VCF file containing all samples may be provided with the argument `--vcf`; if only a subset of the individuals are to be used, they can be specified with the `--subsites` option. Or, if the genotypes are in multiple VCF files, a list of these files can be given with the option --vcf-files. In our example, there is a single VCF file for each individual, so we use the second option.

It is worth noting that VCF input has some limitations when used with ARGweaver. Namely, ARGweaver will ignore any phasing information in the VCF, and sites with insertion/deletion polymorphisms or more than two alleles. If the data is phased, the SITES format will have to be used (see below).

2.4   SITES Format

ARGweaver has its own sequence data format, called SITES format, which is used both as an alternative input format, as well as an output format for sampling phase. All lines in SITES files are tab-delimited. The first line starts with the string "NAMES" and then lists the name of every haploid genome (two per individual). The second line starts with the string "REGION" and is followed by the chromosome name, start position, and end position. All subsequent lines contain two columns: the position of a variant site and a string giving the observed alleles at this site in each of the genomes, in the order given on the first line. Importantly, any position not listed in the file is considered invariant across samples. Here is a short example of a sites file with the two Neandertal individuals and two variant sites on chromosome 2:

```
NAMES    Altai_1 Altai_2 Vindija_1    Vindija_2
REGION   2       24000001       24010000
24000417         GGCG
24008883    .    TTTA
```

2.4.1   Phasing Options

For an ARG to be fully defined for diploid organisms, the genotypes at heterozygous positions must be "phased" into two distinct haploid genome sequences. The low-cost, short-read sequencing methods most widely in use, however, generally produce unphased data, in which the chromosomal origin of each allele at a heterozygous site is unknown. Thus, an important consideration in running ARGweaver is how to address phasing.

ARGweaver can either accept predefined haplotype phases or it can treat the phase as unknown and sample possible phasings as it samples ARGs. The program's default behavior is to assume haploid genome sequences are fully specified (phased input). The option --unphased causes ARGweaver to sample the phase instead. In unphased mode, whenever ARGweaver re-threads a leaf branch, it does so by integrating over possible phasings of the individual corresponding to that leaf. After the threading is complete, it resamples the phase for this individual conditional on the threading choice. This new phase is retained until the next time a leaf for the same individual is re-threaded. The phase sampling step is fast and does not contribute significantly to the run time of each sampling iteration, but it may delay overall convergence of the algorithm.

As noted above, the program does not read phase information from VCF files, so the `--unphased` option is implicit when using VCF input. In this case, the program internally creates two haploid lineages, `<ind>_1` and `<ind>_2`, for each diploid individual, `<ind>`, listed in the VCF file. (Throughout this chapter, we assume diploid input to ARGweaver; in principle, ARGweaver could be used with haploid input, but the assumed model of recombination is best matched to sexually reproducing species.) These haploid labels will be used in some of the output files of ARGweaver. With SITES input and the `--unphased` option, the program requires the user to indicate the haploid pairs corresponding to each individual. They will be detected automatically if the genomes are named with the convention `<ind>_1` and `<ind>_2`. Otherwise, the option `--unphased-file <file.txt>` may be used, where `<file.txt>` has two columns with haploid sample names, and each row corresponds to a single individual.

In unphased mode, ARGweaver will output the current phased data (in SITES.gz format) every time an ARG is sampled. This explicit phasing information makes it possible to map mutations onto branches of the sampled ARGs, among other features. It has the additional benefit of allowing ARGweaver to function as an ARG-based computational phasing method. In practice, however, other existing phasing methods are more efficient, and are likely to achieve greater accuracy, especially if they are able to leverage large reference panels of phased genomes [10].

Nevertheless, even when the data has been pre-phased by another computational method, it may still be worthwhile to use the `--unphased` option. The reason is that the error rates from computational phasing methods can be quite high, and in unphased mode ARGweaver may be able to "correct" phasing choices that are incompatible with the ARGs it samples. In this case, the pre-phased data can still be passed to ARGweaver in SITES format and will be used for initialization, so convergence will be much faster than with unphased input.

2.5 Masked Regions

Whether using VCF or SITES input, ARGweaver assumes that any site which does not appear in the input file is invariant. An "invariant" site in ARGweaver is one in which all individuals have been sequenced and are confidently called homozygous for the same allele (usually the reference allele). This absence of genetic variation is informative about the ARG. For example, a genomic region with many invariant sites will tend to have short branches in the inferred ARG, because fewer mutations are expected on shorter branches.

On the other hand, some sites have unknown genotypes, for example, due to low sequencing depth, poor sequence quality, or poor alignability. For ARGweaver, an unknown genotype means something quite different from an invariant site—it suggests missing data, meaning that no information is provided about the ARG at that genomic location. At a region with many unknown sites, the sampled trees will tend to reflect the neutral coalescent model provided to ARGweaver, because there will be little data available to override this "prior."

Therefore, it is essential to distinguish between "invariant" and "unknown" sites in the input files to ARGweaver. Making this distinction often requires some additional effort, because other population genetic methods often do not depend strongly on it, and many data sets are not processed in a way that tracks this difference. In particular, genotypes that are unknown should be "masked," either by using the genotype NN or by using masking options (described below), so that ARGweaver knows to integrate over all possible genotypes at those positions. Unmasked sites that are not specified in the input can then reasonably be assumed to be invariant.

ARGweaver supports several kinds of masking. For regions with poor alignability (such as repeat regions), it is customary to mask out the entire region across all individuals. This can be done by providing ARGweaver with a BED-formatted file indicating the regions to mask, and using the option --maskmap <mask_file.bed>. (Note that unlike VCF and SITES files, BED files have zero-based start coordinates.) On the other hand, some regions have poor genotype quality only in particular individuals, for example, due to low sequencing coverage or read quality. In these cases, the regions can be delineated in an auxiliary file which is specified using the option --ind-maskmap <ind_mask_file.txt>. This file should have two columns, giving the name of each individual and the name of a file containing a BED-formatted mask specific to that individual. Additional options include --mask-cluster <a,b>, which will mask any region of length b that has $\geq$ a variant sites (possibly indicating alignment errors or mutational hotspots); --vcf-min-qual <Q>, which will mask any genotype with quality less than Q (for VCF files with quality scores); or --vcf-genotype-filter <filter>, which can mask genotypes based on any keys used in the VCF genotype field. For example, --vcf-genotype-filter ˜DP<10;DP>50;GQ<10˜ will mask sites where the depth is less than 10 or greater than 50, or where the genotype quality is less than 10.

In our example analysis, we use a union of several mappability and uniqueness filters developed for the ENCODE project [30]. Details for where these filters were obtained can be found in the online resource. We also use --vcf-min-qual 30 --mask-cluster 2,5. These filters seem sufficient for our illustration, however other filters may be needed for a thorough, careful analysis. In particular, it may be important to mask CpG sites, which have unusually high mutation rates.

2.5.1 Genomic vs Variant VCFs

Many population genomic data sets are now available in VCF format, and it may be tempting to run ARGweaver directly on such a data set. However, VCF is a common file type used for a wide variety of purposes, and it is critical to understand what criteria were used for inclusion or non-inclusion of sites in the files before analyzing them. In particular, VCF files often contain only those positions where high-confidence variants are detected. As discussed above, this convention will make it impossible to distinguish unknown and invariant sites. Thus, more preparation will be necessary for a proper analysis with ARGweaver.

Further processing of such incomplete VCF files generally requires returning to the alignments from which the VCF files were derived. If those alignments are available in the form of a BAM file, then the necessary information can be extracted in a fairly straightforward manner. If they are not available, it may be necessary to regenerate them from the raw reads. Once a BAM file is in hand, the best option is usually to re-run a genotype caller (such as GATK [31]) on the BAM file to generate more complete VCF files including most likely genotypes at every site, as well as quality scores, sequencing depth, and genotype probabilities. A possible shortcut, adequate for many purposes, is to use bamtools [1] to extract the sequencing depth per site from the BAM file, and use the depth as a proxy for genotype quality. For example, if a position is not in the VCF file but has a sequencing depth greater than some cutoff (perhaps 20), then it is very likely invariant. In practice, this thresholding can be accomplished by first creating a BED file for each individual containing the regions with sequencing depth below the desired cutoff, and then using the `--ind-maskmap` option to mask these regions.

In our example, we are lucky to be using VCF files that provide genotype probabilities and confidence scores at every location having aligned reads. However, care must still be taken to deal with these files correctly. ARGweaver assumes that any site absent from a VCF file is invariant, but in this case those sites are actually unknown. To correct this assumption, we create a BED file containing all the regions absent from the VCF for each individual, and use the `--ind-maskmap` option to specify that these regions should be masked in their respective genomes. This is done with the bedops tool [18], and the commands are shown in the example script that comes with this chapter.

2.5.2 Genotype Probabilities

If a genome has only been sequenced at low coverage, there may be too many errors in the genotypes to produce meaningful ARGs. Nevertheless, it is possible to run ARGweaver on low-quality data by having it weight possible genotypes by their probabilities of being correct. There are two ways to specify these probabilities. First, if VCF files are annotated with PL (phred-likelihood) or GL (genotype-likelihood) fields, then the option `--use-genotype-`

`probs` may be used to integrate over the possible genotypes. Second, genotype probabilities can be encoded directly into the SITES file. In this case, each row (following the header) may have an additional $4n$ columns, in the order $p_{1,A}, p_{1,C}, p_{1,G}, p_{1,T}, p_{2,A}, \ldots, p_{n,T}$, where $p_{i,b}$ is the probability of the ith haploid genome having base b. If these columns are present, then `--use-genotype-probs` is implied.

The use of genotype probabilities will slow down ARGweaver, and therefore this feature should only be used if absolutely necessary. There is a modest computational cost, of course, in taking the genotype probabilities into account. The larger issue, however, is that the use of genotype probabilities causes almost every site to be considered "variant," which prohibits the use of site compression (*see* Subheading 4.3). Nevertheless, genotype probabilities may be useful for low-coverage data when the scale of the analysis is not too large.

3   Choosing Model Parameters

ARGweaver assumes fixed rates of mutation, recombination, and coalescence (based on population sizes), which must be specified by the user. As noted above, these parameters can be thought of as defining a "prior" distribution for ARGs, which can be overcome by consideration of the data in determining the "posterior" samples produced by the program. However, these prior estimates can have an appreciable influence on the sampled ARGs, so they should be set as accurately as possible.

3.1   Mutation Rates

ARGweaver can accept a single mutation rate to be used across the entire genomic region that is being analyzed (with the option `--mutrate <rate>`), or it can use a specified map of mutation rates (`--mutmap <ratefile.bed>`). If using a rate map, the file specifying the map should have four columns: chromosome, start coordinate (0-based), end coordinate, and the rate. The rates should be specified in units of expected mutations per base pair per generation.

The mutation rate is particularly important for calibrating the timing of ancestral events. If the given mutation rate is off by a factor of m, then the estimated ages of events will tend to be off by a factor of $1/m$, so that too high a mutation rate will make events seem to have happened much more recently than they actually did. After a period of controversy [26], estimates of mutation rates for humans have stabilized in recent years, but there is still considerable debate about the best average rates to use for evolutionary analyses [16, 25]. In addition, mutation rates are known to vary across species and along the genome in each species, which further complicates their use in ARGweaver.

For our example analysis, we address these issues by using levels of divergence between several closely related primates (human, chimpanzee, gorilla, orangutan, and gibbon) to estimate relative mutation rates in sliding 100 kb windows. We first mask out conserved regions of the genome in order to estimate the neutral substitution rate, which should be proportional to the average mutation rate. Then, we scale all the relative rates so that the average rate is 1.45e − 8 mutations per base pair per generation [17]. The online resource for this chapter contains the full script for obtaining these rates, as well as the rates themselves.

3.2 Recombination Rates

Recombination rates are specified in ARGweaver in units of the probability of a recombination between two neighboring bases per generation. As with mutation rates, the rate may be assumed constant across the region (`--recombrate <rate>`), or a map of rates may be provided to the program (`--recombmap <ratefile.bed>`).

As discussed earlier, there is an important interplay between the mutation and recombination processes in the ARGweaver model. ARGweaver always tries to find ARGs that best fit the data given the model. If too high a recombination rate is used, then ARGweaver may overfit the data, so that it samples as many recombination events as needed to produce trees that allow for minimal numbers of mutations at polymorphic sites. Conversely, an unrealistically low rate will lead to ARGs containing too few recombination events, resulting in local trees that are incompatible with the site patterns in the data (this will be seen as a large number of "noncompats" in the output stats file). Thus, it is important to specify the recombination rates as accurately as possible.

In humans, there are many estimates of recombination rates; for our example analysis we will use one based on a collection of African-American genomes [5]. For other species that may not have existing maps, a genome-wide estimate from the closest model organism should be sufficient.

It is worth noting that, for reasons of computational efficiency, the mutation and recombination map should generally only be specified at modest levels of resolution along the genome sequence. The reason is that ARGweaver must recompute the transition and emission probabilities of its HMM at all positions at which the mutation or recombination rates change, as well as at those at which the local tree changes. Thus, if the rates change more frequently than the local trees, there will be a considerable increase in the run time of the algorithm. For this reason, in our example analysis we smooth out the recombination rates by taking the average across 5 kb windows. The mutation rates were calculated in larger windows so they are already smooth.

3.3 Population Size   The prior model for ARGweaver assumes that all samples are drawn from a single, panmictic population. The panmixia aspect of the prior is weak in the sense that actual structure in the population is usually evident in the data and will be reflected in the sampled ARGs. At the same time, the population size does determine the prior rate at which branches coalesce, so an incorrect prior may skew certain features of the sampled ARGs, such as the relative coalescence times between samples, and the relative rates of candidate recombination events.

The simplest way to specify population size is as a constant-sized population. A quick estimate for the population size can be obtained using Watterson's estimator. If S is the number of segregating sites over L nucleotides in n haploid genomes, and μ is the mutation rate per base pair per generation, then Watterson's estimator for the diploid effective population size is:

$$N = \frac{S}{4\mu L} \left(\sum_{i=1}^{n-1} \frac{1}{i} \right)^{-1}. \tag{1}$$

A slightly better method, for the purposes of ARGweaver, is to use the nucleotide diversity (otherwise known as pi), which is computed as the average number of pairwise differences between any two haploid genomes per base-pair. If using VCF input, this calculation can be accomplished using VCFTools [2] with the command `vcftools --site-pi`. The nucleotide diversity can be divided by 4μ to obtain an estimate of the diploid effective population size, N.

For most populations, a more realistic model allows for a changing population size over time. Programs such as SMC++ [29], MSMC [27, see also Chapter 7], PSMC [7], G-PhoCS [3], or diCal [28]) can be used to obtain estimates of a demographic history that includes such changes. The option `--popsize-file <popsize_file.txt>` can then be used to specify the corresponding history in ARGweaver. The specified popsize file should have two columns: a time (in generations) and a diploid population size. The times should be increasing, with the first time being zero. An example is shown below:

```
0       10000
1500    200
2500    20000
```

This file would represent a bottleneck scenario where there is a population of size 10,000 for the past 1500 generations, but between generations 1500 and 2500 ago, the population size was only 200. Before 2500 generations ago, the population size was

20,000. Note that ARGweaver will round all times in the file to the nearest discrete time point used by the model.

Our example analysis illustrates some of the inherent shortcomings in assuming a single population. The sequence data we will analyze consists of humans sampled from across the globe, including non-Africans (whose populations endured a severe bottleneck associated with the out-of-Africa event, followed by a rapid recent expansion), Africans (whose effective population size is larger, and more stable over time), and ancient hominins (whose effective population sizes are much smaller than those of humans). There simply is no single history which would be a good fit for our data set. In this case, we will simply forge ahead with ARGweaver's default population size of 10,000, which is about the correct order of magnitude for most of our lineages over most of their shared history. However, it is critical that we keep our model misspecification in mind as we interpret our results. In a real analysis, we would probably eventually want to use simulations to understand the implications of our over-simplifying assumptions. At the same time, the fact that we do estimate ARGs that appear to be reasonable in most respects demonstrates that ARGweaver is somewhat robust to the choice of population size.

3.4 Time Discretization

Before running ARGweaver, it is worth thinking about the time discretization scheme and adjusting it to ensure it is appropriate for the analysis at hand. There are three options to consider. First, the number of time points can be changed with the option `--ntimes <ntime>`, with the default number of points being 20. While more resolution may be desired, the running time will increase proportionally to the square of this number.

Second, the option `--maxtime <time>` indicates the maximum time point in the model, in units of generations. All lineages are forced to coalesce by this time, so it usually makes sense to choose a very ancient time. The default in ARGweaver of 200,000 generations is about 20 times the effective human population size and a reasonable choice for most human analyses.

The third relevant option is `--delta <delta>`. The times are distributed on a scale so that recent time points are more close together than distant points. This convention allows for greater resolution on recent time scales, when there are usually more coalescence events (since there are more distinct lineages). The distribution is controlled by the `delta` (δ) parameter, where which bring the points closer together at recent times when it is set to larger values. The exact formula for setting the time points is: $t(i) = (\exp(\frac{i}{K-1}\log(1 + \delta t_{max})) - 1)/\delta$, for K time points and $i \in \{0, 1, \ldots, K - 1\}$. Very small values of $\delta(< 1/t_{max})$ will yield roughly linear distribution of times, whereas very large values δ will place the first few time points so close together that they represent fractions of a generation. The default value of $\delta = 0.01$ produces a reasonable distribution for the default t_{max} of 200,000 generations. The discrete times are written to the terminal and log file at the start of an ARGweaver run, and it is advisable to inspect these values and possibly adjust δ (by restarting the run) as necessary.

It is important to carefully consider your goals when deciding how to set `maxtime` and `delta`. If you are interested in recent history, then you may want to increase `delta`, and your choice of `maxtime` may not be crucial. If you are interested in balancing selection or deep coalescences, it will be important to choose a larger `maxtime` and possibly a smaller `delta` as well.

4   Other Options

4.1   Sampling Frequency

Another decision to be made is how frequently to sample from the MCMC chain. The option `--sample-step <n>` tells ARGweaver to output the ARG sampled on every nth step of the MCMC algorithm. If any of the individuals are unphased, then the program will also output the corresponding phased samples.

As mentioned above, it is customary to "thin" MCMC samples to reduce autocorrelation between the final samples, but it is difficult to know how much thinning will be required prior to an ARGweaver run. Some have argued that thinning is inefficient and unnecessary, and that most properties of the distribution can be better estimated using the full sample [11]. In the case of ARGweaver, however, there is a substantial cost associated with storing and processing each sampled ARG, and adjacent ARGs in the chain are very highly correlated, so some degree of thinning is justified. We will use the default sampling frequency of 10, and return to the issue of autocorrelation as we interpret the results.

4.2   Ancient Samples

If the samples are not all from present day (as with the Neandertals and Denisovan in our example), their ages (in generations before the present) can be specified to ARGweaver. The computed ARGs will then have shorter branches for these samples than for the modern samples. The ages can be specified in a file with two columns—sample name and age in generations—and passed to ARGweaver with the option `--age-file <age_file.txt>`. The ages will be rounded to the nearest time point in the model. Any sample not found in the file is assumed to have age zero, and at least one sample must have age zero (so sample ages should be given relative to the youngest sample).

In our example, we use ages of 4206 generations for the Altai Neandertal, 2482 generations for the Denisovan sample, and 1793 generations for the Vindija Neandertal, corresponding to ages of

122 kya, 72 kya, and 52 kya, respectively [22], and an assumed generation time of 29 years.

4.3 Site Compression

Site compression is usually important for keeping ARGweaver's run time manageable. The option `--compress-seq <c>` will compress blocks of c sites together, resulting in a speed-up of the code of approximately a factor of c. The breakpoints between each block are chosen in a flexible manner so that there is no more than one variant site in the same block. If a block contains a variant site, then the new "compressed site" takes on the site pattern of its variant site; otherwise the compressed site is invariant. The mutation and recombination rates are also increased by a factor of c, since they reflect per-site rates. (This rate inflation is done internally by `arg-sample`; the user should provide rates per uncompressed base pair.)

There are several issues to consider with this option. Compression will fail (with a program abort soon after stating) if variant sites are too close together to compress at the requested level. Compression also causes a loss of resolution; if `--compress-seq 50` is used, then the breakpoints in the ARGs will occur at most every 50 base pairs. Most importantly, recall that ARGweaver assumes that only one recombination event can occur between any two sites. This assumption applies to compressed sites also, so its effect is magnified as the compression factor increases. Therefore, if compression is too high, ARGweaver may not be able to place enough recombination events between variant sites, resulting in poor ARG estimates.

One way to think about how much compression is allowable is to consider the distribution of distances between variant sites, ignoring singleton sites, which contain no topological information. In our data set, non-singleton variant (NSV) sites occur, on average, about every 300 bases. Therefore, one might be tempted to select a compression factor of 50, which would allow an average of ~6 recombinations between each pair of NSV sites. However, bear in mind that the distance between NSV sites is approximately exponentially distributed, which means that at an average distance is 300, only ~15% of NSV sites are more than 50 base pairs apart. We therefore will use a compression factor of 10 in our example, which should allow about 97% of pairs of NSV sites to have more than one recombination between them.

In general, we recommend using compression conservatively, so that multiple recombinations are still possible between most pairs of adjacent NSVs. Too high compression will lead to a high number of "noncompats" in the *.stats* file (described in the next section), so the compression factor may need to be adjusted if this is observed.

5   Running ARGweaver

With all preliminaries addressed, we are now ready to run ARG-weaver and work through our example analysis. The command to run ARGweaver is as follows:

```
arg-sample --vcf-files vcf_files.txt \
 --region $region \
 --vcf-min-qual 30 \
 --subsites inds.txt \
 --maskmap filter.bed.gz \
 --mask-cluster 2,5 \
 --ind-maskmap ind_mask_files.txt \
 --age-file sample_ages.txt \
 --mutmap subst_rate_autosome.bed.gz \
 --recombmap recomb_rate_autosome.bed.gz \
 --compress-seq 10 \
 -o $outdir/out
```

As ARGweaver runs, it produces several output files, all with names having the prefix specified by the -o option. These files include:

- *A log file* (<outroot>.log) , which records the same output that is written to the terminal, including details about the model and data, progress, time, and memory usage, *etc.*, as the iterations continue.

- *A stats file* (<outroot>.stats), consisting of one row per MCMC iteration, with columns including:

  - prior: log probability of the sampled ARG given the model

  - likelihood: log probability of the data given the sampled ARG

  - joint: total log probability of the ARG and the data (prior + likelihood)

  - recombs: number of recombination events in the sampled ARG

  - arglen: total length of all branches summed across sites

  - noncompats: the number of variant sites that cannot be explained by a single mutation under the sampled ARG

- *Sampled ARGs* (<outroot>.<iter>.smc.gz) are written at the sampling frequency requested by the option --sample-step. These are in ARGweaver's SMC format, which is text-readable and lists non-recombining genomic intervals, the tree in each interval (using Newick format), and the recombination events that occur between intervals. The recombination events are described as subtree pruning and regrafting (SPR) events, which define where a particular branch breaks and recoalesces back onto the tree.

- *Phased sites files* (`<outroot>.<iter>.sites.gz`) If the data is unphased, then it will also print the current phased sequence data (in SITES format) for each ARG that is printed. These phased SITES files do not contain information about positions that are masked in all individuals; those regions are written to a file named `<outroot>.masked_regions.bed`.

5.1 Time/Memory Requirements

ARGweaver requires substantial computation time, but the memory usage is low. In our example, a 1 Mbase region took 6 h to complete 1000 MCMC iterations, with a maximum memory usage of 135 Mbytes. The program does not support multithreading. Rather, parallelization is usually achieved by running many genomic segments at the same time on different CPU cores.

5.2 Monitoring Convergence

The stats file produced by ARGweaver can be used to monitor ARGweaver's convergence. Figure 2 shows an example of how

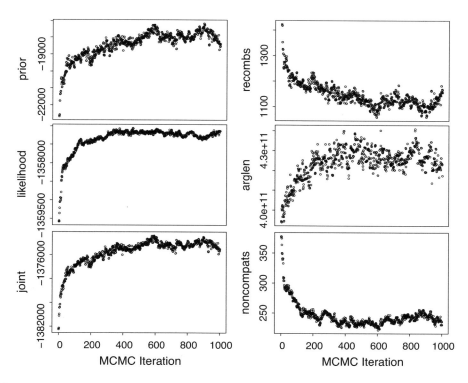

Fig. 2 Traces of various statistics over MCMC iterations. The values should stabilize as the chain reaches convergence; earlier iterations should be discarded as "burn-in." Values related to the probability of the data, including the prior, likelihood, and joint probability, should increase as the MCMC converges from a poor initial guess to higher probability ARGs. The number of "noncompats" tends to decrease until stabilization, since the true ARG usually explains observed site patterns without requiring multiple mutations. The number of recombinations and length of the ARG may increase or decrease before convergence, depending on the data and the model. In this example, a burn-in of 600 iterations seems adequate

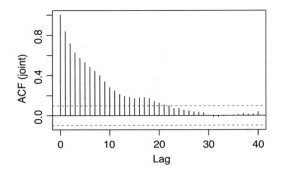

Fig. 3 The autocorrelation in the "joint" likelihood statistic for the run shown in Fig. 2, as plotted by the R function "acf." The region between the blue lines represents the 95% confidence interval for no correlation

statistics change as the sampler reaches equilibrium (known as "stationarity" in the MCMC literature). In this example, the relevant statistics seem to become fairly stable after about 600 iterations, so we will use 600 for the "burn-in" for this run.

It is also useful to examine the autocorrelation of the statistics in the stats file (after removing the burn-in samples). The autocorrelation of the "joint" statistic is shown in Fig. 3, and it appears that in this case autocorrelation reaches insignificant levels in roughly 20 iterations. Autocorrelations for the other statistics look similar and are insignificant between 20–30 iterations (not shown). Therefore, the sampled ARGs should be thinned to at least every 20th MCMC iteration in order to achieve a sample of effectively independent ARGs. It is not necessary to perform thinning in order to estimate the mean or quantiles of ARG statistics, so we will not do that here. Still, it is useful to keep the thinning interval in mind, in order to compute how many effectively independent samples we have. If we run 1000 MCMC iterations, and discard the first 600 as burn-in, and use a thinning interval of 20, we are left with an effective sample size of only 20. This size may be sufficient for inspecting example trees, but is not enough to obtain good estimates of derived quantities such as coalescence times. So, in this case we may want to continue running the MCMC chain for (at least) an additional 1600 iterations in order to end up with 100 effectively independent samples.

5.2.1 Resuming a Run

In our original command, we ran the MCMC for the default number of iterations, which is 1000. It is easy to resume a run, by adding the option --resume and specifying the final number of iterations desired, for example, --iters 2600. In this case it will start from iteration 1000 and perform 1600 additional iterations.

6 Interpreting Results

When enough iterations have been completed, the fun begins! It is time to look at the ARGs and see what might be learned from them. There are countless ways that one might parse and examine a set of ARGs, and custom code may eventually be required for a specific analysis. However, there are some tools in ARGweaver that can be used to get started.

Many of the plots we will show in this and following sections were created by the R package that comes with ARGweaver. We will not show the code to create the plots in this chapter, but it is in the companion material that accompanies this book. The R package Gviz [4] was used to create the gene annotation plots.

6.1 Leaf Trace Plots

Leaf trace plots were introduced in the ARGweaver paper [23] as way to visualize the ARG. In these diagrams, the leaves of the local trees, as they change along the genome sequence, are drawn as horizontal lines, with the vertical distance between neighboring lines proportional to the distance between adjacent leaves in the local tree (see top panel of Fig. 4). Leaf traces should be interpreted with caution, because there are many possible leaf trace plots for the same ARG (depending on arbitrary choices made in ordering the leaves), and the distance between non-adjacent lines in the leaf trace plot is not directly interpretable. Furthermore, the leaf trace is drawn for a single ARG, rather than showing the distribution across sampled ARGs. Nevertheless, the leaf trace is an intuitive graphical description of an ARG that can be used to survey its overall structure.

A leaf trace plot can be created by first running the `arg-layout` executable on a single SMC file. Then, the "plotLeafTrace" function in the ARGweaver R package can plot the resulting file.

Leaf traces around the $DARC$ gene are shown in the top panel of Fig. 4. The first feature that is apparent is that the plot changes quickly along the x-axis in some regions, and more slowly in others, reflecting the posterior estimate of the local recombination rate. The vertical height of the plot also gives a quick indication of the total height of the local trees along the region. That is, a large "spread" of the traces indicates a deep (ancient) time to most recent common ancestry (TMRCA), whereas a small spread indicates a shallow TMRCA. If the traces are colored by population of origin, the leaf trace can also provide an idea of the level of population structure in the data. The leaf trace in our example suggests that the $DARC$ gene is in the middle of a low-diversity region with a relatively high recombination rate.

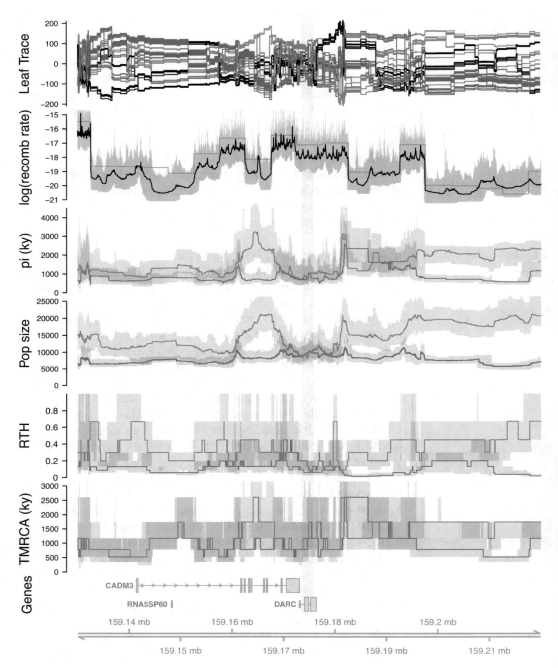

Fig. 4 Leaf trace plots and key statistics for the region surrounding the *DARC* gene. The highlighted region shows the location of SNPs that define different haplotypes, some of which provide resistance to malaria. The recombination rate is shown in units of log(recombs/bp/generation) and is computed across all individuals; the solid gray line shows the prior recombination rates, and the black line shows posterior expected values. The other statistics are shown for different population subsets. TMRCA gives the time to the most recent common ancestor; RTH is the "relative TMRCA half-time," which describes the minimum time for half the lineages to coalesce, normalized by the TMRCA. Blue lines plot African human genomes; red lines plot non-African humans, and black lines in the leaf trace (top panel) represent the ancient hominins. Shading around all lines represent 90% confidence intervals

6.2 Computing Basic ARG Statistics

After examining leaf trace plots, the next way to explore the ARG is to look at various statistics across the genomic region. The first step is to convert the ARGs into a more convenient format. We currently have an SMC file for every sampled ARG, named `<outprefix>.0.smc.gz`, ..., `<outprefix>.<num_iter>.smc.gz`. We use the command `smc2bed-all <outprefix>`, which will combine all the information into a single sorted and indexed BED file, with columns: chromosome, start coordinate (0-based), end coordinate, MCMC iteration number, and Newick tree (representing the local tree for the ARG sampled in this region and iteration).

Now, the executable `arg-summarize` can be used to extract statistics. Some of the more useful options to `arg-summarize` include:

- `--tmrca`: time to the most recent ancestor

- `--pi`: average distance between any two leaves

- `--branchlen`: total tree length

- `--popsize`: estimate of diploid population size based on coalescence rates in the local tree

- `--tmrca-half`: time at which half the samples find a current ancestor

- `--rth`: Relative TMRCA Half life (RTH), defined as the ratio of tmrca-half to tmrca. Unusually low values of this statistic suggest a recent "clustering" of coalescent events, possibly indicating a partial selective sweep [23]

- `--node-dist <leaf1,leaf2>`: Distance between leaf1 and leaf2 on the tree

- `--node-dist-all`: Like node-dist, but for all pairs of leaves

- `--min-coal-time <ind1,ind2>`: Return minimum coalescence time between two individuals (over all four haploid pair combinations)

- `--subset-inds <ind_list.txt>`: Before computing statistics, prune all individuals not listed in given file

- `--mean`: Rather than reporting statistics for every MCMC iteration, report the mean across iterations for all non-recombining intervals. This applies to all statistics requested (i.e., `--tmrca`, `--pi`, *etc.*)

- `--quantile <q1,q2,...>`: Same as `--mean`, but instead report one or more quantiles of statistics across samples

For example, the command:

```
arg-summarize -a <outprefix>.bed.gz --tmrca \
  --subset africans_inds.txt \
  --quantile 0.05,0.5,0.95
```

would compute the 5%, 50% (median), and 95% quantiles of the time to the most recent common ancestor, in the subset of the ARG only containing individuals listed in the file `african_inds.txt`.

ARGweaver supports many additional options not described here. A full list can be obtained using the command `arg-summarize --help`.

Figure 4 shows a plot of several of these statistics (recombination rate, pi, popsize, RTH, TMRCA) in the region surrounding the *DARC* gene. The highlighted region is known to harbor variants which have reached near-fixation throughout Africa and are thought to provide resistance to malaria; however, this region tends not to be detected by most tests for selective sweeps [13]. Looking at Fig. 4, there are some suggestions of possible positive selection in Africa, such as low estimates for pi, population size, and RTH in the highlighted region. It is possible that the relatively high recombination rate in this region has led to a fast breakdown of haplotype structure, which would make a selective sweep difficult to detect.

It is important when looking at these plots to remember the underlying population structure; a low RTH seems to be fairly common in non-African populations due to the population bottleneck, but is more rare in African populations. Similarly, we expect the local population size estimates for the African genomes to be higher than for the non-Africans. Overall, while the RTH in Africa at the highlighted region is low, it is doubtful that this region would be a significant outlier in a genome-wide scan.

6.2.1 Examining Local Trees

Often one of the most useful ways to gain insight into the ARG is to look directly at the estimated local trees. ARGweaver's R package comes with some tools to visualize the trees (the package internally makes use of the "ape" package [19]). For example, Fig. 5 shows one

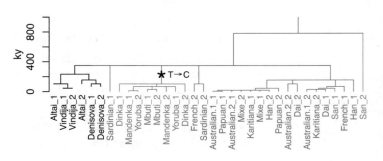

1:159173860–159174688 rep=2000

Fig. 5 A sampled tree at the FY*O mutation locus. The FY*O T → C mutation is mapped to the branch as shown

of the sampled trees at the position of rs2814778, the SNP which defines the African haplotype.

This tree is interesting in several ways. First, it suggests a tight grouping among the African and non-African populations. While this is not particularly unusual for non-Africans (due to the out-of-Africa bottleneck), it is quite rare to observe this level of clustering among Africans. Importantly, the San individual does not cluster with the rest of the Africans, providing a hint as to why the African TMRCA was not unusually low in Fig. 4. In fact, it is known that the FY*O mutation shown in Fig. 5 is not common in the San population, whereas it has reached near-fixation throughout most of the rest of Africa [13].

These observations suggest that perhaps the statistics shown in Fig. 4 do not fully capture the interesting aspects of this tree. Figure 6 shows the same statistics, but also calculated using a subset of Africans that excludes San. This subset (shown in dark blue) is much more of a regional outlier, with unusually low pi, population size, and TMRCA. Notably, the RTH statistic is no longer low in this case; this is because RTH is designed to detect partial sweeps, and the putative sweep is complete in the subset excluding San.

One must of course be cautious about the biases that might result from inspecting the sampled trees, deciding how they are interesting, and then revising the statistical tests to detect these same interesting features. Nevertheless, it is sometimes only by visual inspection and exploration of the sampled trees that the patterns in the data start to become clear. The basic statistics that can be computed with `arg-summarize` are often too crude to give a clear understanding of the ARG features. For example, TMRCA is unaffected by any feature besides the final coalescence time. The pi statistic is also very heavily dominated by long branches, especially due to the logarithmic timescale used by ARGweaver. Finally, RTH is sometimes informative of partial sweeps; however, it fails to detect complete sweeps or partial sweeps that have not yet hit 50% frequency, and it also will give many false positives if the underlying population is expanding or has experienced a bottleneck.

It is therefore sometimes necessary to "browse" through the trees to get an idea of what interesting signals may exist. In our example, we revised our tests both because of the structure of the observed trees, and because this structure was concordant with the known geographical distribution of the DARC haplotypes. In order to determine if this locus is truly special, it would be necessary to compare its trees/statistics to ones from across the genome, or generated from neutral simulations based on a more realistic demographic model.

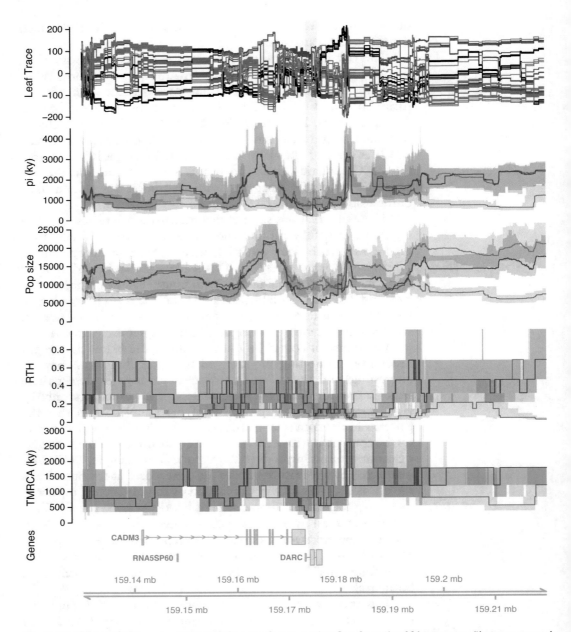

Fig. 6 Modified statistics around the *DARC* gene after removing San from the African group (the new group is shown in dark blue)

6.2.2 Allele Age

The sampled ARGs allow mutations to be mapped to the branches of local trees, which in turn allows the times at which those mutations occurred to be estimated. The arg-summarize program supports computing such "allele ages," based on a single sites file. However, this program is not designed to perform allele age computations when the data is unphased, as it would need to use the sampled phase corresponding with each sampled ARG to properly map the mutations. This problem can be addressed using a script called `allele_age`, which will call `arg-summarize` for each pair of sampled ARG and phase, and report the age for each MCMC iteration.

There are some caveats to interpreting the allele age. First of all, it is typically the case that, at a small fraction of sites, the local tree will not be able to explain a particular variant site with a single mutation (indicating a violation of the "infinite sites" model), so that the mutation time is poorly defined. It is also possible that the assignment of derived/ancestral alleles will not be consistent across all the ARGs; some ARGs may explain an observed variant as a young, low-frequency allele, whereas other samples may flip the ancestral and derived alleles and describe the same variant as an old, high-frequency allele. Also, when a mutation is mapped to a branch, it is equally likely to have arisen anywhere along the branch, so mutations mapped to longer branches will have much more uncertainty in their time estimates than those mapped to short branches.

The `allele_age` function outputs a number of extra columns to help clarify these issues. For each MCMC sample, it outputs the identity of the inferred derived and ancestral alleles, as well as a flag indicating whether the mutation can be explained by the infinite sites model under the sampled tree. It also outputs both the mean allele age (the midpoint of the branch where it was mapped) and the minimum allele age (the most recent point on the branch).

In our DARC example, we find that on average 4.5% of sites do not obey the infinite sites model. The infinite sites violations tend to be concentrated in the same sites across all MCMC replicates (with 2.6% of sites requiring multiple mutations in $> 95\%$ of replicates, and 92% of sites requiring multiple mutations in $< 5\%$ of replicates). This rate of infinite-site violations is not unexpected, and is likely due to a number of factors, including low levels of genotyping error accumulating over 17 samples, true instances of multiple mutations (especially at sites with high mutation rates such as CpG sites), model misspecification, or uncertainty in phasing or the ARG.

Looking at the allele underlying the tree in Fig. 5, we estimate that the allele is between 100 and 300 ky old. This large amount of uncertainty is not surprising, as the mutation is mapped to a fairly long branch above the African subtree.

6.2.3 Neandertal Introgression

Careful inspection of the local trees in our example region reveals another interesting feature. Several of the trees downstream from *DARC*, such as the one shown in Fig. 7, exhibit an atypical placement of one of the Han haplotypes (Han_2), which is clustered tightly with the Neandertal genomes. In other respects, this tree is

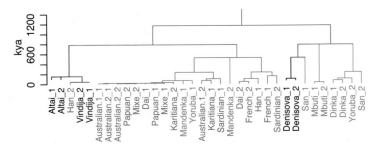

1:159203410–159206979 rep=2000

Fig. 7 Plot of a local tree which shows putative introgression from Neandertal into the Han genome

quite typical. It is therefore possible that this is a region of Neandertal introgression into the Han individual included in our data set.

To investigate further, we can use `arg-summarize` to compute the minimum coalescence time between the Han individual and each of the ancient hominins (for every sampled ARG, at every genomic location), and look at the distribution of these times. The result is plotted in Fig. 8. We see that there is a region between roughly 159.20Mb and 159.21Mb where the Han genome coalesces with both the Neandertal and the Altai genomes significantly more recently than 500 kya, roughly the minimum to be expected in the absence of introgression. (The threshold of 500 kya generously allows for uncertainty in dating coalescence events; the actual population divergence time is closer to 600 kya.) No other individual besides Han has coalescence times significantly below 500 kya in this region (not shown), indicating that this observation is not likely to be an artifact of using an incorrect local mutation rate. Thus, it seems likely that this is a short introgressed region in the Han.

Another way to examine this region is to look at the site patterns within it. The `subsites` program can retrieve sites for particular individuals or genomic intervals; we can use the sampled phased SITES file from the ARGweaver run to examine the site patterns. The R package also has a function (`plotSites()`) to visualize site patterns. Figure 9 shows the site patterns in this putatively introgressed region. It confirms that there are four sites shared by one of the Han haplotypes and at least one Neandertal, which are not shared by any other modern human, making the region a very good candidate for introgression. Most other software for detecting introgression would not confidently find such a short region.

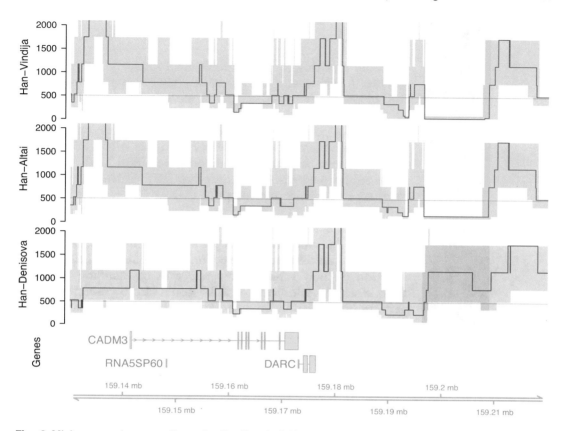

Fig. 8 Minimum coalescence times for the Han individual to each of the three ancient individuals (red: Neandertal, blue: Denisovan; minimum taken across all four haplotype combinations Han_1, Han_2, ancient_1, ancient_2). Times are in ky; shaded colors show 95% confidence intervals across MCMC samples, solid lines give the median. The horizontal line at t=500 ky represents an approximate minimum expected time of coalescence in the absence of introgression, given the divergence time between humans and Neandertal/Denisovans. The highlighted region is putatively introgressed from Neandertal into the Han genome

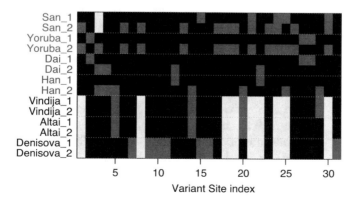

Fig. 9 Variant sites in a subset of individuals in a region of putative Neandertal introgression into Han near the *DARC* gene (chr1:159196000-159209000). Red=minor allele, black=major allele, gray=missing data

7   Discussion

We hope that this chapter is sufficient to help new users decide whether ARGweaver is an appropriate tool to apply to their data, and to get started with an initial analysis. In our example, we explored a few relatively simple statistics that suggested instances of positive selection and Neandertal introgression near the *DARC* locus. In other work, we have also shown that ARGweaver can be used to detect balancing and negative selection [23], as well as more subtle patterns of introgression [6]. However, there are many more evolutionary questions that ARGweaver could potentially shed light upon; in fact, one of the most exciting aspects of ARGweaver is that its possible uses have yet to be fully explored. One could imagine delving deeper into the study of natural selection, for example estimating selection coefficients, or distinguishing selection on new variants from selection on standing variation. Beyond natural selection, the ARG could also be useful to detect patterns of population structure, explore genotype/phenotype correlation, phase haplotypes, estimate recombination rates, etc. As we discussed in the introduction, almost all population genetics questions can be framed as questions about ARG structure. In some cases it may be sufficient to compute summary statistics from the ARG using `arg-summarize`, but in others, it will be necessary to write custom code for analyzing the ARG.

There is no doubt that an ARGweaver analysis requires more time and effort than an analysis based on a typical "off-the-shelf" tool for population genetics. For this reason, ARGweaver generally should not be the first tool that one reaches for when analyzing a new data set. Nevertheless, as we have shown, this additional effort can prove worthwhile for certain kinds of analyses. One such instance is when the genomic data are rare and precious, as is the case with the Neandertal, Denisovan, and other ancient genomes, which is why we highlight the use of those genomes in this chapter. Another circumstance in which ARGweaver may be especially useful is when simpler methods fall short, for example, by being underpowered, or inappropriate in some way for the data or the question at hand. Because ARGweaver utilizes all the genome data, without reducing it to summary statistics, and models the full recombination and coalescence process (within the limitations of the SMC$'$), it should generally have more statistical power than other methods. In addition, its ability to work on low-quality and/or unphased genomes, and produce full evolutionary histories, makes it a uniquely flexible approach that may fill in gaps left behind by more traditional methods. For these reasons, ARGweaver is a "power tool" many population geneticists may wish to add to their toolboxes.

Acknowledgements

The ARGweaver software was initially written by Matt Rasmussen, and later substantially extended by Melissa Hubisz. We thank Aaron Stern, Amy Williams, and Ritika Ramani for helpful comments and suggestions. This material is based upon work supported by the National Science Foundation Graduate Research Fellowship to Melissa Hubisz under Grant No. DGE-1650441. Adam Siepel was supported in part by grant GM102192 from the National Institutes of Health.

References

1. Barnett DW, Garrison EK, Quinlan AR, Strömberg MP, Marth GT (2011) BamTools: a C++ API and toolkit for analyzing and managing BAM files. Bioinformatics 27 (12):1691–1692. https://doi.org/10.1093/bioinformatics/btr174

2. Danecek P, Auton A, Abecasis G, et al (2011) The variant call format and VCFtools. Bioinformatics 27(15):2156–2158. https://doi.org/10.1093/bioinformatics/btr330

3. Gronau I, Hubisz MJ, Gulko B, Danko CG, Siepel A (2011) Bayesian inference of ancient human demography from individual genome sequences. Nat Genet 43(10):1031–1034

4. Hahne F, Ivanek R (2016) Visualizing genomic data using Gviz and bioconductor. Statistical genomics: methods and protocols. Springer, New York, pp 335–351

5. Hinch AG, Tandon A, Patterson N, et al (2011) The landscape of recombination in African Americans. Nature 476 (7359):170–175

6. Kuhlwilm M, Gronau I, Hubisz MJ, et al (2016) Ancient gene flow from early modern humans into Eastern Neanderthals. Nature 530(7591):429–433

7. Li H, Durbin R (2011) Inference of human population history from individual whole-genome sequences. Nature 475:493–496

8. Li N, Stephens M (2003) Modeling linkage disequilibrium and identifying recombination hotspots using single-nucleotide polymorphism data. Genetics 165:2213–2233

9. Li H, Handsaker B, Wysoker A, et al (2009) The sequence alignment/map format and SAMtools. Bioinformatics 25:2078–2079

10. Loh PR, Palamara PF, Price AL (2016) Fast and accurate long-range phasing in a UK Biobank cohort. Nat Genet 48:811–816

11. MacEachern SN, Berliner LM (1994) Subsampling the Gibbs sampler. Am Stat 48 (3):188–190

12. Marjoram P, Wall JD (2006) Fast "coalescent" simulation. BMC Genet 7:16

13. McManus KF, Taravella AM, Henn BM, et al (2017) Population genetic analysis of the DARC locus (Duffy) reveals adaptation from standing variation associated with malaria resistance in humans. PLOS Genet 13(3):1–27. https://doi.org/10.1371/journal.pgen.1006560

14. McVean GA, Cardin NJ (2005) Approximating the coalescent with recombination. Philos Trans R Soc Lond B Biol Sci 360:1387–1393

15. Meyer M, Kircher M, Gansauge MT, et al (2012) A high-coverage genome sequence from an archaic Denisovan individual. Science 338(6104):222–226. https://doi.org/10.1126/science.1224344

16. Moorjani P, Gao Z, Przeworski M (2016) Human germline mutation and the erratic evolutionary clock. PLoS Biol 14(10):e2000744

17. Narasimhan VM, Rahbari R, Scally A, et al (2017) Estimating the human mutation rate from autozygous segments reveals population differences in human mutational processes. Nat Commun 8(1):303. https://doi.org/10.1038/s41467-017-00323-y

18. Neph S, Kuehn MS, Reynolds AP, et al (2012) BEDOPS: high-performance genomic feature operations. Bioinformatics 28 (14):1919–1920. https://doi.org/10.1093/bioinformatics/bts277

19. Paradis E, Claude J, Strimmer K (2004) APE: analyses of phylogenetics and evolution in R language. Bioinformatics 20:289–290

20. Pollard KS, Hubisz MJ, Rosenbloom KR, Siepel A (2010) Detection of nonneutral

substitution rates on mammalian phylogenies. Genome Res 20:110–121

21. Prüfer K, Racimo F, Patterson N, *et al* (2014) The complete genome sequence of a Neanderthal from the Altai Mountains. Nature 505 (7481):43–49

22. Prüfer K, de Filippo C, Grote S, *et al* (2017) A high-coverage Neandertal genome from Vindija Cave in Croatia. Science 358 (6363):655–658. https://doi.org/10.1126/science.aao1887

23. Rasmussen MD, Hubisz MJ, Gronau I, Siepel A (2014) Genome-wide inference of ancestral recombination graphs. PLoS Genet 10(5): e1004342

24. Sankararaman S, Mallick S, Dannemann M, *et al* (2014) The genomic landscape of Neanderthal ancestry in present-day humans. Nature 507:354–357

25. Scally A (2016) The mutation rate in human evolution and demographic inference. Curr Opin Genet Dev 41:36–43

26. Scally A, Durbin R (2012) Revising the human mutation rate: implications for understanding human evolution. Nat Rev Genet 13 (10):745–753

27. Schiffels S, Durbin R (2014) Inferring human population size and separation history from multiple genome sequences. Nat Genet 46 (8):919–925

28. Sheehan S, Harris K, Song YS (2013) Estimating variable effective population sizes from multiple genomes: a sequentially Markov conditional sampling distribution approach. Genetics 194(3):647–662

29. Terhorst J, Kamm JA, Song YS (2016) Robust and scalable inference of population history from hundreds of unphased whole genomes. Nat Genet 49:303–309

30. The ENCODE Project Consortium (2012) An integrated encyclopedia of DNA elements in the human genome. Nature 489:57–74

31. Van der Auwera GA, Carneiro MO, Hartl C, *et al* (2013) From FastQ data to high-confidence variant calls: the genome analysis toolkit best practices pipeline. Curr Protoc Bioinformatics 11 (1110):11.10.1–11.10.33. https://doi.org/10.1002/0471250953.bi1110s43

Part III

Advances in Population Genomics

Chapter 11

Population Genomics of Transitions to Selfing in Brassicaceae Model Systems

Tiina M. Mattila, Benjamin Laenen, and Tanja Slotte

Abstract

Many plants harbor complex mechanisms that promote outcrossing and efficient pollen transfer. These include floral adaptations as well as genetic mechanisms, such as molecular self-incompatibility (SI) systems. The maintenance of such systems over long evolutionary timescales suggests that outcrossing is favorable over a broad range of conditions. Conversely, SI has repeatedly been lost, often in association with transitions to self-fertilization (selfing). This transition is favored when the short-term advantages of selfing outweigh the costs, primarily inbreeding depression. The transition to selfing is expected to have major effects on population genetic variation and adaptive potential, as well as on genome evolution. In the Brassicaceae, many studies on the population genetic, gene regulatory, and genomic effects of selfing have centered on the model plant *Arabidopsis thaliana* and the crucifer genus *Capsella*. The accumulation of population genomics datasets have allowed detailed investigation of where, when and how the transition to selfing occurred. Future studies will take advantage of the development of population genetics theory on the impact of selfing, especially regarding positive selection. Furthermore, investigation of systems including recent transitions to selfing, mixed mating populations and/or multiple independent replicates of the same transition will facilitate dissecting the effects of mating system variation from processes driven by demography.

Key words Self-fertilization, *Arabidopsis*, *Capsella*, Self-incompatibility, Mating system evolution, Heterozygosity, Effective population size, Recombination rate, Transposable element, Efficacy of selection

1 Introduction

Flowering plants harbor a great variety of mating systems and associated floral and reproductive adaptations [1], and there is a rich empirical and theoretical literature on the causes of this diversity [2–6]. About half of all flowering plants harbor genetic self-incompatibility (SI), a molecular recognition system that allows plants to recognize and reject self pollen, and that has arisen multiple times in the history of flowering plants [7]. Despite the fact that molecular SI systems are widespread, loss of SI, often accompanied by a shift to higher selfing rates, has occurred even more frequently,

Julien Y. Dutheil (ed.), *Statistical Population Genomics*, Methods in Molecular Biology, vol. 2090,
https://doi.org/10.1007/978-1-0716-0199-0_11, © The Author(s) 2020

in many independent plant lineages [8]. This transition can be favored under conditions when the benefits of selfing, such as reproductive assurance [2] and the 3:2 inherent genetic transmission advantage of selfing [9], outweigh the costs of inbreeding depression and reduced opportunities for outcrossing through pollen (pollen discounting). As the favorability of the transition hinges on ecological factors including access to mates and pollinators that may vary greatly spatially or temporally, it is perhaps not surprising that the transition to selfing has occurred repeatedly [10]. Over a longer term, however, the loss of SI is associated with a reduction in the net diversification rate [11], a finding that provides tentative support for Stebbins's suggestion that selfing is an evolutionary dead end [12]. While the underlying ecological and evolutionary mechanisms behind this observation remain unclear, it was suggested already by Stebbins [12] that decreased adaptive potential in selfers would lead to higher extinction rates, a suggestion that is supported by theoretical modeling [13]. However, selfing does not only affect adaptation but also the impact of purifying selection [14, 15], and the relative importance of accumulation of deleterious mutations vs. reduced potential for adaptation in selfing lineages currently remains unclear. To fully understand the impact of mating system shifts on evolutionary processes, it is necessary to combine theoretical and empirical investigations, and ideally to study several parallel transitions from outcrossing to selfing. Molecular population genetics has proven to be a powerful tool to shed light on the role of natural selection in shaping the patterns of variation in selfing species. Here we will give an outline of recent work in this area, with a focus on two main model systems in the Brassicaceae.

2   The Molecular Basis of the Loss of SI and Evolution of Self-Fertilization in Brassicaceae

The effects of the transition to self-fertilization on population genomic variation and molecular evolution have been extensively studied in two systems from the Brassicaceae family, *Arabidopsis* and *Capsella*. Both of these genera have outcrossing SI as well as SC species with high selfing rates, and thus serve as good models to study this evolutionary transition [16–18]. The most widely studied SC species are *Arabidopsis thaliana* and *Capsella rubella*, which have both been estimated to be highly selfing [19–22]. The patterns of variation and molecular evolution in these selfers are often contrasted with those in their diploid sister species *Arabidopsis lyrata* and *Capsella grandiflora*, which are both SI and outcrossing. Investigations of the other SC species from these genera such as allopolyploid *Arabidopsis suecica* [23] and *Arabidopsis kamchatica*

[24–26], diploid *Capsella orientalis* and allopolyploid *Capsella bursa-pastoris* [18] have given further insight into the evolution of selfing. In Fig. 1, we provide an overview of the evolutionary relationships among the best-studied *Arabidopsis* and *Capsella* species.

Knowledge on the molecular basis of the breakdown of SI is at the center of studies investigating the early genetic causes of the transition to selfing in the Brassicaceae. In Brassicaceae, the SI recognition system includes two key genes at the nonrecombining self-incompatibility locus (*S*-locus) as well as modifier genes. The gene *SRK* encodes an *S*-locus receptor kinase that is located on the stigma surface and acts as the female specificity determinant, whereas the gene *SCR* encodes a pollen ligand that is deposited on the pollen surface and acts as the male specificity determinant [27]. This reaction is a key-lock protein interaction between the female determinant on the stigma and the male determinant on the pollen coat [27]. When SRK on the stigma binds to SCR from the same *S*-haplotype, a downstream reaction is triggered which culminates in the prevention of pollen tube growth and fertilization [28]. The evolution of selfing proceeds by disruptions of the SI reaction, for example due to loss-of-function mutations in key *S*-locus genes or in unlinked modifier genes (e.g., [29]), after which the selection at these loci is relaxed and the genes may be degraded further.

There has been intense interest in the role of parallel molecular changes underlying repeated shifts to selfing associated with the loss of SI (reviewed in [30]). In particular, theory predicts that mutations that disrupt the function of the male specificity determinant might spread more easily than those that disrupt the female specificity determinant [31, 32], and there is accumulating support for this prediction. For instance, in *A. thaliana*, Tsuchimatsu et al. [33] showed that an inversion in *SCR* underlies SC in many European accessions. In some accessions, SI can be restored by introduction of functional SRK-SCR allele from self-incompatible sister species *A. lyrata* but variability between accessions exists [34, 35]. In *Capsella* homologous machinery has been shown to underlie the SI reaction [36] and there is widespread transspecific shared polymorphism between *C. grandiflora* and SI *Arabidopsis* species at the *S*-locus [37]. The loss of SI in *C. rubella* is due to changes at the *S*-locus [36, 38], and experiments suggest that breakdown of the male specificity function is responsible for this loss [36]. However, the molecular basis of the breakdown of SI in *C. rubella* remains unclear [39] and this is also true for the other SC *Capsella* species. With recent progress in long-read sequencing, which facilitates assembly of the *S*-locus, this area is ripe for further investigation.

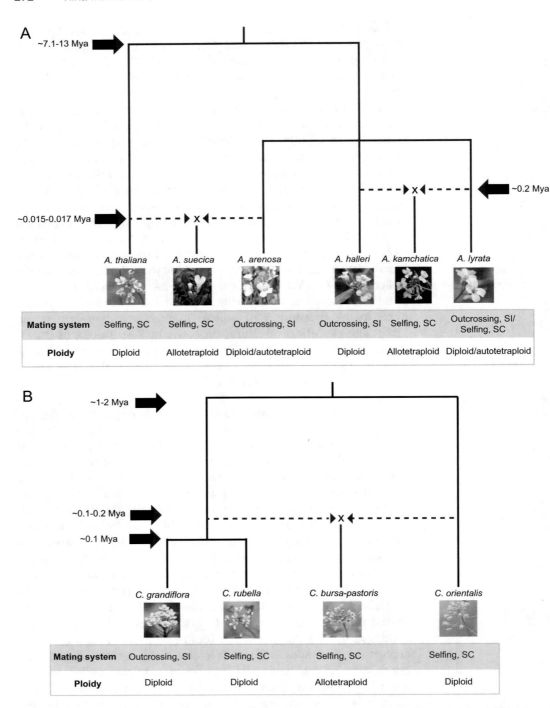

Fig. 1 Schematic drawing of evolutionary relationships among the most well-studied *Arabidopsis* (**a**) and *Capsella* species (**b**). Mating system (selfing or outcrossing) and self-incompatibility status (*SI* self-incompatible, *SC* self-compatible) and ploidy level is indicated for each species. Approximate estimates of split times are indicated by arrows. For *A. thaliana* and *A. lyrata*, we show two recent estimates based on Guo et al. [124] and Beilstein et al. [125]. The estimate of the timing of the origin of *A. suecica* is based on

3 Population Genetics Consequences of Selfing

3.1 Theoretical Expectations

Self-fertilization has drastic consequences on the patterns and distribution of genetic variation, and for the impact of natural selection. The level of selfing is therefore an important factor to consider in population genetics. Here we summarize the expected population genetic consequences of selfing (Fig. 2) and then present empirical results from *Arabidopsis* and *Capsella* that illustrate the theoretical expectations.

Selfing has two major key effects; it results in reduced heterozygosity and a reduced effective population size (N_e). Under complete selfing, heterozygosity is halved every generation. Hence, the heterozygosity of a completely selfing population is almost fully eliminated already after six generations of complete selfing. A side effect is a rapid generation of isolated lines of different genotypes [40] which is expected to result in stronger population structure in selfing species compared to outcrossers [41]. Moreover, selfers are expected to exhibit a reduction in N_e for several reasons. First, selfing immediately results in a twofold reduction of the number of independently sampled gametes, and this is expected to reduce the N_e by a factor of two [42, 43]. Even greater reductions in N_e are expected if selfers undergo more frequent extinction and recolonization dynamics than outcrossers [44], or if the origin of selfing species is often associated with bottlenecks [14]. Furthermore, because selfing results in a rapid decrease in heterozygosity, recombination is less efficient at breaking up linkage disequilibrium in selfers than in outcrossers [45]. In this situation, background selection or recurrent hitchhiking (linked selection) will have a greater impact, reducing neutral genetic diversity beyond what would be expected in an outcrosser [46]. Together, these factors decrease the overall genetic diversity [41] and increase the linkage disequilibrium (LD) of selfing populations [43].

The combined effect of reduced N_e and effective recombination rate will also affect the efficacy of selection genome-wide. On the one hand, when N_e is reduced, a higher proportion of the genome behaves neutrally and alleles that were slightly deleterious in large populations become effectively neutral [47]. In addition, as an effect of the reduced effective recombination rate in selfers, Hill-Robertson interference [48] will increase and therefore limit

---

Fig. 1 (continued) Novikova et al. [106]. The timing of the population split between *C. rubella* and *C. grandiflora* is based on Slotte et al. [76] and the timing of the origin of *C. bursa-pastoris* and the split between *C. orientalis* and the *C. grandiflora/C. rubella* lineage is based on Douglas et al. [18]. Photographs of Arabidopsis species were taken by Jon Ågren (*A. thaliana*), Robin Burns (*A. suecica*), Johanna Leppälä (*A. arenosa*), Tiina Mattila (*A. lyrata*), Vincent Castric (*A. halleri*), and Rie Shimizu-Inatsugi (*A. kamchatica*). All *Capsella* photographs were taken by Kim Steige

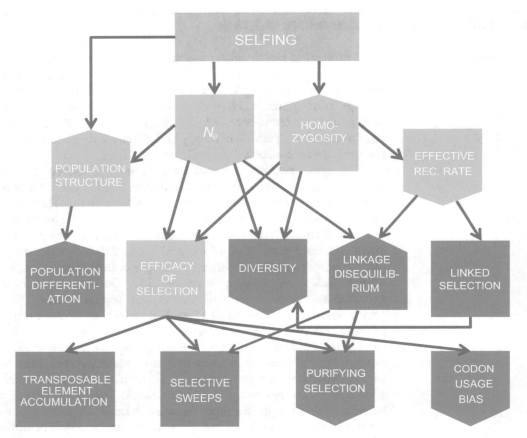

Fig. 2 A network showing the effect of selfing on population parameters (orange) and population genetics statistics (pink). The box shape indicates the predicted effect (increase/decrease) of selfing on each factor (square boxes indicate that the effect can be either an increase or a decrease, depending on the exact combination of parameter values)

selection efficacy further. As a consequence, one may expect selfing lineages to have an excess of nonsynonymous divergence compared to synonymous divergence (d_N/d_S or K_a/K_s) as well as polymorphisms (π_N/π_S), mainly due to weakened selection against slightly deleterious variants [41]. Further, reduced efficacy of selection and recombination rate may also decrease the level of codon usage bias [49, 50]. However, it should be noted that spurious signals of relaxed purifying selection can result as a result of recent demographic change [51], because the time to reach equilibrium after a bottleneck is longer for nonsynonymous than for synonymous polymorphism. Ideally, forward population genetic simulations incorporating selection and demography should therefore be undertaken to validate inference of relaxed purifying selection.

The dynamics of alleles with different levels of dominance will also be affected by the mating system [52], in ways that can sometimes counteract the effect of reduced N_e. For instance, in selfing species, increased homozygosity renders recessive alleles visible to selection, and as a result, fixation probabilities of recessive advantageous alleles are expected to be higher in selfers than in outcrossers [53]. Harmful recessive alleles can also be removed more efficiently, resulting in purging of recessive deleterious alleles, unless the reduction in N_e in selfers is severe enough that genetic drift overpowers the homozygosity effect [54].

Selfing and outcrossing populations have also been shown to differ in the dynamics of adaptive alleles. The reduced efficacy of selection will decrease the probability of slightly beneficial mutation fixation but also the fixation time of beneficial mutation in selfers is faster in comparison with the outcrossing species regardless of the dominance level [13, 55]. Further, in selfing populations, adaptation is more likely to result from new mutations (hard sweeps) while in outcrossers adaptation from standing variation (soft sweeps) is predicted to be more frequent [13, 15]. On the other hand, the effect of linked selection is expected to be stronger in selfing species due to reduced effective recombination rate which will increase the fixation probability of linked harmful mutations [56] and potentially limiting the adaptive potential of selfers.

Mating system has also been hypothesized to affect the transposable element (TE) content of the genome. TEs are mobile genetic elements that make up a large yet variable proportion of many plant genomes [57, 58]. Theory predicts that both mating system variation and differences in the effective population size (N_e) should affect TE content, because these factors affect possibilities for TEs to spread, the potential for evolution of self-regulation of transposition, and the efficacy of selection against deleterious TE insertions. For instance, outcrossing enhances opportunities for TE spread [59], and transposition rates should evolve to be highest in outcrossers, which should therefore be expected to have higher TE content than highly selfing species [60]. Likewise, if the harmful effects of TEs are mostly recessive or codominant, increased homozygosity in selfers leads to more efficient purifying selection against TEs in selfers [61]. On the other hand, natural selection against slightly deleterious TE insertions could be compromised in selfing species, because of their reduced effective population size [41]. Increased homozygosity in selfers might also decrease the deleterious effect of TEs, because this decreases the probability of ectopic recombination [62]. Under this scenario, a transition to selfing would be expected to lead to an increase in TE content. Different models therefore yield contrasting predictions regarding the expected effect of mating system variation on TE content.

3.2 Empirical Results   Several empirical results have confirmed the theoretical predictions regarding the population genetics effects of selfing. Figure 3 summarizes some empirical population genetics results from selfing and outcrossing *Arabidopsis* and *Capsella* species. First, while it is difficult to disentangle the effect of past demographic events from the effect of selfing, both *C. rubella* and *A. thaliana* present high levels of population structure [21, 22, 63]. Indeed, population structure is stronger in the selfer *C. rubella* than in the outcrossing *C. grandiflora*, consistent with theoretical expectations for selfers [22] (Fig. 3).

Furthermore, evidence for a reduction in N_e has been found in natural populations of *C. rubella* and *A. thaliana*. In *C. rubella* both synonymous diversity and population recombination rate are significantly lower than in the outcrossing sister species *C. grandiflora* [22, 64]. Similarly, the selfer *A. thaliana* shows lower synonymous polymorphism in comparison with the outcrossing *A. lyrata* [65, 66] (Fig. 3). An early study also found evidence for a faster decay of linkage disequilibrium (LD) in *A. lyrata* than in *A. thaliana* [66]. However, more recent work has shown that there is also high variation in nucleotide diversity and LD patterns in *A. thaliana*, both between different parts of the genome and across different geographic regions and habitats [67–70]. Patterns in *A. lyrata* are also complicated by strong population size decrease in several extensively studied *A. lyrata* populations [71], which also decreases the population recombination rate. Hence, it is worth noticing that both diversity and LD are both highly dependent on the past demographic history and the local variation in the recombination rate across the genome. The decay of LD also depends on whether estimates are based on local or global population samples [72]. In *A. thaliana* LD decays faster in a world-wide sample in comparison with local populations [68, 70] which may be due to local populations having a low number of founders [72].

Empirical evidence for the impact of selfing on the efficacy of selection started with early investigations of divergence and polymorphism in *A. thaliana* and its outcrossing sister species *A. lyrata*, using a limited number of loci. This study found very limited evidence for relaxed selection in *A. thaliana* [73]. However, using genome-wide polymorphism data, Slotte et al. [74] found evidence for weaker purifying selection on nonsynonymous sites in *A. thaliana* relative to the outcrosser *C. grandiflora*. Further studies confirmed decreased codon usage bias in both *A. thaliana* and *C. rubella* in comparison with the outcrossing *A. lyrata* and *C. grandiflora* [50] (Fig. 3e). Analyses of population genomic data from *C. rubella* and *C. grandiflora* further found evidence for a higher ratio of nonsynonymous to synonymous polymorphism in *C. rubella* [75, 76]. Forward population genomic simulations demonstrated that this was likely primarily a result of the reduced N_e in *C. rubella*, and not due to a major shift in the distribution of fitness effects (DFE) in association with the shift to selfing [76]. More recently, a study that directly estimated the DFE based on analyses of site frequency spectra found evidence for a higher proportion of nearly neutral nonsynonymous mutations in

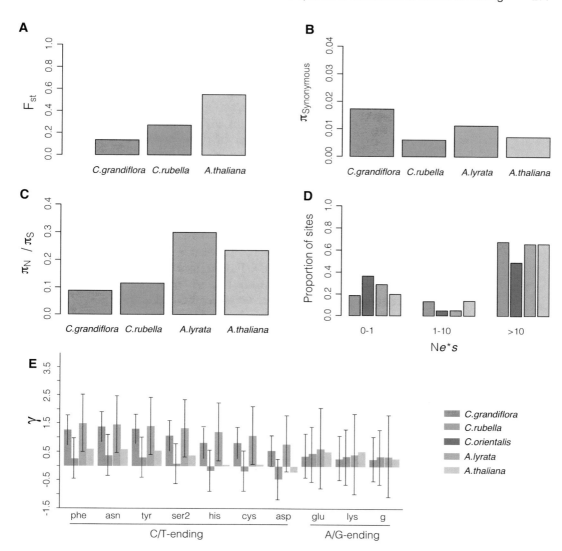

Fig. 3 Empirical results illustrating the impact of a mating system shift in the *Capsella* and *Arabidopsis* genus. (**a**) *Population structure*: Global F_{ST} among populations is higher in the selfer *C. rubella* than in the outcrosser *C. grandiflora* [22]. Elevated values of F_{ST} are also found genome-wise among populations of *Arabidopsis thaliana* [70]. (**b**) *Neutral genetic diversity*: nucleotide diversity at synonymous sites is lower in selfers compared to their outcrossing relatives in both *Capsella* and *Arabidopsis* [74, 76, 126]. (**c**) *Strength of purifying selection*: a higher ratio of nonsynonymous to synonymous nucleotide diversity suggests relaxed purifying selection in the selfing *C. rubella* [74, 76]. In contrast, in *Arabidopsis*, the outcrosser *A. lyrata* has a higher ratio of nonsynonymous to synonymous nucleotide diversity than *A. thaliana*. (**d**) *Distribution of fitness effect* (DFE) in bins of $N_e s$ (the product of the effective population size and the selection coefficient) for new nonsynonymous mutations [18, 74, 83, 126]. (**e**) *Codon usage bias*: maximum likelihood estimates of selection coefficient, Υ, for the ten amino acids with twofold degenerate codons between the selfing and outcrossing species. Whiskers are 95% confidence intervals obtained by the MCMC analysis (*see* Qiu et al. [50] for details, with permission from GBE)

the selfers *C. orientalis* and *C. bursa-pastoris* relative to the outcrosser *C. grandiflora* [18]. These results are in general agreement with the theoretical expectation that purifying selection should be relaxed in selfers.

While some models predict that a shift to selfing should lead to a reduced prevalence of TEs, other models predict the opposite. So far, empirical evidence from *Arabidopsis* and *Capsella* do not unequivocally support either of these predictions. On one hand, the comparison of structure of the *A. thaliana* and *A. lyrata* genome sequences suggested that the *A. thaliana* genome contained large numbers of small deletions, especially in TEs [77]. This result is consistent with the hypothesis that in selfing species TEs are more efficiently removed [61]. One the other hand, Lockton and Gaut [78] found that many TE families are in higher frequency and are subjected to weaker selection in *A. thaliana* in comparison with *A. lyrata*. Furthermore, comparison of *C. rubella*, *C. grandiflora*, *A. thaliana* and *A. lyrata* revealed that the TE frequency and density in *Capsella* showed a stronger resemblance to *A. thaliana* than to *A. lyrata* [76]. This may indicate that the reason for the TE abundance difference between the *Arabidopsis* species is accumulation of TEs in the *A. lyrata* genome rather than decline in the selfing *A. thaliana* lineage.

Focusing on TE content in selfing and outcrossing *Capsella* species, Ågren et al. [79, 80] found an increase in TE number in *C. rubella* but a slight decrease in the selfer *C. orientalis*, in comparison with the outcrossing *C. grandiflora*. In the polyploid selfer *C. bursa-pastoris*, no evidence for a difference in TE dynamics was found in comparison with its parental species, *C. grandiflora* and *C. orientalis* [80]. Thus, while there is some evidence for a reduced prevalence of TEs in selfers, the results are not unequivocal, and further work is needed to clarify whether the contrasting findings might be related to the timing of the shift to selfing and demographic history. Indeed, there is evidence for an effect of demographic history on selection against TEs in *A. lyrata*, where large refugial populations exhibited a signature of purifying selection against TE insertions, whereas in bottlenecked populations, TEs were evolving neutrally [81]. In addition to broad comparative genomic studies contrasting species that differ in their mating system, studies of intraspecific variation can thus provide insight into population genomics and selection on TEs [82]. Because TEs are important contributors to variation in plant genome size and TE silencing can also affect gene regulation [83–85], it is of considerable interest to improve our understanding of the impact of mating system on variation in TE content.

4 Discovering the Geographic Origin and the Timing of the Mating System Shift

Understanding the timing, mode, and geographic location of the shift to selfing is of key importance for proper interpretation of population genomic data from selfing species [86]. For instance, improved understanding of the timing and geographical location of the shift can be key for interpretation of genetic structure, and can allow one to account for underlying neutral (demography induced) processes when investigating changes in the efficacy of selection.

In *A. thaliana*, several studies have estimated the timing of the emergence of selfing based on patterns of polymorphism and demographic modeling. *A. thaliana* is native to the Eurasia and Africa [87–89] and widely spread especially in Europe. It has also recently spread into North America in association with humans [21]. All the currently known accessions are SC, indicating that the evolution of this trait preceded the worldwide spread of the species. Linkage disequilibrium patterns suggested that the transition to selfing in *A. thaliana* occurred as early as 1,000,000 years ago [90] while coalescent modeling using *S*-locus diversity suggested a younger origin with an upper estimate of approximately 400,000 years ago [91].

The recent development of large-scale population genomics datasets from 1135 *A. thaliana* accessions [63], offers one of the best population genomics resource for studying plant population genetics and molecular evolution. A recent demographic modeling study exploited this resource and included additional accessions covering roughly the African distribution of the species to investigate the timing and geographic origin of the shift to selfing in *A. thaliana* [89]. Using a combination of the MSMC method that infers cross-coalescent times and fluctuations in the effective population size using a whole-genome data from multiple populations [92] and the site frequency spectrum based diffusion approximation method δaδi [93], Durvasula et al. [89] inferred the demographic history of the different groups from Africa and Europe. They suggest that partial loss of SI occurred 500,000–1,000,000 years ago subsequent to the migration of the ancestral founding *A. thaliana* population to Africa approximately 800,000–1,200,000 years ago. Although the existence of multiple nonfunctional *S*-haplotypes suggests that the final loss of SI occurred multiple times independently [94, 95], the new study including African accessions shows that all the currently known *S*-haplotypes are found coexisting in Morocco, suggesting that selfing originated in this geographic region [89]. This result was further supported by the higher estimated N_e in African populations, and they estimated that the species spread out-of-Africa some 90,000–140,000 years ago.

The postglacial colonization of *A. thaliana* within Europe has likely proceeded through acquirement of a weedy lifestyle [63]. The first European colonists likely occurred in the southern and eastern part of Europe [63, 96, 97] and the massive spread over central and northern parts of Europe occurred later, possibly associated with human action [63]. Population structure analysis revealed that there are two distinct groups in Central Europe and these two groups have admixed in Central Europe [98–100]. Two distinct lineages are also present in Scandinavia with accessions from Finland as well as from the northern parts of Sweden and Norway forming their own cluster while the southern Swedish accessions cluster with the southern accessions [99, 101]. Using a whole-genome population-genomics approach Lee et al. [102] found five different clusters within European *A. thaliana* accessions and they suggest that that these groups have given rise to the current distribution of the species within Europe.

Another selfing example from *Arabidopsis*, where the demographic and colonization history has been studied, is the polyploid species *A. suecica*, which is a selfing allopolyploid between *A. thaliana* and *A. arenosa* (Fig. 1). The species is spread in central Sweden and southern Finland. Early investigation explored 52 microsatellites and four nuclear sequences [103] and inferred a single and recent origin of the species approximately 12,000–300,000 YA followed by northward spread using a Bayesian coalescent population modeling. This single origin has also been supported by other studies based on the amount of variation in chloroplast sequence data [104, 105]. However, recent investigation of whole-genome resequencing data from 15 *A. suecica* accessions concluded that the multiple origins hypothesis cannot be ruled out [106]. Based on the *S*-locus haplotype dominance patterns in these accessions they suggest that *A. suecica* could have been SC, at least to some degree, immediately after the species emergence approximately 15,100–16,600 years ago (Fig. 1).

In *Capsella* it has been estimated that the timing of the loss of SI is much more recent. Isolation-migration analyses based on 39 gene fragments and assuming a mutation rate of 1.5×10^{-8} suggested that the shift to selfing was concomitant with speciation of *C. rubella* from an outcrossing ancestor similar to present-day *C. grandiflora*, and that this occurred some 20,000 years ago [64]. Likewise, an investigation of the diversity patterns at the *S*-locus across the European range of the species suggested that loss of SI took place in Greece approximately 40,000 years ago [37], since the presumably ancestral long form of *SRK* allele is present in this region while the other accessions harbor a shorter form. Divergence estimates based on analysis of genome-wide founding haplotypes suggested that selfing evolved later, approximately 50,000–100,000 years ago [75]. Assuming a different mutation rate of 7.1×10^{-9}, Slotte et al. [76] analyzed genome-wide site frequency spectra using δaδi, estimated that the timing of the split between *C. grandiflora* and *C. rubella* occurred <200,000 years ago (Fig. 1). Later on, coalescent-based analyses of genome-wide joint site frequency spectra from *C. grandiflora*, *C. orientalis* and *C. bursa-pastoris* were used to infer the timing of allopolyploid speciation resulting in the selfing species *C. bursa-pastoris* ([18], Fig. 1). Together these analyses have provided an evolutionary framework for further genomic studies of the consequences of selfing and allopolyploidy in *Capsella* (Fig. 1).

5   Some Caveats

As many authors point out, estimates of population split times are usually based on assumptions that contain considerable uncertainty (*see*, e.g., ref. 103). For example, a fixed mutation rate or a distribution around a mean is often assumed. The direct mutation rate estimate from *A. thaliana* mutation accumulation lines of 7.1×10^{-9} [107] is commonly used for *Arabidopsis*, whereas, as pointed out above, early estimates in *Capsella* were based on a mutation rate of 1.5×10^{-8} [108] while later studies (e.g., Slotte et al. [76] and Douglas et al. [18]) have used the Ossowski et al. [107] mutation rate estimate. Other assumptions may include constant generation time, recombination rate and analyses may further be restricted to a limited number of demographic models. These factors may be variable between studies and it is important to pay attention to these details when evaluating the results of demographic modeling.

A more severe limitation concerns the effect of selection on linked sites on demographic inference. Simulations have shown that the increased intensity of background selection in selfers can rapidly lead to a strong reduction in neutral diversity [109]. Based on these findings, some authors have questioned whether it is possible to reliably infer demographic changes associated with the shift to selfing based on neutral polymorphism [109]. The impact of linked selection on patterns of neutral polymorphism is indeed a general problem for demographic inference [110–112]. To some degree, it may be possible to circumvent this problem by judiciously choosing which sites to use for demographic inference [46], and by using forward population genetic simulations in software such as SLiM [113, 114] to assess whether results are robust to the effect of selection on linked sites. As an example, a recent study on *Arabis alpina* used site frequency spectra for sites in genomic regions with high recombination rates and low gene density, which should be least affected by linked selection, to infer the demographic history of selfing Scandinavian populations [115]. The reliability of this inference was further checked with forward simulations incorporating background selection and a shift to selfing [115]. This study therefore demonstrated one way to assess the effect of linked selection on demographic inference in selfing populations.

6 Future Directions

In this chapter, we have given a brief overview of population genomic studies of the transition to selfing, focusing on the two model systems *Arabidopsis* and *Capsella*. One limitation of most of the studies presented here is that they have focused on comparing pairs of species with contrasting mating systems. Ideally, to be able to distinguish between idiosyncrasies of one particular contrast and effects of selfing *per se*, future comparative population genomic studies of the effect of selfing should include multiple phylogenetically independent contrasts.

An additional limitation is that most studies have focused on contrasting only highly selfing and obligate outcrossing species, although there is a lot more diversity to plant mating systems. For instance, a substantial proportion (approximately 42%) of flowering plants undergoes a mix of outcrossing and self-fertilization [116]. Despite this fact, and despite the existence of theoretical and simulation-based work on the expected population genomic consequences of partial selfing [15, 54, 56, 117], there is a dearth of empirical population genomic studies including mixed mating populations and species (but *see*, e.g., 115, 118). One exception is a recent study which tested for a difference in the impact of purifying selection among obligate outcrossing, mixed-mating, and highly selfing populations of *Arabis alpina* [115]. This study found no major detectable difference in purifying selection between mixed mating and outcrossing populations, whereas purifying selection efficacy was significantly lower in Scandinavian selfing populations, most likely as a result of a postglacial colonization bottleneck [115]. These results are consistent with the expectation that a low level of outcrossing may be sufficient to prevent accumulation of deleterious alleles [117]. However, further empirical studies in more mixed mating species and populations, ideally including larger sample sizes, are required to establish the generality of this pattern.

Another less empirically studied issue is how the prevalence of positive selection and especially how selective sweeps are impacted by mating system. In *A. thaliana*, the proportion of amino acid substitutions driven by positive selection has been estimated to be close to 0% [74, 119] while for example in the outcrossing relative *C. grandiflora* this proportion has been estimated to be as high 40% [74]. However, making inferences on the underlying cause of this difference is not straightforward. For example, positive selection has also been shown to be rare in the outcrossing *A. lyrata* [120] suggesting that other factors, such as for instance demographic history, may also result in low rates of adaptive substitutions. The theoretical work on the effect of dominance on fixation probabilities has yielded results that could be tested by taking advantage of genome-wide population genetics datasets and methodological developments regarding sweep detection (e.g., [121, 122]). For example, with such methodology, it is possible to test whether there is a difference in the relative occurrence of hard and soft sweeps in selfers and outcrossing populations as predicted by theoretical work [13]. Further advances in the classification of mutations into different dominance classes [123] will allow for testing hypotheses related to the behavior of recessive, additive, and dominant alleles in selfing vs. outcrossing species.

References

1. Barrett SC (2002) Evolution of sex: the evolution of plant sexual diversity. Nat Rev Genet 3:274

2. Darwin C (1876) The effects of cross and self fertilisation in the vegetable kingdom. J Murray, London

3. Darwin C (1877) The different forms of flowers on plants of the same species. J Murray, London

4. Lloyd DG (1984) Gender allocation in outcrossing cosexual plants. In: Dirzo R, Sarukhán J (eds) Perspectives on plant population ecology. Sinauer, Sunderland, MA, pp 277–300

5. Charlesworth D, Charlesworth B (1987) Inbreeding depression and its evolutionary consequences. Annu Rev Ecol Syst 18:237–268

6. Charnov EL (1982) The theory of sex allocation. Princeton University Press, Princeton, NJ

7. Raduski AR, Haney EB, Igić B (2012) The expression of self-incompatibility in angiosperms is bimodal. Evolution 66:1275–1283

8. Igic B, Lande R, Kohn J (2008) Loss of self-incompatibility and its evolutionary consequences. Int J Plant Sci 169:93–104

9. Fisher RA (1941) Average excess and average effect on an allelic substitution. Ann Eugenics 11:53–63

10. Charlesworth D (2006) Evolution of plant breeding systems. Curr Biol 16:R735

11. Goldberg EE, Kohn JR, Lande R et al (2010) Species selection maintains self-incompatibility. Science 330:493–495

12. Stebbins GL (1957) Self fertilization and population variability in the higher plants. Am Nat 91:337–354

13. Glémin S, Ronfort J (2013) Adaptation and maladaptation in selfing and outcrossing species: new mutations versus standing variation. Evolution 67:225–240

14. Wright SI, Kalisz S, Slotte T (2013) Evolutionary consequences of self-fertilization in plants. Proc R Soc B 280:20130133

15. Hartfield M (2016) Evolutionary genetic consequences of facultative sex and outcrossing. J Evol Biol 29:5–22

16. Clauss MJ, Koch MA (2006) Poorly known relatives of *Arabidopsis thaliana*. Trends Plant Sci 11:449–459

17. Hurka H, Friesen N, German DA et al (2012) 'Missing link' species *Capsella orientalis* and *Capsella thracica* elucidate evolution of model plant genus *Capsella* (Brassicaceae). Mol Ecol 21:1223–1238

18. Douglas GM, Gos G, Steige KA et al (2015) Hybrid origins and the earliest stages of diploidization in the highly successful recent polyploid *Capsella bursa-pastoris*. PNAS 112:2806–2811

19. Abbott RJ, Gomes MF (1989) Population genetic structure and outcrossing rate of *Arabidopsis thaliana* (L.) Heynh. Heredity 62:411

20. Bergelson J, Stahl E, Dudek S et al (1998) Genetic variation within and among populations of *Arabidopsis thaliana*. Genetics 148:1311–1323

21. Platt A, Horton M, Huang YS et al (2010) The scale of population structure in *Arabidopsis thaliana*. PLoS Genet 6:e1000843

22. St. Onge KR, Källman T, Slotte T et al (2011) Contrasting demographic history and population structure in *Capsella rubella* and *Capsella grandiflora*, two closely related species with different mating systems. Mol Ecol 20:3306–3320

23. Säll T, Lind-Halldén C, Jakobsson M et al (2004) Mode of reproduction in *Arabidopsis suecica*. Hereditas 141:313–317

24. Shimizu KK, Fujii S, Marhold K et al (2005) *Arabidopsis kamchatica* (Fisch. ex DC.) K. Shimizu & Kudoh and *A. kamchatica* subsp. *kawasakiana* (Makino) K. Shimizu & Kudoh, new combinations. Acta Phytotax Geobot 56:163–172

25. Tsuchimatsu T, Kaiser P, Yew C et al (2012) Recent loss of self-incompatibility by degradation of the male component in allotetraploid *Arabidopsis kamchatica*. PLoS Genet 8: e1002838

26. Shimizu-Inatsugi R, LihovÁ J, Iwanaga H et al (2009) The allopolyploid *Arabidopsis kamchatica* originated from multiple individuals of *Arabidopsis lyrata* and *Arabidopsis halleri*. Mol Ecol 18:4024–4048

27. Takayama S, Isogai A (2005) Self-incompatibility in plants. Annu Rev Plant Biol 56:467–489

28. Nasrallah JB (2017) Plant mating systems: self-incompatibility and evolutionary transitions to self-fertility in the mustard family. Curr Opin Genet Dev 47:54–60

29. Mable BK, Hagmann J, Kim ST et al (2017) What causes mating system shifts in plants? *Arabidopsis lyrata* as a case study. Heredity 118:52–63

30. Shimizu KK, Tsuchimatsu T (2015) Evolution of selfing: recurrent patterns in molecular adaptation. Annu Rev Ecol Evol Syst 46:593–622

31. Uyenoyama MK, Zhang Y, Newbigin E (2001) On the origin of self-incompatibility haplotypes: transition through self-compatible intermediates. Genetics 157:1805–1817

32. Tsuchimatsu T, Shimizu KK (2013) Effects of pollen availability and the mutation bias on the fixation of mutations disabling the male specificity of self-incompatibility. J Evol Biol 26:2221–2232

33. Tsuchimatsu T, Suwabe K, Shimizu-Inatsugi R et al (2010) Evolution of self-compatibility in *Arabidopsis* by a mutation in the male specificity gene. Nature 464:1342–1346

34. Nasrallah ME, Liu P, Nasrallah JB (2002) Generation of self-incompatible *Arabidopsis thaliana* by transfer of two S locus genes from *A. lyrata*. Science 297:247–249

35. Nasrallah ME, Liu P, Sherman-Broyles S et al (2004) Natural variation in expression of self-incompatibility in *Arabidopsis thaliana*: implications for the evolution of selfing. PNAS 101:16070–16074

36. Nasrallah JB, Liu P, Sherman-Broyles S et al (2007) Epigenetic mechanisms for breakdown of self-incompatibility in interspecific hybrids. Genetics 175:1965–1973

37. Guo YL, Bechsgaard JS, Slotte T et al (2009) Recent speciation of *Capsella rubella* from *Capsella grandiflora*, associated with loss of self-incompatibility and an extreme bottleneck. PNAS 106:5246–5251

38. Slotte T, Hazzouri KM, Stern D et al (2012) Genetic architecture and adaptive significance of the selfing syndrome in *Capsella*. Evolution 66:1360–1374

39. Vekemans X, Poux C, Goubet PM et al (2014) The evolution of selfing from outcrossing ancestors in Brassicaceae: what have we learned from variation at the S-locus? J Evol Biol 27:1372–1385

40. Hedrick PW (2005) Genetics of populations. Jones and Bartlett, Boston (MA)

41. Charlesworth D, Wright SI (2001) Breeding systems and genome evolution. Curr Opin Genet Dev 11:685–690

42. Pollak E (1987) On the theory of partially inbreeding finite populations. I. Partial selfing. Genetics 117:353–360

43. Nordborg M (2000) Linkage disequilibrium, gene trees and selfing: an ancestral recombination graph with partial self-fertilization. Genetics 154:923–929

44. Ingvarsson P (2002) A metapopulation perspective on genetic diversity and differentiation in partially self-fertilizing plants. Evolution 56:2368–2373

45. Roze D, Lenormand T (2005) Self-fertilization and the evolution of recombination. Genetics 170:841–857

46. Slotte T (2014) The impact of linked selection on plant genomic variation. Brief Funct Genomics 13:268–275

47. Ohta T (1973) Slightly deleterious mutant substitutions in evolution. Nature 246:96–98

48. Hill WG, Robertson A (1966) The effect of linkage on limits to artificial selection. Genet Res 8:269–294

49. Marais G, Charlesworth B, Wright SI (2004) Recombination and base composition: the case of the highly self-fertilizing plant *Arabidopsis thaliana*. Genome Biol 5:R45

50. Qiu S, Zeng K, Slotte T et al (2011) Reduced efficacy of natural selection on codon usage bias in selfing *Arabidopsis* and *Capsella* species. Genome Biol Evol 3:868–880

51. Brandvain Y, Wright SI (2016) The limits of natural selection in a nonequilibrium world. Trends Genet 32:201–210

52. Glémin S (2007) Mating systems and the efficacy of selection at the molecular level. Genetics 177:905–916

53. Charlesworth B (1992) Evolutionary rates in partially self-fertilizing species. Am Nat 140:126–148

54. Glémin S (2003) How are deleterious mutations purged? Drift versus nonrandom mating. Evolution 57:2678–2687

55. Caballero A, Hill WG (1992) Effects of partial inbreeding on fixation rates and variation of mutant genes. Genetics 131:493–507

56. Hartfield M, Glémin S (2014) Hitchhiking of deleterious alleles and the cost of adaptation in partially selfing species. Genetics 196:281–293

57. Tenaillon MI, Hollister JD, Gaut BS (2010) A triptych of the evolution of plant transposable elements. Trends Plant Sci 15:471–478

58. Ågren JA (2014) Evolutionary transitions in individuality: insights from transposable elements. Trends Evol Ecol 29:90–96

59. Boutin TS, Le Rouzic A, Capy P (2012) How does selfing affect the dynamics of selfish transposable elements? Mob DNA 3(1):5

60. Charlesworth B, Langley CH (1986) The evolution of self-regulated transposition of transposable elements. Genetics 112:359–383

61. Wright SI, Schoen DJ (1999) Transposon dynamics and the breeding system. Genetica 107:139–148

62. Charlesworth D, Charlesworth B (1995) Transposable elements in inbreeding and outbreeding populations. Genetics 140:415–417

63. 1001 Genomes Consortium (2016) 1,135 genomes reveal the global pattern of polymorphism in *Arabidopsis thaliana*. Cell 166:481–491

64. Foxe JP, Slotte T, Stahl EA et al (2009) Recent speciation associated with the evolution of selfing in *Capsella*. PNAS 106:5241–5245

65. Savolainen O, Langley CH, Lazzaro BP et al (2000) Contrasting patterns of nucleotide polymorphism at the alcohol dehydrogenase locus in the outcrossing *Arabidopsis lyrata* and the selfing *Arabidopsis thaliana*. Mol Biol Evol 17:645–655

66. Wright SI, Foxe JP, DeRose-Wilson L et al (2006) Testing for effects of recombination rate on nucleotide diversity in natural populations of *Arabidopsis lyrata*. Genetics 174:1421–1430

67. Nordborg M, Hu TT, Ishino Y et al (2005) The pattern of polymorphism in *Arabidopsis thaliana*. PLoS Biol 3:e196

68. Kim S, Plagnol V, Hu TT et al (2007) Recombination and linkage disequilibrium in *Arabidopsis thaliana*. Nat Genet 39:1151–1155

69. Bomblies K, Yant L, Laitinen RA et al (2010) Local-scale patterns of genetic variability, outcrossing, and spatial structure in natural stands of *Arabidopsis thaliana*. PLoS Genet 6:e1000890

70. Cao J, Schneeberger K, Ossowski S et al (2011) Whole-genome sequencing of multiple *Arabidopsis thaliana* populations. Nat Genet 43:956–963

71. Mattila TM, Tyrmi J, Pyhäjärvi T et al (2017) Genome-wide analysis of colonization history and concomitant selection in *Arabidopsis lyrata*. Mol Biol Evol 34:2665–2677

72. Nordborg M, Borevitz JO, Bergelson J et al (2002) The extent of linkage disequilibrium in *Arabidopsis thaliana*. Nat Genet 30:190–193

73. Wright SI, Lauga B, Charlesworth D (2002) Rates and patterns of molecular evolution in inbred and outbred *Arabidopsis*. Mol Biol Evol 19:1407–1420

74. Slotte T, Foxe JP, Hazzouri KM et al (2010) Genome-wide evidence for efficient positive and purifying selection in *Capsella grandiflora*, a plant species with a large effective population size. Mol Biol Evol 27:1813–1821

75. Brandvain Y, Slotte T, Hazzouri KM et al (2013) Genomic identification of founding haplotypes reveals the history of the selfing species *Capsella rubella*. PLoS Genet 9:e1003754

76. Slotte T, Hazzouri KM, Ågren JA et al (2013) The *Capsella rubella* genome and the genomic consequences of rapid mating system evolution. Nat Genet 45:831–835

77. Hu TT, Pattyn P, Bakker EG et al (2011) The *Arabidopsis lyrata* genome sequence and the basis of rapid genome size change. Nat Genet 43:476–481

78. Lockton S, Gaut B (2010) The evolution of transposable elements in natural populations of self-fertilizing *Arabidopsis thaliana* and its outcrossing relative *Arabidopsis lyrata*. BMC Evol Biol 10:10

79. Ågren JA, Wang W, Koenig D et al (2014) Mating system shifts and transposable element evolution in the plant genus *Capsella*. BMC Genomics 15:602

80. Ågren JA, Huang H, Wright SI (2016) Transposable element evolution in the allotetraploid *Capsella bursa-pastoris*. Am J Bot 103:1197–1202

81. Lockton S, Ross-Ibarra J, Gaut BS (2008) Demography and weak selection drive patterns of transposable element diversity in natural populations of *Arabidopsis lyrata*. PNAS 105:13965–13970

82. Horvath R, Slotte T (2017) The role of small RNA-based epigenetic silencing for purifying selection on transposable elements in *Capsella grandiflora*. Genome Biol Evol 9:2911–2920

83. Steige KA, Laenen B, Reimegård J et al (2017) Genomic analysis reveals major determinants of *cis*-regulatory variation in *Capsella grandiflora*. PNAS 114:1087–1092

84. Wang X, Weigel D, Smith LM (2013) Transposon variants and their effects on gene expression in *Arabidopsis*. PLoS Genet 9: e1003255

85. Steige KA, Reimegård J, Koenig D et al (2015) Cis-regulatory changes associated with a recent mating system shift and floral adaptation in *Capsella*. Mol Biol Evol 32:2501–2514

86. Charlesworth D, Vekemans X (2005) How and when did *Arabidopsis thaliana* become highly self-fertilising. BioEssays 27:472–476

87. Hoffmann MH (2005) Evolution of the realized climatic niche in the genus *Arabidopsis* (Brassicaceae). Evolution 59:1425–1436

88. Brennan AC, Méndez-Vigo B, Haddioui A et al (2014) The genetic structure of *Arabidopsis thaliana* in the south-western Mediterranean range reveals a shared history between North Africa and southern Europe. BMC Plant Biol 14:17

89. Durvasula A, Fulgione A, Gutaker RM et al (2017) African genomes illuminate the early history and transition to selfing in *Arabidopsis thaliana*. PNAS 114(20):5213–5218

90. Tang C, Toomajian C, Sherman-Broyles S et al (2007) The evolution of selfing in *Arabidopsis thaliana*. Science 317:1070–1072

91. Bechsgaard JS, Castric V, Charlesworth D et al (2006) The transition to self-compatibility in *Arabidopsis thaliana* and evolution within S-haplotypes over 10 Myr. Mol Biol Evol 23:1741–1750

92. Schiffels S, Durbin R (2014) Inferring human population size and separation history from multiple genome sequences. Nat Genet 46:919–925

93. Gutenkunst RN, Hernandez RD, Williamson SH et al (2009) Inferring the joint demographic history of multiple populations from multidimensional SNP frequency data. PLoS Genet 5:e1000695

94. Boggs NA, Nasrallah JB, Nasrallah ME (2009) Independent S-locus mutations caused self-fertility in *Arabidopsis thaliana*. PLoS Genet 5(3):e1000426

95. Shimizu KK, Shimizu-Inatsugi R, Tsuchimatsu T et al (2008) Independent origins of self-compatibility in *Arabidopsis thaliana*. Mol Ecol 17(2):704–714

96. Beck JB, Schmuths H, Schaal BA (2008) Native range genetic variation in *Arabidopsis thaliana* is strongly geographically structured and reflects Pleistocene glacial dynamics. Mol Ecol 17:902–915

97. Picó FX, Méndez-Vigo B, Martinez-Zapater J et al (2008) Natural genetic variation of Arabidopsis thaliana is geographically structured in the Iberian peninsula. Genetics 180:1009–1021

98. Sharbel TF, Haubold B, Mitchell-Olds T (2000) Genetic isolation by distance in *Arabidopsis thaliana*: biogeography and postglacial colonization of Europe. Mol Ecol 9:2109–2118

99. François O, Blum MGB, Jakobsson M et al (2008) Demographic history of European populations of *Arabidopsis thaliana*. PLoS Genet 4:e1000075

100. Long Q, Rabanal FA, Meng D et al (2013) Massive genomic variation and strong selection in *Arabidopsis thaliana* lines from Sweden. Nat Genet 45:884–890

101. Huber CD, Nordborg M, Hermisson J et al (2014) Keeping it local: evidence for positive selection in Swedish *Arabidopsis thaliana*. Mol Biol Evol 31:3026–3039

102. Lee C, Svardal H, Farlow A et al (2017) On the post-glacial spread of human commensal *Arabidopsis thaliana*. Nat Commun 8:14458

103. Jakobsson M, Hagenblad J, Tavare S et al (2006) A unique recent origin of the allotetraploid species *Arabidopsis suecica*: evidence from nuclear DNA markers. Mol Biol Evol 23:1217–1231

104. Säll T, Jakobsson M, Lind-Halldén C et al (2003) Chloroplast DNA indicates a single origin of the allotetraploid *Arabidopsis suecica*. J Evol Biol 16:1019–1029

105. Jakobsson M, Säll T, Lind-Halldén C et al (2007) The evolutionary history of the

common chloroplast genome of *Arabidopsis thaliana* and *A. suecica*. J Evol Biol 20:104–121

106. Novikova PY, Tsuchimatsu T, Simon S et al (2017) Genome sequencing reveals the origin of the allotetraploid *Arabidopsis suecica*. Mol Biol Evol 34:957–968

107. Ossowski S, Schneeberger K, Lucas-Lledó JI et al (2010) The rate and molecular spectrum of spontaneous mutations in *Arabidopsis thaliana*. Science 327:92–94

108. Koch MA, Haubold B, Mitchell-Olds T (2000) Comparative evolutionary analysis of chalcone synthase and alcohol dehydrogenase loci in *Arabidopsis, Arabis*, and related genera (Brassicaceae). Mol Biol Evol 17:1483–1498

109. Barrett SC, Arunkumar R, Wright SI (2014) The demography and population genomics of evolutionary transitions to self-fertilization in plants. Philos Trans R Soc Lond B Biol Sci 369(1648):20130344

110. Schrider DR, Shanku AG, Kern AD (2016) Effects of linked selective sweeps on demographic inference and model selection. Genetics 204:1207–1223

111. Messer PW, Petrov DA (2013) Frequent adaptation and the McDonald-Kreitman test. PNAS 110:8615–8620

112. Ewing GB, Jensen JD (2016) The consequences of not accounting for background selection in demographic inference. Mol Ecol 25(1):135–141

113. Messer PW (2013) SLiM: Simulating evolution with selection and linkage. Genetics 194 (4):1037–1039

114. Haller BC, Messer PW (2016) SLiM 2: flexible, interactive forward genetic simulations. Mol Biol Evol 34:230–240

115. Laenen B, Tedder A, Nowak MD et al (2018) Demography and mating system shape the genome-wide impact of purifying selection in *Arabis alpina*. Proc Natl Acad Sci U S A 115(4):816–821

116. Goodwillie C, Kalisz S, Eckert CG (2005) The evolutionary enigma of mixed mating systems in plants: occurrence, theoretical explanations, and empirical evidence. Annu Rev Ecol Evol Syst 36:47–79

117. Kamran-Disfani A, Agrawal AF (2014) Selfing, adaptation and background selection in finite populations. J Evol Biol 27:1360–1371

118. Salcedo A, Kalisz S, Wright SI (2014) Limited genomic consequences of mixed mating in the recently derived sister species pair, *Collinsia concolor* and *Collinsia parryi*. J Evol Biol 27:1400–1412

119. Bustamante CD, Nielsen R, Sawyer SA et al (2002) The cost of inbreeding in *Arabidopsis*. Nature 416:531–534

120. Foxe JP, Dar V, Zheng H et al (2008) Selection on amino acid substitutions in *Arabidopsis*. Mol Biol Evol 25:1375–1383

121. Garud NR, Messer PW, Buzbas EO et al (2015) Recent selective sweeps in North American *Drosophila melanogaster* show signatures of soft sweeps. PLoS Genet 11:1–32

122. Schrider DR, Kern AD (2016) S/HIC: robust identification of soft and hard sweeps using machine learning. PLoS Genet 12: e1005928

123. Huber CD, Durvasula A, Hancock AM et al (2017) Gene expression drives the evolution of dominance. Nature Communications 9:2750

124. Guo X, Liu J, Hao G et al (2017) Plastome phylogeny and early diversification of Brassicaceae. BMC Genomics 18:176

125. Beilstein MA, Nagalingum NS, Clements MD et al (2010) Dated molecular phylogenies indicate a Miocene origin for *Arabidopsis thaliana*. PNAS 107:18724–18728

126. Mattila TM (2017) Post-glacial colonization, demographic history, and selection in *Arabidopsis lyrata*: genome-wide and candidate gene based approach. University of Oulu. PhD thesis

Chapter 12

Genomics of Long- and Short-Term Adaptation in Maize and Teosintes

Anne Lorant, Jeffrey Ross-Ibarra, and Maud Tenaillon

Abstract

Maize is an excellent model for the study of plant adaptation. Indeed, post domestication maize quickly adapted to a host of new environments across the globe. And work over the last decade has begun to highlight the role of the wild relatives of maize—the teosintes *Zea mays* ssp. *parviglumis* and ssp. *mexicana*—as excellent models for dissecting long-term local adaptation.

Although human-driven selection associated with maize domestication has been extensively studied, the genetic basis of natural variation is still poorly understood. Here we review studies on the genetic basis of adaptation and plasticity in maize and its wild relatives. We highlight a range of different processes that contribute to adaptation and discuss evidence from natural, cultivated, and experimental populations. From an applied perspective, understanding the genetic bases of adaptation and the contribution of plasticity will provide us with new tools to both better understand and mitigate the effect of climate changes on natural and cultivated populations.

Key words Maize, Teosinte, Adaptation, Plasticity, Convergence

1 Introduction

A combination of archeobotanical records and genetic data has established that maize (*Zea mays* ssp. *mays*) was domesticated around 9000 years ago in the Balsas river valley of Mexico from the wild teosinte *Zea mays* ssp. *parviglumis* [1–3]. Unlike complex domestication scenarios involving multiple domestication events in the common bean (*Phaseolus vulgaris* L.) and the lima bean (*Phaseolus lunatus* L.) [4] or multiple progenitors from different regions in barley (*Hordeum vulgare*; [5], maize stands a relatively simple scenario involving only a single domestication event resulting in a moderate decrease of genetic diversity of roughly 20% [6].

With the rise of coalescent simulation tools since the late 1990s [7], researchers have repeatedly attempted to establish demographic scenarios of maize domestication. All concur with a simple bottleneck model, that is, a reduction of effective population size

Julien Y. Dutheil (ed.), *Statistical Population Genomics*, Methods in Molecular Biology, vol. 2090,
https://doi.org/10.1007/978-1-0716-0199-0_12, © The Author(s) 2020

(N_e), with <10% of the teosinte population contributing to the maize gene pool [8–11]. A recent investigation indicates that this bottleneck was followed by a major expansion resulting in an N_e for modern maize much larger than that of teosinte [11]. However, the complexity of the forces acting to shape diversity at a genome-wide scale makes it difficult to disentangle them. On the one hand, domestication has likely promoted strong positive selection at ~2% to 4% of loci [10] producing one of the most famous textbook example of selective sweeps at *tb1*, a gene responsible for the reduced branching phenotype in maize [12]. On the other hand, purifying selection has also reduced neutral genetic diversity [11]. Such selection may lead to an excess of rare variants, a footprint easily confounded with both positive selection and population expansion [13].

After its initial domestication, the geographic range of maize has rapidly exceeded that of its wild relatives, with documented routes of diffusion northward and southward out of Mexico [14, 15] and to the European continent [16]. Today the maize gene pool worldwide consists of locally adapted open-pollinated populations (landraces) as well as modern inbred lines, derived from landraces, that are used in hybrid production for modern breeding. Such spatial movement has exerted a diversity of selective pressures, triggering changes in the phenology of individuals that ultimately determines the completion of the annual cycle and individual fitness [17, 18].

In the last decade, the annual teosintes *Zea mays* ssp. *parviglumis* and ssp. *mexicana* have emerged as models for dissecting long-term adaptation to natural selection [19]. While their distribution is rather limited geographically, teosintes span extremely various environmental conditions in terms of temperatures, precipitations and elevations. Migration is also somewhat limited by the complex landscape of Mexico [20, 21]. Moreover, both teosinte taxa display a high level of nucleotide diversity [22] consistent with large estimates of effective population sizes from 120k to 160k [23]. Together, these conditions set the stage for extensive local adaptation.

Populations respond to environmental changes in three ways: (1) by shifting their range via migration to environments whose conditions are similar to their original conditions; (2) by genetic adaptation through the recruitment of preexisting or new alleles that increase the fitness of individuals carrying them; or (3) by phenotypic adjustments without genetic alterations, a mechanism called phenotypic plasticity.

Recent range shifts driven by global warming have been reported in tree species distributed in California, Oregon and Washington with an average shift compared to mature trees of about 27 m in altitude and 11 kms northward, toward colder environments [24]. Likewise, rising temperatures have likely caused the upslope migration reported for vascular plants species across European boreal-to-temperate mountains [25].

Such measurement in natural populations of teosintes are currently unavailable making the assessment of recent migration in response to climate change unknown. However, a niche modeling study showed that the range of annual teosintes appears to be quite similar to what it was at the time of domestication [26]. From the same study, relatively minor shifts of the niche have occurred even over the dramatic changes of the last glacial maximum, suggesting that migration over long ranges was not necessary.

In this chapter, we focus on adaptation and phenotypic plasticity. We review methods used to explore genetic adaptation and the factors constraining it. Next, we review empirical reports of short- and long-term adaptation in maize and teosintes. Finally, we discuss the role genetic convergence and phenotypic plasticity have played during adaptation.

2   How to Explore Adaptation?

Genetic adaptation can be defined as the modulation of allele frequencies through natural and/or artificial selection. Natural selection is imposed by changes in environmental conditions, or artificial selection by humans. Identification of adaptive loci (Fig. 1a, b) and/or traits (Fig. 1c, d) uses spatial or temporal variation in conjunction with quantification of traits in native environments (Fig. 1f) or in common gardens (Fig. 1g) [27–30]. While the temporal approach includes retrospective studies that follow the phenotypic and genetic composition of populations through time (for instance [31] to infer past selective events, the spatial approach relies on samples of populations that are geographically separated [30, 32].

In *Zea*, experimental approaches have been coupled with genotyping of sampled/evolved populations to identify the genomic bases of observed phenotypic changes. More often, however, studies have focused only on species-wide population genomic analyses tracing patterns of variation. These include searches for (1) spatial associations of allele frequencies with environmental factors or phenotypes (Fig. 1a); (2); shifts in allele frequencies across genetic groups (e.g., comparing wild and cultivated samples) using genome scans (Fig. 1b); and (3) differential gene expressions related to population/subspecies differentiation. An increasingly popular approach that was initiated in 2003 by Jaenicke-Despres [33] is the use of ancient DNA, as maize cobs are often well preserved making them an attractive source for ancient DNA studies. Such studies provide access to temporal samples to address past selective events that shaped genomes.

A. Allele frequencies among populations correlate with environmental variation at a candidate locus

C. Phenotypic variation among populations at a candidate trait correlates with environmental variation

E. Genotypic variation at a candidate locus correlates with phenotypic variation at a candidate trait across populations

B. Genetic differentiation between extreme populations exceeds neutral expectation at a candidate locus

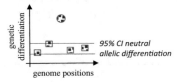

D. Phenotypic differentiation among populations exceeds neutral expectation at a candidate trait

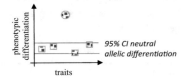

F. Genotypic variation at a candidate locus associates with phenotypic variation at a candidate trait within populations

Fig. 1 Experimental approaches to detect potentially adaptive polymorphisms and traits using population genetic (**a**, **b**) or phenotypic (**c**, **d**) data, or combining both (**e**, **f**). A candidate polymorphism whose allele frequency among populations varies with spatial or temporal variation can be detected using correlation-based methods (**a**) or genome-wide scans, where it displays an elevated differentiation of allele frequencies compared with neutral (squares) loci (**b**). A candidate trait that covaries with spatial or temporal variation among populations can be detected using correlation-based methods (**c**) or when phenotypic differentiation measured in common environment(s) exceeds genotypic differentiation at neutral (squares) loci (**d**). A link between candidate loci and traits can be established by correlating genotypic and phenotypic variation measures in common environment(s) across populations (**e**), and within populations (**f**)

3   What Constraints Adaptation?

Genetic adaptation can proceed through a single beneficial mutation that occurs after the onset of selection pressure, in which case the classical genetic footprint of a "hard" selective sweep is observed. Alternatively, it can proceed through a single mutation segregating in the population before the onset of selection (standing genetic variation), or through recurrent beneficial mutations. In these latter cases, adaptation produces a "soft" sweep footprint [34].

Hard sweeps are characterized by local shifts in allele frequencies due to the hitchhiking of neutral sites around a selected de novo variant occurring on a specific haplotype. Such changes in allele frequencies can easily be detected by genome scans. In contrast, soft sweeps, which derive from multiple adaptive alleles sweeping in the population, are substantially harder to detect at a genome-wide scale.

The relative contribution of hard and soft sweeps has been a long-standing debate and ultimately raises the important question of what limits adaptation. Experimental evolution in model organisms with short generation time such as *Escherichia coli*, yeast and *Drosophila melanogaster* have provided insights into those questions [35–40]. What emerges from these studies is that relevant parameters include the mutation rate, drift and selection [41, 42]. We surveyed these parameters in eight divergent selection experiments undertaken in maize (Table 1) and detail below our interpretations. By applying continuous directional selection on a given quantitative trait, such experiments aim to quantify and understand the limits of selection. However, it should be noted none of the cited work has included multiple replicates.

One of the most puzzling observations across experiments is that the response to selection is generally steady over time. In the Golden Glow (GG) experiment, the response varies from 4.7% to 8.7% of the original phenotypic value per cycle of selection across 24 cycles [48]. In the Krug Yellow Dent (KYD), it was estimated at 1.6% and 2.5% per cycle respectively, for high and low seed size direction [59]. In the Iowa Stalk Synthetic (BSSS), the response was of 3.9% per cycle for higher grain yield [50]. In the Iowa Long Ear Synthetic (BSLE), an increase of 1.4% and a decrease of 1.9% per cycle for high and low ear length were observed [59]. The results were more equivocal for Burn's White (BW), for which the response is much stronger and steadier toward high (between 0.1% and 0.3%) than low values (between 0% and 0.32%) for both protein and oil content. This pattern of shift between a strong and steady response to a plateau-like response for the low trait values is explained by physiological limits. Hence after 65 generations a lower limit for protein content is reached where the percentage of oil in the grain (close to 0% in the late generations) is no longer detectable [46, 47]. A similar situation has been reported for some of the late flowering families of MBS847 and F252 that are not able to produce seeds in the local climate conditions where they are selected, while the early still display a significant response after 16 generations [43]. Overall, mutations do not appear limiting regardless of the design, whether it started from highly inbred material or a diverse set of intercrossed landraces (Table 1).

What differs from one experiment to another, however, is the genomic footprint of the response to selection. Such footprints have been investigated in all but the BW and BSLE design. In GG, in which the mutational target size—the number of sites affecting the trait—was restricted, the effective population size was the highest of all and the selection was intense. The signal is consistent with genome-wide soft sweeps [48, 49]. In KYD, characterized by a larger mutational target, stronger drift (smaller effective population size), but weaker selection, both hard and soft sweeps are observed [45]. In BSSS, in which the mutational target

Table 1
Description of eight long-term (>16 generations) Divergent Selection (DS) experiments in maize with groups of features primarily (but not exclusively) related to Mutations (3), Drift (1), Selection (2) and Power to detect selection targets (5) highlighted by groups

DS experiments	F252 (F252)	MBS847 (MBS)	Krug Yellow Dent (KYD)	Burn's White (BW)	Burn's White (BW)	Golden Glow (GG)	Iowa Stiff Stalk Synthetic (BSSS)	Iowa Long Ear Synthetic (BSLE)
References[a]	[43, 44]	[43, 44]	[45]	[46, 47]	[46, 47]	[48, 49]	[50–52]	[53]
Directions (High/low)[b]	H/L	H/L	H/L	H/L	H/L	H	H	H/L
Trait[c]	Flowering	Flowering	Seed size	Protein	Oil	Ears/plant	Grain yield	Ear length
Material type[d]	Inbred	Inbred	OP variety	OP variety	OP variety	OP variety	Synthetic population	Synthetic population
Mutational target[e]	>60 QTLs [54]	>60 QTLs [54]	>300 loci [55]	102–178 factors	14–69 factors	Limited [48]	Large [56]	25 QTLs [57]
Standing variation[f]	1.9%	0.19%	Pervasive	Pervasive	Pervasive	Pervasive	Pervasive	Pervasive
Census population size[g]	1000	1000	1200–1500	60–120	60–120	4250 (1–12) 14,250 (13–30)	>1240	4000
N_e[h]	3.1–20.2	5.8–13.5	369	4–12	4–12	667	10–20	14
Selection coefficient (%)	1	1	8	20	20	0.5–5	5	7.5
Heritability[j]	0.14/0.13	0.13/0.16	–	0.21/0.07	0.23/0.23	0.88	0.4	0.05

Number of founders[j]	2 haplotypes	2 haplotypes	100 founders	24 ears (H) 12 ears (L)	24 ears (H) 12 ears (L)	~300 founders	16 founders	12 founders
Reproductive mode	Selfing	Selfing	Outcrossing	Outcrossing	Outcrossing	Outcrossing	Outcrossing	Outcrossing
Sampling[k]	All/ind	All/ind	All/bulk	All/bulk	All/bulk	All/bulk	All/bulk	All/bulk
Number of generations	16	16	30	114	114	30	17	27

[a]References from which values were taken for each DS experiment are indicated in superscript

[b]Direction of selection toward higher and/or lower values than the initial material

[c]Protein and Oil designate protein and oil content of the grain, Ears/plan relates to prolificacy

[d]Inbred: Inbred line; OP variety: Open Pollinated population

[e]Number of factors in BW was estimated from the trait value, predicted gain and additive genetic variance

[f]Standing variation was estimated from 50k SNP array for F252 and MBS

[g]For GG, 4250 individuals were evaluated from cycles 1–12, and 14,250 in the following cycles

[h]Effective population size (N_e) estimates given from the variance of offspring number [58], range is given when N_e was estimated at each generation

[i]Broad-sense heritability estimated from genetic variation between progenies of the same family. Average values across generations is reported here

[j]Expressed either as number of haplotypes (a single founder = individual bears 2 haplotypes), number of founders, or number of ears (all individuals of a given ear share identical mother but different fathers). For GG, most selection cycles used 300 founders

[k]Seeds from all time points (All) are available, and were either collected separately on each selected individual (/ind) or in bulk (/bulk)

size is the largest, the effective population size small and the selection intense, the signal is consistent with hard sweeps [51]. The F252 and MBS populations display the most limited standing variation and at the same time the strongest drift and selection of all experiments; in these a rapid fixation of new mutations explains the response to selection [43, 44]. Effective population size primarily determines the likelihood of soft sweeps. Hence, when θ (four times the product of effective population size and the beneficial mutation rate) is equal or above 1, and selection is strong enough, adaptation proceeds from multiple de novo mutations or standing variation [60]. Below 1, soft sweeps' contribution diminishes with θ. In the experiments from Table 1, selection is strong but $\theta \ll 1$ in all cases. Nevertheless, hard and soft sweeps were associated respectively with the lowest (F252 and MBS) and highest (GG) effective population size, consistent with N_e being a key player. Comparisons among experiments thus contribute to understanding the parameters of importance and their interactions that together shape the genomic patterns of the response to selection.

An additional layer of complexity that may substantially impact evolutionary trajectories is that of genetic correlations among traits. Such correlations may emerge from genes with pleiotropic effects, epistatic interactions among genes, and/or loci in tight linkage affecting various traits. While some studies have found that covariance between traits rarely affect adaptation [61], several examples instead suggest that they may either constrain or facilitate adaptation as predicted by Lande [62]. For instance, in *Arabidopsis thaliana* a recent study indicates that polymorphisms with intermediate degrees of pleiotropy favored rapid adaptation to microhabitats in natura [63]. In the case of domestication, tight linkage between genes conferring the so-called domestication syndrome has been invoked as a mechanism facilitating adaptation to the cultivated environment in allogamous species, preventing gene flow from wild relatives to break coadapted suites of alleles [64]. It turns out that rather than clustering, plant domestication genes identified so far are single locus which are mainly transcription factors (reviewed in [65]) most of which likely display strong epistatic interactions. *tb1* in maize, for instance, interacts with another locus on a different chromosome to alter the sex of maize inflorescences. The introgression of the *tb1* teosinte allele alone changes only ~20% of the inflorescence sex but the introgression of both alleles converts 90% of maize's female flowers to male [66]. The maize *tb1* allele segregates at low frequency in teosinte populations but is rarely found associated with the domesticated allele of chromosome 3, as both are likely to evolve under negative selection in teosinte [12, 66]. Their association in maize has however facilitated the acquisition of the domesticated phenotype.

4   Mechanisms of Genetic Adaptation in Maize and Teosintes

Populations of teosinte have long evolved under natural selection. In contrast, maize populations have been under artificial human selection that moved phenotypes toward optimal traits tailored to agriculture during a shorter time frame of ~9000 years [1, 2, 22]. These time scales have left distinct genetic signatures. In theory, traits fixed by domestication should involve genes with larger effect sizes, and standing variation should be a major contributor to domestication [67]. This is supported by crosses between maize and teosintes that led to the discovery of six main QTLs responsible for major phenotypic differences between them, notably vegetative architecture and inflorescence sexuality ([68, 69], reviewed in [70]). Among these QTLs, genes with major phenotypic effects have been discovered such as *tb1* and *tga1* (*teosinte glume architecture1*). In addition to these major genes, a collection of targets (2–4% of the genome according to [6, 10]) have likely contributed to the domesticated phenotype. In contrast, Genome Wide Association (GWA) studies on traits selected over much longer time scale such as drought tolerance or flowering time have highlighted only minor effect loci that rarely contribute to more than 5% of the phenotypic variation [54, 67, 71, 72].

In addition to the time frame over which adaptation occurs, another important factor for evolution is the nature of variation for selection to act on. Maize and teosintes are genetically very diverse, with as much nucleotide diversity in coding regions between two maize lines as there are between humans and chimpanzees [73]. This diversity is even higher in intergenic regions [74, 75]. Some adaptive mutations are found in coding sequences. Examples include nonsynonymous changes in the *tga1* gene responsible for the "naked kernel" maize phenotype, and in the *diacylglycerol acyltransferase* (*DGAT1–2*) gene resulting in elevated kernel oil content in maize lines [76, 77]. But most observations support adaptation from regulatory noncoding sequences. Indeed, in comparison with *Arabidopsis*, where adaptive variants are enriched in coding sequences [78], in maize and teosintes these are predominantly found in noncoding region: estimates in *Zea* show that noncoding variants may explain as much phenotypic variation as those in coding regions [79, 80]. Selection on regulatory sequences drive important expression changes; hence, genes displaying footprints of selection in maize are usually more expressed than in teosintes [6], and are associated with modified coexpression networks [81].

Adaptive variation also results from structural variants. In contrast to the *Arabidopsis* or rice genomes where Transposable Elements (TEs) account for 20–40% of sequence, the maize genome is composed of about 85% TEs [82, 83]. Genome size varies considerably within *Zea* resulting in over 30% differences among maize lines or landraces [79, 84, 85]. Because of their deleterious effect, TEs are often negatively selected and silenced by DNA methylation [86]. But some may also impact gene expression and function in a beneficial manner by various mechanisms such as gene inactivation or differential expression caused by insertion in regulatory regions [87] or TE-mediated genomic rearrangements causing gene insertion, deletion or duplication (reviewed in [88]). A handful of examples of their beneficial impact has been reported in *Zea*. A classic example in maize is at the *tb1* locus, where a transposon inserted in the cis-regulatory region, doubling expression [89]. Teosinte, like most grasses, produces numerous branches tipped by a male inflorescence. In contrast, maize has only one main stalk terminated by a single tassel with repressed development of lateral branches. The increased expression level of *tb1* is the major contributor to this apical dominance [89]. Beyond TEs, Copy Number Variants (CNVs) are also common in the maize genome [90] and they contribute significantly to phenotypic variation [79, 91].

Another important player in adaptation in *Zea* is gene flow. Indeed, teosinte populations are found in sympatry with maize and hybridization between them is common [92]. Highland maize shows up to 20% *mexicana* introgression, which has likely facilitated their adaptation to high elevations [3, 93]. An ancient DNA study revealed that ancestral highland maize already showed evidence of introgression from *mexicana* [15]. Introgressed regions found at high frequency in highland maize overlap with previously identified QTLs driving adaptive traits [93, 94], emphasizing the importance of introgression during post-domestication adaptation. Similarly, recent results suggest that admixture between distinct genetic groups has facilitated adaptation to mid-latitudes in North America and Europe [16].

5   Local Adaptation in Maize and Teosintes

Strictly defined, a genotype can be considered locally adapted if it has a higher fitness at its native site than any other nonnative genotypes [95]. Locally adapted alleles can be either neutral or deleterious in other environments. Two models depict those situations, namely conditional neutrality and antagonistic pleiotropy [96]. Despite numerous studies, the genetic processes underlying local adaptation in natural populations are still poorly understood. This is mainly due to traits driving local adaptation being mostly quantitative [29]. This complex determinism may involve numerous, but not necessarily substantial, allele frequency changes.

Studies showed that highland maize landraces outperform lowland maize populations in their native environment but perform worse than any other population at lower elevation sites [97], suggesting strong adaptation for high altitude.

Natural selection acts on phenotypic traits, changing the frequency of underlying alleles and shifting population phenotypes toward local optima. Since these optima rely on local conditions, genes ecologically important usually differ between sub-populations in heterogeneous environments, resulting in divergence in allele frequencies over time. This characteristic has been utilized in genome scans to mine correlations between allele frequencies and environmental variables (Fig. 1a). Such studies have revealed that, in teosintes, these loci impact flowering time and adaptation to soil composition [20, 98, 99]. Flowering time was also a key component of maize's local adaptation to higher latitudes during post-domestication. Maize evolved a reduced sensitivity to photoperiod, in part due to a CACTA-like TE insertion in the promoter region of the *ZmCCT* gene that drives photoperiod response in early flowering maize [100, 101]. An example of adaptation driven by soil interactions is the tolerance of maize and teosintes to aluminum in highly acidic soils. In these lines, the adaptation is linked to tandem duplications of the *MATE1* gene involved in the extrusion of toxic compounds [91].

Numerous other biotic and abiotic factors are likely involved in adaptation in maize and teosintes, including predation, parasitism, moisture, and herbicide [102, 103]. For example, a study on *parviglumis* has shown that in response to herbivory, immunity genes involved in the inhibition of insects' digestive proteases experienced a recent selective sweep in a region of Mexico, probably reflecting local adaptation [104].

Interestingly, four large inversion polymorphisms seem to play an important role in local adaptation. Among them, a 50 Mb inversion on chromosome 1 is found at high frequency in *parviglumis* (20–90%), low frequency in *mexicana* (10%), and is absent in maize. This inversion is highly correlated with altitude and significantly associated with temperature and precipitation [20, 105]. Inversions on chromosomes 3, 4 and 9 also displayed environmental association in teosintes and maize landraces for the first two and in teosintes for the last one [20, 72]. Local adaptation to different habitats or niches is a gradual process that can promote divergence and, in the long run, ecological speciation [106]. Genotyping of a broad sample of 49 populations covering the entire geographic range of teosintes has recently provided some evidence of this. Aguirre-Liguori et al. [98] showed that both within *parviglumis* and *mexicana*, populations distributed at the edge of the ecological niche experience stronger local adaptation, suggesting that local adaptation may have contributed to divergence between these two subspecies.

6   How Convergent Is Adaptation?

Convergent adaptation is the result of independent events of similar phenotypic changes to adapt to analogous environmental constraints [107]. In this review, we concentrated on genetic convergence in populations of the same, or closely related, species which are the result of convergent evolution at the molecular level. By molecular convergence, we include convergence at the same nucleotide positions, genes or orthologues. Several studies illustrate this, suggesting that genomes may respond in predictable ways to selection [108–112]. The selected alleles can originate from independent mutation events in different lineages, from shared ancestral variation or by introgression [111].

A classical way to study convergence is experimental evolution. During these experiments, replicates of the same genotype are grown for many generations in new environments. Such studies have often shown that convergent evolution is common [37, 113]. Domestication can be thought of as an example of long-term experimental evolution, and domesticates provide striking examples of phenotypic convergence, with common traits usually referred to as the domestication syndrome. These phenotypes include, but are not limited to, larger fruits or gains, less branching, loss of shattering, and loss of seed dormancy [114]. QTL mapping can be performed to identify the genes controlling these phenotypes in different species. As an example, seeds on wild grasses shed naturally at maturity. During domestication this trait was rapidly selected against since it causes inefficient harvesting [115]. QTL mapping of sorghum, rice and maize reveals that the *Shattering1* genes are involved in the loss of the dispersal mechanism and were under convergent evolution during their domestication [116].

But genetic convergence can also be observed over much shorter evolutionary time, at the intraspecific level across populations. Here genome scans for extreme differentiation in allele frequency between multiple pairs of diverged populations along gradients, for instance, are typically employed. This method has been used to test for convergent adaptation in highland maize landraces and teosintes. Fustier et al. [99] found several instances (24/40) of convergence involving the same haplotype in two gradients of adaptation to high altitude in teosintes. In maize, the Mesoamerican and South American populations independently adapted from distinct lowland populations to high elevation conditions [14]. These populations exhibit several similar phenotypic characteristics not observed in lowland populations such as changes in inflorescence morphology and stem coloration. A study found that highland adaptation is likely due to a combination of introgression events, selection on standing genetic variation and independent de novo mutations [117]. These studies also showed that convergent evolution involving identical nucleotide changes is uncommon and most selected loci arise from standing genetic variation present in lowland populations. This is not surprising given the relative short time frame of highland adaptation in maize compared to teosinte subspecies.

Recently, a new method has been developed to infer modes of convergence [118], using covariance of allele frequencies in windows around a selected site to explicitly compare different models of origin for a selected variant. This novel method should give a better insight on the genetic mechanisms underlying convergence.

7   What Is the Role of Phenotypic Plasticity?

Phenotypic plasticity is defined as the capacity of a genotype to produce a range of expressed phenotypes in distinct environments. This is achieved through differential developmental pathways in response to changing conditions [119, 120]. Plasticity can be an important process during adaptation. Indeed, populations with flexible phenotypes are predicted to better cope with environmental changes and to display a greater potential for expansion [121]. This process is particularly important for plants as they are fixed in a specific location and not sheltered from the environment [122].

When the environment changes, the phenotypic optimum of a population is likely altered as well. As a result, individuals that show a plastic response in the direction of the new optimum will have a fitness advantage. In contrast, individuals that exhibit no plasticity or that produce phenotypes too far from the optimum will be selected against.

Plasticity has limits, however, and may entail a fitness cost. For instance, compared to developmentally fixed phenotypes, plastic individuals in constant environments may display lower fitness or produce a less adapted phenotype. Possible reasons include sensory mechanisms that have a high energetic cost, the epistatic effects of regulatory genes involved in the plastic response, lag time between the perception and the phenotypic response and genetic correlations among traits [123–125].

Phenotypic plasticity is difficult to study as it arises from genetic and environmental interactions which are often hard to disentangle. After a number of generations of constant selection, for example, the fixation of genetic variation that constitutively expresses the trait can lead to a loss of plasticity via a process called genetic assimilation [126–128]. Hence an initially plastic phenotype may result in genetic adaptation after genetic assimilation. Some examples of plastic responses are well documented in plants, for example, the response to vernalization in *Arabidopsis* regulating flowering time in some ecotypes [122]. Another example is the change in seed dormancy in response to the environment which prevents germination when conditions are unlikely to lead to the survival of the plant [124].

Taxa in *Zea* are good models to investigate plasticity as maize is grown worldwide and adapted to a diversity of environments. In addition, studies of teosintes allow comparison to ancestral levels of plasticity. A recent experiment evaluated plasticity in maize by studying Genotype by Environment interactions (GxE) for a number of phenotypes in 858 inbred lines across 21 locations across North America [129]. Results demonstrated that genes selected for high yield in temperate climates in North America correlated with low variance in GxE. This suggests a loss of plasticity accompanying selection for stable crop performance across environments, a major goal for breeders. In addition, GxE was mainly explained by regulatory regions [129], an observation in agreement with previous findings indicating that most phenotypic variation in maize is due to gene regulation [130].

Recent work on maize and *parviglumis* growing under environmental conditions mimicking those encountered at the time of maize domestication (comparatively lower CO_2 atmospheric concentration and lower temperatures) gives better insights into this phenomenon. The results showed that teosintes grown in these conditions exhibit contemporary maize-like phenotypes [131]. In contrast, modern maize has lost this plastic response. Over 2000 candidate loci associated with phenotypic changes showed altered expression in teosintes but not in maize, implying that they are no longer environmentally responsive (Fig. 2; [132]). Such loss of phenotypic plasticity may limit the ability of maize to cope with environmental variability in the face of current climate changes.

8   Conclusion

Ongoing global warming has drastic effects on maize production, with an estimated impact of temperature and precipitation on yield of 3.8% worldwide between 1980 and 2008 [133]. Predicted changes that include further increases in temperatures and decline in rainfall, as well as shifts of pests and diseases, represent a huge challenge. There is thus a pressing need to better understand the dynamics and genomic basis of adaptation. Future climate projections predict that changes in temperature will impact the distribution and survival of both cultivated maize and its wild relatives [26, 134]. Most modeling studies, however, have focused on the climate tolerance of species, while the response to climate can depend on other factors such as plasticity and local adaptation. This suggests that the response should be studied at the level of individual populations to better understand the basis of adaption.

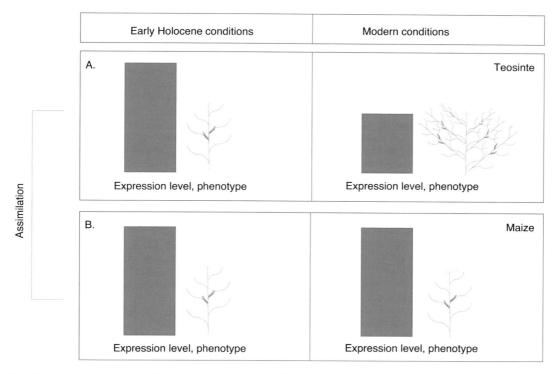

Fig. 2 Schematic representation of differences in plastic responses between maize and teosintes in Early-Holocene (EH) conditions. (**a**) *Parviglumis* plants exhibit maize-like phenotypes in the EH conditions (vegetative architecture, inflorescence sexuality and seed maturation). Phenotypes of *parviglumis* in modern conditions are typical of today's plants. These changes in phenotypes are associated with altered expression levels of over 2000 candidate loci in teosinte; here we represent the schematic expression of one gene between the two environments in teosinte. (**b**) In contrast, these same traits and underlying gene expression remain unchanged in maize between EH and modern conditions

Acknowledgments

Maud Tenaillon acknowledges the Kavli Institute for Theoretical Physics at UCSB, supported in part by the National Science Foundation under Grant No. NSF PHY-1125915. J.R.I. would like to acknowledge support from the USDA Hatch project (CA-D-PLS-2066-H). The authors wish to acknowledge funding support from the US National Science Foundation (IOS-0922703 and IOS-1238014) as well as Michelle Stitzer, Mirko Ledda, and Domenica Manicacci for comments. Finally, the authors would like to acknowledge Yves Vigouroux for his helpful reviews.

References

1. Piperno DR, Flannery KV (2001) The earliest archaeological maize (Zea mays L.) from highland Mexico: new accelerator mass spectrometry dates and their implications. Proc Natl Acad Sci U S A 98:2101–2103. https://doi.org/10.1073/pnas.98.4.2101

2. Matsuoka Y, Vigouroux Y, Goodman MM, Sanchez GJ, Buckler E, Doebley J (2002) A single domestication for maize shown by multilocus microsatellite genotyping. Proc Natl Acad Sci U S A 99:6080–6084. https://doi.org/10.1073/pnas.052125199

3. van Heerwaarden J, Doebley J, Briggs WH, Glaubitz JC, Goodman MM, de Jesus Sanchez Gonzalez J, Ross-Ibarra J (2011) Genetic signals of origin, spread, and introgression in a large sample of maize landraces. Proc Natl Acad Sci 108:1088–1092. https://doi.org/10.1073/pnas.1013011108

4. Kwak M, Toro O, Debouck DG, Gepts P (2012) Multiple origins of the determinate growth habit in domesticated common bean (Phaseolus vulgaris). Ann Bot 110:1573–1580. https://doi.org/10.1093/aob/mcs207

5. Poets AM, Fang Z, Clegg MT, Morrell PL (2015) Barley landraces are characterized by geographically heterogeneous genomic origins. Genome Biol 16:173. https://doi.org/10.1186/s13059-015-0712-3

6. Hufford MB, Xu X, van Heerwaarden J, Pyhäjärvi T, Chia J-M, Cartwright RA, Elshire RJ, Glaubitz JC, Guill KE, Kaeppler SM, Lai J, Morrell PL, Shannon LM, Song C, Springer NM, Swanson-Wagner RA, Tiffin P, Wang J, Zhang G, Doebley J, McMullen MD, Ware D, Buckler ES, Yang S, Ross-Ibarra J (2012) Comparative population genomics of maize domestication and improvement. Nat Genet 44:808–811. https://doi.org/10.1038/ng.2309

7. Hudson RR (2002) Generating samples under a Wright-Fisher neutral model of genetic variation. Bioinformatics 18:337–338. https://doi.org/10.1093/bioinformatics/18.2.337

8. Eyre-Walker A, Gaut RL, Hilton H, Feldman DL, Gaut BS (1998) Investigation of the bottleneck leading to the domestication of maize. Proc Natl Acad Sci U S A 95:4441–4446. https://doi.org/10.1073/pnas.95.8.4441

9. Tenaillon MI, U'Ren J, Tenaillon O, Gaut BS (2004) Selection versus demography: a multilocus investigation of the domestication process in maize. Mol Biol Evol 21:1214–1225. https://doi.org/10.1093/molbev/msh102

10. Wright SI, Bi IV, Schroeder SG, Yamasaki M, Doebley JF, McMullen MD, Gaut BS (2005) The effects of artificial selection on the maize genome. Science 308:1310–1314. https://doi.org/10.1126/science.1107891

11. Beissinger TM, Wang L, Crosby K, Durvasula A, Hufford MB, Ross-Ibarra J (2016) Recent demography drives changes in linked selection across the maize genome. Nat Plants 2:16084. https://doi.org/10.1038/nplants.2016.84

12. Doebley J, Stec A, Gustus C (1995) Teosinte branched1 and the origin of maize: evidence for epistasis and the evolution of dominance. Genetics 141:333–346

13. Cvijović I, Good BH, Desai MM (2018) The effect of strong purifying selection on genetic diversity. Genetics 209(4):1235–1278. https://doi.org/10.1534/genetics.118.301058

14. Vigouroux Y, Glaubitz JC, Matsuoka Y, Goodman MM, Sánchez GJ, Doebley J (2008) Population structure and genetic diversity of new world maize races assessed by DNA microsatellites. Am J Bot 95:1240–1253. https://doi.org/10.3732/ajb.0800097

15. Da Fonseca RR, Smith BD, Wales N, Cappellini E, Skoglund P, Fumagalli M, Samaniego JA, Carøe C, Avila-Arcos MC, Hufnagel DE, Korneliussen TS, Vieira FG, Jakobsson M, Arriaza B, Willerslev E, Nielsen R, Hufford MB, Albrechtsen A, Ross-Ibarra J, Gilbert MTP (2015) The origin and evolution of maize in the Southwestern United States. Nat Plants 1:14003. https://doi.org/10.1038/nplants.2014.3

16. Brandenburg JT, Mary-Huard T, Rigaill G, Hearne SJ, Corti H, Joets J, Vitte C, Charcosset A, Nicolas SD, Tenaillon MI (2017) Independent introductions and admixtures have contributed to adaptation of European maize and its American counterparts. PLoS Genet 13(3):e1006666. https://doi.org/10.1371/journal.pgen.1006666

17. Chuine I (2010) Why does phenology drives species distribution? Philos Trans R Soc London B 365:3149–3160. https://doi.org/10.1098/rstb.2010.0142

18. Swarts K, Gutaker RM, Benz B, Blake M, Bukowski R, Holland J, Kruse-Peeples M, Lepak N, Prim L, Romay MC, Ross-Ibarra J, Sanchez-Gonzalez JJ, Schmidt C, Schuenemann VJ, Krause J, Matson RG, Weigel D, Buckler ES, Burbano HA (2017) Genomic estimation of complex traits reveals ancient

maize adaptation to temperate North America. Science 357:512–515. https://doi.org/10.1126/science.aam9425

19. Hufford MB, Bilinski P, Pyhäjärvi T, Ross-Ibarra J (2012) Teosinte as a model system for population and ecological genomics. Trends Genet 28:606–615

20. Pyhäjärvi T, Hufford MB, Mezmouk S, Ross-Ibarra J (2013) Complex patterns of local adaptation in teosinte. Genome Biol Evol 5:1594–1609. https://doi.org/10.1093/gbe/evt109

21. De Jesús Sánchez González J, Corral JAR, García GM, Ojeda GR, De La Cruz LL, Holland JB, Medrano RM, Romero GEG (2018) Ecogeography of teosinte. PLoS One 13(2): e0192676. https://doi.org/10.1371/journal.pone.0192676

22. Fukunaga K, Hill J, Vigouroux Y, Matsuoka Y, Sanchez J, Liu KJ, Buckler ES, Doebley J (2005) Genetic diversity and population structure of teosinte. Genetics 169:2241–2254. https://doi.org/10.1534/genetics.104.031393

23. Ross-Ibarra J, Tenaillon M, Gaut BS (2009) Historical divergence and gene flow in the genus Zea. Genetics 181:1399–1413. https://doi.org/10.1534/genetics.108.097238

24. Monleon VJ, Lintz HE (2015) Evidence of tree species' range shifts in a complex landscape. PLoS One 10(1):e0118069. https://doi.org/10.1371/journal.pone.0118069

25. Pauli H, Gottfried M, Dullinger S, Abdaladze O, Akhalkatsi M, Alonso JLB, Coldea G, Dick J, Erschbamer B, Calzado RF, Ghosn D, Holten JI, Kanka R, Kazakis G, Kollar J, Larsson P, Moiseev P, Moiseev D, Molau U, Mesa JM, Nagy L, Pelino G, Puscas M, Rossi G, Stanisci A, Syverhuset AO, Theurillat J-P, Tomaselli M, Unterluggauer P, Villar L, Vittoz P, Grabherr G (2012) Recent plant diversity changes on Europe's mountain summits. Science 336:353–355. https://doi.org/10.1126/science.1219033

26. Hufford MB, Martínez-Meyer E, Gaut BS, Eguiarte LE, Tenaillon MI (2012) Inferences from the historical distribution of wild and domesticated maize provide ecological and evolutionary insight. PLoS One 7(11): e47659. https://doi.org/10.1371/journal.pone.0047659

27. Anderson JT, Geber MA (2010) Demographic source-sink dynamics restrict local adaptation in Elliott's blueberry (Vaccinium elliottii). Evolution 64:370–384. https://

28. Fournier-Level A, Korte A, Cooper MD, Nordborg M, Schmitt J, Wilczek AM (2011) A map of local adaptation in Arabidopsis thaliana. Science 334:86–89. https://doi.org/10.1126/science.1209271

29. Savolainen O, Lascoux M, Merilä J (2013) Ecological genomics of local adaptation. Nat Rev Genet 14:807–820

30. Endler JA (1986) Natural selection in the wild. Princeton University Press, Princeton, NJ

31. Thompson J, Charpentier A, Bouguet G, Charmasson F, Roset S, Buatois B, Vernet P, Gouyon P-H (2013) Evolution of a genetic polymorphism with climate change in a Mediterranean landscape. Proc Natl Acad Sci U S A 110:2893–2897. https://doi.org/10.1073/pnas.1215833110

32. Tiffin P, Ross-Ibarra J (2014) Advances and limits of using population genetics to understand local adaptation. Trends Ecol Evol 29:673–680

33. Jaenicke-Despres V (2003) Early allelic selection in maize as revealed by ancient DNA. Science 302:1206–1208. https://doi.org/10.1126/science.1089056

34. Hermisson J, Pennings PS (2017) Soft sweeps and beyond: understanding the patterns and probabilities of selection footprints under rapid adaptation. Methods Ecol Evol 8:700–716. https://doi.org/10.1111/2041-210X.12808

35. Bell G, Collins S (2008) Adaptation, extinction and global change. Evol Appl 1:3–16. https://doi.org/10.1111/j.1752-4571.2007.00011.x

36. Burke MK, Dunham JP, Shahrestani P, Thornton KR, Rose MR, Long AD (2010) Genome-wide analysis of a long-term evolution experiment with Drosophila. Nature 467:587–590. https://doi.org/10.1038/nature09352

37. Tenaillon O, Rodriguez-Verdugo A, Gaut RL, McDonald P, Bennett AF, Long AD, Gaut BS (2012) The molecular diversity of adaptive convergence. Science 335:457–461. https://doi.org/10.1126/science.1212986

38. Burke MK, Liti G, Long AD (2014) Standing genetic variation drives repeatable experimental evolution in outcrossing populations of saccharomyces cerevisiae. Mol Biol Evol 31:3228–3239. https://doi.org/10.1093/molbev/msu256

39. Graves JL, Hertweck KL, Phillips MA, Han MV, Cabral LG, Barter TT, Greer LF, Burke

doi.org/10.1111/j.1558-5646.2009.00825.x

MK, Mueller LD, Rose MR, Singh N (2017) Genomics of parallel experimental evolution in drosophila. Mol Biol Evol 34:831–842. https://doi.org/10.1093/molbev/msw282

40. Good BH, McDonald MJ, Barrick JE, Lenski RE, Desai MM (2017) The dynamics of molecular evolution over 60,000 generations. Nature 551:45–50. https://doi.org/10.1038/nature24287

41. Schlötterer C, Kofler R, Versace E, Tobler R, Franssen SU (2014) Combining experimental evolution with next-generation sequencing: a powerful tool to study adaptation from standing genetic variation. Heredity (Edinb) 114:1–10. https://doi.org/10.1038/hdy.2014.86

42. Franssen SU, Kofler R, Schlötterer C (2017) Uncovering the genetic signature of quantitative trait evolution with replicated time series data. Heredity (Edinb) 118:42–51. https://doi.org/10.1038/hdy.2016.98

43. Durand E, Tenaillon MI, Raffoux X, Thépot S, Falque M, Jamin P, Bourgais A, Ressayre A, Dillmann C (2015) Dearth of polymorphism associated with a sustained response to selection for flowering time in maize. BMC Evol Biol 15:103. https://doi.org/10.1186/s12862-015-0382-5

44. Durand E, Tenaillon MI, Ridel C, Coubriche D, Jamin P, Jouanne S, Ressayre A, Charcosset A, Dillmann C (2010) Standing variation and new mutations both contribute to a fast response to selection for flowering time in maize inbreds. BMC Evol Biol 10:2. https://doi.org/10.1186/1471-2148-10-2

45. Odhiambo MO, Compton WA (1987) Twenty cycles of divergent mass selection for seed size in corn. Crop Sci 27:1113–1116

46. Moose SP, Dudley JW, Rocheford TR (2004) Maize selection passes the century mark: a unique resource for 21st century genomics. Trends Plant Sci 9:358–364

47. Dudley JW, Lambert RJ (2010) 100 generations of selection for oil and protein in corn. Plant Breed Rev 24:79–110

48. De Leon N, Coors JG (2002) Twenty-four cycles of mass selection for prolificacy in the Golden Glow maize population. Crop Sci 42:325–333. https://doi.org/10.2135/cropsci2002.3250

49. Maita R, Coors JG (1996) Twenty cycles of biparental mass selection for prolificacy in the open-pollinated maize population Golden Glow. Crop Sci 36:1527–1532

50. Lamkey KR (1992) Fifty years of recurrent selection in the Iowa Stiff Stalk Synthetic maize population. Maydica 37:19

51. Gerke JP, Edwards JW, Guill KE, Ross-Ibarra J, McMullen MD (2015) The genomic impacts of drift and selection for hybrid performance in maize. Genetics 201:1201–1211. https://doi.org/10.1534/genetics.115.182410

52. Hallauer AR, Carena MJ, Filho JBM (2010) Quantitative genetics in maize breeding. In: Carena MJ (ed) Handbook of plant breeding. Springer, New York, NY, pp 1–680

53. Lopez-Reynoso JJ, Hallauer AR (1998) Twenty-seven cycles of divergent mass selection for ear length in maize. Crop Sci 38:1099–1107

54. Buckler ES, Holland JB, Bradbury PJ, Acharya CB, Brown PJ, Browne C, Ersoz E, Flint-Garcia S, Garcia A, Glaubitz JC, Goodman MM, Harjes C, Guill K, Kroon DE, Larsson S, Lepak NK, Li H, Mitchell SE, Pressoir G, Peiffer JA, Rosas MO, Rocheford TR, Romay MC, Romero S, Salvo S, Villeda HS, Sofia da Silva H, Sun Q, Tian F, Upadyayula N, Ware D, Yates H, Yu J, Zhang Z, Kresovich S, McMullen MD (2009) The genetic architecture of maize flowering time. Science 325:714–718. https://doi.org/10.1126/science.1174276

55. Liu J, Huang J, Guo H, Lan L, Wang H, Xu Y, Yang X, Li W, Tong H, Xiao Y et al (2017) The conserved and unique genetic architecture of kernel size and weight in maize and rice. Plant Physiol 175:774–785

56. Yang J, Mezmouk S, Baumgarten A, Buckler ES, Guill KE, McMullen MD, Mumm RH, Ross-Ibarra J (2017) Incomplete dominance of deleterious alleles contributes substantially to trait variation and heterosis in maize. PLoS Genet 13(9):e1007019. https://doi.org/10.1371/journal.pgen.1007019

57. Ross AJ, Hallauer AR, Lee M (2006) Genetic analysis of traits correlated with maize ear length. Maydica 51:301–313

58. Crow JF, Kimura M et al (1970) An introduction to population genetics theory. Harper and Row, New York

59. Lopez-Reynoso JDJ, Hallauer AR (1998) Twenty-seven cycles of divergent mass selection for ear length in maize. Crop Sci 38:1099–1107. https://doi.org/10.2135/cropsci1998.0011183X003800040035x

60. Messer PW, Petrov DA (2013) Population genomics of rapid adaptation by soft selective sweeps. Trends Ecol Evol 28:659–669.

https://doi.org/10.1016/j.tree.2013.08.003

61. Agrawal AF, Stinchcombe JR (2009) How much do genetic covariances alter the rate of adaptation? Proc R Soc B Biol Sci 276:1183–1191. https://doi.org/10.1098/rspb.2008.1671

62. Lande R (1979) Quantitative genetic analysis of multivariate evolution, applied to brain: body size allometry. Evolution 33:402. https://doi.org/10.2307/2407630

63. Frachon, L., Libourel, C., Villoutreix, R. et al. (2017) Intermediate degrees of synergistic pleiotropy drive adaptive evolution in ecological time. Nat Ecol Evol 1, 1551–1561 https://doi.org/10.1038/s41559-017-0297-1

64. Le Thierry D'Ennequin M, Toupance B, Robert T, Godelle B, Gouyon PH (1999) Plant domestication: a model for studying the selection of linkage. J Evol Biol 12:1138–1147. https://doi.org/10.1046/j.1420-9101.1999.00115.x

65. Martínez-Ainsworth NE, Tenaillon MI (2016) Superheroes and masterminds of plant domestication. C R Biol 339:268–273. https://doi.org/10.1016/j.crvi.2016.05.005

66. Lukens LN, Doebley J (1999) Epistatic and environmental interactions for quantitative trait loci involved in maize evolution. Genet Res 74:291–302. https://doi.org/10.1017/S0016672399004073

67. Wallace JG, Larsson SJ, Buckler ES (2014) Entering the second century of maize quantitative genetics. Heredity (Edinb) 112:30–38

68. Beadle GW (1972) Mystery of maize. F Museum Nat Hist Bull 43:2–11

69. Briggs WH, McMullen MD, Gaut BS, Doebley J (2007) Linkage mapping of domestication loci in a large maize teosinte backcross resource. Genetics 177:1915–1928. https://doi.org/10.1534/genetics.107.076497

70. Stitzer MC, Ross-Ibarra J (2018) Maize domestication and gene interaction. New Phytol 220(2):395–408. https://doi.org/10.1111/nph.15350

71. Cook JP, McMullen MD, Holland JB, Tian F, Bradbury P, Ross-Ibarra J, Buckler ES, Flint-Garcia SA (2012) Genetic architecture of maize kernel composition in the nested association mapping and inbred association panels. Plant Physiol 158:824–834. https://doi.org/10.1104/pp.111.185033

72. Romero Navarro JA, Willcox M, Burgueño J, Romay C, Swarts K, Trachsel S, Preciado E, Terron A, Delgado HV, Vidal V, Ortega A, Banda AE, Montiel NOG, Ortiz-Monasterio-I, Vicente FS, Espinoza AG, Atlin G, Wenzl P, Hearne S, Buckler ES (2017) A study of allelic diversity underlying flowering-time adaptation in maize landraces. Nat Genet 49:476–480. https://doi.org/10.1038/ng.3784

73. Tian F, Stevens NM, Buckler ES (2009) Tracking footprints of maize domestication and evidence for a massive selective sweep on chromosome 10. Proc Natl Acad Sci 106:9979–9986. https://doi.org/10.1073/pnas.0901122106

74. Tenaillon MI, Sawkins MC, Long AD, Gaut RL, Doebley JF, Gaut BS (2001) Patterns of DNA sequence polymorphism along chromosome 1 of maize (Zea mays ssp. mays L.). Proc Natl Acad Sci U S A 98:9161–9166. https://doi.org/10.1073/pnas.151244298

75. Buckler ES, Gaut BS, McMullen MD (2006) Molecular and functional diversity of maize. Curr Opin Plant Biol 9:172–176

76. Wang H, Nussbaum-Wagler T, Li B, Zhao Q, Vigouroux Y, Faller M, Bomblies K, Lukens L, Doebley JF (2005) The origin of the naked grains of maize. Nature 436:714–719. https://doi.org/10.1038/nature03863

77. Zheng P, Allen WB, Roesler K, Williams ME, Zhang S, Li J, Glassman K, Ranch J, Nubel D, Solawetz W, Bhattramakki D, Llaca V, Deschamps S, Zhong GY, Tarczynski MC, Shen B (2008) A phenylalanine in DGAT is a key determinant of oil content and composition in maize. Nat Genet 40:367–372. https://doi.org/10.1038/ng.85

78. Hancock AM, Brachi B, Faure N, Horton MW, Jarymowycz LB, Sperone FG, Toomajian C, Roux F, Bergelson J (2011) Adaptation to climate across the Arabidopsis thaliana genome. Science 334:83–86. https://doi.org/10.1126/science.1209244

79. Chia JM, Song C, Bradbury PJ, Costich D, De Leon N, Doebley J, Elshire RJ, Gaut B, Geller L, Glaubitz JC, Gore M, Guill KE, Holland J, Hufford MB, Lai J, Li M, Liu X, Lu Y, McCombie R, Nelson R, Poland J, Prasanna BM, Pyhäjärvi T, Rong T, Sekhon RS, Sun Q, Tenaillon MI, Tian F, Wang J, Xu X, Zhang Z, Kaeppler SM, Ross-Ibarra J, McMullen MD, Buckler ES, Zhang G, Xu Y, Ware D (2012) Maize HapMap2 identifies extant variation from a genome in flux. Nat Genet 44:803–807. https://doi.org/10.1038/ng.2313

80. Rodgers-Melnick E, Vera DL, Bass HW, Buckler ES (2016) Open chromatin reveals the functional maize genome. Proc Natl Acad Sci 113:E3177–E3184. https://doi.org/10.1073/pnas.1525244113

81. Swanson-Wagner R, Briskine R, Schaefer R, Hufford MB, Ross-Ibarra J, Myers CL, Tiffin P, Springer NM (2012) Reshaping of the maize transcriptome by domestication. Proc Natl Acad Sci U S A 109:11878–11883. https://doi.org/10.1073/pnas.1201961109

82. Schnable PS, Ware D, Fulton RS, Stein JC, Wei F, Pasternak S, Liang C, Zhang J, Fulton L, Graves TA, Minx P, Reily AD, Courtney L, Kruchowski SS, Tomlinson C, Strong C, Delehaunty K, Fronick C, Courtney B, Rock SM, Belter E, Du F, Kim K, Abbott RM, Cotton M, Levy A, Marchetto P, Ochoa K, Jackson SM, Gillam B, Chen W, Yan L, Higginbotham J, Cardenas M, Waligorski J, Applebaum E, Phelps L, Falcone J, Kanchi K, Thane T, Scimone A, Thane N, Henke J, Wang T, Ruppert J, Shah N, Rotter K, Hodges J, Ingenthron E, Cordes M, Kohlberg S, Sgro J, Delgado B, Mead K, Chinwalla A, Leonard S, Crouse K, Collura K, Kudrna D, Currie J, He R, Angelova A, Rajasekar S, Mueller T, Lomeli R, Scara G, Ko A, Delaney K, Wissotski M, Lopez G, Campos D, Braidotti M, Ashley E, Golser W, Kim H, Lee S, Lin J, Dujmic Z, Kim W, Talag J, Zuccolo A, Fan C, Sebastian A, Kramer M, Spiegel L, Nascimento L, Zutavern T, Miller B, Ambroise C, Muller S, Spooner W, Narechania A, Ren L, Wei S, Kumari S, Faga B, Levy MJ, McMahan L, Van Buren P, Vaughn MW, Ying K, Yeh C-T, Emrich SJ, Jia Y, Kalyanaraman A, Hsia A-P, Barbazuk WB, Baucom RS, Brutnell TP, Carpita NC, Chaparro C, Chia J-M, Deragon J-M, Estill JC, Fu Y, Jeddeloh JA, Han Y, Lee H, Li P, Lisch DR, Liu S, Liu Z, Nagel DH, McCann MC, SanMiguel P, Myers AM, Nettleton D, Nguyen J, Penning BW, Ponnala L, Schneider KL, Schwartz DC, Sharma A, Soderlund C, Springer NM, Sun Q, Wang H, Waterman M, Westerman R, Wolfgruber TK, Yang L, Yu Y, Zhang L, Zhou S, Zhu Q, Bennetzen JL, Dawe RK, Jiang J, Jiang N, Presting GG, Wessler SR, Aluru S, Martienssen RA, Clifton SW, McCombie WR, Wing RA, Wilson RK (2009) The B73 maize genome: complexity, diversity, and dynamics. Science 326:1112–1115. https://doi.org/10.1126/science.1178534

83. Tenaillon MI, Hollister JD, Gaut BS (2010) A triptych of the evolution of plant transposable elements. Trends Plant Sci 15:471–478

84. Muñoz-Diez C, Vitte C, Ross-Ibarra J, Gaut BS, Tenaillon MI (2012) Using nextgen sequencing to investigate genome size variation and transposable element content. Top Curr Genet 24:41–58

85. Diez CM, Meca E, Tenaillon MI, Gaut BS (2014) Three groups of transposable elements with contrasting copy number dynamics and host responses in the maize (Zea mays ssp. mays) genome. PLoS Genet 10(4):e1004298. https://doi.org/10.1371/journal.pgen.1004298

86. Hollister JD, Gaut BS (2009) Epigenetic silencing of transposable elements: a trade-off between reduced transposition and deleterious effects on neighboring gene expression. Genome Res 19:1419–1428. https://doi.org/10.1101/gr.091678.109

87. Waters AJ, Makarevitch I, Noshay J, Burghardt LT, Hirsch CN, Hirsch CD, Springer NM (2017) Natural variation for gene expression responses to abiotic stress in maize. Plant J 89:706–717. https://doi.org/10.1111/tpj.13414

88. Vitte C, Fustier MA, Alix K, Tenaillon MI (2014) The bright side of transposons in crop evolution. Briefings Funct Genomics Proteomics 13:276–295. https://doi.org/10.1093/bfgp/elu002

89. Studer A, Zhao Q, Ross-Ibarra J, Doebley J (2011) Identification of a functional transposon insertion in the maize domestication gene tb1. Nat Genet 43:1160–1163. https://doi.org/10.1038/ng.942

90. Springer NM, Ying K, Fu Y, Ji T, Yeh CT, Jia Y, Wu W, Richmond T, Kitzman J, Rosenbaum H, Iniguez AL, Barbazuk WB, Jeddeloh JA, Nettleton D, Schnable PS (2009) Maize inbreds exhibit high levels of copy number variation (CNV) and presence/absence variation (PAV) in genome content. PLoS Genet 5(11):e1000734. https://doi.org/10.1371/journal.pgen.1000734

91. Maron LG, Guimarães CT, Kirst M, Albert PS, Birchler JA, Bradbury PJ, Buckler ES, Coluccio AE, Danilova TV, Kudrna D, Magalhaes JV, Piñeros MA, Schatz MC, Wing RA, Kochian LV (2013) Aluminum tolerance in maize is associated with higher MATE1 gene copy number. Proc Natl Acad Sci U S A 110:5241–5246. https://doi.org/10.1073/pnas.1220766110. Ronald Sederoff by R

92. Baltazar BM, Sánchez-Gonzalez JDJ, De La Cruz-Larios L, Schoper JB (2005) Pollination

between maize and teosinte: an important determinant of gene flow in Mexico. Theor Appl Genet 110:519–526. https://doi.org/10.1007/s00122-004-1859-6

93. Hufford MB, Lubinksy P, Pyhäjärvi T, Devengenzo MT, Ellstrand NC, Ross-Ibarra J (2013) The genomic signature of crop-wild introgression in maize. PLoS Genet 9(5): e1003477. https://doi.org/10.1371/journal.pgen.1003477

94. Lauter N, Gustus C, Westerbergh A, Doebley J (2004) The inheritance and evolution of leaf pigmentation and pubescence in teosinte. Genetics 167:1949–1959. https://doi.org/10.1534/genetics.104.026997

95. Kawecki TJ, Ebert D (2004) Conceptual issues in local adaptation. Ecol Lett 7:1225–1241. https://doi.org/10.1111/j.1461-0248.2004.00684.x

96. Anderson JT, Lee CR, Rushworth CA, Colautti RI, Mitchell-Olds T (2013) Genetic trade-offs and conditional neutrality contribute to local adaptation. Mol Ecol 22:699–708

97. Mercer K, Martínez-Vásquez Á, Perales HR (2008) Asymmetrical local adaptation of maize landraces along an altitudinal gradient. Evol Appl 1:489–500. https://doi.org/10.1111/j.1752-4571.2008.00038.x

98. Aguirre-Liguori JA, Tenaillon MI, Vázquez-Lobo A, Gaut BS, Jaramillo-Correa JP, Montes-Hernandez S, Souza V, Eguiarte LE (2017) Connecting genomic patterns of local adaptation and niche suitability in teosintes. Mol Ecol 26:4226–4240. https://doi.org/10.1111/mec.14203

99. Fustier MA, Brandenburg JT, Boitard S, Lapeyronnie J, Eguiarte LE, Vigouroux Y, Manicacci D, Tenaillon MI (2017) Signatures of local adaptation in lowland and highland teosintes from whole-genome sequencing of pooled samples. Mol Ecol 26:2738–2756. https://doi.org/10.1111/mec.14082

100. Hung H-Y, Shannon LM, Tian F, Bradbury PJ, Chen C, Flint-Garcia SA, McMullen MD, Ware D, Buckler ES, Doebley JF, Holland JB (2012) ZmCCT and the genetic basis of day-length adaptation underlying the postdomestication spread of maize. Proc Natl Acad Sci 109:E1913–E1921. https://doi.org/10.1073/pnas.1203189109

101. Yang Q, Li Z, Li W, Ku L, Wang C, Ye J, Li K, Yang N, Li Y, Zhong T, Li J, Chen Y, Yan J, Yang X, Xu M (2013) CACTA-like transposable element in ZmCCT attenuated photoperiod sensitivity and accelerated the postdomestication spread of maize. Proc Natl Acad Sci 110:16969–16974. https://doi.org/10.1073/pnas.1310949110

102. Linhart YB, Grant MC (1996) Evolutionary significance of local genetic differentiation in plants. Annu Rev Ecol Syst 27:237–277. https://doi.org/10.1146/annurev.ecolsys.27.1.237

103. Valverde BE (2007) Status and management of grass-weed herbicide resistance in Latin America. Weed Technol 21:310–323. https://doi.org/10.1614/WT-06-097.1

104. Moeller DA, Tiffin P (2008) Geographic variation in adaptation at the molecular level: a case study of plant immunity genes. Evolution 62:3069–3081. https://doi.org/10.1111/j.1558-5646.2008.00511.x

105. Fang Z, Pyhäjärvi T, Weber AL, Dawe RK, Glaubitz JC, Sánchez González JJ, Ross-Ibarra C, Doebley J, Morrell PL, Ross-Ibarra J (2012) Megabase-scale inversion polymorphism in the wild ancestor of maize. Genetics 191:883–894. https://doi.org/10.1534/genetics.112.138578

106. Schluter D (2009) Evidence for ecological speciation and its alternative. Science 323:737–741

107. Wood TE, Burke JM, Rieseberg LH (2005) Parallel genotypic adaptation: when evolution repeats itself. Genetica 123:157–170

108. Colosimo PF, Hosemann KE, Balabhadra S, Villarreal G, Dickson H, Grimwood J, Schmutz J, Myers RM, Schluter D, Kingsley DM (2005) Widespread parallel evolution in sticklebacks by repeated fixation of ectodysplasin alleles. Science 307:1928–1933. https://doi.org/10.1126/science.1107239

109. Pearce RJ, Pota H, Evehe MSB, Bâ EH, Mombo-Ngoma G, Malisa AL, Ord R, Inojosa W, Matondo A, Diallo DA, Mbacham W, Van Den Broek IV, Swarthout TD, Getachew A, Dejene S, Grobusch MP, Njie F, Dunyo S, Kweku M, Owusu-Agyei S, Chandramohan D, Bonnet M, Guthmann JP, Clarke S, Barnes KI, Streat E, Katokele ST, Uusiku P, Agboghoroma CO, Elegba OY, Cissé B, A-Elbasit IE, Giha HA, Kachur SP, Lynch C, Rwakimari JB, Chanda P, Hawela M, Sharp B, Naidoo I, Roper C (2009) Multiple origins and regional dispersal of resistant dhps in African Plasmodium falciparum malaria. PLoS Med 6(4):e1000055. https://doi.org/10.1371/journal.pmed.1000055

110. Chan YF, Marks ME, Jones FC, Villarreal G, Shapiro MD, Brady SD, Southwick AM, Absher DM, Grimwood J, Schmutz J, Myers RM, Petrov D, Jónsson B, Schluter D, Bell

MA, Kingsley DM (2010) Adaptive evolution of pelvic reduction in sticklebacks by recurrent deletion of a pitxl enhancer. Science 327:302–305. https://doi.org/10.1126/science.1182213

111. Stern DL (2013) The genetic causes of convergent evolution. Nat Rev Genet 14:751–764

112. Roesti M, Gavrilets S, Hendry AP, Salzburger W, Berner D (2014) The genomic signature of parallel adaptation from shared genetic variation. Mol Ecol 23:3944–3956. https://doi.org/10.1111/mec.12720

113. Riehle MM, Bennett AF, Long AD (2001) Genetic architecture of thermal adaptation in Escherichia coli. Proc Natl Acad Sci 98:525–530. https://doi.org/10.1073/pnas.98.2.525

114. Gaut BS (2015) Evolution is an experiment: assessing parallelism in crop domestication and experimental evolution. Mol Biol Evol 32:1661–1671. https://doi.org/10.1093/molbev/msv105

115. Fuller DQ, Denham T, Arroyo-Kalin M, Lucas L, Stevens CJ, Qin L, Allaby RG, Purugganan MD (2014) Convergent evolution and parallelism in plant domestication revealed by an expanding archaeological record. Proc Natl Acad Sci 111:6147–6152. https://doi.org/10.1073/pnas.1308937110

116. Lin Z, Li X, Shannon LM, Yeh CT, Wang ML, Bai G, Peng Z, Li J, Trick HN, Clemente TE, Doebley J, Schnable PS, Tuinstra MR, Tesso TT, White F, Yu J (2012) Parallel domestication of the Shattering1 genes in cereals. Nat Genet 44:720–724. https://doi.org/10.1038/ng.2281

117. Takuno S, Ralph P, Swart K, Elshire RJ, Glaubitz JC, Buckler ES, Hufford MB, Ross-Ibarra J (2015) Independent molecular basis of convergent highland adaptation in maize. Genetics 200:1297–1312. https://doi.org/10.1534/genetics.115.178327

118. Lee KM, Coop G (2017) Distinguishing among modes of convergent adaptation using population genomic data. Genetics 207(4):1591–1619. https://doi.org/10.1534/genetics.117.300417

119. Gilbert SF, Epel D (2009) Ecological developmental biology: integrating epigenetics, medicine, and evolution. Yale J Biol Med 82(4):231–232

120. Beldade P, Mateus ARA, Keller RA (2011) Evolution and molecular mechanisms of adaptive developmental plasticity. Mol Ecol 20:1347–1363. https://doi.org/10.1111/j.1365-294X.2011.05016.x

121. Wennersten L, Forsman A (2012) Population-level consequences of polymorphism, plasticity and randomized phenotype switching: a review of predictions. Biol Rev 87:756–767

122. Des Marais DL, Hernandez KM, Juenger TE (2013) Genotype-by-environment interaction and plasticity: exploring genomic responses of plants to the abiotic environment. Annu Rev Ecol Evol Syst 44:5–29

123. DeWitt TJ, Sih A, Wilson DS (1998) Costs and limits of phenotypic plasticity. Trends Ecol Evol 13:77–81

124. Nicotra AB, Atkin OK, Bonser SP, Davidson AM, Finnegan EJ, Mathesius U, Poot P, Purugganan MD, Richards CL, Valladares F, van Kleunen M (2010) Plant phenotypic plasticity in a changing climate. Trends Plant Sci 15:684–692

125. Auld JR, Agrawal AA, Relyea RA (2010) Re-evaluating the costs and limits of adaptive phenotypic plasticity. Proc R Soc B Biol Sci 277:503–511. https://doi.org/10.1098/rspb.2009.1355

126. Kuzawa CW, Bragg JM (2012) Plasticity in human life history strategy. Curr Anthropol 53:S369–S382

127. Diggle PK, Miller JS (2013) Developmental plasticity, genetic assimilation, and the evolutionary diversification of sexual expression in Solanum. Am J Bot 100:1050–1060. https://doi.org/10.3732/ajb.1200647

128. Standen EM, Du TY, Larsson HCE (2014) Developmental plasticity and the origin of tetrapods. Nature 513:54–58. https://doi.org/10.1038/nature13708

129. Gage JL, Jarquin D, Romay C, Lorenz A, Buckler ES, Kaeppler S, Alkhalifah N, Bohn M, Campbell DA, Edwards J, Ertl D, Flint-Garcia S, Gardiner J, Good B, Hirsch CN, Holland J, Hooker DC, Knoll J, Kolkman J, Kruger G, Lauter N, Lawrence-Dill CJ, Lee E, Lynch J, Murray SC, Nelson R, Petzoldt J, Rocheford T, Schnable J, Schnable PS, Scully B, Smith M, Springer NM, Srinivasan S, Walton R, Weldekidan T, Wisser RJ, Xu W, Yu J, De Leon N (2017) The effect of artificial selection on phenotypic plasticity in maize. Nat Commun 8(1):1348. https://doi.org/10.1038/s41467-017-01450-2

130. Wallace JG, Bradbury PJ, Zhang N, Gibon Y, Stitt M, Buckler ES (2014) Association mapping across numerous traits reveals

patterns of functional variation in maize. PLoS Genet 10(12):e1004845. https://doi.org/10.1371/journal.pgen.1004845

131. Piperno DR, Holst I, Winter K, McMillan O (2015) Teosinte before domestication: experimental study of growth and phenotypic variability in late Pleistocene and early Holocene environments. Quat Int 363:65–77

132. Lorant A, Pedersen S, Holst I, Hufford MB, Winter K, Piperno D, Ross-Ibarra J (2017) The potential role of genetic assimilation during maize domestication. PLoS One 12(9):

e0184202. https://doi.org/10.1371/journal.pone.0184202

133. Lobell DB, Schlenker W, Costa-Roberts J (2011) Climate trends and global crop production since 1980. Science 333:616–620. https://doi.org/10.1126/science.1204531

134. Ureta C, Martínez-Meyer E, Perales HR, Álvarez-Buylla ER (2012) Projecting the effects of climate change on the distribution of maize races and their wild relatives in Mexico. Glob Chang Biol 18:1073–1082. https://doi.org/10.1111/j.1365-2486.2011.02607.x

Chapter 13

Neurospora from Natural Populations: Population Genomics Insights into the Life History of a Model Microbial Eukaryote

Pierre Gladieux, Fabien De Bellis, Christopher Hann-Soden, Jesper Svedberg, Hanna Johannesson, and John W. Taylor

Abstract

The ascomycete filamentous fungus *Neurospora crassa* played a historic role in experimental biology and became a model system for genetic research. Stimulated by a systematic effort to collect wild strains initiated by Stanford geneticist David Perkins, the genus *Neurospora* has also become a basic model for the study of evolutionary processes, speciation, and population biology. In this chapter, we will first trace the history that brought *Neurospora* into the era of population genomics. We will then cover the major contributions of population genomic investigations using *Neurospora* to our understanding of microbial biogeography and speciation, and review recent work using population genomics and genome-wide association mapping that illustrates the unique potential of *Neurospora* as a model for identifying the genetic basis of (potentially adaptive) phenotypes in filamentous fungi. The advent of population genomics has contributed to firmly establish *Neurospora* as a complete model system and we hope our review will entice biologists to include *Neurospora* in their research.

Key words Ascomycete, Filamentous fungi, Population genomics, Biogeography, Speciation, Reverse ecology, Introgression, Self–nonself recognition, Selective sweep

1 Introduction: Fungi and Population Genomics

Among complex eukaryotes, fungi have excellent potential as models for population studies at diverse levels, and in particular at the genomic level [1–3]. Population genetics as a discipline has long been largely concerned with plants and animals [4], but this trend is currently being tempered by the massive production of fungal genomic data. The first eukaryote to have its genome sequenced was fungal (the baker's yeast *Saccharomyces cerevisiae*) and the rate of genomic sequences production is higher in the fungal kingdom than in any other eukaryotic kingdom. For instance, as of mid-2017, an estimated 2000 fungal genomes have been sequenced and assembled, and several thousand resequenced genomes are available for population genomic investigations [5]. Fungi

Julien Y. Dutheil (ed.), *Statistical Population Genomics*, Methods in Molecular Biology, vol. 2090,
https://doi.org/10.1007/978-1-0716-0199-0_13, © The Author(s) 2020

have relatively small and low-complexity genomes by eukaryotic standards (typically 30–40 Mb, ~10,000 genes), many fungi have haploid genetics, and these genomic advantages have contributed to making fungi the leading kingdom for eukaryotic genome sequencing [6]. It also follows that fungal population genomics is the only variety of eukaryotic population genomics that is truly "genomic," given that, unlike fungi, most plant and animals genomes cannot be sequenced telomere-to-telomere in relatively large numbers within a reasonable time. However, in the context of evolutionary and ecological genetics, what has long been lacking is access to—and essential information on—fungal natural populations [4]. The genus *Neurospora* stood out early as an outstanding model for fungal population studies, with large numbers of isolates that could be sampled in a predictable manner in various ecosystems [2]. In this chapter we will begin by briefly summarizing the biological features, human and historical factors that have contributed to bring *Neurospora* into the realm of evolutionary biology and ecology. We will then cover the major contributions of *Neurospora* to our understanding of fungal biogeography, fungal speciation and the permeability of barriers to gene flow. Finally, we will review the early contributions of *Neurospora* to our current knowledge of the genetic basis of (potentially adaptive) phenotypes in filamentous fungi.

2   The Rise of *Neurospora* as a Model for Evolutionary and Ecological Genetics

Neurospora is one of the most easily recognized of filamentous ascomycetes (Fig. 1). Originally described as a contaminant in French army bakeries [7], *Neurospora* is most often encountered as powdery masses of bright, carotenoid-colored mycelium and mitospores (=conidia, *see* **Note 1**) on the surface of burned or heated substrates. Visible, pink to orange colonies on scorched vegetation or cooked foodstuffs form the primary source of *Neurospora* collections, but aconidial noncolored species can also be isolated from heat-treated soil. The ecological components of the life cycle of *Neurospora* are not fully understood and might involve close association with plants, such as endophytism [8]. Sex, however, is well understood, beginning with the discovery of sexual fruiting bodies (perithecia, *see* **Note 1**) with meiotic products aligned in linear tetrads by mycologists Cornelius Shear and Bernard Dodge [9]. These ordered tetrads stimulated the use of *Neurospora* as the fungal rival of *Drosophila* and maize as a model for genetic research [10–12]. The fact that *Neurospora* is an haplont (which facilitates recognition of recessive loss-of-function mutations) and that it can be grown on simple minimal media (making it possible to impose further nutritional requirements by mutation) were other salient biological features that popularized *Neurospora*

Fig. 1 *Neurospora* colonies growing on the surface of coffee ground (**a**) and burned shrub (**b**)

for genetic investigations [12, 13]. In 1941, George Beadle and Edward Tatum used *Neurospora* to obtain the first biochemical mutants and to show that genes control metabolic reactions, which was referred to as the "one gene–one enzyme" hypothesis [14, 15]. Beadle and Tatum's experiments helped "convince many skeptical biologists that genes control the fundamental processes of life, and not just the final touches of development, such as wing shape or eye pigment" and "started a new era by bringing genetics and biochemistry together" [13].

In parallel with the adoption of *Neurospora* as a model for molecular and cell biology, David Perkins introduced *Neurospora* into the realm of evolutionary biology and ecology in 1968 by putting in place the long-term study of wild populations [11]. The initial objective of the systematic sampling initiative set up by Perkins was to provide genetic variants for laboratory investigations. Although the mode of primary colonization of *Neurospora* and other fundamental aspects of the ecology of these organisms remain a particular mystery, a distinctive advantage of *Neurospora* relative to many microbes was the relative ease of sampling in diverse ecosystems [2, 16]. Perkins, Dave Jacobson and other scholars eventually gathered a collection of >5000 isolates, access to which is still provided by the Fungal Genetics Stock Center (University of Kansas). Surveys of the wild strains have continued to provide genetic variants for a variety of laboratory investigations on mitochondria and senescence plasmids, genes governing vegetative incompatibility or mating types, and meiotic drive or transposable elements [4, 17, 18]. *Neurospora* entered the genomic era in 2003 with the release of its genomic sequence [19], and the Perkins collection was quickly perceived as a boon by functional, ecological, and evolutionary genomicists. Recently, wild isolates proved valuable for variation-guided functional analyses of cell–cell recognition, via either genome-wide association studies or QTL mapping [20–22]. Resequencing data for wild isolates has also been used to identify genes that underlie variation in phenotypes related to self-recognition and cold tolerance [22, 23]; reviewed below in the final section 5). More generally, the collection also advanced knowledge of the systematics, biogeography, population biology, and evolutionary history of *Neurospora* and ascomycetes in general [2, 17, 24]; reviewed in next two sections 3 and 4).

3   *Neurospora* Population Genomics Has Revealed Cryptic Species with Large Variation in the Extent of Their Geographical Distribution

3.1   Nothing Is Generally Everywhere

Microbes have long been thought to have large geographic distributions, in contrast to the highly restricted ranges of larger organisms. Fungi were not immune to this misconception, and the idea that dispersal ability per se does not limit the geographic distribution of these organisms remains quite widespread today, even among biologists. The misconception that many fungi had global distributions is largely based on two factors. First, the observation that almost all fungi produce tiny, powder-like propagules on structures promoting their dissemination by wind [32] and second, reliance on morphological species recognition criteria that have proved to be too broad for fungi, and have given an inaccurate picture of fungal diversity, distributions and ecologies [33, 34].

3.2   Geographic Endemicity Within Globally Distributed Neurospora Morphospecies

Studies on wild *Neurospora* isolates have altered our understanding of fungal biogeography, providing a perfect illustration that the inferred geographic range of a fungal species depends upon the method of species recognition. More generally, studies on wild *Neurospora* isolates have shown that fungal species are highly structured and that fungal distribution have been shaped by geological and climatic events the same way as macrobes have [34, 35]. Conventional criteria based on morphology are of little use to ascertain taxonomic status in *Neurospora* as most conidiating species cannot be distinguished from one another by the size, color and shape of their vegetative and reproductive organs [9, 16, 17]. Hence, by morphological species recognition, only two species of conidiating *Neurospora* are found: one with eight ascospores per ascus, and one with four ascospores per ascus (*see* **Note 1**). The two morphospecies are both cosmopolitan in temperate and tropical latitudes (Table 1). In vitro mating compatibility tests and phylogenetic analyses, however, reveal that the two morphological species have their own biogeography and encompass multiple endemic species: under biological species recognition seven species are found, while under phylogenetic species recognition at least twenty-six species are identified (Table 1). Similar to many plant

Table 1
Conidiating species of *Neurospora*

Morphological species	Biological species	Phylogenetic species	Distribution	References
Neurospora with four-ascospores per ascus	*N. tetrasperma*[a]	*N. tetrasperma* sp.[b]	Western Europe, Pacific Islands, Oceania, America, South and Southeast Asia, Western tropical Africa	Perkins database; [16, 24]
		Lineages 1, 2, 7, 8	Gulf of Mexico	[25–28]
		Lineage 3	Gulf of Mexico, Eastern North America	[25–27]
		Lineage 4	South-East Asia, America	[25–28]
		Lineage 5	Oceania	[25–28]
		Lineage 6	Polynesia, Mexico, Southeast Asia	[25–28]
		Lineage 9	Western tropical Africa	[25, 27, 28]
		Lineage 10	Western Europe	[27, 28]
		Lineage 11	Canary Islands	[16, 26]
Neurospora with eight-ascospores per ascus	*N. sitophila*	ND[c]	Western Europe, Asia, Turkey, Polynesia, Oceania, America, Western tropical Africa	Perkins database; [16, 24]
	N. crassa	*N. perkinsii*	Western Tropical Africa	[29, 30]
		N. crassa	Western Europe, South Asia, Gulf of Mexico, Western Tropical Africa	[16, 24, 29, 30]
	N. intermedia	*N. intermedia*	South and Southeast Asia, Polynesia, Western Tropical Africa, Gulf of Mexico	[29, 30]
	N. hispaniola	*N. hispaniola*	Hispaniola	[29, 30]
	N. metzenbergii	*N. metzenbergii*	Madagascar, Gulf of Mexico	[29, 30]
	N. discreta	*N. discreta* sensu stricto	Gulf of Mexico	[31]
		N. discreta PS4	Western Europe, North America, Papua New Guinea, Western tropical Africa	[16, 24, 31]
		N. discreta PS5, PS6, PS8	Western tropical Africa	[31]
		N. discreta PS7	Gulf of Mexico, Central America	[31]
		N. discreta PS9	New Zealand	[31]

(continued)

Table 1
(continued)

Morphological species	Biological species	Phylogenetic species	Distribution	References
		N. discreta PS10	New Zealand, Brazil	[31]

[a]Subdivision within the biological species, and high congruence between the phylogenetic and biological species recognition are found when using a quantitative measurement of the reproductive success, incorporating characters such as viability and fertility of offspring [25]
[b]Many strains in the Perkins database, or listed in publications, remain to be phylogenetically identified
[c]ND: No data on the existence of cryptic species within *N. sitophila*

and animal genera, the distribution of some of the phylogenetic species appears to be quite limited (e.g., *N. hispaniola* was reported in the Hispaniola island only) while others have very broad distributions (e.g., globally distributed *N. discreta* PS4). Genealogies of multiple genes have been widely used in fungi, including *Neurospora* [24, 25, 29–31], improving our understanding of the structure of fungal biodiversity. The observation that fungal species defined by morphology typically harbor several to many endemic species can be explained by the relative paucity of morphological characters and the slower rate of morphological change for organisms with less elaborate development and fewer cells, allowing genetic isolation to precede recognizable morphological changes [34]. Just as genetic isolation can precede morphological change, phylogenetic divergence can precede reproductive isolation, such that one biological species can embrace several phylogenetic species, as was shown when biological and phylogenetic species recognition were compared in *Neurospora* [36]. Species recognition by genealogical concordance, popular as it has been, has limitations, related to heterogeneity in the congruence of sequenced loci with the species tree, and inadequate sampling of substrates throughout the geographic range of species; *Neurospora* is not an exception.

3.3 On the Difficulty of Species Diagnosis in Neurospora and Fungi

Morphology is of little use to identify species in *Neurospora*, like in many microscopic filamentous fungi. Perkins and collaborators have published a bountiful collection of protocols to induce and assess mating in vitro in *Neurospora*, but systematic analyses of pre- and postmating barriers in large collection of isolates is challenging to implement in modern labs. Species recognition based on genealogical concordance among gene trees became the gold-standard in *Neurospora*, revealing cryptic species diversity [34]. Species recognition by genealogical concordance, however, suffers from two limitations. The first limitation is the requirement for the use of the same sets of sequenced loci across studies. The second limitation is that the resolving power of sequenced loci is most often not

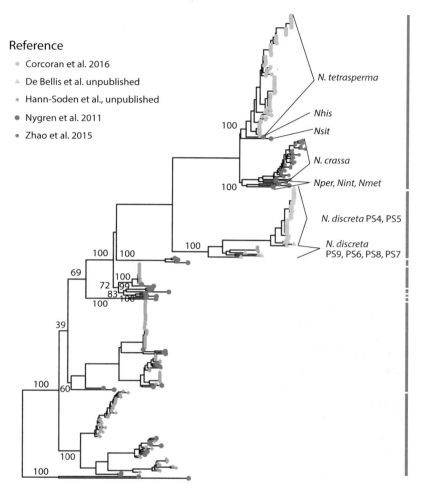

Reference
- Corcoran et al. 2016
- De Bellis et al. unpublished
- Hann-Soden et al., unpublished
- Nygren et al. 2011
- Zhao et al. 2015

N. tetrasperma

Nhis
Nsit

N. crassa

Nper, Nint, Nmet

N. discreta PS4, PS5

N. discreta
PS9, PS6, PS8, PS7

Fig. 2 Total evidence genealogy inferred using RAxML v8 [37] based on the concatenation of sequences at six loci published by [38], including *Bml, mak-2, nik-1, ORF1, pkc,* and *tef-1*. Phylogenetic species identified by genealogical concordance across multiple gene genealogies are indicated by shaded areas and bootstrap supports in the total evidence phylogeny. Taxon names are reported only for best-sampled conidiating species. *Nhis, N. hispaniola; Nsit, N. sitophila; Nper, N. perkinsii; Nint, N. intermedia; Nmet, N. metzenbergii.* Marker sequences were retrieved from genomic sequences using blastn for all datasets but the ref. [38] dataset which was downloaded from NCBI. Samples from Hann-Soden et al. (Unpublished) and some of the samples from Nygren et al. [38] were isolated from heat-treated soil, while the remaining samples were isolated from burned vegetation

known *a priori,* and markers not chosen based prior knowledge of the species tree.

Figure 2 provides an illustration of the phylogenetic diversity encompassed by the genus *Neurospora,* but also of the limitations of species recognition by genealogical concordance. Sequences at six loci previously identified by Nygren et al. [38] were retrieved from GenBank, extracted from publicly available genomes [22, 28, 39] and extracted from the genomes of isolates originating from multiple sites in North America (De Bellis et al., unpublished) and isolated from the soil spore bank treated with heat (Hann-Soden et al., unpublished). Sequences at the six loci were concatenated and the resulting total evidence tree recapitulates the phylogenetic species identified by applying species recognition by genealogical concordance sensu Dettman et al. [29]. Two important insights emerge from this analysis. The first striking feature is that the species identified are different from the species previously identified by Dettman et al. [29] using three loci not included in the Nygren et al. [38] loci. Many of the phylogenetic and biological species previously identified do not stand as distinct species in this analysis. Another important result is the relatively large number of species that can coexist within the same spore bank (not shown here). In summary, using half a dozen loci for phylogenetic species delineation makes it more operational but also limits its ability to resolve population subdivision. Genomic data should have increased power to resolve species limits and evolutionary relationships, but standard species recognition by genealogical concordance remains useful until the cost of sequencing a set of markers is higher than sequencing a full genome.

3.4 Population Structure Within Neurospora Phylogenetic Species

Phylogenetic analyses and mating compatibility tests have great potential to augment knowledge of the taxonomy, ecology and biogeography of fungal genera and species, but these approaches are not operational when the goal is to infer fine-scale population genetic structure [34, 40]. High-throughput sequencing technologies have made it possible for individual laboratories to acquire whole-genome sequences across populations and test hypotheses of geographic endemicity or genome evolution previously formulated based on sequence diversity and reproductive biology. Population genomic studies of eight-spore biological species of *Neurospora* are illustrative of the finer resolution afforded by genomic information to characterize population structure. For instance, in selecting populations of *Neurospora* for genome wide association studies, researchers relied on populations previously identified by concordance of gene or microsatellite genealogies that included a strain with a reference genome and that spanned significant environmental variation. In the case of *N. crassa*, the population boarded the Gulf of Mexico and for *N. discreta* PS4, the population displayed a remarkably large latitudinal distribution along western North America. However, in *N. crassa* phylogenomic analyses and model-based Bayesian clustering of transcriptomic data for 50 isolates revealed not 1 population, but multiple divergent lineages, with the two best sampled lineages found in Louisiana and the Caribbean [23] (Fig. 3). Subsequent phylogenomic analyses of SNPs from the resequenced transcriptomes of 112 *N. crassa* individuals from the same geographic area as the Louisiana population showed no population subdivision, providing an ideal setting for a

(A) Population genomic structure of *Neurospora discreta* PS4

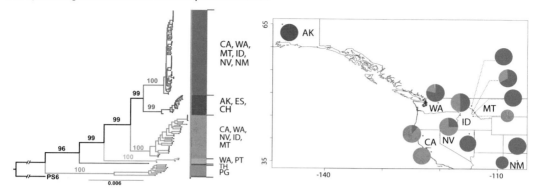

(B) Population genomic structure of *Neurospora crassa*

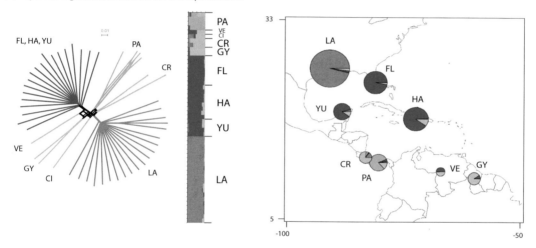

Fig. 3 (**a**) Population genomic structure of *Neurospora discreta* PS4 as inferred from whole genome resequencing of 128 isolates (ref. [39] and De Bellis et al., Unpublished). *Left*: RAxML8 whole genome genealogy; center: ancestry proportions in $K = 7$ clusters as inferred using sparse Nonnegative Matrix Factorization [41]; *right*: sampling sites in North America and proportions of isolates assigned to four European/North American clusters as pie charts. (**b**) Population genomic structure of American *Neurospora crassa* as inferred from reanalysis of previously published transcriptome resequencing data [23]. *Left*: Neighbor-Net inferred from biallelic SNPs without missing data using Splitstree [42]; *center*: ancestry proportions in $K = 4$ cluster as inferred using Structure 2.3.4 [43–45] based on a random 10% of the 134 k SNPs without missing data; right: sampling sites in America and sum of membership proportions in four Structure clusters as pie charts. US/Mexico State Abbreviations: *AK* Alaska, *CA* California, *FL* Florida, *ID* Idaho, *LA* Louisiana, *MT* Montana, *NV* Nevada, *WA* Washington, *YU* Yucatan. ISO country codes: *CH* Switzerland, *CI* Côte d'Ivoire, *CR* Costa Rica, *ES* Spain, *GY* Guyana, *HA* Haiti, *PA* Panama, *PG* Papua New Guinea, *PT* Portugal, *TH* Thailand, *VE* Venezuela

genome-wide association study (*see* last section). In the same way, phylogenomic analyses and model-based Bayesian clustering of whole genome information for 128 *N. discreta* PS4 isolates revealed not one population, but six divergent lineages [39]

(De Bellis et al. unpublished) (Fig. 3). Demographic inference based on a diffusion approximation to the site frequency spectrum [46] estimated relatively recent divergence times (≈0.4 M generations) between the lineages discovered within *N. crassa* and *N. discreta* PS4 [23, 39]. Relating these divergence time estimates with historical events requires a generation time, which is a perennial question in *Neurospora*, and in fungi in general. Extensive observations of *Neurospora* colonies at burn sites led to only a few observations of fruiting bodies in nature, possibly owing to the difficulty in the recognition of black perithecia on/within burned substrate, or the delay of sexual reproduction after conidial blooms [18]. Perithecia were observed on maize cobs [47], under the bark of fire-injured trees [2, 48], or protruding through cracked tissues of scorched sugar cane [49]. However, it remains unknown whether sexual cycles in *Neurospora* are synchronized with wild-fires. As proposed by Turner et al. [18], "Perhaps the dramatic conidiating blooms seen on burned or scorched vegetation are exceptional sporadic events punctuating a mode of growth that is otherwise inconspicuous or invisible." Given these uncertainties, a plausible scenario is that divergence within North American *N. crassa* and *N. discreta* PS4 (≈0.4 M generations) has been driven by climate oscillations of the Pleistocene (2.6 Mya–11 kya), but alternatives cannot be ruled out without further information on the generation time and other aspects of the population biology of *Neurospora*.

3.5 Comparative Population Genomics of Selfing and Outcrossing Neurospora Species

Like plant evolutionary biologists, the evolutionary causes and consequences of self-fertilization is a question of long-standing interest to research scholars working with *Neurospora* [50, 51]. It was recognized at an early stage that the variety of lifestyles in *Neurospora* shows promise for comparative studies [12]. In contrast to the eight-spored *Neurospora* species, meiotic products in the morphospecies *N. tetrasperma* are packaged into four relatively large ascospores [52]. In nature and in the laboratory, the mycelium that emerges from germinating *N. tetrasperma* ascospores is normally heterokaryotic with component nuclei of opposite mating type (A and a; *see* **Note 2**). *Neurospora tetrasperma* therefore superficially resembles true homothallic (*see* **Note 3**) species [53] in that each ascospore can usually produce a self-fertile mycelium and complete the sexual cycle without needing to find a compatible mate [52]. As a consequence, the breeding system (*see* **Note 4**) of *N. tetrasperma* is referred to as "pseudohomothallic" (*see* **Note 3**) [54]. Such reproductive systems (*see* **Note 4**) based on heterokaryosis (*see* **Note 5**) are unique to fungi, and they can be described as "a form of heterothallism with provisions to allow prolonged inbreeding" (*see* **Note 3**) [55]. In the laboratory, heterokaryotic (*see* **Note 5**) wild-type strains self, but self-sterile homokaryotic (A or a) conidia (*see* **Note 5**) derived from wild-type strains are functionally heterothallic (*see* **Note 3**), and can be outcrossed [4, 17, 52, 56]. Recently the frequency of homokaryotic conidium production by pseudohomothallic *N. tetrasperma* strains collected from New Zealand and the UK was systematically studied [27]. Large differences in the number of homokaryotic conidia produced by different populations were observed, with the rate of homokaryotic conidia being twice as high as previously reported [56], suggesting ample opportunities for outcrossing sex. Laboratory crosses employing homokaryotic isolates of opposite mating types obtained from different strains, however, revealed a high frequency of sexual dysfunction caused by vegetative incompatibility between interacting mycelia [55, 57]. These mating compatibility studies suggest that outcrossing may be limited in nature and that large difference in self-sterile spore production across *N. tetrasperma* populations is not necessarily associated with large difference in outcrossing rates.

More recently, population genomics was used to test if vegetative incompatibility is effectively blocking outcrossing in *N. tetrasperma* populations, by quantifying the level of outcrossing in situ and correlating this factor to population structure and genome evolution [28]. Phylogenomic analysis and model-based clustering of whole genome information for large set of strains confirmed nine of the ten cryptic phylogenetic species previously identified [25–27] and revealed an additional lineage in Europe, but no subdivision within species was detected [28, 58]. Nucleotide diversity was of the same order of magnitude in populations of *N. tetrasperma* as seen in populations of outcrossers *N. crassa* and *N. discreta*, suggesting no strong reduction of within-population diversity in *N. tetrasperma* as would be expected under inbreeding (Fig. 4). Analyses of linkage disequilibrium (Chapter 1) were consistent also consistent with selfing [28] (Fig. 4). The ratio of nonsynonymous to synonymous nucleotide diversity (π_N/π_S) was relatively high in *N. tetrasperma* (>0.7), suggesting a relatively high proportion of slightly deleterious mutation consistent with selfing, although π_N/π_S ratios were comparable to those observed in the outcrosser *N. crassa* (Fig. 4). There is a difference between the 8-spored, heterothallic and 4-spored pseudohomothallic species, however, when it comes to species recognition by concordance of gene genealogies. With *N. discreta* and *N. crassa*, 400 k year old lineages were not diagnosed as species, but the 700 k year old lineages (not appreciably from 400 k in this instance) within *N. tetrasperma* were diagnosed as distinct phylogenetic species. This difference is consistent with theoretical predictions, because population differentiation should increase as consequence of lower within-deme diversity and the combined action of other reproductive and life-history traits which tend to increase isolation [61].

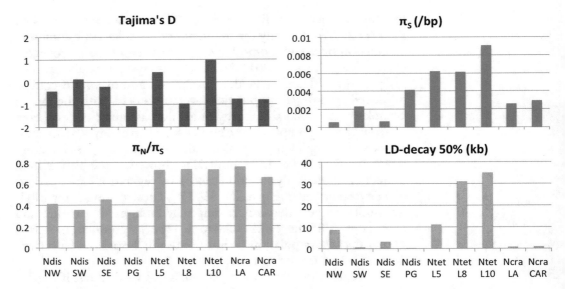

Fig. 4 Summary statistics of population genomic variation in best-sampled *Neurospora* lineages. For *N. discreta* unpublished whole-genome resequencing data (De Bellis et al., Unpublished) were aligned against the reference genome FGSC8579. For *N. crassa*, transcriptome resequencing were downloaded from Short Read Archive (accession SRA026962; [23]) and aligned against reference OR74A v2.0 (hosted at Ensembl Fungi). For *N. tetrasperma*, VCF files were downloaded from the Dryad Digital Repository (https:/datadryad. org/resource/doi: https:/doi.org/10.5061/dryad.162mh; [28]). LD decay values for *N. tetrasperma* and *N. crassa* were previously published by [23, 28]. Computations were carried out in Egglib v3 [59], excluding codon-coding nucleotide triplets with missing data. π_S is the nucleotide diversity at synonymous sites. LD-decay 50% is the distance over which linkage disequilibrium (LD) decays to half its maximum. For *N. tetrasperma* L8 and L10, actual values of LD-decay 50% are 31 kb and >500 kb and bars were truncated for clarity for these two lineages. π_N/π_S is the ratio of nonsynonymous to synonymous nucleotide diversity, which gives, under near neutrality, an estimate of the proportion of effectively neutral mutations that are strongly dependent on the effective population size N_e [60]. The mating-type chromosome of *N. tetrasperma* was excluded from calculations. *Ndis NW* Northwestern *Neurospora discreta* lineage (purple in Fig. 2), *Ndis SW* Southwestern *Neurospora discreta* lineage (orange in Fig. 2), *Ndis SE* Southeastern *Neurospora discreta* lineage (dark green in Fig. 2), *Ndis PG Neurospora discreta* lineage from Papua New Guinea (light blue in Fig. 2), *Ntet L5 Neurospora tetrasperma* lineage 5, *Ntet L8 Neurospora tetrasperma* lineage 8, *Ntet L10 Neurospora tetrasperma* lineage 10, *Ncra LA Neurospora crassa* Louisiana lineage, *Ncra CAR Neurospora crassa* Caribbean lineage

4   *Neurospora* Population Genomics Has Refined Our Views on the Permeability of Barriers to Gene Flow

Population genomics offers great potential for enhanced characterization of cryptic population subdivision in *Neurospora*, revealing that the genus, which is ubiquitous in temperate, subtropical and tropical regions, is structured into a variety of species and divergent lineages within species. The observed richness in *Neurospora* species and lineages does not result in complete geographic separation (i.e., allopatry) and many lineages or species have overlapping ranges (i.e., sympatry or parapatry) (Table 1). Not only multiple species [4, 62], but also multiple lineages within species [39] (Fig. 3), may be collected from the same sites, just centimeters from each other, raising the question of hybridization and admixture [63, 64]. Although true hybrids have not been found in nature [29], the production in the lab of small but significant numbers of viable hybrid progeny has been known since the description of the genus ([65] cited by [58]) and repeatedly confirmed since then (e.g., [18, 36, 66]). For instance, crosses between allopatric *N. intermedia* and *N. crassa* in the lab can typically yield 1–15% black (i.e., potentially viable) ascospores (*see* **Note 1**) [36]. The problem with studying hybridization in *Neurospora* and in fungi in general has been that neither a discerning eye nor simple mating tests can reliably identify hybrids among strains in situ; this is where genomic approaches come to the rescue of the mycologist.

Genomic studies of *Neurospora* populations provided indirect evidence for hybridization and introgression. The first element supporting the existence of hybridization came from the outcrossing heterothallic *N. crassa* (*see* **Note 3**). A genomic island of elevated relative and absolute divergence identified between the Louisiana and Caribbean populations of *N. crassa* [23] showed an unusually large number of fixed differences, suggesting that divergence between haplotypes was older that the splitting of populations. Haplotypic structure at the genomic island was also different between the two populations, with less haplotype diversity and nonuniform haplotype boundaries in the Louisiana population. Together these observations point to the introgression and selective sweep of a single migrant tract in Louisiana from a more genetically diverged population or species that remains to be identified. Further indirect evidence for hybridization came from comparative and population genomic investigations of the pseudohomothallic *N. tetrasperma* (*see* **Note 3**), providing a good illustration of the impact of mating system on population structure in fungi. As described in the previous section, the meiotic pathway of *N. tetrasperma* was reprogrammed so that each ascospore receives mat A and mat a haploid nuclei produced from a single diploid nucleus, which favors selfing. This particular meiotic process is dependent on the segregation of mating-type alleles at the first division of meiosis, which is assured by the suppression of recombination between the centromere and the mating-type locus [67]. Because recombination is suppressed, a large part of the mating-type chromosomes of *N. tetrasperma* degenerates, accumulating nonsynonymous polymorphisms/substitutions and nonoptimal codons [28]. However, genome-wide analyses of *N. tetrasperma* revealed patterns of divergence consistent with a history of introgressive hybridization with several heterothallic relatives, and in particular large tracts of DNA of allospecific origin restricted to the mating-type chromosome. It is also worth noting that these hybridization events occurred despite the morphologically enforced, preferentially selfing, mating system of *N. tetrasperma*. The introgressed tracts have been fixed within *N. tetrasperma* lineages and some of them carry signatures of selective sweeps, suggesting that they confer an adaptive advantage in natural populations. An hypothesis is that the introgression of nondegenerated mating type chromosomes from related species may contribute to maintain integrity of mating type chromosome and constitute a process of genomic reinvigoration acting to reduce the mutational load. These findings corroborate a prediction made by Metzenberg and Randall, and reported by ref. [18] as a personal communication: "periodically during evolution, a deteriorated mating type chromosome is replaced following a cross between *N. tetrasperma* and one of the heterothallic species."

The population genomic studies presented so far show evidence of introgressive hybridization, but none has caught populations in the act of mixing. The most recent evidence for population admixture comes from the study of the population genomic structure of a single species, the outcrossing *N. discreta PS4*. Phylogenomic analyses and model-based clustering of whole-genome data showed four, well-diverged lineages: Papua New-Guinea (PNG), Alaska and Europe (AK-EU), California and Washington state (CA-WA), and New Mexico and Washington state (NM-WA). Admixture analyses using Frappe and a genome scan for lineage-diagnostic SNPs revealed an Alaskan strain that possesses 12% of the genome of the apparently allopatric, NM-WA lineage [39]. Yet, at the same time, there was no evidence of admixture at one collecting site in Washington where the CA-WA and NM-WA lineages are clearly sympatric. The finding of one admixed individual suggested that reproductive isolation is not complete between all pairs of lineages within *N. discreta* and that there might have been opportunities for gene flow between them. Analyses of postdivergence gene flow using the *dadi* package [46], which infers demographic parameters based on a diffusion approximation to the site frequency spectrum, supported models of gene flow following secondary contact, both between North American and PNG lineages, and among lineages in North America/Europe. The finding of nonzero migration rates between all lineages suggests that their geographic distributions have been overlapping to some extent. In North America/Europe, lineages may have diverged following repeated periods of isolation in separate glacial refugia [68], interspersed with periods of secondary contact potentially permitting gene flow that only began relatively recently, as suggested by the finding of a late generation hybrid in Alaska indicative of ongoing admixture.

The best supported models of secondary contact between North American/European lineages assumed heterogeneous gene flow across the genome, and parameter estimates indicated that only a small fraction of the genome experienced relatively higher migration rates between lineages. This heterogeneity in introgression rates may reflect the indirect effects of selection, with limited gene flow in the neighborhood of barrier loci (genetic incompatibilities or genomic features under divergent ecological selection), higher gene flow at adaptively introgressed regions, and basal gene flow in regions not affected by selection [69]. The widespread compatibility of crosses attempted in vitro between North American/European indicated a lack of intrinsic premating barriers (i.e., assortative mating by mate choice), intrinsic postmating prezygotic barriers (i.e., gametic incompatibility) or a form of intrinsic early postzygotic isolation (i.e., zygotic mortality), and the absence of these barriers may have contributed to facilitate gene flow following secondary contact. The lack of intrinsic premating barriers appears to be general among *Neurospora* lineages and species, and may result from constraints on the evolution of pheromone-receptor systems involved in mating [70]. The peptide pheromones that mediate premating attraction between mating-type compatible isolates are identical between *Neurospora* and outgroups [71], consistent with a lack of pheromone-based mate choice. Together these findings and observations suggest that the barriers that limit gene flow following secondary contact and account for the observed heterogeneity in migration rates are mostly extrinsic barriers (e.g., immigrant inviability) or late intrinsic barriers (e.g., hybrid inviability or sterility). This work also illustrates the great potential of speciation genomics for increasing our understanding of fungal biogeography, revealing features such as sympatry in the recent past or admixture between apparently allopatric species, that would not be accessible without genomic data, especially given the scarcity of exploitable fossil records in fungi. Further work on the fine-scale population genetic structure of *Neurospora* is required to quantify lineage diversity at the local scale and the extent of interlineage admixture, and more detailed investigations of barriers to gene flow (e.g., by measuring hybrid inviability and sterility) should provide more insights on the factors contributing to the maintenance or mixing of lineages in sympatry.

5 Studies *Neurospora* Provide Insights into the Genetic Basis of (Potentially Adaptive) Phenotypes in Wild Microbial Eukaryotes

The difficulty of defining the boundaries of populations has been a major impediment for studying the genetic basis of differences between individuals within populations and of adaptive differences between populations. However, by revealing geographic endemism, genomic approaches to characterizing fungal populations have offered new opportunities to identify the molecular underpinnings of key biological features in *Neurospora*. In cases where segregating phenotypic traits could be scored in a *Neurospora* population, quantitative trait locus (QTL) mapping or genome-wide association studies (GWAS) could be used successfully to identify the underlying genomic features. The rates of LD decay observed in *N. crassa* or *N. discreta* for instance (Table 1) are suitable for linkage mapping. QTL mapping was used to identify loci in *N. crassa* associated with a reinforced, post-mating, female mate choice barrier between *N. crassa* and *N. intermedia* [72]. However, due to the presence of large regions of linkage inherent to the QTL approach, the authors could not identify the genes responsible for the quantitative trait. When combined with dense, genome-wide marker coverage and lack of population structure, GWAS can overcome the limitations of QTL studies and offer higher resolution by increasing the range of genetic and phenotypic variation surveyed, by avoiding the generation of time-consuming crosses and taking advantage of many more generations of recombination. GWAS has been used successfully in *N. crassa* to identify the genetic basis of the complex trait of germling communication, the process by which conidia germinating near each other can sense each other, reorient their growth toward one another, and fuse (*see* **Note 6**). By quantifying the proportion of communicating germlings in germinating populations of conidia representing 24 Louisiana *N. crassa* isolates, and using RNAseq to genotype isolates, Palma-Guerrero et al. [21] successfully associated a calcium sensor with the cell-to-cell communication trait.

Another strategy used for rapid trait mapping in segregating populations or crosses is bulked segregant analysis, that is, the genotyping or sequencing of bulked pools of segregating individuals with the most extreme phenotypes. The advantage of bulked segregant analysis over standard QTL analysis or GWAS is that there is no necessity for genotyping all individuals in the segregating population or progeny, which can increase the number of progeny surveyed and hence decrease the size of linked regions surrounding quantitative trait loci compared to classical approaches. As a proof of principle for mutation mapping by bulked segregant analysis, Pomraning et al. [73] used bulked segregant analysis to map the mutation(s) underlying temperature-responsive cell cycle regulation in the classic *N. crassa ndc-1* (*nuclear division cycle-1*) mutant. Two hundred progeny from a wild x mutant cross were tested for the arrest of the nuclear division cycle just prior to DNA synthesis when grown at 37 °C, and a subset of 63 progeny with extreme phenotypes was sequenced en masse in two pools, allowing identification of a single mutation in a single gene.

In cases where bulked segregant analysis does not pinpoint a single gene, population genomic approaches can be used to identify the genes associated with traits. This approach has been demonstrated by studies of the genetic basis of cell communication and fusion in *Neurospora* (*see* **Note 6**). Heller et al. [20, 74] used bulked segregant analysis to map the genes responsible for kind discrimination among germlings and rapid cell death following germling fusion (i.e., "germling regulated death"), identifying 100 kb and 180 kb regions of linkage associated with the self-recognition traits. The genes within these regions could be filtered down to just two genes using genome scans for polymorphism and divergence.

Although *Neurospora* has favorable characteristics for linkage mapping studies, such as a small genome, the ability to cheaply maintain large immortal populations in lab, and haploid genetics, the difficulty in working with *Neurospora*—or any microbe—has been the identification of relevant phenotypes beyond simple traits. However, high-throughput sequencing has made it possible to use an unbiased "reverse ecology" approach to identifying adaptive phenotypes [75], in which genes with functions related to ecologically relevant traits are identified by examining genomic signatures of natural selection. An American *N. crassa* population bordering the Gulf of Mexico had been chosen for GWAS but when 50 of the isolates had been genotyped using RNAseq, analyses of population subdivision revealed multiple clusters and no cluster had sufficient sample size for a GWAS [23]. Fortunately, two clusters had at least 20 members, one in Louisiana and one further south in the Caribbean. Demographic inference based on a diffusion approximation to the site frequency spectrum [46] estimated a relatively higher population migration rate from Louisiana into the Caribbean (0.77 effective migrants per generation) than in the other direction, and a relatively recent divergence time (≈0.4 Mya) in agreement with the small proportion of fixed differences (9.4% of total SNPs; [23]). The Louisiana and Caribbean areas differ by 2–10° of latitude and winter temperatures are on average 9 °C cooler in Louisiana. The resulting hypothesis, that the Louisiana lineage had adapted to life at lower temperature, was not disproved by measuring fitness (i.e., growth) of isolates from each lineage at low (10 °C) and medium (25 °C) temperature. Genome scans to detect regions of extreme divergence in coding sequences, using measures of both relative and absolute divergence [76], revealed more than 30 such regions, but only 2 were identified by all divergence metrics. These two regions encompassed genes known to protect against cold temperatures: a cold shock RNA helicase and a prefoldin chaperone that, in yeast, protects actin from cold temperatures. To test the association between genotypes at candidate genes and fitness in cold temperature, the authors took advantage of the comprehensive *N. crassa* gene deletion collection to devise growth tests [77]. Among the eight genes found in the two regions of extreme genomic divergence, only the RNA helicase and prefoldin knockouts showed loss of cold tolerance.

The identification of genes underlying the ability of mycelia to distinguish self from nonself (i.e., allorecognition genes; *see* **Note**

6) in *N. crassa* by genomics and evolutionary approaches provides another example in which genes, and not phenotypes, were identified first. In filamentous fungi, allorecognition can result in somatic incompatibility (*see* **Note 6**), which is a type of programmed cell death that is triggered following fusion of genetically different cells. In *N. crassa*, somatic allorecognition controls both the fusion of germlings (as in the example of "germling regulated death" cited above) and the fusion of hyphae. Hyphal allorecognition is genetically controlled by the so-called *het* loci, and the *het* loci that have been characterized to date encode proteins carrying a HET domain (encoding a cell death effector) and show signatures of long-term balancing selection [78, 79]. Mining resequencing data from 26 Louisiana isolates for HET domain loci displaying elevated levels of variability, excess of intermediate frequency alleles, and deep gene genealogies identified 34 HET domain loci out of 69. Transformation, incompatibility assays, and genetic analyses revealed that one of the 34 candidates functioned as a *het* locus (*het-e*) that had been identified almost 50 years ago but awaited cloning [80]. The remaining 33 loci are, of course, prime targets for future investigations. These findings are encouraging and the collection of additional and more precise information on the biological and geographical origin of the samples, but also on relevant phenotypes (e.g., related to interactions with other microbes), should make it possible to exploit even better the potential of such reverse ecology approaches in the future.

6   Conclusion

Almost a century has passed since Shear and Dodge [9] described *Neurospora*, 70 years since Beadle and Tatum made it genetically conspicuous [14], and 50 years since Perkins et al. [17] initiated the global collections that fueled a host of research programs that continue to this day. More than 25 years ago, Perkins cautiously wrote that "Research on molecular, cellular and genetic mechanisms is certain to continue. It remains to be seen whether the promise of *Neurospora* for population genetics will be fulfilled." Perkins would surely be happy to witness the tight integration between the two areas of research that now prevails, and to observe that the ways wild strains have proved useful for functional and evolutionary studies far exceeding what he and his collaborators anticipated when systematic collection was initiated in 1968. Much remains to be uncovered about the species richness of *Neurospora* and the distribution of known *Neurospora* species, although our understanding of *Neurospora* biogeography and speciation history is more advanced than it is for the vast majority of free-living microbes. The frequency of wild fires is not a limiting factor, providing a plethora of opportunities to collect new strains, and ongoing work also suggests that both conidial and aconidial species can be retrieved from soil. In the Perkins and Jacobson collection, the phenotype of F1 crosses to *N. crassa* is reported for more than 3600 isolates, and other phenotypes such as meiotic drive phenotype (i.e., spore killer type [66]), color, burned/nonburned substrate for more than 5000. Only a small fraction of this collection has been sequenced, and even with the many genomes already at hand, there is little indication of diminishing returns as additional species or populations are sequenced. The will be no lack of ecological or evolutionary questions. Most of the questions listed by Turner et al. [18] in their conclusion remain to be answered, and much remains to be done to dispel our ignorance about many aspects of the biology of microbial eukaryotes. We predict that, in the future, the message of hope of Perkins and Turner [4] "We are hopeful that the *Neurospora* work reviewed here will encourage wider studies in the genetics of fungal populations and will contribute to an increased appreciation of the potential contribution of the fungi." will not need to be reiterated in any future review about the population genomics of *Neurospora*.

7 Notes

1. An *ascospore* is a sexual spore of an ascomycete fungus, generated through meiosis (=meiospore). An *ascus* is a cell bearing ascospores. A *perithecium* is a spherical type of fruiting body in ascomycete fungi, containing ascus. A *conidium* is an asexual spore of a fungus, generated through mitosis (=mitospore).

2. *Mating type loci* are genes that control sexual compatibility. *Neurospora* has two mating type alleles, referred to as Mat A et al.

3. *Homothallism* defines situations where the successful fusion of gametes does not require functionally different mating-type alleles. *Heterothallism* defines situations where the successful fusion of gametes can occur only between haploids carrying functionally different mating-type alleles. *Pseudohomothallism* qualifies heterothallic species for which self-fertility is enforced by a modified program of meiosis that maintains a constant state of heterokaryosis, where nuclei of opposite mating type share a mycelium and are transmitted together in sexual or asexual spores.

4. The *reproductive system* is the combination of the reproductive mode, the breeding system and the mating system [81]. The *reproductive mode* qualifies the process by which genes are transmitted across generation; reproductive mode can be asexual, sexual, or mixed when there is an alternation of sexual and asexual reproduction during the life cycle. The *breeding system*

refers to the physiologic determinants of mating compatibility, regulated strictly, in fungi, in the haploid stage by mating-type loci. The breeding system of fungi can be heterothallic or homothallic, pseudohomothallism being a specific case of heterothallism (*see* **Note 3**). The *mating system* refers to the degree of genetic relatedness between mates. Outcrossing corresponds to the mating between cells derived from meioses in two different unrelated individuals, whereas inbreeding corresponds to the mating between related individuals. Inbreeding can be caused by selfing, the mating between meiotic products of the same diploid genotype, and several types of selfing can be distinguished in fungi [82]. Contrary to persistent misconceptions in the fungal literature, the breeding system has little influence on the mating system. For instance heterothallism, does not prevent selfing [27, 83].

5. *Heterokaryotic* refers to multinucleate fungal cells that have two or more genetically different (but somatically compatible) nuclei. *Homokaryotic* refers to multinucleate fungal cells where all nuclei are genetically identical.

6. *Allorecognition* refers to self–nonself recognition between conspecific individuals, while *xenorecognition* refers to self–nonself recognition between heterospecific individuals. *Somatic incompatibility* refers to the possible outcome of allorecognition processes, which limit successful somatic fusion to very closely related individuals or tissues [84].

References

1. Gladieux P, Ropars J, Badouin H, Branca A, Aguileta G, De Vienne DM et al (2014) Fungal evolutionary genomics provides insight into the mechanisms of adaptive divergence in eukaryotes. Mol Ecol 23(4):753–773. https://doi.org/10.1111/Mec.12631

2. Jacobson DJ, Powell AJ, Dettman JR, Saenz GS, Barton MM, Hiltz MD et al (2004) Neurospora in temperate forests of western North America. Mycologia 96(1):66–74

3. Leducq JB (2014) Ecological genomics of adaptation and speciation in fungi. Adv Exp Med Biol 781:49

4. Perkins DD, Turner BC (1988) Neurospora from natural populations: toward the population biology of a haploid eukaryote. Exp Mycol 12(2):91–131

5. Taylor JW, Branco S, Gao C, Hann-Soden C, Montoya L, Sylvain I et al (2017) Sources of fungal genetic variation and associating it with phenotypic diversity. Microbiol Spectr 5(5)

6. Stajich JE, Berbee ML, Blackwell M, Hibbett DS, James TY, Spatafora JW et al (2009) The fungi. Curr Biol 19(18):R840–R8R5

7. Ad P (1843) Extrait d'un rapport adresse a M. Le Marechal Duc de Dalmatie, Ministre de la Guerre, President du Conseil, sur une alteration extraordinaire du pain de munition. Ann Chim Phys 9:5–21. 3rd Ser

8. Kuo H-C, Hui S, Choi J, Asiegbu FO, Valkonen JPT, Lee Y-H (2014) Secret lifestyles of Neurospora crassa. Sci Rep 4:5135. https://doi.org/10.1038/srep05135. http://www.nature.com/srep/2014/140530/srep05135/abs/srep05135.html-supplementary-information

9. Shear CL, Dodge BO (1927) Life histories and heterothallism of the red bread-mold fungi of the Monilia sitophila group. US Government Printing Office, Washington, DC

10. Davis RH (2007) Tending neurospora: David Perkins, 1919–2007, and Dorothy Newmeyer

Perkins, 1922–2007. Genetics 175 (4):1543–1548

11. Perkins DD, Davis RH (2002) Neurospora chronology 1843-2002. Fungal Genet Reports 49(1):4–8

12. Perkins DD (1992) Neurospora: the organism behind the molecular revolution. Genetics 130 (4):687

13. Davis RH, Perkins DD (2002) Neurospora: a model of model microbes. Nat Rev Genet 3 (5):397–403

14. Beadle GW, Tatum EL (1941) Genetic control of biochemical reactions in Neurospora. Proc Natl Acad Sci 27(11):499–506

15. Horowitz NH (1985) Roots: the origins of molecular genetics: one gene, one enzyme. BioEssays 3(1):37–39

16. Luque EM, Gutiérrez G, Navarro-Sampedro L, Olmedo M, Rodríguez-Romero J, Ruger-Herreros C et al (2012) A relationship between carotenoid accumulation and the distribution of species of the fungus Neurospora in Spain. PLoS One 7(3):e33658

17. Perkins DD, Turner BC, Barry EG (1976) Strains of neurospora collected from nature. Evolution 30(2):281–313. https://doi.org/10.2307/2407702

18. Turner BC, Perkins DD, Fairfield A (2001) Neurospora from natural populations: a global study. Fungal Genet Biol 32(2):67–92

19. Galagan JE, Calvo SE, Borkovich KA, Selker EU, Read ND, Jaffe D et al (2003) The genome sequence of the filamentous fungus Neurospora crassa. Nature 422 (6934):859–868

20. Heller J, Zhao J, Rosenfield G, Kowbel DJ, Gladieux P, Glass NL (2016) Characterization of greenbeard genes involved in long-distance kind discrimination in a microbial eukaryote. PLoS Biol 14(4):e1002431

21. Palma-Guerrero J, Hall CR, Kowbel D, Welch J, Taylor JW, Brem RB et al (2013) Genome wide association identifies novel loci involved in fungal communication. PLoS Genet 9(8):e1003669

22. Zhao J, Gladieux P, Hutchison E, Bueche J, Hall C, Perraudeau F et al (2015) Identification of allorecognition loci in neurospora crassa by genomics and evolutionary approaches. Mol Biol Evol 32(9):2417–2432

23. Ellison CE, Hall C, Kowbel D, Welch J, Brem RB, Glass NL et al (2011) Population genomics and local adaptation in wild isolates of a model microbial eukaryote. Proc Natl Acad Sci 108(7):2831–2836

24. Jacobson DJ, Dettman JR, Adams RI, Boesl C, Sultana S, Roenneberg T et al (2006) New

findings of Neurospora in Europe and comparisons of diversity in temperate climates on continental scales. Mycologia 98(4):550–559

25. Menkis A, Bastiaans E, Jacobson DJ, Johannesson H (2009) Phylogenetic and biological species diversity within the Neurospora tetrasperma complex. J Evol Biol 22 (9):1923–1936. https://doi.org/10.1111/j.1420-9101.2009.01801.x

26. Corcoran P, Dettman JR, Sun Y, Luque EM, Corrochano LM, Taylor JW et al (2014) A global multilocus analysis of the model fungus Neurospora reveals a single recent origin of a novel genetic system. Mol Phylogenet Evol 78:136–147. https://doi.org/10.1016/j.ympev.2014.05.007

27. Corcoran P, Jacobson DJ, Bidartondo MI, Hickey PC, Kerekes JF, Taylor JW et al (2012) Quantifying functional heterothallism in the pseudohomothallic ascomycete Neurospora tetrasperma. Fungal Biol 116:962–975. https://doi.org/10.1016/j.funbio.2012.06.006

28. Corcoran P, Anderson JL, Jacobson DJ, Sun Y, Ni P, Lascoux M et al (2016) Introgression maintains the genetic integrity of the mating-type determining chromosome of the fungus Neurospora tetrasperma. Genome Res 26 (4):486–498

29. Dettman J, Jacobson D, Taylor J (2003) A multilocus genealogical approach to phylogenetic species recognition in the model eukaryote *Neurospora*. Evolution 57(12):2703–2720

30. Villalta CF, Jacobson DJ, Taylor JW (2009) Three new phylogenetic and biological Neurospora species: N. hispaniola, N. metzenbergii and N. perkinsii. Mycologia 101(6):777–789

31. Dettman JR, Jacobson DJ, Taylor JW (2006) Multilocus sequence data reveal extensive phylogenetic species diversity within the Neurospora discreta complex. Mycologia 98 (3):436–446

32. Pringle A, Baker D, Platt J, Wares J, Latgé J, Taylor J (2005) Cryptic speciation in the cosmopolitan and clonal human pathogenic fungus *Aspergillus fumigatus*. Evolution 59 (9):1886–1899

33. Gladieux P, Feurtey A, Hood ME, Snirc A, Clavel J, Dutech C et al (2015) The population biology of fungal invasions. Mol Ecol 24 (9):1969–1986. https://doi.org/10.1111/mec.13028

34. Taylor JW, Turner E, Townsend JP, Dettman JR, Jacobson D (2006) Eukaryotic microbes, species recognition and the geographic limits of species: examples from the kingdom Fungi.

Philos Trans R Soc Lond Ser B Biol Sci 361 (1475):1947–1963. https://doi.org/10.1098/rstb.2006.1923

35. Taylor JW, Turner E, Pringle A, Dettman J, Johannesson H (2006) Fungal species: thoughts on their recognition, maintenance and selection. Fungi Environ:313–339

36. Dettman JR, Jacobson DJ, Turner E, Pringle A, Taylor JW (2003) Reproductive isolation and phylogenetic divergence in Neurospora: comparing methods of species recognition in a model eukaryote. Evolution 57(12):2721–2741

37. Stamatakis A (2014) RAxML version 8: a tool for phylogenetic analysis and post-analysis of large phylogenies. Bioinformatics 30 (9):1312–1313

38. Nygren K, Strandberg R, Wallberg A, Nabholz B, Gustafsson T, García D et al (2011) A comprehensive phylogeny of Neurospora reveals a link between reproductive mode and molecular evolution in fungi. Mol Phylogenet Evol 59(3):649–663

39. Gladieux P, Wilson BA, Perraudeau F, Montoya LA, Kowbel D, Hann-Soden C et al (2015) Genomic sequencing reveals historical, demographic and selective factors associated with the diversification of the fire-associated fungus Neurospora discreta. Mol Ecol 24 (22):5657–5675

40. Taylor JW, Fisher MC (2003) Fungal multilocus sequence typing—it's not just for bacteria. Curr Opin Microbiol 6(4):351–356. https://doi.org/10.1016/s1369-5274(03)00088-2

41. Frichot E, Mathieu F, Trouillon T, Bouchard G, François O (2014) Fast and efficient estimation of individual ancestry coefficients. Genetics 196(4):973–983

42. Bryant D, Moulton V (2004) Neighbor-Net: an agglomerative method for the construction of phylogenetic networks. Mol Biol Evol 21 (2):255–265. https://doi.org/10.1093/molbev/msh018

43. Pritchard JK, Stephens M, Donnelly P (2000) Inference of population structure using multilocus genotype data. Genetics 155(2):945–959

44. Falush D, Stephens M, Pritchard JK (2003) Inference of population structure using multilocus genotype data: linked loci and correlated allele frequencies. Genetics 164(4):1567–1587

45. Hubisz MJ, Falush D, Stephens M, Pritchard JK (2009) Inferring weak population structure with the assistance of sample group information. Mol Ecol Res 9(5):1322–1332. https://doi.org/10.1111/j.1755-0998.2009.02591.x

46. Gutenkunst RN, Hernandez RD, Williamson SH, Bustamante CD (2009) Inferring the joint demographic history of multiple populations from multidimensional SNP frequency data. PLoS Genet 5(10):e1000695

47. Pandit A, Dubey PS, Mall S (2000) Sexual reproduction of yellow ecotype of Neurospora intermedia in nature. Fungal Genet Reports 47 (1):81–82

48. Kitazima K (1925) On the fungus luxuriantry grown on the bark of the trees injured by the great fire of Tokyo on September 1, 1923. Japan J Phytopathol 1(6):15–19

49. Pandit A, Maheshwari R (1996) Life-history of Neurospora intermedia in a sugar cane field. J Biosci 21(1):57–79

50. Perkins DD (1994) Neurospora tetrasperma bibliography. Fungal Genet Reports 41 (1):72–78

51. Perkins DD (2003) Neurospora tetrasperma bibliography—additions. Fungal Genet Reports 50(1):24–26

52. Dodge BO (1927) Nuclear phenomena associated with heterothallism and homothallism in the ascomycete Neurospora. J Agric Res 35:289–305

53. Glass NL, Metzenberg RL, Raju NB (1990) Homothallic Sordariaceae from nature: the absence of strains containing only theA mating type sequence. Exp Mycol 14(3):274–289

54. Dodge BO (1957) Rib formation in ascospores of Neurospora and questions of terminology. Bull Torrey Bot Club 84(3):182–188

55. Jacobson DJ (1995) Sexual dysfunction associated with outcrossing in Neurospora tetrasperma, a pseudohomothallic ascomycete. Mycologia:604–617

56. Raju NB (1992) Functional heterothallism resulting from homokaryotic conidia and ascospores in Neurospora tetrasperma. Mycol Res 96(2):103–116

57. Saenz GS, Stam JG, Jacobson DJ, Natvig DO (2001) Heteroallelism at the het-c locus contributes to sexual dysfunction in outcrossed strains of Neurospora tetrasperma. Fungal Genet Biol 34(2):123–129

58. Corcoran P (2013) Neurospora tetrasperma from natural populations: toward the population genomics of a model fungus. Acta Universitatis Upsaliensis, Uppsala

59. De Mita S, Siol M (2012) EggLib: processing, analysis and simulation tools for population genetics and genomics. BMC Genet 13(1):1

60. Akashi H, Osada N, Ohta T (2012) Weak selection and protein evolution. Genetics 192 (1):15–31

61. Charlesworth D, Pannell J (2001) Mating systems and population genetic structure in the light of coalescent theory. In: Silvertown J, Westwood AJ (eds) Plants stand still but their genes don't British ecological society special symposium 2000. Blackwell, Oxford, pp 73–96

62. Powell AJ, Jacobson DJ, Salter L, Natvig DO (2003) Variation among natural isolates of Neurospora on small spatial scales. Mycologia 95(5):809–819

63. Skupski MP, Jackson DA, Natvig DO (1997) Phylogenetic analysis of heterothallic Neurospora species. Fungal Genet Biol 21 (1):153–162

64. Taylor JW, Natvig DO (1989) Mitochondrial DNA and evolution of heterothallic and pseudohomothallic Neurospora species. Mycol Res 93(3):257–272

65. Dodge BO (1928) Unisexual conidia from bisexual mycelia. Mycologia 20(4):226–234

66. Svedberg J (2017) Catching the spore killers: genomic conflict and genome evolution in neurospora, Digital comprehensive summaries of Uppsala Dissertations from the Faculty of Science and Technology. Acta Universitatis Upsaliensis, Uppsala, p 51

67. Jacobson DJ (2005) Blocked recombination along the mating-type chromosomes of Neurospora tetrasperma involves both structural heterozygosity and autosomal genes. Genetics 171(2):839–843

68. Swenson NG, Howard DJ (2005) Clustering of contact zones, hybrid zones, and phylogeographic breaks in North America. Am Nat 166 (5):581–591. https://doi.org/10.1086/491688

69. Roux C, Tsagkogeorga G, Bierne N, Galtier N (2013) Crossing the species barrier: genomic hotspots of introgression between two highly divergent Ciona intestinalis species. Mol Biol Evol 30(7):1574–1587

70. Turner E, Jacobson DJ, Taylor JW (2010) Reinforced postmating reproductive isolation barriers in Neurospora, an Ascomycete microfungus. J Evol Biol 23(8):1642–1656

71. Poggeler S, Masloff S, Jacobsen S, Kuck U (2000) Karyotype polymorphism correlates with intraspecific infertility in the homothallic ascomycete Sordaria macrospora. J Evol Biol 13(2):281–289

72. Turner E, Jacobson DJ, Taylor JW (2011) Genetic architecture of a reinforced, postmating, reproductive isolation barrier between Neurospora species indicates evolution via natural selection. PLoS Genet 7(8):e1002204

73. Pomraning KR, Smith KM, Freitag M (2011) Bulk segregant analysis followed by high-throughput sequencing reveals the Neurospora cell cycle gene, ndc-1, to be allelic with the gene for ornithine decarboxylase, spe-1. Eukaryot Cell 10(6):724–733

74. Heller J, Clavé C, Gladieux P, Saupe SJ, Glass NL (2018) NLR surveillance of essential SEC-9 SNARE proteins induces programmed cell death upon allorecognition in filamentous fungi. PNAS 115(10):E2292–E2301

75. Li YF, Costello JC, Holloway AK, Hahn MW (2008) "Reverse ecology" and the power of population genomics. Evolution 62 (12):2984–2994

76. Cruickshank TE, Hahn MW (2014) Reanalysis suggests that genomic islands of speciation are due to reduced diversity, not reduced gene flow. Mol Ecol 23(13):3133–3157

77. Colot HV, Park G, Turner GE, Ringelberg C, Crew CM, Litvinkova L et al (2006) A high-throughput gene knockout procedure for Neurospora reveals functions for multiple transcription factors. Proc Natl Acad Sci 103 (27):10352–10357

78. Glass NL, Dementhon K (2006) Non-self recognition and programmed cell death in filamentous fungi. Curr Opin Microbiol 9 (6):553–558

79. Muirhead CA, Glass NL, Slatkin M (2002) Multilocus self-recognition systems in fungi as a cause of trans-species polymorphism. Genetics 161(2):633–641

80. Wilson JF, Garnjobst L (1966) A new incompatibility locus in Neurospora crassa. Genetics 53(3):621–631

81. Neal PR, Anderson GJ (2005) Are 'mating systems' 'breeding systems' of inconsistent and confusing terminology in plant reproductive biology? Or is it the other way around? Plant Syst Evol 250(3-4):173–185. https://doi.org/10.1007/s00606-004-0229-9

82. Billiard S, López-Villavicencio M, Devier B, Hood ME, Fairhead C, Giraud T (2011) Having sex, yes, but with whom? Inferences from fungi on the evolution of anisogamy and mating types. Biol Rev 86(2):421–442

83. Giraud T, Yockteng R, Lopez-Villavicencio M, Refregier G, Hood ME (2008) Mating system of the anther smut fungus Microbotryum violaceum: selfing under heterothallism. Eukaryot Cell 7(5):765–775

84. Aanen DK, Debets AJM, de Visser J, Hoekstra RF (2008) The social evolution of somatic fusion. BioEssays 30(11-12):1193–1203

Population Genomics of Fungal Plant Pathogens and the Analyses of Rapidly Evolving Genome Compartments

Christoph J. Eschenbrenner, Alice Feurtey, and Eva H. Stukenbrock

Abstract

Genome sequencing of fungal pathogens have documented extensive variation in genome structure and composition between species and in many cases between individuals of the same species. This type of genomic variation can be adaptive for pathogens to rapidly evolve new virulence phenotypes. Analyses of genome-wide variation in fungal pathogen genomes rely on high quality assemblies and methods to detect and quantify structural variation. Population genomic studies in fungi have addressed the underlying mechanisms whereby structural variation can be rapidly generated. Transposable elements, high mutation and recombination rates as well as incorrect chromosome segregation during mitosis and meiosis contribute to extensive variation observed in many species. We here summarize key findings in the field of fungal pathogen genomics and we discuss methods to detect and characterize structural variants including an alignment-based pipeline to study variation in population genomic data.

Key words Fungal pathogens, Genome compartments, Transposable elements, De novo assembly, Multiple genome alignments

1 Introduction

The kingdom Fungi comprises a diverse group of pathogens that infect animals and plants. Understanding the evolution and infection biology of fungal pathogen species is evidently necessary to know how to combat the diseases caused by these organisms. Primary objectives to be addressed in population genomic studies of fungal pathogens relate to the origin of the pathogen, routes of migration, and epidemiology. Moreover, genome data can shed light on the underlying determinants of pathogenicity, which may be new targets in disease control. Finally, as we will outline in this chapter, fungal pathogens provide interesting model systems to study the evolution of genome architecture.

In this chapter, our focus will be on fungi that cause disease on plants. Genome data permitted the reconstruction of the

Julien Y. Dutheil (ed.), *Statistical Population Genomics*, Methods in Molecular Biology, vol. 2090,
https://doi.org/10.1007/978-1-0716-0199-0_14, © The Author(s) 2020

evolutionary histories of some of the most important fungal plant pathogens. For example, the speciation history of the ascomycete wheat pathogen *Zymoseptoria tritici* has been reconstructed by whole genome coalescence analyses revealing that this pathogen emerged with the onset of wheat domestication in the Middle East during the Neolithic Revolution 10–12,000 years ago [1, 2]. Population genetic analyses of isolates representing a world-wide collection of *Z. tritici* was applied to infer the migration history of the pathogen and showed a subsequent dispersal of the pathogen with the spread of wheat cultivation to Europe, Asia and later to New World countries [3]. Another important and recently emerged wheat pathogen is the wheat blast fungus *Magnaporthe oryzae*. The wheat blast disease first emerged in South America and strict quarantine strategies were employed to contain the pathogen within one region and avoid dispersal to other continents. However, the disease was recently reported in Bangladesh. Islam and colleagues were able to track the origin of the wheat blast outbreak in Bangladesh to South America using a genome-wide SNP dataset from 20 isolates collected from different host species in Brazil and Bangladesh [4]. This type of phylogenomic studies and "genomic surveillance" has proven of great relevance to monitor plant disease outbreaks and support the design of improved disease management strategies.

Genome data from fungal plant pathogens has also been a resource for the discovery of genes encoding virulence determinants. In particular quantitative trait locus (QTL) mapping and genome-wide association studies (GWAS) have proven powerful in this field. QTL mapping, based on phenotypic analyses and marker segregation in progeny populations, have been applied to identify the avirulence gene *AvrStb6* in *Z. tritici* [5, 6]. However, QTL analysis has several drawbacks: it relies on the analyses of crosses between two strains. This limits the resolution of the study, depending on the amount of variation between the two strains. Moreover, many fungi propagate primarily by asexual reproduction and many sexual species cannot be crossed under laboratory conditions excluding the possibility of QTL analysis. GWAS on the other hand uses outbred population and polymorphisms that represent the standing genetic variation in a population, providing a higher resolution along the genome [7]. GWAS analyses have been used to identify polymorphisms associated with fungicide sensitivity, mycotoxin production and aggressiveness of the wheat pathogen *Fusarium graminearum* [8], virulence determinants of the pine tree pathogen *Heterobasidion annosum* [9], and toxin production of another wheat pathogen *Parastagnospora nodorum* [10].

Another way to detect genes relevant for pathogenicity in fungi, is to apply evolutionary predictions to identify signatures of recent or past selection. Genes involved in host–pathogen interactions are expected to evolve by antagonistic selection, either following an "arms-race" or a "trench-warfare" scenario of coevolution [11, 12]. The "arms-race" scenario refers to positive selection that repeatedly fixes new advantageous alleles at the locus under selection. The trench-warfare scenario on the other hand refers to the continuous maintenance of different alleles in the population by balancing and diversifying selection. Thus, identifying genes with signatures of positive or balancing selection in pathogen genome will likely uncover genes playing a role in host–pathogen interaction. Evolutionary predictions have been used to identify a number of virulence determinants in fungal plant pathogens and confirm the prediction that virulence determinants indeed often exhibit a signature of positive selection and accelerated evolution [13, 14].

Genome sequencing of hundreds of pathogenic fungal species has revealed extensive variation in genome structure and size [12]. Sequenced genomes range in size from 2 Mb in the Microsporidia to 2 Gb in Pucciniales species, and comprise different levels of ploidy and in some species even aneuploidy [15]. A consistent finding from comparative studies of pathogenic fungi is an extreme extent of genome plasticity whereby closely related species or individuals of the same species can have highly different genome structure and size, and vary in gene content and gene organization [12]. There is evidence that this genome plasticity is crucial for the pathogenic lifestyle. Indeed, variation is essential for pathogens to rapidly adapt to changes in their environment, in particular changes in host resistances, and a highly flexible genome composition appears to be an adaptive mechanism for pathogens to rapidly generate new genetic variation.

The field of fungal pathogen genomics has focused on the sources and patterns of genomic variation, and the contribution of this variation to gene evolution, in particular the evolution of virulence related genes, so called effectors. Effector genes encode secreted proteins that are involved in the suppression of host defenses and these genes are located in genomic segments exhibiting structural variation, including accessory chromosomes and islands of repetitive DNA (e.g., [16–19]). The challenge of studying patterns of evolution in these regions lies in the difficulty of assembling and comparing structurally different sequences.

Population genomic analyses, taking structural variation into account, have been instrumental in determining the underlying drivers of rapid evolution and genome variation in most pathogenic fungal species. This chapter will summarize some of the key discoveries from population genomics analyses of fungal pathogens.

2   Key Discoveries from Population Genomics in Plant Fungal Pathogens

2.1 High Recombination Rates and Population Admixture Contribute to Rapid Adaptation of Fungal Plant Pathogen Genomes

Population genomic data has been applied in a few studies to address the rapid evolution of fungal plant pathogens (reviewed in [12, 20]). Mechanisms that generate genetic variation in a population include mutational processes, recombination and gene flow. The fate of this variation is then determined by selection, genetic drift and the effective population size of the organism. Many aspects make it difficult to study the population genetics and demography of fungi and to assess the contribution of different evolutionary mechanisms to evolution. Most population genetic analyses rely on evolutionary models that make assumptions about the underlying genetic structure of the population (e.g., random mating, infinite site model, a low and constant recombination rate, clonality, skewed offspring, and constant population size). In fungi, many species reproduce both asexually and sexually. More generally, the reproductive mode of fungi can be considered as a continuum ranging from predominantly clonal to strictly out-crossing. Furthermore, the reproductive mode of a particular taxa may change over time. For example, a species may propagate asexually for a certain time followed by a time of more frequent sexual reproduction. Extensive differences in the content of transposable elements between closely related species may support the occurrence of prolonged periods of asexual reproduction in many individual lineages [21, 22].

In population genetic analyses, it is often necessary to have a clear definition of generation time, in order to convert relative time to actual years. However, the generation time of a fungal individual that produces both by asexual and sexual reproduction is difficult to define. In these organisms, not only sexual generations can contribute to novel genetic variation but also asexual generations where high mutations rates generate clonal variation. Furthermore, little is known about the variation in sexual or asexual generations per year. However, the frequency of sexual mating and spore formation may vary from year to year according to environmental conditions and the availability of compatible sexual partners. In summary, analyses of fungal population genomic data, based on existing population models involve many uncertainties caused by our limited understanding of the population biology of fungal pathogen species and by the inadequacy of classic population genetic assumptions to the life history traits of these organisms.

Despite these limitations, population genomic data has, for a few model species, provided new insight into genome the evolution and population biology of the plant pathogens. For example, the impact of recombination on genome evolution has been studied in both ascomycete and basidiomycete pathogens. Badouin and coworkers used population genomic data to infer linkage disequilibrium (LD) along the genome of two closely related species of the anther smut fungus *Microbotryum* [23]. Using information about the extent of LD and the site frequency spectrum (SFS), the authors could determine the distribution and frequency of selective sweeps along the genomes and thereby demonstrate the recent impact of natural selection on gene and genome evolution in the two species. While recombination in *Microbotryum* has been crucial to fix adaptive mutations, suppression of recombination in other parts of the genome has shaped evolution of mating type chromosomes. On these chromosomes recombination suppression has contributed to the generation and maintenance of "super genes" comprising the genes responsible for pre- and postmating compatibility [24].

The impact of recombination has also been studied in *Z. tritici* and its close relative *Zymoseptoria ardabiliae* using population genomic data. These analyses revealed exceptionally high rates of recombination, including recombination hotspots localizing in protein coding genes [25]. Furthermore a strong correlation of recombination with both positive and negative selection was recently demonstrated [26]. Thereby, a negative correlation of recombination and pN/pS, the proportion of nonsynonymous to synonymous polymorphisms, demonstrates an important role of recombination in removing nonadaptive mutations. On the other hand, a positive correlation of recombination with the rate of adaptive nonsynonymous mutations, ω_A, was reported, showing that recombination likewise contributes to the efficient fixation of advantageous mutations in this species.

The impact of intra-specific gene flow on the population genetic structure and dynamic was elegantly demonstrated by a transcriptome sequencing of wheat leaves infected with the yellow stripe rust pathogen *Puccinia striiformis* [27]. *P. striiformis* is an obligate pathogen and difficult to culture on artificial media. Direct sequencing of infected leaf material thus provides a powerful approach to capture the genetic diversity of isolates in the field. Bueano-Sancho and colleagues used data from 246 infected leaves of wheat, triticale, and rye collected in 2 years and at different geographical locations. They used population genetic analyses to infer the population structure and recent patterns of gene flow and admixture of the European rust population and demonstrate extremely diverse populations and rapid seasonal shifts of the rust populations [27]. A significant impact of gene flow on the population genetic structure of fungal pathogens has been demonstrated in other studies also using population genomic data, for example, in the rice blast pathogen *Magnaporthe oryzae* [28] and the ash dieback pathogen *Hymenoscyphus fraxineus* [29].

The impact of new mutations has also been extensively studied in fungal plant pathogen genomes. This is because many species show exceptionally high rates of mutational changes in some segments of their genomes, and the ability to rapidly generate new genetic variation by mutations likely represents an adaptive trait. In the next section we outline the peculiarity of many plant pathogen genomes with respect to genome architecture of the distribution of mutation-prone genome regions.

3 Fungal Plant Pathogen Genomes Are Often Compartmentalized, A Trait Driven by Transposable Elements

The origin of genome compartments in fungal pathogens is still poorly understood, but can only be studied with well-assembled and aligned genome sequences that allow us to study patterns of nucleotide variation within and around these particular genomic regions. Improved genome assemblies have provided insight into the repetitive fraction of fungal pathogen genomes. Repeat contents can vary from less than 1% in *Fusarium graminearum* to more than 80% in some rust and mildew species [30, 31]. The factors determining repeat accumulation are poorly understood, but can include sexual versus asexual reproduction and different genome defense mechanisms such as DNA methylation and Repeat Induced Point mutations (RIP). Transposable elements may accumulate during prolonged asexual reproduction in the absence of recombination; however, some of the sequenced species with the highest repeat content, such as many rust fungi, are sexual, suggesting that other factors likewise are important determinants of transposable element activities.

In some fungal pathogen species a large portion of the repetitive elements are found in particular accessory segments or entire chromosomes that are nonessential but in some species important for virulence. The genome of the asexual fungus *Verticillium dahliae* comprises particular islands enriched with transposable element and encoding effector genes [16, 32]. These islands are present in different lineages of the pathogen and contribute to variation in virulence. Interestingly, these genomic islands harbor little nucleotide variation among individuals that share a particular island, possibly reflecting the strong impact of natural selection on the genes encoded by these regions. Variation in virulence phenotypes is thus given by the presence–absence polymorphism of an entire genomic fragment.

The genome of the fungus *Leptosphaeria maculans* infecting oil seed rape also comprises repeat rich compartments that encode effector proteins [17]. These regions show a particular mutation pattern conferred by RIP. RIP acts to inactivate transposable elements by introducing mutations in repetitive sequence. RIP produces cytosine to thymine (C to T) mutations and can thereby locally impacts the GC content of the sequence [33]. This is the case for *L. maculans* where the repeat-rich islands have become AT isochores with highly distinct GC content compared to the remaining genome.

Genome compartments can also be contained in the genome as accessory chromosomes. The wheat pathogen *Zymoseptoria tritici* has a large number of such accessory chromosomes, eight of them have been sequenced in the reference isolate. These chromosomes can be lost and rearranged during mitosis as well as meiosis [34, 35]. Beside this large complement of accessory chromosomes, *Z. tritici* also exhibits a considerable amount of chromosome length polymorphisms of the core chromosomes as demonstrated by electrophoretic separation of chromosomes and PacBio sequencing [36, 37]. In the soil-borne pathogen *Fusarium oxysporum* lineage-specific chromosomes encode virulence determinants that enable the fungus to be pathogenic on specific host species by the defeat of host defenses [22].

How are accessory chromosomes lost and how are they maintained in populations? A few studies mainly focusing on *F. oxysporum* and *Z. tritici* have started to address these questions. These studies have demonstrated the exceptionally fast rate of accessory chromosomes loss during mitosis [38, 39]. In *F. oxysporum* amplification and maintenance of the chromosomes likely depend on the horizontal exchange of these chromosomes by vegetative fusion of hyphae. In *Z. tritici* however, the accessory chromosomes can be amplified during meiosis by a meiotic drive mechanism [40]. In both species, mechanisms that allow the loss of chromosomes as well as mechanisms that reamplify the chromosomes may have evolved to rapidly generate new genetic variation in the populations of pathogens.

4   Interspecific Hybridization Contributes to Genome Evolution of Fungal Plant Pathogens

Reproductive barriers between fungal species are in many cases poor predictors of species boundaries. Sexual mating and fusion of hyphae between nonconspecific individuals have been frequently described and demonstrate a pathway of gene exchange across species boundaries in the kingdom Fungi. We have recently reviewed the literature on fungal hybridization [41] and will here only mention a few prominent examples of hybridization and gene exchange between fungal species.

Hybridization has been shown to be responsible for the rapid emergence of new virulent lineages of different fungal plant pathogens, including *Ophiostoma nova-ulmi*, the causal agent of Dutch Elm disease and the powdery mildew pathogen *Blumeria graminis-triticale* on crop species Triticale [42, 43]. For the Dutch Elm disease fungus, occasional hybridization events have played a role in the exchange of virulence determinants between otherwise distinct lineages. *B. graminis-triticale*, on the other hand, is the product of few hybridization events between powdery mildew species infecting wheat and rye, respectively. The evidence for a hybridization event is a particular mosaic distribution of genetic variation that clearly reflects the two parental genomes recombined in one genome [43]. The two examples demonstrate very different outcome of hybridization ranging from a few signatures of introgression to entirely mixed parental genomes and hybrid speciation.

The exchange of genetic material can also occur as horizontal gene transfer where only a fragment of DNA is integrated into the genome of one species from another organism. The wheat pathogens *Parastagonospora nodorum* and *Pyrenophora tritici-repentis* are two distantly related ascomycete pathogens. However, their genomes comprise one region of exceptionally high sequence identity [44]. This region that is flanked by transposable elements includes a gene that encodes a proteinaceous toxin, ToxA. ToxA is a virulence factor that confers necrosis in susceptible wheat cultivars and the acquisition of the *ToxA* gene by *P. tritici-repentis* from *P. nodorum* by horizontal gene transfer, allowed the emergence a new virulent lineage of *P. tritici-repentis* infecting wheat. Interestingly, genome sequencing revealed that the *ToxA* gene also is present in another wheat pathogen *Bipolaris sorokiniana* suggesting that this gene may be carried by a bacterial or viral vector frequently associated with wheat [45].

Multiple signatures of hybridization and interspecific gene exchange supports a high extent of flexibility in terms of genome content and structure in fungal plant pathogens. The finding that introgression and horizontal gene transfer in some cases involve virulence determinants underlines the importance of studying not only these regions, but also the processes whereby they occur. However, hybridization events between more distantly related species may be challenging to identify with population genomic data. This is because outlier loci in the genome that comprise highly diverged haplotypes can be difficult to assemble by reference-based assembly approaches. Below we discuss how to circumvent this issue by alignment of de novo assembled genomes.

5   Discovering Variation in Population Genomic Data

5.1   Variant Calling Through Short-Read Mapping: Methods and Limits

Most population genomics approaches are based on the mapping of short sequencing reads, using software such as bwa or bowtie to a well-assembled reference genome [46–48]. Tools such as *GATK*, *SAMtools mpileup*, or *FreeBayes* can be used to call single nucleotide variants and small indels from the mapping file and output this information in a Variant Call Format (VCF) file [49–52]. Here, we will not go further into details about these methods for SNP discovery as these have been extensively reviewed elsewhere (e.g., [53–55]).

Variant discovery through mapping of short reads to a reference is supported by a large number of well-documented tools. However, these methods have drawbacks, some of them especially relevant in nonmodel organisms such as most fungal pathogens. As mentioned above, many pathogenicity-related genes locate in repeat rich compartments of fungal pathogen genomes, and mapping based approaches may not be ideal for the characterization of genetic variation in these regions. Alignment in low-complexity or repetitive regions, although facilitated by paired or mated reads, is often challenging due to the difficulty of correctly mapping the sequence to the reference [55]. Dependence on a reference genome can also be an issue in nonmodel organisms for which a complete reference genome is not always available. Indeed, any misassembly or single nucleotide error in the reference genome could be reflected in the final variants. Poor assembly quality would also lead to structural variation being impossible to discover. Finally, mapping of short reads will not perform efficiently in presence of high genetic variability. Such high variability may be found locally in genomes that have experienced introgression or in some regions have a higher mutation rate. In either case, reads containing multiple alternative alleles might not map correctly, resulting in the under-representation of the diverging haplotypes [55].

Another limitation to mapping-based approaches is the detection of structural variation. To detect translocations or inversions, genomes can be de novo assembled and compared in a multiple genome alignment (Fig. 1). Fungal genomes are convenient for this approach as they often are relatively small and can be sequenced in the haploid phase, therefore preventing issues with heterozygosity and phasing.

Sequencing technologies based on longer reads (e.g., PacBio SMRT or Nanopore sequencing) provide improved resources for de novo assembly. These technologies have proven valuable in the improved detection of structural variation in plant pathogen genomes, including repeat-rich accessory segments on core chromosomes [56, 57]. Below, we describe methods to use de novo genome assembly based on both long and short read sequencing and give the details of a pipeline which allows variants calling from these assembled genomes.

6   De Novo Assembly and the Rise of Long-Read Sequencing

A number of assemblers are available for the different types of sequencing reads available including short reads produced by Illumina sequencing and long reads produced, for example, by SMRT

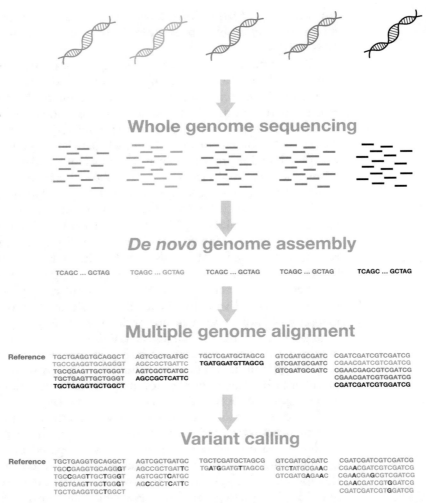

Fig. 1 Generation of population genomic datasets using multiple genome alignment (MGA). Genomes of multiple individuals are generated by short or long read sequencing and assembled de novo. De novo genome assemblies are aligned to generate a MGA. The alignment consists of alignment blocks of different sizes (number of sequences) and lengths (base-pair of alignment). The MGA is projected against a single reference sequence (here shown in red). The projection rearranges each alignment block so that the reference sequence represents the positive strand of the genome. Variable positions can be called directly from the MGA and summarized in a VCF file

sequencing. For de novo assembly of short read data programs like *SPAdes* [58], *SOAPdenovo2* [59] or *IDBA-UD* [60] based on de Brujn graph assemblies are available [61]. De Brujn graph-based assemblers work by splitting the short reads into even shorter units of uniform size, the so-called k-mers. These k-mers provide the basis for the reconstruction of the genome sequence based on overlap of different k'mers while information about the local connectivity of each k'mer is preserved by a De Bruijn graph structure (*see*, e.g., ref. 62, 63). To properly handle repetitive regions the De

Bruijn graph assembler masks repetitive and low-complexity regions and assemble the remaining genome into many contigs and scaffolds.

De novo genome assemblies of long read data is based on other algorithms and build on the alignment of overlapping reads [64, 65]. Long read sequencing with SMRT technology provides an average read length of 10 kb that can be assembled with assemblers like *Canu* [64], *Falcon* [65], or *SMRTAssembly* (©Pacific Biosciences). Nanopore sequencing is producing even longer reads, mostly dependent on the length of the extracted DNA fragment, by a MinION instrument. Methods to assemble genomes based on this technology partially overlap with the ones used with SMRT technology, for example, with the *Canu* assembler [64] and are reviewed in de Lannoy et al. [66]. Nanopore reads have been used to improve the N50 of the maize pathogen *Rhizoctonia solani* by an order of magnitude compared to previous efforts [67]. This improvement is even more pronounced in genomes with high repetitive content (e.g., [56, 68]). In the oat crown rust fungus *Puccinia coronata* f. sp. *avenae* with a genome-wide repeat content of more than 50%, long read sequencing has enabled detailed characterization of structural variants [69]. Moreover, assembly of long read data provided a map of SNPs not only between individuals but also between nuclei in the dikaryotic hyphae of *P. coronata* f. sp. *avenae*, a level of variation so far poorly studied in fungi.

The main inconvenience with long-read sequencing methods so far is the high error rate. To circumvent this issue, it is necessary to either increase the sequencing depth or to combine the advantages of short and long reads. Indeed, assemblies of long and short read data from the same genome is also possible with "hybrid assemblers" like *hybridSPAdes* whereby the long read data ensures the assembly of long scaffolds and the short read data provides high coverage of individual nucleotides in the assembly [70]. Instead of using both types of reads during the assembly process, it is also possible to correct long-read de novo assembly with short-read data, using software like *Pilon* [71]. Such an approach was recently used to assemble genomes of the species *Leptosphaeria* and *Zymoseptoria* [37, 72].

It is important to note that different assemblers (for short-read data as well as long-read data) may perform differently with different genome datasets depending on the repeat content and sequence complexity. Moreover, the long-read technologies are improving at a very rapid rate and new tools and methods are constantly developed. We therefore advise reviewing the latest methods, testing different assemblers with a given dataset and comparing the resulting assemblies with tools such as *Quast* to determine the best performance [73]. To evaluate the quality of the assemblies, key parameters to compare are the total length of the assembly, the number of contigs and the overall size of the assembled fragments which can be summarized by the N50 value (defined as the largest contig length, L, whereby contigs of length superior or equal to L accounts for at least 50% of the bases of the assembly).

For population genomics analyses of the fungal wheat pathogen *Z. tritici* we have developed a pipeline based on de novo genome assembly and multi genome alignments (Fig. 1). This method has allowed us to quantify and characterize accessory regions in the genome of *Z. tritici* and to identify hitherto overseen signatures of introgression along the genome of the pathogen [74]. Following de novo assembly of either short or long read sequencing, the next step in our pipeline is the generation of a multiple genome alignment with a multiple genome aligner such a *TBA* [75], *Mugsy* [76], or *progessiveMauve* [77]. These aligners first generate pairwise alignments of all genomes and next combine these into a multiple genome alignment. The resulting alignment, for example, in "multiple alignment format" *maf* file consists of a large number of local alignment blocks, that differ in their length and the number of sequences included in the block (*see* Chapter 2). The variation in sequence numbers per block along the genome may reflect actual presence/absence variation in genome segments, but can also reflect the parts of the genome that is prone to assembly and/or alignment errors. A thorough filtering and realignment of the alignment blocks is therefore necessary to ensure that the observed patterns are biologically relevant. Programs like *Mafft* or *T-Coffee* are available for realignment of alignment blocks to ensure the optimal comparison of sequences [78, 79].

Filtering and variant detection from a multiple genome alignment can be done with programs like *Maffilter* [80] (*see* also Chapter 2). *Maffilter* allows the list of variant sites identified across the aligned genomes to be outputted as a VCF file. This format is identical to the one used by classic variant calling following a mapping approach. This is especially convenient, as it will allow for these variants to be used as input by any population genomics programs designed to work on this well-known format. Another advantage of this pipeline is that it allows to detect variants simultaneously using sequencing data produced by different technologies, for example, in the case here some genomes are obtained by Illumina sequencing and other by PacBio SMRT sequencing (Feurtey et al. unpublished).

7   Detection of Structural Variation in Genomes

Structural variation is increasingly being recognized as an important level of genetic variation to study. In a study of a single human genome, Pang and colleagues found that the genome differed from the reference human genome by only 0.1% when considering SNPs but by approximately 1.2% when considering other source of genetic variation such as insertions, deletions, or copy number variations [81]. In fungal plant pathogens, structural variation is recognized as an important type of variation as highlighted in some of the examples summarized above.

Methods based on read mapping can be applied to characterize structural variation along genomes [82]. These methods rely on several types of information to detect structural variants including read-depth, and the distribution of paired-end and split reads. Read depth in a mapping, that is, the number of sequencing reads aligning to a specific locus, can give information about copy number variations and deletions. For example, a locus with a higher depth than expected could indicate a duplication and a lower depth (close to 0 in a haploid genome, half the expected depth in a diploid genome) a deletion [83].

Deletions in the resequenced genome compared to the reference genome will cause the insert size of paired-end reads (the DNA fragment including the sequenced reads and the gap sequence between the reads) to appear larger than expected, while an insertion will make the insert look smaller than expected. Furthermore, pairs in which one read aligns to the genome while the other does not may reflect an insertion of a TE if the second read aligns to a repeated element somewhere else in the genome. Likewise translocations, inversions, and other kinds of structural variants can be inferred from pairs of reads. Aligning DNA genomic sequencing reads using an aligner created for RNAseq and thus able to split a read sequence (usually, due to intronic sequences being spliced out of the read) would allow detecting deletions since the deleted sequence will look like a splicing junction site. Software that can detect such structural variants include *Pindel*, *Delly*, or *LUMPY* [84–86]. More details about these methods and software can be found for instance in [83, 87, 88].

Although these methods can uncover many structural variants from short and long reads, they do have their limits. Some of these methods make strict assumptions about the sequencing data, which are not always met in real data. Methods based on read depth assume that the sequencing depth is uniform across the genome and that variation mainly is explained by structural variants. However, variation in GC content and sequence composition along the genome can also cause variation in sequencing efficiency and thereby sequencing depth [83, 89]. Moreover, genomic segments such as accessory chromosomes or large insertions which do not always exist in a reference genome cannot be detected by mapping of short reads to a reference [88].

Whole genome assembly is able to uncover all types of structural variation, including large DNA fragments, which are not present in the reference genome. Another advantage of whole genome assembly is that, if the quality of the assembly is good, it provides strong evidence that no structural variant has gone unde-tected [88]. When the number of genomes is low, structural variants can be identified visually using, for instance, *Symap* or *circos*, which provide easily interpretable visualization of genome alignments [90, 91]. Specific software able to detect structural variants from de novo assemblies have also been developed such as Assemblytics and AsmVar, an automatization step that accounts for structural variants a population level [92, 93]. In summary several tools are available to detect and characterize structural variants in population genomic data. In organisms, like fungal pathogens, with highly variable genomes, accounting for structural variants is essential in order to understand genome evolution and the impact of mutation and recombination along the genome.

8   Conclusion

Analyses of genetic variation in fungal plant pathogen genomes have to a large extent focused on highly variable regions, on species-specific traits and presence–absence variation. More detailed analyses in a few species point to these regions being of particular interest as they can encode important pathogenicity factors. Variation in these regions is therefore considered to be adaptive in accordance with rapid host–pathogen coevolution. Population genomic studies that aim to characterize genetic variation in highly variable regions rely on high quality assemblies and alignments. De novo assemblies of long read sequence data provide an important new resource to capture variation in these regions, including variation in transposable element sequences.

The processes that drive genome evolution in fungal pathogen genomes is still poorly understood. We have demonstrated exceptionally high rates of recombination and particular mechanisms that introduce new mutations at high a rate. Furthermore, we know that fungal pathogens can exchange genes with other species either by sexual mating or fusion of asexual structures. However, the underlying mechanism of these processes, as well as the impact of natural selection on genetic variation generated by these is still to be unraveled.

With their small genome size and in many cases particular genome architecture, fungal pathogens, however, provide excellent models organisms for fundamental studies of genome evolution. Moreover, a better understanding of evolutionary processes occurring in pathogen populations is crucial for the development of agricultural ecosystems with higher disease resistance [94].

Acknowledgments

Research in the lab of EHS is funded by the German Research Council, the Max Planck Society and the State of Schleswig Holstein. The preparation of this book chapter was, furthermore, supported by Deutsche Forschungsgemeinschaft (DFG), in the context of the priority program SPP1819—"Rapid evolutionary adaptation—Potential and constraints."

References

1. Stukenbrock EH, Banke S, Javan-Nikkhah M, McDonald BA (2007) Origin and domestication of the fungal wheat pathogen Mycosphaerella graminicola via sympatric speciation. Mol Biol Evol 24:398–411

2. Stukenbrock EH, Bataillon T, Dutheil JY, Hansen TT, Li R, Zala M, McDonald BA, Wang J, Schierup MH (2011) The making of a new pathogen: insights from comparative population genomics of the domesticated wheat pathogen Mycosphaerella graminicola and its wild sister species. Genome Res 21:2157–2166

3. Banke S, McDonald BA (2005) Migration patterns among global populations of the pathogenic fungus Mycosphaerella graminicola. Mol Ecol 14:1881–1896

4. Islam MT, Croll D, Gladieux P et al (2016) Emergence of wheat blast in Bangladesh was caused by a South American lineage of Magnaporthe oryzae. BMC Biol 14:84

5. Zhong Z, Marcel TC, Hartmann FE, Ma X, Plissonneau C, Zala M, Ducasse A, Confais J, Compain J, Lapalu N (2017) A small secreted protein in Zymoseptoria tritici is responsible for avirulence on wheat cultivars carrying the Stb6 resistance gene. New Phytol 214 (2):619–631

6. Kema GHJ, Mirzadi Gohari A, Aouini L et al (2018) Stress and sexual reproduction affect the dynamics of the wheat pathogen effector AvrStb6 and strobilurin resistance. Nat Genet 50:375–380

7. Genissel A, Confais J, Lebrun M-H, Gout L (2017) Association genetics in plant pathogens: minding the gap between the natural variation and the molecular function. Front Plant Sci 8:1301

8. Talas F, McDonald BA (2015) Genome-wide analysis of Fusarium graminearum field populations reveals hotspots of recombination. BMC Genomics 16:996

9. Dalman K, Himmelstrand K, Olson Å, Lind M, Brandström-Durling M, Stenlid J (2013) A genome-wide association study identifies genomic regions for virulence in the non-model organism Heterobasidion annosum s.s. PLoS One 8:e53525

10. Gao Y, Liu Z, Faris JD, Richards J, Brueggeman RS, Li X, Oliver RP, McDonald BA, Friesen TL (2016) Validation of genome-wide association studies as a tool to identify virulence factors in Parastagonospora nodorum. Phytopathology 106:1177–1185

11. Tellier A, Moreno-Gámez S, Stephan W (2014) Speed of adaptation and genomic footprints of host–parasite coevolution under arms race and trench warfare dynamics. Evolution 68:2211–2224

12. Möller M, Stukenbrock EH (2017) Evolution and genome architecture in fungal plant pathogens. Nat Rev Microbiol 15(12):756–771

13. Poppe S, Dorsheimer L, Happel P, Stukenbrock EH (2015) Rapidly evolving genes are key players in host specialization and virulence of the fungal wheat pathogen Zymoseptoria tritici (Mycosphaerella graminicola). PLoS Pathog 11:e1005055

14. Schweizer G, Münch K, Mannhaupt G, Schirawski J, Kahmann R, Dutheil JY (2018) Positively selected effector genes and their contribution to virulence in the smut fungus Sporisorium reilianum. Genome Biol Evol 10:629–645

15. Stajich JE (2017) Fungal genomes and insights into the evolution of the kingdom.

16. de Jonge R, Bolton MD, Kombrink A, van den Berg GCM, Yadeta KA, Thomma BPHJ (2013) Extensive chromosomal reshuffling drives evolution of virulence in an asexual pathogen. Genome Res 23:1271–1282

17. Rouxel T, Grandaubert J, Hane JK et al (2011) Effector diversification within compartments of the Leptosphaeria maculans genome affected by repeat-induced point mutations. Nat Commun 2:202

18. Enkerli J, Bhatt G, Covert SF (1997) Nht1, a transposable element cloned from a

dispensable chromosome in Nectria haematococca. Mol Plant Microbe Interact 10 (6):742–749. https://doi.org/10.1094/MPMI.1997.10.6.742

19. Schirawski J, Mannhaupt G, Münch K et al (2010) Pathogenicity determinants in smut fungi revealed by genome comparison. Science 330:1546–1548

20. McDonald BA, Linde C (2002) Pathogen population genetics, evolutionary potential, and durable resistance. Annu Rev Phytopathol 40:349–379

21. Grandaubert J, Lowe RGT, Soyer JL et al (2014) Transposable element-assisted evolution and adaptation to host plant within the Leptosphaeria maculans-Leptosphaeria biglobosa species complex of fungal pathogens. BMC Genomics 15:891

22. Ma L-J, van der Does HC, Borkovich KA et al (2010) Comparative genomics reveals mobile pathogenicity chromosomes in Fusarium. Nature 464:367–373

23. Badouin H, Gladieux P, Gouzy J, Siguenza S, Aguileta G, Snirc A, Le Prieur S, Jeziorski C, Branca A, Giraud T (2016) Widespread selective sweeps throughout the genome of model plant pathogenic fungi and identification of effector candidates. Mol. Ecol. 26 (7):2041–2062

24. Branco S, Carpentier F, Rodríguez de la Vega RC et al (2018) Multiple convergent supergene evolution events in mating-type chromosomes. Nat Commun 9:2000

25. Stukenbrock EH, Dutheil JY (2018) Fine-scale recombination maps of fungal plant pathogens reveal dynamic recombination landscapes and intragenic hotspots. Genetics 208:1209–1229

26. Grandaubert J, Dutheil JY, Stukenbrock EH (2019) The genomic determinants of adaptive evolution in a fungal pathogen. Evolution Letters 3(3):299–312

27. Hubbard A, Lewis CM, Yoshida K, Ramirez-Gonzalez RH, de Vallavieille-Pope C, Thomas J, Kamoun S, Bayles R, Uauy C, Saunders DGO (2015) Field pathogenomics reveals the emergence of a diverse wheat yellow rust population. Genome Biol 16:1–15

28. Gladieux P, Condon B, Ravel S, Soanes D, Maciel JLN, Nhani A, Chen L, Terauchi R, Lebrun MH, Tharreau D (2018) Gene flow between divergent cereal-and grass-specific lineages of the rice blast fungus Magnaporthe oryzae. mBio 9:e01219–e01217

29. McMullan M, Rafiqi M, Kaithakottil G et al (2018) The ash dieback invasion of Europe was founded by two genetically divergent individuals. Nat Ecol Evol 2:1000–1008

30. Wicker T, Oberhaensli S, Parlange F et al (2013) The wheat powdery mildew genome shows the unique evolution of an obligate biotroph. Nat Genet 45:1092–1096

31. Duplessis S, Cuomo CA, Lin Y-C et al (2011) Obligate biotrophy features unraveled by the genomic analysis of rust fungi. Proc Natl Acad Sci 108:9166–9171

32. de Jonge R, Peter van Esse H, Maruthachalam K et al (2012) Tomato immune receptor Ve1 recognizes effector of multiple fungal pathogens uncovered by genome and RNA sequencing. Proc Natl Acad Sci U S A 109:5110–5115

33. Gladyshev E (2017) Repeat-induced point mutation (RIP) and other genome defense mechanisms in fungi. Microbiol Spectr 5(4). https://doi.org/10.1128/microbiolspec.FUNK-0042-2017

34. Wittenberg AHJ, van der Lee TAJ, Ben M'Barek S, Ware SB, Goodwin SB, Kilian A, Visser RGF, Kema GHJ, Schouten HJ (2009) Meiosis drives extraordinary genome plasticity in the haploid fungal plant pathogen Mycosphaerella graminicola. PLoS One 4:e5863

35. Habig M, Quade J, Stukenbrock EH (2017) Forward genetics approach reveals host genotype-dependent importance of accessory chromosomes in the fungal wheat pathogen Zymoseptoria tritici. MBio 8:e01919–e01917

36. Mehrabi R, Taga M, Kema GHJ (2007) Electrophoretic and cytological karyotyping of the foliar wheat pathogen Mycosphaerella graminicola reveals many chromosomes with a large size range. Mycologia 99:868–876

37. Plissonneau C, Hartmann FE, Croll D (2018) Pangenome analyses of the wheat pathogen Zymoseptoria tritici reveal the structural basis of a highly plastic eukaryotic genome. BMC Biol 16:5

38. Vlaardingerbroek I, Beerens B, Schmidt SM, Cornelissen BJC, Rep M (2016) Dispensable chromosomes in Fusarium oxysporum f. sp. lycopersici. Mol Plant Pathol 17:1455–1466

39. Moeller M, Habig M, Freitag M, Stukenbrock EH (2018) Extraordinary genome instability and widespread chromosome rearrangements during vegetative growth. bioRxiv 304915

40. Habig M, Kema G, Holtgrewe Stukenbrock E (2018) Meiotic drive of female-inherited supernumerary chromosomes in a pathogenic fungus. Elife 7:pii: e40251

41. Feurtey A, Stukenbrock EH (2018) Interspecific gene exchange as a driver of adaptive evolution in fungi. Annu Rev Microbiol 72:377–398

42. Brasier CM (2001) Rapid evolution of introduced plant pathogens via interspecific hybridization. Bioscience 51:123

43. Menardo F, Praz CR, Wyder S et al (2016) Hybridization of powdery mildew strains gives rise to pathogens on novel agricultural crop species. Nat Genet 48:201–205

44. Friesen TL, Stukenbrock EH, Liu Z, Meinhardt S, Ling H, Faris JD, Rasmussen JB, Solomon PS, McDonald BA, Oliver RP (2006) Emergence of a new disease as a result of interspecific virulence gene transfer. Nat Genet 38:953–956

45. Mcdonald MC, Ahren D, Simpfendorfer S, Milgate A, Solomon PS (2018) The discovery of the virulence gene ToxA in the wheat and barley pathogen Bipolaris sorokiniana. Mol Plant Pathol 19(2):432–439

46. Li H, Durbin R (2009) Fast and accurate short read alignment with Burrows-Wheeler transform. Bioinformatics 25(14):1754–1760. https://doi.org/10.1093/bioinformatics/btp324

47. Lunter G, Goodson M (2011) Stampy: a statistical algorithm for sensitive and fast mapping of Illumina sequence reads. Genome Res 21:936–939

48. Langmead B (2010) Aligning short sequencing reads with Bowtie. Curr Protoc Bioinforma Chapter 11:Unit 11.7

49. Li H (2011) A statistical framework for SNP calling, mutation discovery, association mapping and population genetical parameter estimation from sequencing data. Bioinformatics 27:2987–2993

50. Garrison E, Marth G (2012) Haplotype-based variant detection from short-read sequencing. arXiv preprint arXiv:1207.3907.

51. DePristo MA, Banks E, Poplin R et al (2011) A framework for variation discovery and genotyping using next-generation DNA sequencing data. Nat Genet 43:491–498

52. McKenna A, Hanna M, Banks E et al (2010) The genome analysis toolkit: a MapReduce framework for analyzing next-generation DNA sequencing data. Genome Res 20:1297–1303

53. Mielczarek M, Szyda J (2016) Review of alignment and SNP calling algorithms for next-generation sequencing data. J Appl Genet 57:71–79

54. Altmann A, Weber P, Bader D, Preuß M, Binder EB, Müller-Myhsok B (2012) A beginners guide to SNP calling from high-throughput DNA-sequencing data. Hum Genet 131:1541–1554

55. Pfeifer SP (2017) From next-generation resequencing reads to a high-quality variant data set. Heredity 118:111–124

56. Plissonneau C, Stürchler A, Croll D (2016) The evolution of orphan regions in genomes of a fungal pathogen of wheat. MBio 7: e01231-16

57. Faino L, Seidl MF, Datema E, Berg GC, Janssen A, Wittenberg AH (2015) Single-molecule real-time sequencing combined with optical mapping yields completely finished fungal genome. MBio. https://doi.org/10.1128/mBio.00936-15

58. Bankevich A, Nurk S, Antipov D et al (2012) SPAdes: a new genome assembly algorithm and its applications to single-cell sequencing. J Comput Biol 19:455–477

59. Luo R, Liu B, Xie Y, Li Z, Huang W, Yuan J, He G, Chen Y, Pan Q, Liu Y (2012) SOAPdenovo2: an empirically improved memory-efficient short-read de novo assembler. Gigascience 1(1):18

60. Peng Y, Leung HCM, Yiu SM, Chin FYL (2012) IDBA-UD: a de novo assembler for single-cell and metagenomic sequencing data with highly uneven depth. Bioinformatics 28:1420–1428

61. Pevzner PA, Tang H, Waterman MS (2001) An Eulerian path approach to DNA fragment assembly. Proc Natl Acad Sci 98:9748–9753

62. Zerbino DR, Birney E (2008) Velvet: algorithms for de novo short read assembly using de Bruijn graphs. Genome Res 18:821–829

63. Chaisson M, Pevzner P, Tang H (2004) Fragment assembly with short reads. Bioinformatics 20:2067–2074

64. Koren S, Walenz BP, Berlin K, Miller JR, Bergman NH, Phillippy AM (2016) Canu: scalable and accurate long—read assembly via adaptive k—mer weighting and repeat separation, pp 1–35

65. Chin C-S, Peluso P, Sedlazeck FJ et al (2016) Phased diploid genome assembly with single-molecule real-time sequencing. Nat Methods 13:1050–1054

66. de Lannoy C, de Ridder D, Risse J (2017) The long reads ahead: de novo genome assembly using the MinION. F1000Research 6:1083

67. Datema E, Hulzink RJM, Blommers L, et al. (2016) The megabase-sized fungal genome of Rhizoctonia solani assembled from nanopore reads only. bioRxiv 084772

68. Faino L, Seidl MF, Shi-Kunne X, Pauper M, van den Berg GCMM, Wittenberg AHJJ, Thomma BPHJHJ (2016) Transposons passively and actively contribute to evolution of the two-speed genome of a fungal pathogen. Genome Res 26:1091–1100

69. Miller ME, Zhang Y, Omidvar V et al (2018) De novo assembly and phasing of dikaryotic

genomes from two isolates of Puccinia coronata f. sp. avenae, the causal agent of oat crown rust. mBio 9

70. Antipov D, Korobeynikov A, McLean JS, Pevzner PA (2016) HybridSPAdes: an algorithm for hybrid assembly of short and long reads. Bioinformatics 32:1009–1015

71. Walker BJ, Abeel T, Shea T et al (2014) Pilon: an integrated tool for comprehensive microbial variant detection and genome assembly improvement. PLoS One 9:e112963

72. Dutreux F, Da Silva C, Couloux A et al (2018) De novo assembly and annotation of three Leptosphaeria genomes using Oxford Nanopore MinION sequencing. Sci Data 5:180235

73. Gurevich A, Saveliev V, Vyahhi N, Tesler G (2013) QUAST: quality assessment tool for genome assemblies. Bioinformatics 29:1072–1075

74. Feurtey A, Stevens DM, Stephan W, Stukenbrock EH (2019) Interspecific gene exchange introduces high genetic variability in crop pathogen. Genome Biol Evol. In Press

75. Blanchette M, Kent WJ, Riemer C et al (2004) Aligning multiple genomic sequences with the threaded blockset aligner. Genome Res 14:708–715

76. Angiuoli SV, Salzberg SL (2011) Mugsy: fast multiple alignment of closely related whole genomes. Bioinformatics 27:334–342

77. Darling AE, Mau B, Perna NT (2010) Progressivemauve: multiple genome alignment with gene gain, loss and rearrangement. PLoS One 5(6):e11147. https://doi.org/10.1371/journal.pone.0011147

78. Katoh K, Asimenos G, Toh H (2009) Multiple alignment of DNA sequences with MAFFT. Bioinforma DNA Seq Anal 537:39–64

79. Notredame C, Higgins DG, Heringa J (2000) T-coffee: a novel method for fast and accurate multiple sequence alignment. J Mol Biol 302:205–217

80. Dutheil JY, Gaillard S, Stukenbrock EH (2014) MafFilter: a highly flexible and extensible multiple genome alignment files processor. BMC Genomics 15:53

81. Pang AW, MacDonald JR, Pinto D et al (2010) Towards a comprehensive structural variation map of an individual human genome. Genome Biol 11:R52

82. Wala JA, Bandopadhayay P, Greenwald NF, et al (2018) SvABA: genome-wide detection of structural variants and indels by local assembly. bioRxiv 105080. https://doi.org/10.1101/gr.221028.117

83. Escaramís G, Docampo E, Rabionet R (2015) A decade of structural variants: description, history and methods to detect structural variation. Brief Funct Genomics 14:305–314

84. Ye K, Schulz MH, Long Q, Apweiler R, Ning Z (2009) Pindel: a pattern growth approach to detect break points of large deletions and medium sized insertions from paired-end short reads. Bioinformatics 25:2865–2871

85. Rausch T, Zichner T, Schlattl A, Stutz AM, Benes V, Korbel JO (2012) DELLY: structural variant discovery by integrated paired-end and split-read analysis. Bioinformatics 28: i333–i339

86. Layer RM, Chiang C, Quinlan AR, Hall IM (2014) LUMPY: a probabilistic framework for structural variant discovery. Genome Biol 15: R84

87. Tattini L, D'Aurizio R, Magi A (2015) Detection of genomic structural variants from next-generation sequencing data. Front Bioeng Biotechnol 3:92. https://doi.org/10.3389/fbioe.2015.00092

88. Sedlazeck FJ, Lee H, Darby CA, Schatz MC (2018) Piercing the dark matter: bioinformatics of long-range sequencing and mapping. Nat Rev Genet 19:329–346

89. Sims D, Sudbery I, Ilott NE, Heger A, Ponting CP (2014) Sequencing depth and coverage: key considerations in genomic analyses. Nat Rev Genet 15:121–132

90. Soderlund C, Bomhoff M, Nelson WM (2011) SyMAP v3.4: a turnkey synteny system with application to plant genomes. Nucleic Acids Res 39:e68

91. Krzywinski M, Schein J, Birol İ, Connors J, Gascoyne R, Horsman D, Jones SJ, Marra MA (2009) Circos: an information aesthetic for comparative genomics. Genome Res 19:1639–1645. https://doi.org/10.1101/gr.092759.109

92. Nattestad M, Schatz MC (2016) Assemblytics: a web analytics tool for the detection of variants from an assembly. Bioinformatics 32:3021–3023

93. Liu S, Huang S, Rao J, Ye W, Krogh A, Wang J (2015) Discovery, genotyping and characterization of structural variation and novel sequence at single nucleotide resolution from de novo genome assemblies on a population scale. Gigascience 4(1):64

94. McDonald BA, Stukenbrock EH (2016) Rapid emergence of pathogens in agro-ecosystems: global threats to agricultural sustainability and food security. Phil Trans R Soc B 371:20160026

Chapter 15

Population Genomics on the Fly: Recent Advances in *Drosophila*

Annabelle Haudry, Stefan Laurent, and Martin Kapun

Abstract

Drosophila melanogaster, a small dipteran of African origin, represents one of the best-studied model organisms. Early work in this system has uniquely shed light on the basic principles of genetics and resulted in a versatile collection of genetic tools that allow to uncover mechanistic links between genotype and phenotype. Moreover, given its worldwide distribution in diverse habitats and its moderate genome-size, *Drosophila* has proven very powerful for population genetics inference and was one of the first eukaryotes whose genome was fully sequenced. In this book chapter, we provide a brief historical overview of research in *Drosophila* and then focus on recent advances during the genomic era. After describing different types and sources of genomic data, we discuss mechanisms of neutral evolution including the demographic history of *Drosophila* and the effects of recombination and biased gene conversion. Then, we review recent advances in detecting genome-wide signals of selection, such as soft and hard selective sweeps. We further provide a brief introduction to background selection, selection of noncoding DNA and codon usage and focus on the role of structural variants, such as transposable elements and chromosomal inversions, during the adaptive process. Finally, we discuss how genomic data helps to dissect neutral and adaptive evolutionary mechanisms that shape genetic and phenotypic variation in natural populations along environmental gradients. In summary, this book chapter serves as a starting point to *Drosophila* population genomics and provides an introduction to the system and an overview to data sources, important population genetic concepts and recent advances in the field.

Key words *Drosophila melanogaster*, Population genetics, Demography, Recombination, Selection, Background selection, Selective sweeps, Inversions, Transposable elements, Clines

1 Introduction

The fruit fly *Drosophila melanogaster* is a small Dipteran that originates from sub-Saharan Africa [1] and has since then colonized all continents except for Antarctica as a human commensal [2, 3]. Within the last 15–20,000 years it expanded its range to Europe and Asia and was only recently introduced to Australia and the Americas (~200 years ago according to [1, 4]). Because of its short life cycle and its simple maintenance, it was first adopted as a laboratory model organism by William Castle and later by Thomas

Julien Y. Dutheil (ed.), *Statistical Population Genomics*, Methods in Molecular Biology, vol. 2090,
https://doi.org/10.1007/978-1-0716-0199-0_15, © The Author(s) 2020

Hunt Morgan at the beginning of the twentieth century [3, 5]. At a time when the basic principles of heredity were still under heavy debate, Morgan used the *Drosophila* system to experimentally prove and extend the fundamental predictions of Mendelian genetics, which led to the discovery of genes and their location on chromosomes. This early work was rewarded with Nobel prizes to Morgan and several of his former students and research assistants and forms the basis of our present day understanding of genetic mechanisms [6]. Subsequently, the *Drosophila* system was further exploited, and resulted in the development of numerous genetic tools such as balancer chromosomes, gene-specific knockout mutants and other transgenic constructs, including the Gal4/UAS system to study gene expression or more recently, the CRISPR/Cas9 system for site-specific genome engineering. Moreover, with its condensed genome of ~180 Mb, *D. melanogaster* was among the first eukaryotic organisms whose genome was fully sequenced, assembled and annotated [7].

Beside major advances in functional genetics *Drosophila* has also proven powerful for population genetic inference. Accordingly, numerous major population genetics discoveries have first been made in flies. Theodosius Dobzhansky, together with coworkers and students, was one of the first to systematically investigate genetic variation in *Drosophila*—particularly by focusing on chromosomal inversions. His groundbreaking work gave a first insight into the evolutionary processes that shape genetic variation and subsequently paved the ground for the modern synthesis of evolutionary biology (*see* **Note 1**) [8, 9]. By sequencing the *Adh* gene in 11 lines collected in 5 natural populations, Hudson generated the first fruit fly DNA sequence polymorphism data, identifying only one nonsynonymous polymorphism out of 43 SNPs [10]. As early as the 1980s, methods based on restriction enzymes were applied to *D. melanogaster* to quantify natural genetic variation across multiple loci [11, 12], followed by the first analyses of Sanger sequenced DNA fragments from dozens of genes [13]. These studies provided the first insights into genome-wide patterns of variation in DNA sequences, revealing abundant silent nucleotide site diversity, less abundant nonsynonymous diversity and rarer small insertions and deletions and transposable element insertions [14]. Based on the null hypothesis of neutral evolution, Hudson et al. proposed a first statistical test of selection based on comparing polymorphism and divergence: the Hudson–Kreitman–Aguadé (HKA) test [15], which postulates that genes should all exhibit the same ratio of within-species variability (polymorphism) to between-species divergence at neutral sites. As an extension of the HKA test, McDonald and Kreitman developed a novel test to specifically detect positive selection on protein sequences, first used to detect positive selection at the *Adh* locus in *Drosophila*, and which has since become a ubiquitous test of neutrality [16]. The ratio of nonsynonymous to synonymous divergence is expected to be equal to the ratio of nonsynonymous to synonymous polymorphism if nonsynonymous sites are neutral or deleterious, but higher if they are adaptive. Some of the strongest evidence for adaptive molecular evolution documented in all organisms has come from application of the McDonald–Kreitman test and methods based on it (reviewed in [17, 18]). Finally, a major discovery made in *D. melanogaster* was that the level of nucleotide variability is positively correlated with the local recombination rate [19], suggesting that selection may constitute a major constraint on levels of genomic diversity.

In summary, the fruit fly *D. melanogaster* is an ideal model for studying neutral and adaptive genome evolution in outbred, sexual organisms since it is characterized by a long history as a genetic model organism [5], exhibits well-documented, rapid, and widespread adaptations over short (<20 generations) timescales in natural populations [20, 21], has powerful genetic tools [5, 22] a well-annotated genome [23], and genome-wide polymorphisms data for several populations (*see* Subheading 2.2 for details). Moreover, the genomes of over 25 of its congeners have been recently sequenced [24]. In particular, comparative genomics analyses on 12 species provided fundamental new insights into genome evolution [25] and led to the ModENCODE Project [26], which aims at identifying functional elements in the *D. melanogaster* and *Caenorhabditis elegans* genomes. In this chapter, we will focus on population genomics studies (*see* **Note 2**), mostly based on next generation sequencing data, and review different aspects of both neutral and selective evolution based on the *Drosophila* system.

2 Data Sources

2.1 Data Acquisition Techniques

One particular strength of the *Drosophila* system is its simple maintenance under laboratory conditions. *Drosophila* is commonly propagated as isofemale lines which originate from a single wild-caught and inseminated female. This allows researchers to conduct molecular and phenotypic measurements across several years using the same genetic material and to preserve natural genetic variation under laboratory conditions. In this paragraph we briefly review the nature of the genetic material that has been sequenced in large genome sequencing projects and how these different approaches potentially affect patterns of variation and missing data.

2.1.1 Isofemale Inbred Lines

Isofemale inbred lines are started from single gravid females whose progeny are allowed to interbreed. These lines can be maintained for several years as long as flies are regularly transferred to new vials with fresh fly-food (a well-known task for any student in a *Drosophila* lab having worked in a fly-room). A high degree of inbreeding due to small population sizes leads to a rapid reduction of genetic variation and heterozygosity at every generation within each iso-female line. Inbred lines are often referred to as *F* (Filial genera-tions) followed by the number of the generations of full-sib mating (*F3*, *F10*, *F20*, ...). Due to their near-complete homozygosity, every line should be considered as contributing a single genome to the total sample (and not two, as it could be assumed for an outbred sample). Since isofemale lines are propagated separately and are not allowed to interbreed, they are a versatile tool to preserve genetic variation under laboratory conditions, given that sufficient isofemale lines per population are maintained [27]. One significant issue with this approach is that lines derived from equa-torial populations have shown to be particularly resistant to inbreeding, a problem that has been linked to the presence of inversion polymorphisms hosting recessive lethal mutations. In these lines, large regions (>500 kb) of residual heterozygosity can be observed [28] which complicates the determination of patterns of polymorphism and divergence in this population [29, 30]. More-over, given the small population sizes at which isofemale lines are usually propagated, novel mutations that appeared after the capture of the wild-caught ancestors are likely to accumulate in each line over time. Isofemale lines that are maintained in the laboratory for long periods of time will thus slightly deviate from their ancestors and be poorer indicators of natural variation compared to recently established lines.

2.1.2 Haploid Embryo Sequencing

To circumvent problems caused by residual heterozygosity, Langley et al. proposed to sequence the amplified genome of a single haploid embryo [29]. Most eggs fertilized by recessive male sterile mutants *ms(3)K81* fail to develop [31]. The few that do, however, only contain one haploid maternal genome. Such a single haploid embryo derived from a cross between a female from any line of interest and an *ms(3)K81* male provides enough genomic DNA for whole-genome amplification and sequencing [29]. Although whole-genome amplification increases variance in coverage and the frequency of chimeric reads, this technique provides a powerful approach to uniquely generate high-quality sequencing data using standard paired-end sequencing protocols. Similar to isofemale inbred lines, this technique provides a single genome per sequenced individual (female) but allows for obtaining phased DNA sequences even in the presence of inbreeding-resistant polymorphic inversions.

2.1.3 Genomic Sequencing and Phasing of Hemiclones

Whole-genome sequencing of hybrid F1 crosses—the so-called hemiclones—which share one common parent [32], represents an alternative approach to generate phased haplotype sequencing data. Wild-type *Drosophila* strains are therefore crossed with the same highly inbred or fully isogenic lab-strain that acts as a reference. The resulting F1 hemiclones are then sequenced as single individuals alongside their lab-strain parent to bioinformatically distinguish between the reference and the unknown wild-type allele. This method has been recently employed in *D. melanogaster* and allowed to combine cytological screens with whole-genome sequencing to generate and analyze fully phased genomes with known inversion polymorphisms [33]. Additionally, this approach was used to sequence and characterize a panel of more than 200 wild-type chromosomes from a North American *D. melanogaster* population [34].

2.1.4 Pooled Sequencing (Pool-Seq)

Pool-Seq is a sequencing technique, where tissues or whole bodies of multiple individuals are pooled prior to DNA extraction, library preparation, and whole-genome sequencing. In contrast to single individual sequencing, Pool-Seq is very cost-efficient and has proven powerful to accurately estimate population-wide allele frequencies [35–37]. However, Pool-Seq also comes at the cost of losing information about individual genotypes and haplotype structure. Moreover, it remains very difficult to distinguish low-frequency variants from sequencing errors, which further complicates population genetics inference [38, 39] and precludes calculating classic population genetic estimators without statistical adjustments (*see* for example [40–43]).

It is important to note that these approaches neither allow to measure genotype variation in natural populations, which is the proportion of heterozygote individuals within a population nor the proportion of heterozygote sites within a single diploid individual.

2.2 Consortia and Available Datasets

The first finished genome draft of *D. melanogaster* was published more than 17 years ago, and was among the very first fully sequenced eukaryotic genomes [7]. Since then, the quality of the reference sequence has further improved, and the number of functional annotations, such as gene models or regulatory elements, keeps increasing continuously. Both sequence and annotation data are publicly available at www.flybase.org, a bioinformatics database that is the main repository of genetic and molecular information for *D. melanogaster* (and other species from the Drosophilidae family). *D. melanogaster* was also one of the first species for which full-genome intraspecific variation data was collected. The first whole-genome population genetics study in *D. melanogaster* surveyed natural variation in three African (Malawi) and six North American (North Carolina) strains using low-coverage sequencing [44].

2.2.1 Drosophila Genetic Reference Panel (DGRP) and Drosophila Population Genomics Project (DPGP)

The first two projects to systematically investigate the genomic variability in natural *D. melanogaster* populations were the DGRP [45] and DPGP [46] initiatives. Both consortia independently sequenced more than 160 isofemale inbred lines (F20), all sampled in Raleigh, North Carolina, USA; a sample that was later extended to 205 lines [47]. The major aim was to generate whole-genome sequencing data that can be used for genome-wide association studies. The genetic and phenotypic data are available from http://dgrp2.gnets.ncsu.edu. While the DGRP data are well suited for quantitative genetics analyses (using stable, well-described, and homogeneous genetic material), they only provide information about the genetic variation at a single location (North-Eastern USA) although a large portion of the genetic diversity of the species is known to reside in its ancestral range in sub-Saharan Africa [48, 49]. The DGRP data is thus neither suitable for investigating the demographic history of worldwide populations nor the patterns and processes leading to local adaptations that likely facilitated the range expansion and ultimately led to a cosmopolitan distribution of *D. melanogaster*.

2.2.2 Drosophila Population Genomics Projects

The *Drosophila* population genomic project (DPGP, http://www.dpgp.org) is an ongoing major population genomic sequencing effort: beside the Raleigh population, the DPGP sequenced a population of Malawi (Africa) that exhibited >40% more polymorphism genome-wide compared to the North-American one [46]. Then, the DPGP2 sequenced 139 wild-derived strains representing 22 populations from sub-Saharan Africa [50]. The analyses of the DPGP2 data confirmed that the most genetically diverse populations are located in Southern Africa (e.g., Zambia). Afterward, the DPGP3 increased the sample size for a Zambian population (Siavonga) up to 197 lines [51]. Most DPGP2 and all DPGP3 lines were sequenced from haploid embryos as described above.

2.2.3 The Drosophila Genome Nexus

The Drosophila Genome Nexus is a population genomic resource that integrates single-individual *D. melanogaster* genomes from multiple published sources [51, 52], including DPGP and DGRP among others [30, 53–55]. The aim was to generate a comprehensive dataset using the same bioinformatics methods to facilitate comparisons among them. The latest iteration (DGN v.1.1 [52]), contains a total of 1121 genomes, from 83 populations in Africa, Europe, North America, and Australia. It especially highlighted differences in levels of heterozygosity among the different datasets. The genome browser PopFly allows for the visualization and retrieval of numerous population genomics statistics, such as estimates of nucleotide diversity, linkage disequilibrium, recombination rates [56].

2.2.4 Dros-RTEC and DrosEU

Complementary to previous efforts, which aim at sequencing single individual genomes in large numbers from a single population (DGRP, DPGP, DPGP3) or in small numbers from multiple locations (DPGP2 [30]), two consortia in North America (Dros-RTEC

[57]) and in Europe (DrosEU [43]) recently started to generate Pool-Seq data from wild-caught flies from numerous sampling sites to quantitatively assess genetic variation and differentiation through time and space in natural populations. To date, DrosEU has sequenced and analyzed 48 samples from more than 30 localities all across Europe, which revealed strong and previously unknown population structure—mostly along the longitudinal axis—in Europe. Moreover, population genetic analyses of these data allowed for a description of novel candidates for selective sweeps, to detect previously unknown clines of mitochondrial haplotypes, inversions and transposable elements (TE) and to isolate novel viral species in the microbiome. The Dros-RTEC consortium similarly sequenced 72 samples of *D. melanogaster* collected from 23 localities mostly in North America [57]. Due to their focus on rapid seasonal adaptation, many localities were sampled at different time points over the course of 1–6 years, which allows for a quantitative investigation of genome-wide seasonal fluctuations in SNPs and inversion polymorphisms. These analyses revealed that previous candidates for seasonality exhibit highly predictable annual allele frequency fluctuations and those signatures of seasonal adaptation parallel spatial differentiation along latitudinal gradients.

2.2.5 Other Data

In addition to these concerted sampling and sequencing efforts, there is a rapidly growing number that similarly sequenced pools of flies from natural populations. For example, Pool-Seq data of populations from the temperature gradients along the North American and Australia were generated [58–60]. Large pools of flies collected from Vienna/Austria and Bolzano/Italy were sequenced by [61]. More recently, Kofler and colleagues [62] generated and analyzed Pool-Seq data from more than 550 South African flies. In combination with the aforementioned Pool-Seq data from large consortia, these data represent highly valuable resources to tackle fundamental questions about the adaptive process on complex spatial and temporal scales.

3 Neutral Evolution

3.1 Demographic Analyses

D. melanogaster is one of eight species described in the melanogaster subgroup of the subgenus *Sophophora*. Within this group, two species are cosmopolitan (*D. melanogaster* and *D. simulans*), while the remaining six are endemic to the Afrotropical region (*D. sechellia*, *D. mauritania*, *D. erecta*, *D. orena*, *D. teissieri*, *D. yakuba*). This has led early studies to suggest an Afrotropical origin of *D. melanogaster* and *D. simulans* and is now widely accepted [4]. As expected under this hypothesis, the genome-wide average diversity measured in Afrotropical populations of *D. melanogaster* is higher than in non-African populations

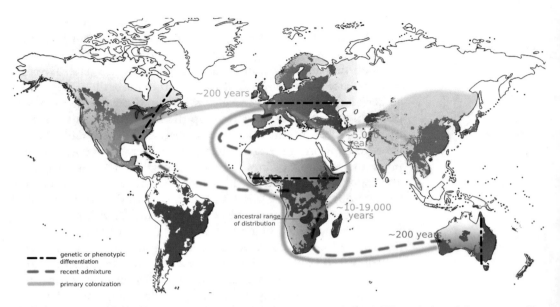

Fig. 1 Map illustrating worldwide distribution, migration routes and clinal differentiation of the cosmopolitan species *D. melanogaster*. Populations are separated in ancestral (red), ancient (orange) and newly introduced (blue) populations, according to the categorization in David and Capy [4]. The expected ancestral range (Zambia) is highlighted in dark red. Primary colonization routes across populations are shown by colored arrows: the European colonization started approximately ~10–19,000 years ago [66, 67], followed by a spread to Asia ~5000 years ago [66] and a more recent range expansion to Australia and North America within the last 200 years [2]. Patterns of recent admixture (dotted grey arrows) were documented from European alleles in Africa [50], and from African alleles to North America and Australia [68, 69]. Clinal genetic and phenotypic differentiation (dash-dotted black arrows and color gradient) are documented along latitudinal gradients in North America [58, 68] and Australia [60, 68], and along longitudinal gradients in Africa [70] and in Europe [43]. At the time of the review, no information about demography of South-American populations was available. The dark grey areas depict expected habitable geographic regions and were modeled from 4951 unique worldwide sampling-points in the TaxoDros database (http:/taxodros.uzh.ch) and climatic data from the WorldClim database (http:/worldclim.org) using the *R*-package *dismo* (http:/rspatial.org/sdm/). Note that distribution models can be confounded by unequal sampling and may thus explain the missing predicted distribution in South-Western Africa, Central Asia, and Russia

[44, 48, 63, 64]. In addition and similarly to *Homo sapiens*, the genetic variation outside sub-Saharan Africa represents a subset of the diversity found within sub-Saharan African populations, which further suggest that South-Eastern tropical Africa represents the ancestral range of the species [65].

In an influential review summarizing the results of early population variation surveys in *D. melanogaster* [4], David and Capy categorized worldwide natural populations into three groups: ancestral, ancient, and new populations (Fig. 1). Ancestral populations are located in sub-Saharan Africa, where they probably have split from the sister species *D. simulans* approximately 2.3 million years ago [71]. Ancient populations are located in Eurasia and migrated out of their ancestral range presumably at the end of the last ice age. The third group, new populations, is located in America and Australia, and represents a blend of ancestral and ancient populations that recently colonized these two continents along European shipping routes during the last centuries. Although these early insights were based on a small number of loci, they have proven to be surprisingly robust and 30 years later, the categorization of David and Capy is still widely accepted. Several studies, however, took advantage of the increasing amount of genetic data and the rapidly developing field of model-based inference in population genetics, to investigate the demographic history of the species within a probabilistic framework. These studies evaluated the likelihood of competing demographic scenarios and provided estimates for demographic parameters such as the age of the split between African and non-African populations, and population sizes at different times of the colonization process. In the next paragraph we review how genome-wide data and statistical modeling updated the insights formulated by David and Capy [4].

Early population genetics surveys identified East and South African populations to be closer to mutation-drift equilibrium compared to West African populations, which were characterized by higher linkage disequilibrium and lower diversity levels [63–65]. These findings suggest that East and South Africa include the ancestral range of the species. Demographic inference using samples from sub-Saharan Africa indicated that the ancestral population has experienced a population size expansion approximately 60,000 years ago (ranging from 26,000 to 95,000 [72]). This ancestral expansion is found in all published models incorporating the African population and is necessary to fit the excess of rare variants measured in samples from the ancestral range (e.g., Zambia, Zimbabwe). These models, however, assume that all sampled mutations are neutral, which is unlikely because of putatively unknown regulatory elements and the presence of background selection [73]. A simulation showed that ignoring background selection in demographic inference leads to an overestimation of growth models [74]. Estimations of the coalescence rate through time using smc++ indicated that the rate of coalescence in a sample from the Zambian population (Siavonga, DPGP3) has been constantly decreasing in the last 100,000 years [75], which is in line with the population expansion scenario suggested by previous studies. Furthermore, Terhorst et al. [75] measured a strong reduction in the coalescent rate for times older than 100,000 years, suggesting either a very large ancestral population size or substantial population structure in the ancestral population [76]. Neither of these two processes is accounted for in current demographic models for D. *melanogaster* and more work is needed to evaluate whether the decreased ancestral rate measured by [75] is reflecting true ancestral processes or rather aspects of the genomic data that are not accounted by the method. More specifically, it remains to be clarified whether this approach can correctly recover neutral demographic processes when applied to small compact genomes with a high proportion of nonneutral regions. More recently, Kapopoulou et al. [77] estimated the age of the split between ancestral (Zambia) and West African populations to be approximately 72,000 years, which suggests that the population expansion reported by earlier studies could well reflect a genuine early range expansion of the species on the African continent.

3.1.1   Out of Africa

Analysis of European samples revealed that the time of split between African and European populations occurred around 13,000 years [66, 72]. These early studies, however, did not include gene flow between populations in their models and therefore predicted that their estimates were probably younger than the true age of divergence between African and European lineages. Indeed, Kapopoulou et al. [78] recently confirmed this prediction using genome-wide polymorphism data and by explicitly accounting for the effect of gene flow in their inference procedure. Their demographic results identified gene flow as an important factor in the recent history of European and African populations and reported divergence time estimates of approximately 48,000 years. Independently, Pool et al. [50] reported pervasive influence of European admixture in many African populations with greater admixture proportion in urban locations. The "ancient" status of Southeast Asian populations has also been confirmed by [66, 79]. Similarly to the European case described above, divergence time estimates between Asian and European populations strongly depend on whether or not gene flow is taken into account in the inference method (22,000 vs. 5000 years, respectively).

North-American populations are considered as newly introduced because the colonization process has been observed directly by entomologists in the second half of the nineteenth century [2]. Strikingly, *D. melanogaster* was identified as the most common species across the USA only 25 years after its introduction, suggesting a dramatic population expansion after colonization [2]. A genome-wide analysis of 39 flies sampled as part of the DGRP project [45], using an approximate Bayesian computation method (ABC) revealed the admixed nature of this population with European and African admixture proportions of 85% and 15%, respectively [67]. These estimations confirmed similar conclusions reached earlier using microsatellite data [80]. This very recent secondary contact between African and European lineages is likely responsible for the North-South clinal genetic variation observed in Northern America and Australia (Fig. 1), but local adaptation could contribute to the maintenance of this clinal variation by opposing itself to the homogenizing effect of gene flow [54, 68]. Arguello et al. [79] recently confirmed the importance of Afro-European admixture in the ancestry of North American and

Australian flies using a larger dataset and a more precise inference procedure. The mosaic ancestry of American and Australian fly populations therefore represents an exciting opportunity to study how migration and selection interact along a clinal heterogeneous environment. Methods based on hidden Markov models were developed to estimate patterns of local ancestry in samples of North-American populations (where the term local refers to an arbitrary subgenomic unit) [69, 81]. In samples with a predominant European genetic background, their results identify significant differences in the proportion of African ancestry between functional classes of genomic loci.

3.2 Recombination

In most sexually reproducing eukaryotes, recombination ensures both the proper segregation of homologous chromosomes during meiosis and the creation of new combinations of alleles at each generation. During meiosis, a substantial number of double-strand breaks result in meiotic recombination between homologs. These double-strand breaks are repaired either as crossover (CO) or non-crossover (NCO) gene conversions: COs imply reciprocal exchange between flanking regions, whereas NCOs do not. Both forms of recombination are key factors in genome evolution as their rates determine the probability to which extent genomic sites are linked or evolve independently and hence affect the evolutionary fate of the alleles. A fundamental understanding of recombination rates is thus crucial in population genomic studies. In *Drosophila*, meiotic recombination only occurs in females, but not in males [82], a dimorphism known as "achiasmy" (an extreme case of hetero-chiasmy observed in many species [83]).

In the 1990s, several studies revolutionized population genetics by showing that the level of genetic diversity in populations of *Drosophila* species was lower in regions of low recombination [19, 84, 85]. Recombination itself seems to be the major factor determining patterns of nucleotide diversity along the genome. Indeed, mutation associated with recombination can be excluded as the cause of this correlation, at least in *Drosophila*, given the lack of correlation between recombination and divergence [19, 45]. The frequent occurrence of these patterns [86] has motivated further exploration and estimation of genome-wide patterns of recombination and diversity.

Classically, the estimation of recombination rates generally relies on the "Marey approach" that compares a genetic map, which quantifies distances as CO frequency (in cM) to a physical map (distances in base pairs). A user-friendly web service called MareyMap Online [87] allows to get recombination rate estimates based on such an approach. In their landmark study, Begun and Aquadro [19] found a strong positive correlation between nucleotide diversity estimated at 20 genes and local rates of COs in natural populations of *D. melanogaster*. They used the coefficient of exchange as a measure of recombination rate, based on the physical distance among cytological markers in combination with DNA content estimates from densities of polytene chromosomes [88]. The fully sequenced *Drosophila* genome, which became available in 2000 [7], represents a highly accurate physical map that was necessary to generate detailed recombination maps. Marais et al. [89] fitted a third-order polynomial, which provided a first overview of the distribution of COs along each chromosomal arm. They showed that CO rates decline in proximity to telomeres and centromeres. Accounting for specific recombination patterns of the telomeric and centromeric regions, Fiston-Lavier and colleagues provided corrected estimates of local recombination rates in *D. melanogaster* [90].

Besides classical recombination maps based on crosses, alternative approaches take population genetic variation into account to estimate CO rates. Patterns of linkage disequilibrium (LD) in a population result from historical recombination events. Recombination (CO) rates across the genome can thus be inferred from linkage disequilibrium, through the population-scaled recombination parameter $\rho = 4 N_e r$ where N_e is the effective population size, and r the CO rate between base pairs per generation [91]. Mc Vean et al. [92] developed a coalescent-based method implemented in the software *LDHat* for the estimation of local recombination rates ($4 N_e r$ per kilobase) using a composite likelihood approximation [93], based on the segregation of a high density of physically mapped SNPs. Originally developed for human populations, this method has been applied to many species including *Drosophila* [46]. Besides providing a recombination map with a higher resolution, Langley et al. showed that r and ρ were strongly positively correlated at a large scale [46], indicating these independent estimates are both capturing heterogeneity in recombination. However, compared to humans, *D. melanogaster* harbors much higher SNP densities, population recombination parameters are an order of magnitude higher and footprints of positive selection are more widespread. Since the *LDhat* method assumes neutral evolution, it can infer spurious recombination hotspots under certain conditions of selection. Chan et al. [94] proposed a corrected method (*LDhelmet*), which is more robust to the effects of selection and computed an improved fine-scale, genome-wide recombination map in *D. melanogaster*, including a handful of hotspots of at least ten times the background recombination rate.

Combining both crosses and population variation approaches, Comeron et al. [95] proposed a method to distinguish between the two possible outcomes of the repairing of double strand breaks associated with meiotic recombination: CO and NCO gene conversions. While COs involve DNA exchange between chromatid arms of homologous chromosomes on a large-scale, NCOs are nonreciprocal recombination events with a swap of small DNA fragment. First described in *Drosophila* [96–98], CO interference prevents the formation of two COs in close proximity and thus, reduces the probability of double CO events (~1 CO per chromosome per meiosis [99]). Based on the size of genetic regions affected by gene conversion, Comeron et al. [95] estimated separately rates of CO and NCO from crosses, making use of the very high density of SNPs in *D. melanogaster* (139 million), which allowed them to design a 2 kb-resolution map of recombination. Unlike COs, NCOs appear to be uniformly distributed throughout the genome [95], insensitive to the centromere effect and without interference [100], and more frequent (rates of NCO: CO could reach values over 100 [95]).

While extrapolated and direct recombination estimates are consistent on a large scale, the latter ones show greater variability at the center of the chromosomal arms [101]. Altogether, these recombination maps provide baseline estimates for population genomic studies, especially to model the expected variation under selection at linked sites (*see* Subheading 4.2 and [102]).

3.3 Biased Gene Conversion

Both CO and NCO recombination involve gene conversion. In particular, the presence of heterozygous sites within heteroduplex DNA results in the formation of mismatches, which lead to the conversion of one allele by the other during the repair. There is evidence, from diverse eukaryotic lineages, that GC:AT mismatches tend to be more often repaired in GC than in AT alleles, a process called GC-biased gene conversion (gBGC [103, 104]). gBGC has been inferred as the main driver of GC-content evolution in vertebrates [103, 105–107] and several other taxa [108–111]. gBGC is a nonadaptive mechanism that mimics natural selection, because it confers a higher transmission probability of GC over AT alleles in heterozygotes. Therefore, gBGC needs to be accounted for in molecular evolution studies to correctly model neutral evolution of the genome [112, 113]. The impact of gBGC in *D. melanogaster* is, however, less clear: GC content is positively correlated with CO rate [89, 114, 115], but not with NCO rate [95]. Globally, whole-genome polymorphism and divergence data did not support a gBGC model in *D. melanogaster* [116], except for the *X* chromosome [117, 118] where it may partly explain the stronger signal of selection on codon usage compared to autosomes [119].

3.4 Population Genetics of Chromosomal Inversions

Chromosomal inversions were first discovered in *D. melanogaster* almost exactly 100 years ago [120]. They represent structural mutations that result in the reversal of genetic order in the affected genomic region relative to the noninverted ("standard") arrangement [121, 122]. Inversions can have strong effects on genome evolution in various different ways: breakpoints may disrupt genes (e.g., [118]) or result in gene duplications due to staggered breaks [123, 124]. Moreover, inversions can trigger positional effects, where expression patterns of genes are altered due to changes in their relative chromosomal position ([125–127] but *see* [128]). However, their most fundamental effect is the strong suppression of recombination in heterozygotes, since crossing-over within the inverted region results in abnormal chromatids [129–131]. As shown in humans where inversions can cause numerous diseases, many of these effects have deleterious consequences [132]; however there are some rare adaptive cases (reviewed in Subheading 4.5). Upon their discovery, inversions have been predominantly studied in species of the genus *Drosophila*. Particularly the pioneering work of Theodosius Dobzhansky and colleagues in *D. pseudoobscura* and *D. persimilis* [8, 9, 133, 134] gave a first insight into the evolutionary processes that shape genetic variation and differentiation in natural populations [135–137]. However, only due to recent advances in whole-genome sequencing technology, it became possible to quantitatively test for different evolutionary models and characterize the genetic effects of inversions on a genome-wide scale. Consistent with the action of spatially varying selection, many inversions in *Drosophila* are commonly found to exhibit steep clines along environmental gradients [138–141]. Several of these, such as the latitudinal gradient of the well-studied *In (3R)Payne* inversion in *D. melanogaster*, are replicated on multiple continents and persist over time ([33, 142] but *see* [143]). Recent large-scale genomic datasets of *D. melanogaster*, for the first time, allow a quantitative assessment of the genetic and evolutionary pattern associated with inversions. Analyses of genome-wide data from African flies allowed for (1) determination of the age and geographic origin of various cosmopolitan and endemic inversions. These analyses revealed that most common cosmopolitan inversions are of African origin and predate the out-of-Africa migration [144]. Furthermore, these data provide (2) a first insight into the amount and distribution of genetic variation and differentiation associated with inversions. Data analyses of the DGRP, for example, found that inversions contribute strongest to genetic differentiation and substructure within a population from Raleigh/North Carolina [47]. Moreover, only with the help of dense genome-wide sequencing, it became possible to show that genetic differentiation is not homogeneously elevated within inversions, but decays toward the inversion center [33, 141, 144–146]. Consistent with theoretical predictions [147–149], these data suggest that there is a limited amount of genetic exchange among karyotypes rather than a complete inhibition of recombination. In addition, local peaks of strong differentiation close to the inversion center suggest that several inversions, for example *In(3R)Payne*, contain various adaptive loci which are in strong linkage with the inversion breakpoints [33, 141, 144, 145]. Analyses of genomic data in combination with long-range PCR further helped (3) to reconstruct the exact genetic composition of inversion breakpoints [150] and (4) facilitated the development of inversion-specific marker SNPs, which now make it possible to reliably estimate inversion abundance and frequency in single-individual and Pool-Seq data, respectively [33, 141, 151]. Together, these analyses highlight that whole-genome data for the first time allows to quantitatively elucidate the mechanisms underlying the evolution of chromosomal inversions.

3.5 Population Genomics of Transposable Elements

Transposable elements (TEs) are mobile, self-replicating, repeated DNA sequences found in every eukaryotic genome at varying proportions among taxa [152, 153], among closely related species [24] and among individuals of the same species ([154] for maize; [155] for *Arabidopsis*; [156] for *Drosophila*). Because of their mutagenic potential (either by inserting into functional regions or by promoting chromosomal rearrangements via ectopic recombination—**Note 3**), TEs are thought to play a significant role in populations' evolution and adaptation [157]. According to the nearly neutral theory, TE insertions are expected to be generally neutral or deleterious to the host genome [158]. However, rare cases of adaptive insertions have also been documented (*see* Subheading 4.6 for examples in *Drosophila*). The general model of TE dynamics is the *transposition-selection balance model* [159]. It assumes that the maintenance of TEs in the population is explained by an equilibrium between (1) the increase in copy number through a constant transposition rate and (2) their removal driven by natural selection, through the combined effect of excision and purifying selection acting against the deleterious effects of inserted TEs [159, 160]. This model predicts that most TEs should be segregating at low TE frequency in *D. melanogaster* populations (*see* [161] for detailed review). The *burst-transposition model* [162] relaxes the assumption of constant transposition rate over time in proposing periods of intense TE transposition (bursts) to explain TE dynamics. According to this model, recent insertions have not yet reached an equilibrium between their transposition rate and negative selection. TEs may thus be at low frequency even under a strictly neutral model. Here, a positive correlation between insertions age and their frequency is expected (recently active TEs should be at low population frequency while long-time inactive TEs could reach fixation).

D. *melanogaster* has been used as a model species for the study of TE population dynamics for more than 25 years [163] and recent whole-genome population data fuelled this area of research allowing for testing of previous hypotheses. A bulk of new programs was recently developed to estimate TE insertion frequency in a population using NGS datasets (*see* [164] for review). On top of the 5434 annotated TEs described in the reference genome, 10,208 and 17,639 insertions were discovered in European [165] and North American DGRP [166] populations, respectively. However, these numbers needs to be considered cautiously as the performance of methods detecting polymorphic TE insertions based on short read data depends on many variables, such as the sequencing coverage, the element family, the age of insertion, the size of the copy, the genomic location (*see* a benchmark in [167]). The large predominance of low frequency insertions along with the scarcity of insertions in exonic regions observed in both datasets supports the *transposition-selection balance model*. In contrast, Kofler, Betancourt and Schlötterer provided evidence that half of the TE families have had transposition rates that vary with time [165], giving support to the *burst-transposition model*. However, they also found an excess of rare variants in young TE insertions compared to neutral expectations which suggests the action of purifying selection [166]. Overall, population genomics analyses of TEs provide empirical support for both hypotheses and indicate that they are not mutually exclusive. This is in agreement with previous in situ analyses suggesting that models of evolution could vary among elements and populations [168]. Although dynamics of some TE families can be explained by a neutral model with transposition rates varying over time, purifying selection is necessary to fully explain the patterns of population distribution of TEs [161, 169].

4   Selection

D. melanogaster has been a model species for many studies aiming at describing the genetic basis of adaptation. Comparisons between theoretical models of positive and negative selection with empirical data have started in the early 1980s, when PCR coupled with Sanger sequencing allowed to directly measure natural variation. The positive correlation between local rates of recombination and genetic diversity [19] was among the most important observations made by these early studies and has been interpreted as evidence for the widespread effect of selection along the genome. This postulate challenged the paradigm of the Nearly Neutral extension of the Neutral Theory [170, 171], which assumes that the large majority of polymorphic and divergent sites are neutral or slightly deleterious. Since then, the search for genes underlying adaptation as well as the quantification of the genome-wide impact of selection has stimulated the development of statistical methods aiming at detecting past adaptive processes from DNA polymorphism data. In 1991, McDonald and Kreitman developed their reference test of selection, and detected adaptation on the *Adh* locus in *Drosophila* [16]. Based on the McDonald and Kreitman test ratios, the fraction α of nonsynonymous substitutions driven to fixation by position selection can be estimated by $1 - (D_s P_n)/(D_n P_s)$, with D_s and D_n the number of synonymous and nonsynonymous substitutions, respectively and P_s and P_n the number of synonymous and nonsynonymous polymorphisms, respectively ([172] and *see* Chapter 6 for more details). Numerous studies have provided evidence for pervasive molecular adaptation in *D. melanogaster*, suggesting that approximately 50% of the amino acid changing substitutions ($\alpha = 0.5$), and similarly large proportions of noncoding substitutions, were adaptive [173–178].

4.1 Hitchhiking Effects

The first mathematical formulation of the effect of a positively selected allele on intra-specific genetic diversity was proposed by Maynard Smith and Haigh in 1974 and coined the "**hitchhiking model**" [179]. Selection reduces diversity not only at selected sites, but also at linked neutral sites, and the number of variants linked together around a single selected target is inversely proportional to the recombination rate. The hitchhiking model summarizes the relation between the strength of selection on a single adaptive mutant allele, the local recombination rate, and the distribution of surrounding neutral alleles across sites and samples. Under such a linkage model, when a beneficial allele establishes itself in the population, the high rate at which this establishment occurs creates an irregularity in the distribution of neutral alleles around the selected allele. This characteristic signature resulting from positive selection has been coined "**selective sweep**" (hard sweep), a terminology used to describe both the adaptive process and the resulting signal in genetic data. This model served as basis for the development of statistical tools designed to capture the signal of a selective sweep in the presence of different confounding factors ([180, 181] and *see* Chapter 5 for more discussion on sweep detection). *D. melanogaster* has been among the first organisms for which this approach has been used to map selective sweeps [72, 182, 183], eventually yielding to the identification of several candidate genes/regions for adaptations (Table 1) that allowed *D. melanogaster* to extend its geographic range to very heterogeneous environments and to recent anthropogenic changes [184, 200, 205]. However, the particular demographic history of the species, and especially the severe founding events followed by population expansion should be considered as a confounding factor, strongly increasing the rate of false positives and thus reducing the performance of sweep detection methods in this specific biological system [206–209]. These insights into the confounding effects of adaptive and neutral processes motivated two lines of research: (a) characterizing neutral models accounting for the major demographic events having affected the genome-wide distribution of neutral alleles (see paragraph above) and (b) more general formulation of the adaptive process initially described by [179].

Soft sweep theory extended the Maynard Smith's and Haigh's hitchhiking model, by including the possibility of (1) recurrent mutations leading to beneficial alleles and (2) segregating neutral alleles becoming positively selected (i.e., selection from standing

Table 1
Documented selective sweeps in African and non-African populations of _D. melanogaster_

Gene(s) involved in the sweep	Sweep size (kb)	Populations	Biological function	Reference
Acetylcholineesterase (Ace)	~1.5	Non-African populations	Insecticide resistance	Karasov et al. [184]; Messer and Petrov [185]; Kapun et al. [43]
Argonaute-2 (AGO2)	>50	_D. melanogaster, D. simulans_ and _D. yakuba_	Resistance to viral infection	Obbard et al. [186]
brinker gene (_brk_)	83–124	European population	Cold tolerance	Glinka et al. [187]; Wilches et al. [188]
_CG18 508_ and _Fcp3C_	14	Non-African populations		DuMont and Aquadro [189]
CHKov1	~25	Non-African populations	Resistance to viral infection	Magwire et al. [190]
Cyp6g1		Non-African populations	Insecticide resistance	Schmidt et al. [191]; Battlay et al. [192, 193]; Kapun et al. [43]
Diminutive (dm)	25	African and non-African populations	Positive regulator of body size	Jensen et al. [194]
Fezzik (fiz)	1.8	European population	Growth	Saminadin-Peter et al. [195] Glaser-Schmitt and Parsch [196]
HDAC6	2.7	African population	Stress surveillance	Svetec et al. [197]
Notch	14	Non-African populations	Development	DuMont and Aquadro [189]
phantom (phm)	12–20	European population	Cytochrome P450 enzyme	Orengo and Aguade [198]
polyhomeotic-proximal (ph-p)	30	European population	Reduced temperature-induced plasticity	Beisswanger and Stephan [199]; Voigt et al. [200]
roughest (rst)	0.361	African population	Apoptosis	Pool et al. [201]
Suppressor of Hairless (Su[H])	1.2	African population	Growth; _Notch_ signalling	Depaulis et al. [202]
wings apart-like (wapl)	~60	European populations	Chromatin organization	Beisswanger et al. [203]
>50 candidates		North-American population		Garud et al. [204]

variation; reviewed in [210]). Both cases predict an association of the beneficial alleles with several background haplotypes (versus a single one in the hard sweep model). Garud et al. [204] scanned the DGRP dataset to capture signature of hard and soft sweeps, and found a significantly higher number of candidate genomic regions than expected under the neutral admixture model previously calibrated for this population [67]. Furthermore, they found that among their top 50 candidates most cases were better explained by soft than hard sweeps, suggesting that standing genetic variation and recurrence of beneficial alleles play an important role in real-life adaptive processes in *D. melanogaster*. However, the statistical significance of their results is highly dependent on an appropriate calibration of neutral demographic models, suggesting that the performance of soft-sweep detection methods still needs to be tested under a large range of demographic models. In the meantime, the results of genome-wide soft-sweep detection studies should be evaluated carefully when used to support claims about adaptive processes [211].

4.2 Recurrent Hitchhiking and Background Selection

Beyond the study of single instances of selective sweeps, *D. melanogaster* and *D. simulans* have also been used to investigate the genome-wide effect of recurring sweeps on genetic variation. The relevant model is the recurrent hitchhiking model [212], which describes genome-wide patterns of variation as a function of the occurrence rate of selective sweeps and the distribution of fitness effects of advantageous mutations. Several studies have developed model-based inference approaches to estimate these two parameters using polymorphism and divergence data, reviewed in [213, 214]. All consistent with a strong impact of selection on the pattern of diversity in this species, a wide range of the strength of selection on beneficial mutations ($N_e s$, where N_e is the effective population size and s the selection coefficient) was estimated, ranging from 1–10 [215, 216], ~12 [217], ~40 [218], 350–3500 [72, 172, 219] to ~10,000 [220]. These studies showed that the rate and strength of positive selection was large enough such that a significant amount of neutral alleles in the genome cannot be seen as evolving independently from adaptive sweeping alleles (the dependencies being caused by genetic linkage between beneficial and neutral alleles). Essentially, the disparate estimates reflect variation in the calibration of the different models, in particular according to (1) the type of selection assumed, (2) the modeled relationship between diversity estimates and selection (strength and frequency) through the action of recombination. These results also revealed the difficulty of telling apart whether genome-wide selection is characterized by a small number of large effect or a large number of small effect adaptive alleles.

The relative importance of positive selection in *Drosophila* has been challenged, however, by studies describing the effect of strictly deleterious alleles on linked neutral variants [73, 221, 222]. This hitchhiking effect caused by selection against recurrent deleterious mutations called **background selection** has been shown to be a valid alternative explanation for low variability in genome regions with low recombination rates [73, 223]. Importantly, Comeron generated a map describing the strength of background selection along the genome as a function of the local recombination and deleterious mutation rate [224]. This study showed that a large proportion (70%) of the observed variation in the level of diversity across autosomes can be explained by background selection alone and therefore called for the inclusion of background selection in further population genomics analyses. Elyashiv et al. recently proposed a method to jointly estimate the parameters of distinct modes of linked selection, accounting for both positive (selective sweeps) and negative background selection [225]. Applied on *D. melanogaster*, they showed that negative selection at linked sites has had an even more drastic effect on diversity patterns in *D. melanogaster* than previously appreciated based on classical selective sweeps models (1.6–2.5-fold). Their results further suggest that 4% of substitutions between *D. melanogaster* and *D. simulans* have experienced strong positive selection ($s \approx 10^{-3.5}$) and that 35% to 45% of substitutions have been weakly selected (s between $10^{-5.5}$ and 10^{-6}).

4.3   Selection on Noncoding DNA

Since the 2000s, whole-genome comparative analyses accumulated evidence that only a small portion of conserved sequences across species (i.e., potentially functional) was composed of protein-coding genes [226, 227]. In the meantime, genomic surveys identified noncoding genomic sequences showing exceptionally high levels of similarity across species, which were termed **conserved noncoding elements** or CNEs (reviewed in [228]). In *Drosophila*, CNEs are estimated to cover ~30–40% of the genome [226, 229]. The high levels of evolutionary conservation observed in these regions are postulated to be the result of functional constraints since many CNEs partially overlap with *cis*-regulatory elements [230] and functional noncoding RNAs [231, 232]. In *Drosophila*, several classes of noncoding DNA evolve considerably slower than synonymous sites yet show an excess of between-species divergence relative to polymorphism when compared with synonymous sites [175]. While the former observation indicates selective constraints, the latter is a signature of adaptive evolution, which resembles patterns of protein evolution in *Drosophila* [173, 174]. To quantify the intensity and the relative importance of selection in shaping the evolution of noncoding DNA, several studies applied extensions of the McDonald–Kreitman approach, combining polymorphism and divergence analyses. When analysing noncoding DNA in a population from Zimbabwe, Andolfatto estimated that ~20% of nucleotide divergence in introns and intergenic

DNA and ~60% in UTRs were driven to fixation by positive selection [175]. Using a hierarchical Bayesian framework, he estimated that significant positive selection acted on noncoding sequences, especially in UTRs [175]. This was recently supported by a whole-genome survey of 50 European populations that showed that UTRs and noncoding RNAs are the noncoding genomic regions most subjected to adaptive selection, with >40% of divergence being driven by positive selection [229]. Specifically focusing on CNEs of the X chromosome, Casillas et al. [233] observed a large excess of low-frequency derived SNP alleles within CNE relative to non-CNE regions in an African and two European populations. While low levels of purifying and positive selection also act outside of CNEs, Casillas et al. [233] estimated that 85% of the CNEs were functional and evolved under moderately strong purifying selection ($N_e s$ ~10–100). Altogether, these studies strongly suggest that CNEs are not solely neutral genomic regions with extremely low mutation rates known as mutation "cold spots" [234] but shaped by both purifying selection and adaptive evolution in *Drosophila*. Moreover, these findings support the important role of noncoding regulatory changes in evolution.

4.4 Selection on Synonymous Codon Usage

The McDonald and Kreitman test and its extensions are built around the hypothesis that synonymous or fourfold degenerate sites (*see* **Note 4**) mostly evolve neutrally, while nonsynonymous sites are under strong purifying or positive selection. However, both synonymous and fourfold degenerate sites might be subject to selection on synonymous codon usage (see original reference for *Drosophila* by [235], and more recent review by [236]). Comparison of polymorphism and divergence patterns suggested that both strong ($4N_e s \gg 1$) and weak ($4N_e s \sim 1$) selection applies to synonymous sites in *D. melanogaster* [237, 238]. In this species, the level of codon bias is positively correlated to the levels of expression [239], but negatively correlated to the levels of divergence [240, 241]. Both findings suggest selection on codon usage bias. As in most *Drosophila* species, all preferred codons are GC-ending [239, 242]; selection on codon bias is therefore expected to increase GC content at synonymous sites. Several attempts to detect selection on codon usage bias in *D. melanogaster* have come to conflicting conclusions. Some studies detected evidence for selection favoring GC-ending codons [119, 243], although the intensity of selection may be weaker in *D. melanogaster* compared to other *Drosophila* species [244]. Other studies did not find support for such on-going selection [245, 246], but rather revealed an excess of substitutions toward AT-ending codons. This may either reflect a reduction in selection efficacy ($4N_e s$) or a shift in the mutational bias in *D. melanogaster* lineage [247]. The population genetics of codon usage bias can however be affected by confounding, nonadaptive processes such as GC-biased gene conversion

([113] but *see* Subheading 3.3). In a recent study, Jackson et al. [248] modeled base composition evolution, and found evidence for selection on fourfold degenerate sites along both *D. melanogaster* and *D. simulans* lineages over a substantial period. They showed that while selection intensity on codon usage was rather stable in *D. simulans* in the recent past, it was declining in *D. melanogaster*. In conclusion, the observed AT-biased substitution pattern could not only result from a mutational bias, but likely partially reflects an ancestral reduction in selection intensity.

4.5 Adaptive Chromosomal Inversions

There is ample evidence that inversions play a pivotal role during adaptive processes and various hypotheses have been developed to explain their evolutionary impact [135–137, 249]: (1) According to the "coadaptation" model, inversions have higher fitness and spread because they suppress maladaptive crossing-over which would unlink coadapted alleles at epistatically interacting loci with high marginal fitness [9, 250]. Genomic analyses in *D. pseudoobscura* support this model and provide evidence that loci in tight linkage with an inversion show epistatic interactions [251]. (2) Under the "local adaptation model," an inversion bears higher fitness because it captures and protects locally adapted loci from recombination with maladaptive migrant haplotypes as initially proposed by [252] and recently revised by [253]. A remarkable conclusion of this model is that the selective advantage of an inversion is determined only by the migration rate of maladapted haplotypes and the amount of linkage among the locally adapted loci. (3) The frequent occurrence of fixed inversions in different species of the genus *Drosophila* [254–257] and in other species groups [258, 259] suggests that many divergent inversions evolved by underdominance and are important components of the speciation process by suppressing gene flow among young sym- or parapatric species [260]. Similarly, inversions play a key role in the evolution of sex chromosomes by keeping together alleles in sex determining factors and sexually antagonistic genes [261]. (4) Conversely, inversions can also be maintained due to overdominance or other types of balancing selection. In line with this model, many inversions, particularly in *Drosophila*, are commonly found to segregate at intermediate frequencies in natural and experimental populations [262].

4.6 Adaptive Insertions of Transposons

Like other type of mutations, TE insertions are expected to be mostly deleterious or evolutionary neutral. However, some transposable elements could be beneficial and positively selected. There are several possible mechanisms by which a TE can be advantageous; either by directly affecting the gene function of individual genes, or by modifying regulatory elements [263, 264]. Due to recent technical advances in sequencing technology (NGS) and due to the rapidly growing number of whole-genome data, the ability to detect selected TE insertions has considerably increased in the past few years. Different methods have been developed to infer selection acting on TE insertions. Villanueva-Cañas et al. [265] provide a detailed overview over the main approaches and their specificities: (1) **DNA sequence conservation** analyses can be used to detect past events of domestication of TEs as regulatory elements, where TE insertions are conserved among closely related species due to purifying selection (*see* for example [229]). (2) Methods developed to detect **selection on linked polymorphisms** from SNPs (*see* Subheading 4.1 and Chapter 5 for more discussion) can also be applied to identify positively selected TEs. Based on either a bias in frequency spectra or haplotype structure, over 35 putatively adaptive TEs were identified in genome-wide studies in *D. melanogaster* to date [161, 165, 266, 267]. (3) A third method is built around **environmental association analyses** that include genome scans for selection performed in parallel in populations from different environments to detect specific adaptation driven by environmental conditions. Using this approach, González et al. [268] discovered several recent TE insertions in *D. melanogaster* that are putatively involved in local adaptation. These TEs exhibit low population frequencies in ancestral population (Africa) but are common in derived populations (North America and Australia). (4) Using a coalescent framework approach, Blumenstiel et al. [169] identified seven additional putative adaptive insertions exhibiting higher population frequency than expected according to their **estimated allele age**. (5) Finally, selection on TE insertions should be validated at the phenotypic level using **functional assays** to identify the molecular and fitness effects. One well-documented example is the insertion *Bari-Jheh* that was found to affect the level of expression of its nearby genes under oxidative stress conditions and to increase resistance to this stress [269, 270].

Beside the impact of single TE insertions, there is growing evidence for a more global effect of TEs on molecular functions. Especially, in *Drosophila*, TEs seem to play a role in a diversity of cellular processes [161], such as the establishment of dosage compensation [271], heterochromatin assembly [272] and brain genomic heterogeneity [273].

4.7 Faster-X Evolution

According to a theory proposed by Charlesworth et al. [274], the rates of evolution of X-linked loci are expected to be faster than autosomal ones if mutations are partially recessive ($0 < h < 1/2$, with h the coefficient of dominance) and expressed in both sexes or males only. In heterogametic males (XY), X-linked mutations are hemizygous and therefore directly exposed to selection, whereas new recessive autosomal mutations are masked from expression in heterozygotes individuals. Moreover, the effective recombination rate is ~1.8-fold greater on the X compared to autosomes [213], which reduces Hill–Robertson interference and increases the

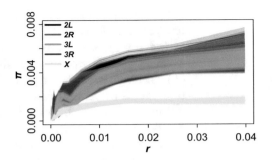

Fig. 2 Correlation between nucleotide diversity and recombination. Nucleotide diversity (π) calculated in 10 kb nonoverlapping windows was estimated for 48 European populations (DrosEU data, [43] and compared to recombination rate (r) obtained from [95] for the four autosomal arms (*2L*, *2R*, *3L*, and *3R*) and the X chromosome. We therefore averaged π in equally sized bins according to discrete log-transformed values of r observed in the corresponding genomic regions. The shaded polygons surrounding average values (central lines) for each of the 48 populations show the 95% confidence intervals

efficiency of selection. The increased selection in hemizygous males together with the higher efficiency of selection due to the increased recombination may act synergistically to account for the "faster-X evolution," which is generally supported by genomic data collected in *Drosophila* populations (reviewed in [275]). Levels of polymorphism are similar on X-linked loci to autosomal ones in African populations, but lower in derived populations [30, 50], which might be a consequence of selective sweeps in response to the adaptation of new environments [276]. However, recombination seems to play a secondary role in determining pattern of diversity along the X-chromosome. Contrary to autosomes, the X-chromosome exhibits global nucleotide diversity only weakly correlated with recombination rate (Fig. 2), and a nonsynonymous diversity completely independent [277].

In contrast with polymorphism, divergence among *Drosophila* relatives is greater for the X than for autosomes (reviewed in [275]). Higher efficiency of selection on X is supported by the estimated higher percentage of sites undergoing both strongly deleterious and adaptive evolution than autosomes, and a lower level of weak negative selection in *D. melanogaster* [45, 46, 277]. Codon usage bias in *Drosophila* is also higher for X-linked genes than for autosomal ones, possibly due to the higher effective recombination rate and their resulting reduced susceptibility to Hill–Robertson effects [119]. In the end, faster-X evolution also implies that genes for reproductive isolation have a higher probability of being X-linked, what is generally true [161].

5 Perspectives: Temporal and Geographical Clines

Organisms with broad geographic distributions, such as various species of the genus *Drosophila*, are commonly found along environmental gradients. Such transects have long been in the focus of evolutionary geneticists, as they provide natural test beds to investigate the evolutionary underpinnings of local adaptation [278]. Studying spatially or temporally changing genotypes and phenotypes, which are commonly referred to as "clines" [279], has a long history in *D. melanogaster* [280–282]. While there is growing evidence for longitudinal clines in Africa [70] and in Europe [43], most data have been collected from latitudinal gradients along the North American and Australian east coasts. A large body of literature documents steep and persistent clines in many fitness-related phenotypes, which are often recapitulated on multiple continents. These include, for example, clines in body-, wing- and organ-size [283–287], lifetime fecundity and lifespan [288] as well as heat and cold resistance [289–291]. Similarly, various genetic polymorphisms such as microsatellites [292], SNPs [293], TEs [266] and inversions [141–143] have been found to vary clinally. Besides these well-defined spatial clines there is growing evidence for rapid adaptation on seasonal timescales leading to temporal clines. These are characterized by predictable annual fluctuations in allele frequencies [20] and variation in life history traits [21] and innate immunity [294].

Ongoing advances in next-generation sequencing technology prompted the development of analytical methods to identify putative targets of local and clinal adaptation (reviewed in [295]) and only recently allowed to extend the hunt for clinal genetic variation from single loci to genome-wide scales. A rapidly increasing number of studies in *D. melanogaster* have started to comprehensively investigate clinal genomic patterns—mostly by comparing the endpoints of latitudinal gradients from the Australian and North American east coasts [58–60, 293]. Many of these pioneering studies identified common patterns in the distribution of genome-wide clinal variation which provide insights, but also raise new questions about the evolutionary mechanisms involved in adaptation: (1) Loci with extensive clinal differentiation are not homogeneously distributed along the genome, but strongly clustered within large inversions [58, 60, 141], which suggests that inversions play an important role during local adaptation—potentially by keeping together coadapted loci associated with polygenic trait variation [252, 253]. However, the identity of these loci and the affected traits remain largely unknown so far. (2) Many clinal polymorphisms, such as the chromosomal inversion *In(3R)Payne* and variants of the alcohol dehydrogenase (*Adh*) locus, are paralleled on multiple continents [33, 58, 59, 293] and change frequencies in a predictable fashion. While parallel adaptive evolution due to spatially varying selection along analogous environmental gradients on different continents may shape many clinal patterns, other non-adaptive evolutionary forces could have similar effects. For example, a handful of studies found independent evidence for varying levels of admixture with African genetic variation both in North America [54, 67, 80] and Australia [68]. These findings highlight that clines, which are often considered to be the prime outcome of spatially or temporally varying selection, are potentially confounded with neutral evolutionary processes such as spatially restricted gene flow or admixture [296]. At last, (3) all aforementioned studies failed to identify large numbers of clinal loci with large or even fixed allele differences at the opposite endpoints of the latitudinal gradients. For example, no more than 0.1% of all SNPs exhibited allele frequency differences >0.5 between Florida and Maine, while not a single SNP exceed an allele frequency difference of 0.92 in the analyses of [58]. These findings are consistent with observations from other *Drosophila* species, which also found moderate and gradual clinal allele frequency changes [140, 297, 298], but in stark contrast to common model expectations for clinal evolution [299]. Together, these first analyses of clinal genomic data clearly show that it still remains challenging to disentangle the evolutionary contribution of selection and demography to clinal variation in natural populations.

Efforts of two large population genomic consortia are currently underway to densely sample natural populations through time and space both in North America [57] and Europe [43]. These comprehensive datasets will markedly extend earlier efforts that focused mostly on the comparison of clinal endpoints. Particularly the analyses of previously largely ignored European *D. melanogaster* populations will allow to make clear predictions about the adaptive process in derived populations from North America and Australia.

6 Notes

1. The mathematical framework that integrated Darwin's theory of evolution and the mechanisms of heredity discovered by Mendel.

2. Defined as genome scale analyses of polymorphism including polymorphism/divergence comparisons, but not analyses strictly based on divergence.

3. Ectopic recombination: recombination between two similar nonhomologous sequences, that is, two TE copies of the same family inserted at different genomic locations. Such DNA exchange between nonorthologous regions leads to chromosomal rearrangement.

4. Fourfold degenerate sites consist in sites for which all four possible nucleotides at this position would encode for the same amino acid, representing a subset of all synonymous sites.

Acknowledgments

We are very grateful to Laurent Duret, Jeffrey Jensen, Gabriel Marais, Alan Moses, Cristina Vieira, and Stephen Wright for their helpful comments on the manuscript. We are very thankful to Julien Dutheil for the invitation to write this chapter.

References

1. Lachaise D, Cariou M-L, David JR et al (1988) Historical biogeography of the Drosophila melanogaster species subgroup. In: Evolutionary biology. Springer, Boston, MA, pp 159–225

2. Keller A (2007) Drosophila melanogaster's history as a human commensal. Curr Biol 17:R77–R81. https://doi.org/10.1016/j.cub.2006.12.031

3. Markow TA (2015) The secret lives of Drosophila flies. elife 4:e06793. https://doi.org/10.7554/eLife.06793

4. David JR, Capy P (1988) Genetic variation of Drosophila melanogaster natural populations. Trends Genet 4:106–111

5. Hales KG, Korey CA, Larracuente AM, Roberts DM (2015) Genetics on the Fly: a primer on the Drosophila model system. Genetics 201:815–842. https://doi.org/10.1534/genetics.115.183392

6. Kohler RE (1994) Lords of the fly. University of Chicago Press, Chicago

7. Adams MD, Celniker SE, Holt RA et al (2000) The genome sequence of Drosophila melanogaster. Science 287:2185–2195

8. Wright S (1982) Dobzhansky's genetics of natural populations, I-XLIII. Evolution (N Y) 36:1102. https://doi.org/10.2307/2408088

9. Dobzhansky T (1970) Genetics of the evolutionary process. Columbia University Press, New York

10. Hudson RR (1983) Properties of a neutral allele model with intragenic recombination. Theor Popul Biol 23:183–201

11. Langley CH, Montgomery E, Quattlebaum WF (1982) Restriction map variation in the Adh region of Drosophila. Proc Natl Acad Sci U S A 79:5631–5635. https://doi.org/10.1073/PNAS.79.18.5631

12. Aquadro CF, Desse SF, Bland MM et al (1986) Molecular population genetics of the alcohol dehydrogenase gene region of Drosophila melanogaster. Genetics 114:1165–1190

13. Powell JR (1997) Progress and prospects in evolutionary biology: the Drosophila model. Oxford University Press, New York

14. Charlesworth B, Charlesworth D (2017) Population genetics from 1966 to 2016. Hered 118:2–9. https://doi.org/10.1038/hdy.2016.55

15. Hudson RR, Kreitman M, Aguade M (1987) A test of neutral molecular evolution based on nucleotide data. Genetics 116:153–159

16. McDonald J, Kreitman M (1991) Adaptive protein evolution at the Adh locus in Drosophila. Nature 351:652–654

17. Booker TR, Jackson BC, Keightley PD (2017) Detecting positive selection in the genome. BMC Biol 15:98. https://doi.org/10.1186/s12915-017-0434-y

18. Kern AD, Hahn MW (2018) The neutral theory in light of natural selection. Mol Biol Evol 35:1366–1371. https://doi.org/10.1093/molbev/msy092

19. Begun DJ, Aquadro CF (1992) Levels of naturally occurring DNA polymorphism correlate with recombination rates in D. melanogaster. Nature 356:519–520

20. Bergland AO, Behrman EL, O'Brien KR et al (2014) Genomic evidence of rapid and stable adaptive oscillations over seasonal time scales in Drosophila. PLoS Genet 10:e1004775. https://doi.org/10.1371/journal.pgen.1004775

21. Behrman EL, Watson SS, O'Brien KR et al (2015) Seasonal variation in life history traits in two Drosophila species. J Evol Biol

28:1691–1704. https://doi.org/10.1111/jeb.12690

22. St Johnston D (2002) The art and design of genetic screens: Drosophila melanogaster. Nat Rev Genet 3:176–188. https://doi.org/10.1038/nrg751

23. Hoskins RA, Carlson JW, Wan KH et al (2015) The release 6 reference sequence of the *Drosophila melanogaster* genome. Genome Res 25:445–458. https://doi.org/10.1101/gr.185579.114

24. Sessegolo C, Burlet N, Haudry A (2016) Strong phylogenetic inertia on genome size and transposable element content among 26 species of flies. Biol Lett 12:20160407. https://doi.org/10.1098/rsbl.2016.0407

25. Clark AG, Eisen MB, Smith DR et al (2007) Evolution of genes and genomes on the Drosophila phylogeny. Nature 450:203–218. https://doi.org/10.1038/nature06341

26. Celniker SE, Dillon LAL, Gerstein MB et al (2009) Unlocking the secrets of the genome. Nature 459:927–930. https://doi.org/10.1038/459927a

27. David JR, Gibert P, Legout H et al (2005) Isofemale lines in Drosophila: an empirical approach to quantitative trait analysis in natural populations. Heredity (Edinb) 94:3–12. https://doi.org/10.1038/sj.hdy.6800562

28. Falconer DS (1989) Introduction to quantitative genetics, 3rd edn. Longman, Scientific & Technical, Burnt Mill, Harlow, Essex/New York

29. Langley CH, Crepeau M, Cardeno C et al (2011) Circumventing heterozygosity: sequencing the amplified genome of a single haploid Drosophila melanogaster embryo. Genetics 188:239–246. https://doi.org/10.1534/genetics.111.127530

30. Grenier JK, Arguello JR, Moreira MC et al (2015) Global diversity lines - a five-continent reference panel of sequenced Drosophila melanogaster strains. G3 5:593–603. https://doi.org/10.1534/g3.114.015883

31. Fuyama Y (1984) Gynogenesis in Drosophila melanogaster1. Jpn J Genet 59:91–96

32. Rice WR, Linder JE, Friberg U et al (2005) Inter-locus antagonistic coevolution as an engine of speciation: assessment with hemiclonal analysis. Proc Natl Acad Sci U S A 102:6527–6534. https://doi.org/10.1073/pnas.0501889102

33. Kapun M, van Schalkwyk H, McAllister B et al (2014) Inference of chromosomal inversion dynamics from Pool-Seq data in natural and laboratory populations of Drosophila melanogaster. Mol Ecol 23:1813–1827. https://doi.org/10.1111/mec.12594

34. Gilks WP, Pennell TM, Flis I et al (2016) Whole genome resequencing of a laboratory-adapted Drosophila melanogaster population sample. F1000Res 5:2644. https://doi.org/10.12688/f1000research.9912.1

35. Zhu Y, Bergland AO, González J, Petrov DA (2012) Empirical validation of pooled whole genome population re-sequencing in Drosophila melanogaster. PLoS One 7:e41901. https://doi.org/10.1371/journal.pone.0041901

36. Fracassetti M, Griffin PC, Willi Y (2015) Validation of pooled whole-genome re-sequencing in Arabidopsis lyrata. PLoS One 10:e0140462. https://doi.org/10.1371/journal.pone.0140462

37. Schlötterer C, Tobler R, Kofler R, Nolte V (2014) Sequencing pools of individuals - mining genome-wide polymorphism data without big funding. Nat Rev Genet 15:749–763. https://doi.org/10.1038/nrg3803

38. Cutler DJ, Jensen JD (2010) To pool, or not to pool? Genetics 186:41–43. https://doi.org/10.1534/genetics.110.121012

39. Lynch M, Bost D, Wilson S et al (2014) Population-genetic inference from pooled-sequencing data. Genome Biol Evol 6:1210–1218. https://doi.org/10.1093/gbe/evu085

40. Futschik A (2010) The next generation of molecular markers from massively parallel sequencing of pooled DNA samples. Genetics 186:207–218. https://doi.org/10.1534/genetics.110.114397

41. Kofler R, Orozco-terWengel P, De Maio N et al (2011) PoPoolation: a toolbox for population genetic analysis of next generation sequencing data from pooled individuals. PLoS One 6:e15925. https://doi.org/10.1371/journal.pone.0015925

42. Kofler R, Pandey RV, Schlötterer C (2011) PoPoolation2: identifying differentiation between populations using sequencing of pooled DNA samples (Pool-Seq). Bioinformatics 27:3435–3436. https://doi.org/10.1093/bioinformatics/btr589

43. Kapun M, Barron M, Staubach F, et al (2018) Genomic analysis of European Drosophila melanogaster populations on a dense spatial scale reveals longitudinal population structure and continent-wide selection. bioRxiv. https://doi.org/10.1101/313759

44. Sackton TB, Kulathinal RJ, Bergman CM et al (2009) Population genomic inferences from

sparse high-throughput sequencing of two populations of Drosophila melanogaster. Genome Biol Evol 1:449–465. https://doi.org/10.1093/gbe/evp048

45. Mackay TF, Richards S, Stone EA et al (2012) The Drosophila melanogaster genetic reference panel. Nature 482:173–178. https://doi.org/10.1038/nature10811

46. Langley CH, Stevens K, Cardeno C et al (2012) Genomic variation in natural populations of Drosophila melanogaster. Genetics 192:533–598. https://doi.org/10.1534/genetics.112.142018

47. Huang W, Massouras A, Inoue Y et al (2014) Natural variation in genome architecture among 205 Drosophila melanogaster genetic reference panel lines. Genome Res 24:1193–1208. https://doi.org/10.1101/gr.171546.113

48. Andolfatto P (2001) Contrasting patterns of X-linked and autosomal nucleotide variation in Drosophila melanogaster and Drosophila simulans. Mol Biol Evol 18:279–290. https://doi.org/10.1093/oxfordjournals.molbev.a003804

49. Hutter S, Li H, Beisswanger S et al (2007) Distinctly different sex ratios in African and European populations of Drosophila melanogaster inferred from chromosomewide single nucleotide polymorphism data. Genetics 177:469–480. https://doi.org/10.1534/genetics.107.074922

50. Pool JE, Corbett-Detig RB, Sugino RP et al (2012) Population genomics of sub-Saharan Drosophila melanogaster: African diversity and non-African admixture. PLoS Genet 8:e1003080. https://doi.org/10.1371/journal.pgen.1003080

51. Lack JB, Cardeno CM, Crepeau MW et al (2015) The Drosophila genome nexus: a population genomic resource of 623 Drosophila melanogaster genomes, including 197 from a single ancestral range population. Genetics 199:1229–1241. https://doi.org/10.1534/genetics.115.174664

52. Lack JB, Lange JD, Tang AD et al (2016) A thousand fly genomes: an expanded Drosophila genome nexus. Mol Biol Evol 33:3308–3313. https://doi.org/10.1093/molbev/msw195

53. Campo D, Lehmann K, Fjeldsted C et al (2013) Whole-genome sequencing of two North American Drosophila melanogaster populations reveals genetic differentiation and positive selection. Mol Ecol 22:5084–5097. https://doi.org/10.1111/mec.12468

54. Kao JY, Zubair A, Salomon MP et al (2015) Population genomic analysis uncovers African and European admixture in Drosophila melanogaster populations from the South-Eastern United States and Caribbean Islands. Mol Ecol 24:1499–1509. https://doi.org/10.1111/mec.13137

55. Bergman CM, Haddrill PR (2015) Strain-specific and pooled genome sequences for populations of Drosophila melanogaster from three continents. F1000Res 4:31. https://doi.org/10.12688/f1000research.6090.1

56. Hervas S, Sanz E, Casillas S et al (2017) Pop-Fly: the Drosophila population genomics browser. Bioinformatics 33:2779–2780. https://doi.org/10.1093/bioinformatics/btx301

57. Machado H, Bergland AO, Taylor R, et al (2018) Broad geographic sampling reveals predictable and pervasive seasonal adaptation in Drosophila. https://doi.org/10.1101/337543

58. Fabian DK, Kapun M, Nolte V et al (2012) Genome-wide patterns of latitudinal differentiation among populations of Drosophila melanogaster from North America. Mol Ecol 21:4748–4769. https://doi.org/10.1111/j.1365-294X.2012.05731.x

59. Reinhardt JA, Kolaczkowski B, Jones CD et al (2014) Parallel geographic variation in Drosophila melanogaster. Genetics 197:361–373. https://doi.org/10.1534/genetics.114.161463

60. Kolaczkowski B, Kern AD, Holloway AK, Begun DJ (2011) Genomic differentiation between temperate and tropical Australian populations of Drosophila melanogaster. Genetics 187:245–260. https://doi.org/10.1534/genetics.110.123059

61. Bastide H, Betancourt A, Nolte V et al (2013) A genome-wide, fine-scale map of natural pigmentation variation in Drosophila melanogaster. PLoS Genet 9:e1003534. https://doi.org/10.1371/journal.pgen.1003534

62. Kofler R, Nolte V, Schlötterer C (2015) Tempo and mode of transposable element activity in Drosophila. PLoS Genet 11:e1005406. https://doi.org/10.1371/journal.pgen.1005406

63. Veuille M, Baudry E, Cobb M et al (2004) Historicity and the population genetics of Drosophila melanogaster and D. simulans. Genetica 120:61–70

64. Haddrill PR, Thornton KR, Charlesworth B, Andolfatto P (2005) Multilocus patterns of nucleotide variability and the demographic

and selection history of Drosophila melanogaster populations. Genome Res 15:790–799

65. Pool JE, Aquadro CF (2006) History and structure of sub-Saharan populations of Drosophila melanogaster. Genetics 174:915–929. https://doi.org/10.1534/genetics.106. 058693

66. Laurent SJY, Werzner A, Excoffier L, Stephan W (2011) Approximate Bayesian analysis of Drosophila melanogaster polymorphism data reveals a recent colonization of Southeast Asia. Mol Biol Evol 28:2041–2051. https://doi.org/10.1093/molbev/msr031

67. Duchen P, Zivkovic D, Hutter S et al (2013) Demographic inference reveals African and European admixture in the North American Drosophila melanogaster population. Genetics 193:291–301. https://doi.org/10.1534/genetics.112.145912

68. Bergland AO, Tobler R, González J et al (2016) Secondary contact and local adaptation contribute to genome-wide patterns of clinal variation in Drosophila melanogaster. Mol Ecol 25:1157–1174. https://doi.org/10.1111/mec.13455

69. Pool JE (2015) The mosaic ancestry of the Drosophila genetic reference panel and the D. melanogaster reference genome reveals a network of Epistatic fitness interactions. Mol Biol Evol 32:3236–3251. https://doi.org/10.1093/molbev/msv194

70. Fabian DK, Lack JB, Mathur V et al (2015) Spatially varying selection shapes life history clines among populations of Drosophila melanogaster from sub-Saharan Africa. J Evol Biol 28:826–840. https://doi.org/10.1111/jeb.12607

71. Li YJ, Satta Y, Takahata N (1999) Paleodemography of the Drosophila melanogaster subgroup: application of the maximum likelihood method. Genes Genet Syst 74:117–127

72. Li H, Stephan W (2006) Inferring the demographic history and rate of adaptive substitution in Drosophila. PLoS Genet 2:e166. https://doi.org/10.1371/journal.pgen. 0020166

73. Charlesworth B, Morgan MT, Charlesworth D (1993) The effect of deleterious mutations on neutral molecular variation. Genetics 134:1289–1303

74. Ewing GB, Jensen JD (2016) The consequences of not accounting for background selection in demographic inference. Mol Ecol 25:135–141. https://doi.org/10.1111/mec.13390

75. Terhorst J, Kamm JA, Song YS (2017) Robust and scalable inference of population history from hundreds of unphased whole genomes. Nat Genet 49:303–309. https://doi.org/10.1038/ng.3748

76. Mazet O, Rodríguez W, Grusea S et al (2016) On the importance of being structured: instantaneous coalescence rates and human evolution—lessons for ancestral population size inference? Heredity (Edinb) 116:362–371. https://doi.org/10.1038/hdy.2015.104

77. Kapopoulou A, Pfeifer SP, Jensen JD, Laurent S (2018) The demographic history of African Drosophila melanogaster. bioRxiv 340406. https://doi.org/10.1101/340406

78. Kapopoulou A, Kapun M, Pavlidis P, et al (2018) Early split between African and European populations of Drosophila melanogaster. bioRxiv 340422. https://doi.org/10.1101/340422

79. Arguello JR, Laurent S, Clark AG (2019) Demographic history of the human commensal Drosophila melanogaster. Genome Biol Evol 11:844–854. https://doi.org/10.1093/gbe/evz022

80. Caracristi G, Schlötterer C (2003) Genetic differentiation between American and European Drosophila melanogaster populations could be attributed to admixture of African alleles. Mol Biol Evol 20:792–799. https://doi.org/10.1093/molbev/msg091

81. Corbett-Detig R, Nielsen R (2017) A hidden Markov model approach for simultaneously estimating local ancestry and admixture time using next generation sequence data in samples of arbitrary ploidy. PLoS Genet 13: e1006529. https://doi.org/10.1371/journal.pgen.1006529

82. Morgan TH (1910) Sex limited inheritance in Drosophila. Science 32:120–122. https://doi.org/10.1126/science.32.812.120

83. Lenormand T, Dutheil J (2005) Recombination difference between sexes: a role for haploid selection. PLoS Biol 3:e63. https://doi.org/10.1371/journal.pbio.0030063

84. Stephan W, Langley CH (1989) Molecular genetic variation in the centromeric region of the X chromosome in three Drosophila ananassae populations. I. Contrasts between the vermilion and forked loci. Genetics 121:89–99

85. Aguade M, Miyashita N, Langley CH (1989) Reduced variation in the yellow-achaete-scute region in natural populations of Drosophila melanogaster. Genetics 122:607–615

86. Cutter AD, Payseur BA (2013) Genomic signatures of selection at linked sites: unifying the disparity among species. Nat Rev Genet

14:262–274. https://doi.org/10.1038/nrg3425

87. Siberchicot A, Bessy A, Guéguen L, Marais GA (2017) MareyMap online: a user-friendly web application and database service for estimating recombination rates using physical and genetic maps. Genome Biol Evol 9:2506–2509. https://doi.org/10.1093/gbe/evx178

88. Lindsley DL, Sandler L (1977) The genetic analysis of meiosis in female Drosophila melanogaster. Philos Trans R Soc Lond Ser B Biol Sci 277:295–312. https://doi.org/10.1098/RSTB.1977.0019

89. Marais G, Mouchiroud D, Duret L (2001) Does recombination improve selection on codon usage? Lessons from nematode and fly complete genomes. Proc Natl Acad Sci U S A 98:5688–5692

90. Fiston-Lavier AS, Singh ND, Lipatov M, Petrov DA (2010) Drosophila melanogaster recombination rate calculator. Gene 463:18–20. https://doi.org/10.1016/j.gene.2010.04.015

91. Hudson RR (1987) Estimating the recombination parameter of a finite population model without selection. Genet Res 50:245–250

92. McVean GAT, Myers SR, Hunt S et al (2004) The fine-scale structure of recombination rate variation in the human genome. Science 304:581–584. https://doi.org/10.1126/science.1092500

93. Hudson RR (2001) Two-locus sampling distributions and their application. Genetics 159:1805–1817

94. Chan AH, Jenkins PA, Song YS (2012) Genome-wide fine-scale recombination rate variation in Drosophila melanogaster. PLoS Genet 8:e1003090. https://doi.org/10.1371/journal.pgen.1003090

95. Comeron JM, Ratnappan R, Bailin S (2012) The many landscapes of recombination in Drosophila melanogaster. PLoS Genet 8:e1002905. https://doi.org/10.1371/journal.pgen.1002905

96. Sturtevant AH (1913) A third group of linked genes in Drosophila ampelophila. Science 37:990–992. https://doi.org/10.1126/science.37.965.990

97. Sturtevant AH (1915) The behavior of the chromosomes as studied through linkage. Z Indukt Abstamm Vererbungsl 13:234–287. https://doi.org/10.1007/BF01792906

98. Muller HJ (1916) The mechanism of crossing-over. Am Nat 50:193–221

99. Carpenter AT (1975) Electron microscopy of meiosis in Drosophila melanogaster females:

II. The recombination nodule--a recombination-associated structure at pachytene? Proc Natl Acad Sci U S A 72:3186–3189. https://doi.org/10.1073/PNAS.72.8.3186

100. Miller DE, Smith CB, Kazemi NY et al (2016) Whole-genome analysis of individual meiotic events in Drosophila melanogaster reveals that noncrossover gene conversions are insensitive to interference and the centromere effect. Genetics 203:159–171. https://doi.org/10.1534/genetics.115.186486

101. Fiston-Lavier A-S, Petrov DA. Drosophila melanogaster recombination rate calculator. http://petrov.stanford.edu/cgi-bin/recombination-rates_updateR5.pl

102. Campos JL, Zhao L, Charlesworth B (2017) Estimating the parameters of background selection and selective sweeps in Drosophila in the presence of gene conversion. Proc Natl Acad Sci 114:E4762–E4771. https://doi.org/10.1073/pnas.1619434114

103. Duret L, Galtier N (2009) Biased gene conversion and the evolution of mammalian genomic landscapes. Annu Rev Genomics Hum Genet 10:285–311. https://doi.org/10.1146/annurev-genom-082908-150001

104. Mancera E, Bourgon R, Brozzi A et al (2008) High-resolution mapping of meiotic crossovers and non-crossovers in yeast. Nature 454:479–485. https://doi.org/10.1038/nature07135

105. Figuet E, Ballenghien M, Romiguier J, Galtier N (2015) Biased gene conversion and GC-content evolution in the coding sequences of reptiles and vertebrates. Genome Biol Evol 7:240–250. https://doi.org/10.1093/gbe/evu277

106. Bolívar P, Mugal CF, Nater A, Ellegren H (2016) Recombination rate variation modulates gene sequence evolution mainly via GC-biased gene conversion, not Hill–Robertson interference, in an Avian system. Mol Biol Evol 33:216–227. https://doi.org/10.1093/molbev/msv214

107. Glémin S, Arndt PF, Messer PW et al (2015) Quantification of GC-biased gene conversion in the human genome. Genome Res 25:1215–1228. https://doi.org/10.1101/gr.185488.114

108. Pessia E, Popa A, Mousset S et al (2012) Evidence for widespread GC-biased gene conversion in eukaryotes. Genome Biol Evol 4:675–682. https://doi.org/10.1093/gbe/evs052

109. Glémin S, Clément Y, David J, Ressayre A (2014) GC content evolution in coding

regions of angiosperm genomes: a unifying hypothesis. Trends Genet 30:263–270. https://doi.org/10.1016/j.tig.2014.05.002

110. Wallberg A, Glémin S, Webster MT (2015) Extreme recombination frequencies shape genome variation and evolution in the honeybee, Apis mellifera. PLoS Genet 11: e1005189. https://doi.org/10.1371/journal.pgen.1005189

111. Haudry A, Cenci A, Guilhaumon C et al (2008) Mating system and recombination affect molecular evolution in four Triticeae species. Genet Res (Camb) 90:97–109. https://doi.org/10.1017/S0016672307009032

112. Romiguier J, Roux C (2017) Analytical biases associated with GC-content in molecular evolution. Front Genet 8:16. https://doi.org/10.3389/fgene.2017.00016

113. Galtier N, Roux C, Rousselle M et al (2018) Codon usage bias in animals: disentangling the effects of natural selection, effective population size and GC-biased gene conversion. Mol Biol Evol 35(5):1092–1103. https://doi.org/10.1093/molbev/msy015

114. Singh ND (2005) Genomic heterogeneity of background Substitutional patterns in Drosophila melanogaster. Genetics 169:709–722. https://doi.org/10.1534/genetics.104.032250

115. Liu G, Li H (2008) The correlation between recombination rate and dinucleotide bias in Drosophila melanogaster. J Mol Evol 67:358–367. https://doi.org/10.1007/s00239-008-9150-0

116. Robinson MC, Stone EA, Singh ND (2014) Population genomic analysis reveals no evidence for GC-biased gene conversion in Drosophila melanogaster. Mol Biol Evol 31:425–433. https://doi.org/10.1093/molbev/mst220

117. Galtier N, Bazin E, Bierne N (2006) GC-biased segregation of noncoding polymorphisms in Drosophila. Genetics 172:221–228

118. Haddrill PR, Waldron FM, Charlesworth B (2008) Elevated levels of expression associated with regions of the Drosophila genome that lack crossing over. Biol Lett 4:758–761. https://doi.org/10.1098/rsbl.2008.0376

119. Campos JL, Zeng K, Parker DJ et al (2013) Codon usage bias and effective population sizes on the X chromosome versus the autosomes in Drosophila melanogaster. Mol Biol Evol 30:811–823. https://doi.org/10.1093/molbev/mss222

120. Sturtevant AH (1917) Genetic factors affecting the strength of linkage in Drosophila. Proc Natl Acad Sci U S A 3:555–558. https://doi.org/10.1073/pnas.3.9.555

121. Sturtevant AH (1919) Contributions to the genetics of Drosophila melanogaster. III. Inherited linkage variations in the second chromosome. Contributions to the genetics of Drosophila melanogaster

122. Sturtevant AH (1921) A case of rearrangement of genes in Drosophila. Proc Natl Acad Sci U S A 7:235–237. https://doi.org/10.1073/pnas.7.8.235

123. Mattei JF, Mattei MG, Ardissone JP et al (1980) Pericentric inversion, inv(9) (p22 q32), in the father of a child with a duplication-deletion of chromosome 9 and gene dosage effect for adenylate kinase-1. Clin Genet 17:129–136. https://doi.org/10.1111/j.1399-0004.1980.tb00121.x

124. Puerma E, Orengo D-J, Aguadé M (2016) Multiple and diverse structural changes affect the breakpoint regions of polymorphic inversions across the Drosophila genus. Sci Rep 6:36248. https://doi.org/10.1038/srep36248

125. Wargent JM, Hartmann-Goldstein IJ (1974) Phenotypic observations on modification of position-effect variegation in Drosophila melanogaster. Heredity (Edinb) 33:317–326. https://doi.org/10.1038/hdy.1974.98

126. Salm MPA, Horswell SD, Hutchison CE et al (2012) The origin, global distribution, and functional impact of the human 8p23 inversion polymorphism. Genome Res 22:1144–1153. https://doi.org/10.1101/gr.126037.111

127. Said I, Byrne A, Serrano V et al (2018) Linked genetic variation and not genome structure causes widespread differential expression associated with chromosomal inversions. Proc Natl Acad Sci U S A 115:5492–5497. https://doi.org/10.1073/pnas.1721275115

128. Lavington E, Kern AD (2017) The effect of common inversion polymorphisms In(2L)t and In(3R)Mo on patterns of transcriptional variation in Drosophila melanogaster. G3 (Bethesda) 7:3659–3668. https://doi.org/10.1534/g3.117.1133

129. Dobzhansky T, Sturtevant AH (1938) Inversions in the chromosomes of Drosophila Pseudoobscura. Genetics 23:28–64

130. Dobzhansky T, Epling C (1948) The suppression of crossing over in inversion heterozygotes of Drosophila Pseudoobscura. Proc

Natl Acad Sci U S A 34:137–141. https://doi.org/10.1073/pnas.34.4.137

131. Garcia C, Valente VLS (2018) Drosophila chromosomal polymorphism: from population aspects to origin mechanisms of inversions. Intech

132. Puig M, Casillas S, Villatoro S, Cáceres M (2015) Human inversions and their functional consequences. Brief Funct Genomics 14:369–379. https://doi.org/10.1093/bfgp/elv020

133. Anderson WW, Arnold J, Baldwin DG et al (1991) Four decades of inversion polymorphism in Drosophila pseudoobscura. Proc Natl Acad Sci U S A 88:10367–10371. https://doi.org/10.1073/pnas.88.22.10367

134. Krimbas CB, Powell JR (1992) The inversion polymorphism of Drosophila subobscura. In: Drosophila inversion polymorphism. CRC Press, Boca Raton, FL, pp 127–220

135. Hoffmann AA, Sgrò CM, Weeks AR (2004) Chromosomal inversion polymorphisms and adaptation. Trends Ecol Evol 19:482–488. https://doi.org/10.1016/j.tree.2004.06.013

136. Hoffmann AA, Rieseberg LH (2008) Revisiting the impact of inversions in evolution: from population genetic markers to drivers of adaptive shifts and speciation? Annu Rev Ecol Evol Syst 39:21–42. https://doi.org/10.1146/annurev.ecolsys.39.110707.173532

137. Kirkpatrick M (2010) How and why chromosome inversions evolve. PLoS Biol 8:e1000501. https://doi.org/10.1371/journal.pbio.1000501

138. Schaeffer SW (2008) Selection in heterogeneous environments maintains the gene arrangement polymorphism of Drosophila pseudoobscura. Evolution (N Y) 62:3082–3099. https://doi.org/10.1111/j.1558-5646.2008.00504.x

139. Rezende EL, Balanyà J (2010) Climate change and chromosomal inversions in Drosophila subobscura. Clim Res 43:103–114. https://doi.org/10.3354/cr00869

140. Rego C, Balanyà J, Fragata I et al (2010) Clinal patterns of chromosomal inversion polymorphisms in Drosophila subobscura are partly associated with thermal preferences and heat stress resistance. Evolution (N Y) 64:385–397. https://doi.org/10.1111/j.1558-5646.2009.00835.x

141. Kapun M, Fabian DK, Goudet J, Flatt T (2016) Genomic evidence for adaptive inversion clines in Drosophila melanogaster. Mol Biol Evol 33:1317–1336. https://doi.org/10.1093/molbev/msw016

142. Knibb WR (1982) Chromosome inversion polymorphisms in Drosophila melanogaster II. Geographic clines and climatic associations in Australasia, North America and Asia. Genetica 58:213–221. https://doi.org/10.1007/BF00128015

143. Umina PA, Weeks AR, Kearney MR et al (2005) A rapid shift in a classic clinal pattern in Drosophila reflecting climate change. Science 308:691–693. https://doi.org/10.1126/science.1109523

144. Corbett-Detig RB, Hartl DL (2012) Population genomics of inversion polymorphisms in Drosophila melanogaster. PLoS Genet 8:e1003056. https://doi.org/10.1371/journal.pgen.1003056

145. Rane RV, Rako L, Kapun M, LEE SF (2015) Genomic evidence for role of inversion 3RP of Drosophila melanogaster in facilitating climate change adaptation. Mol Ecol 24:2423–2432. https://doi.org/10.1111/mec.13161/pdf

146. Fuller ZL, Haynes GD, Richards S, Schaeffer SW (2017) Genomics of natural populations: evolutionary forces that establish and maintain gene arrangements in Drosophila pseudoobscura. Mol Ecol 26:6539–6562. https://doi.org/10.1111/mec.14381

147. Navarro A, Betrán E, Barbadilla A, Ruiz A (1997) Recombination and gene flux caused by gene conversion and crossing over in inversion heterokaryotypes. Genetics 146:695–709

148. Andolfatto P, Depaulis F, Navarro A (2001) Inversion polymorphisms and nucleotide variability in Drosophila. Genet Res (Camb) 77:1–8. https://doi.org/10.1017/S0016672301004955

149. Guerrero RF, Rousset F, Kirkpatrick M (2012) Coalescent patterns for chromosomal inversions in divergent populations. Philos Trans R Soc Lond B Biol Sci 367:430–438. https://doi.org/10.1098/rstb.2011.0246

150. Corbett-Detig RB, Cardeno C, Langley CH (2012) Sequence-based detection and breakpoint assembly of polymorphic inversions. Genetics 192:131–137. https://doi.org/10.1534/genetics.112.141622

151. Navarro A, Faria R (2014) Pool and conquer: new tricks for (c)old problems. Mol Ecol 23:1653–1655. https://doi.org/10.1111/mec.12685

152. Sotero-Caio CG, Platt RN, Suh A, Ray DA (2017) Evolution and diversity of transposable elements in vertebrate genomes. Genome

Biol Evol 9:161–177. https://doi.org/10.1093/gbe/evw264

153. Tenaillon MI, Hollister JD, Gaut BS (2010) A triptych of the evolution of plant transposable elements. Trends Plant Sci 15:471–478. https://doi.org/10.1016/j.tplants.2010.05.003

154. Tenaillon MI, Hufford MB, Gaut BS, Ross-ibarra J (2011) Genome size and transposable element content as. Mol Biol 3:219–229. https://doi.org/10.1093/gbe/evr008

155. Stuart T, Eichten SR, Cahn J et al (2016) Population scale mapping of transposable element diversity reveals links to gene regulation and epigenomic variation. Elife 5. https://doi.org/10.7554/eLife.20777

156. Vieira C, Fablet M, Lerat E et al (2012) A comparative analysis of the amounts and dynamics of transposable elements in natural populations of Drosophila melanogaster and Drosophila simulans. J Environ Radioact 113:83–86. https://doi.org/10.1016/j.jenvrad.2012.04.001

157. Kidwell MG, Lisch DR (2001) Perspective: transposable elements, parasitic DNA, and genome evolution. Evolution 55:1–24

158. Charlesworth B, Langley CH, Sniegowski PD (1997) Transposable element distributions in Drosophila. Genetics 147:1993–1995

159. Charlesworth B, Charlesworth D (1983) The population dynamics of transposable elements. Genet Res 42:1. https://doi.org/10.1017/S0016672300021455

160. Charlesworth B, Jarne P, Assimacopoulos S (1994) The distribution of transposable elements within and between chromosomes in a population of Drosophila melanogaster. III. Element abundances in heterochromatin. Genet Res 64:183. https://doi.org/10.1017/S0016672300032845

161. Barrón MG, Fiston-Lavier A-S, Petrov DA, González J (2014) Population genomics of transposable elements in *Drosophila*. Annu Rev Genet 48:561–581. https://doi.org/10.1146/annurev-genet-120213-092359

162. Bergman CM, Bensasson D (2007) Recent LTR retrotransposon insertion contrasts with waves of non-LTR insertion since speciation in Drosophila melanogaster. Proc Natl Acad Sci U S A 104:11340–11345. https://doi.org/10.1073/pnas.0702552104

163. Charlesworth B, Langley CH (1989) The population genetics of Drosophila transposable elements. Annu Rev Genet 23:251–287. https://doi.org/10.1146/annurev.ge.23.120189.001343

164. Ewing AD (2015) Transposable element detection from whole genome sequence data. Mob DNA 6:24. https://doi.org/10.1186/s13100-015-0055-3

165. Kofler R, Betancourt AJ, Schlötterer C (2012) Sequencing of pooled DNA samples (Pool-Seq) uncovers complex dynamics of transposable element insertions in Drosophila melanogaster. PLoS Genet 8:e1002487. https://doi.org/10.1371/journal.pgen.1002487

166. Cridland JM, Macdonald SJ, Long AD, Thornton KR (2013) Abundance and distribution of transposable elements in two Drosophila QTL mapping resources. Mol Biol Evol 30:2311–2327. https://doi.org/10.1093/molbev/mst129

167. Rishishwar L, Mariño-Ramírez L, Jordan IK (2017) Benchmarking computational tools for polymorphic transposable element detection. Brief Bioinform 18:908–918. https://doi.org/10.1093/bib/bbw072

168. Vieira C, Lepetit D, Dumont S, Biémont C (1999) Wake up of transposable elements following Drosophila simulans worldwide colonization. Mol Biol Evol 16:1251–1255

169. Blumenstiel JP, Chen X, He M, Bergman CM (2014) An age-of-allele test of neutrality for transposable element insertions. Genetics 196:523–538. https://doi.org/10.1534/genetics.113.158147

170. Ohta T (1973) Slightly deleterious mutant substitutions in evolution. Nature 246:96–98. https://doi.org/10.1038/246096a0

171. Kimura M (1983) The neutral theory of molecular evolution. Cambridge University Press, Cambridge

172. Eyre-Walker A (2006) The genomic rate of adaptive evolution. Trends Ecol Evol 21:569–575

173. Fay JC, Wyckoff GJ, Wu C-I (2002) Testing the neutral theory of molecular evolution with genomic data from Drosophila. Nature 415:1024–1026. https://doi.org/10.1038/4151024a

174. Smith NGC, Eyre-Walker A (2002) Adaptive protein evolution in Drosophila. Nature 415:1022–1024

175. Andolfatto P (2005) Adaptive evolution of non-coding DNA in Drosophila. Nature 437:1149–1152

176. Shapiro JA, Huang W, Zhang C et al (2007) Adaptive genic evolution in the Drosophila genomes. Proc Natl Acad Sci 104:2271–2276. https://doi.org/10.1073/pnas.0610385104

177. Eyre-Walker A, Keightley PD (2009) Estimating the rate of adaptive molecular evolution in the presence of slightly deleterious mutations and population size change. Mol Biol Evol 26:2097–2108. https://doi.org/10.1093/molbev/msp119

178. Castellano D, Coronado-Zamora M, Campos JL et al (2016) Adaptive evolution is substantially impeded by Hill–Robertson interference in *Drosophila*. Mol Biol Evol 33:442–455. https://doi.org/10.1093/molbev/msv236

179. Maynard Smith J, Haigh J (1974) The hitchhiking effect of a favourable gene. Genet Res 23:23–55

180. Kim Y, Stephan W (2002) Detecting a local signature of genetic hitchhiking along a recombining chromosome. Genetics 160:765–777

181. Nielsen R, Williamson S, Kim Y et al (2005) Genomic scans for selective sweeps using SNP data. Genome Res 15:1566–1575. https://doi.org/10.1101/gr.4252305

182. Glinka S, Ometto L, Mousset S et al (2003) Demography and natural selection have shaped genetic variation in Drosophila melanogaster: a multi-locus approach. Genetics 165:1269–1278

183. Ometto L, Glinka S, De Lorenzo D, Stephan W (2005) Inferring the effects of demography and selection on Drosophila melanogaster populations from a chromosome-wide scan of DNA variation. Mol Biol Evol 22:2119–2130

184. Karasov T, Messer PW, Petrov DA (2010) Evidence that adaptation in Drosophila is not limited by mutation at single sites. PLoS Genet 6:e1000924. https://doi.org/10.1371/journal.pgen.1000924

185. Messer PW, Petrov DA (2013) Population genomics of rapid adaptation by soft selective sweeps. Trends Ecol Evol 28:659–669. https://doi.org/10.1016/j.tree.2013.08.003

186. Obbard DJ, Jiggins FM, Bradshaw NJ, Little TJ (2011) Recent and recurrent selective sweeps of the antiviral RNAi gene Argonaute-2 in three species of Drosophila. Mol Biol Evol 28:1043–1056. https://doi.org/10.1093/molbev/msq280

187. Glinka S, De Lorenzo D, Stephan W (2006) Evidence of gene conversion associated with a selective sweep in Drosophila melanogaster. Mol Biol Evol 23:1869–1878. https://doi.org/10.1093/molbev/msl069

188. Wilches R, Voigt S, Duchen P et al (2014) Fine-mapping and selective sweep analysis of

QTL for cold tolerance in Drosophila melanogaster. G3 (Bethesda) 4:1635–1645. https://doi.org/10.1534/g3.114.012757

189. DuMont VB, Aquadro CF (2005) Multiple signatures of positive selection downstream of notch on the X chromosome in Drosophila melanogaster. Genetics 171:639–653. https://doi.org/10.1534/genetics.104.038851

190. Magwire MM, Bayer F, Webster CL et al (2011) Successive increases in the resistance of Drosophila to viral infection through a transposon insertion followed by a duplication. PLoS Genet 7:e1002337. https://doi.org/10.1371/journal.pgen.1002337

191. Schmidt JM, Good RT, Appleton B et al (2010) Copy number variation and transposable elements feature in recent, ongoing adaptation at the Cyp6g1 locus. PLoS Genet 6:e1000998. https://doi.org/10.1371/journal.pgen.1000998

192. Battlay P, Green L, Leblanc P, et al (2018) Structural variants and selective sweep foci contribute to insecticide resistance in the Drosophila melanogaster genetic reference panel. bioRxiv 301937. https://doi.org/10.1101/301937

193. Battlay P, Schmidt JM, Fournier-Level A, Robin C (2016) Genomic and Transcriptomic associations identify a new insecticide resistance phenotype for the selective sweep at the *Cyp6g1* locus of *Drosophila melanogaster*. G3 (Bethesda) 6:2573–2581. https://doi.org/10.1534/g3.116.031054

194. Jensen JD, Bauer DuMont VL, Ashmore AB et al (2007) Patterns of sequence variability and divergence at the diminutive gene region of Drosophila melanogaster: complex patterns suggest an ancestral selective sweep. Genetics 177:1071–1085. https://doi.org/10.1534/genetics.106.069468

195. Saminadin-Peter SS, Kemkemer C, Pavlidis P, Parsch J (2012) Selective sweep of a cis-regulatory sequence in a non-African population of Drosophila melanogaster. Mol Biol Evol 29:1167–1174. https://doi.org/10.1093/molbev/msr284

196. Glaser-Schmitt A, Parsch J (2018) Functional characterization of adaptive variation within a cis-regulatory element influencing Drosophila melanogaster growth. PLoS Biol 16:e2004538. https://doi.org/10.1371/journal.pbio.2004538

197. Svetec N, Pavlidis P, Stephan W (2009) Recent strong positive selection on Drosophila melanogaster HDAC6, a gene encoding a stress surveillance factor, as revealed by population genomic analysis. Mol Biol Evol

26:1549–1556. https://doi.org/10.1093/molbev/msp065

198. Orengo DJ, Aguade M (2007) Genome scans of variation and adaptive change: extended analysis of a candidate locus close to the phantom gene region in Drosophila melanogaster. Mol Biol Evol 24:1122–1129. https://doi.org/10.1093/molbev/msm032

199. Beisswanger S, Stephan W (2008) Evidence that strong positive selection drives neofunctionalization in the tandemly duplicated polyhomeotic genes in Drosophila. Proc Natl Acad Sci 105:5447–5452. https://doi.org/10.1073/pnas.0710892105

200. Voigt S, Laurent S, Litovchenko M, Stephan W (2015) Positive selection at the *Polyhomeotic* locus led to decreased thermosensitivity of gene expression in temperate *Drosophila melanogaster*. Genetics 200:591–599. https://doi.org/10.1534/genetics.115.177030

201. Pool JE, Bauer DuMont V, Mueller JL, Aquadro CF (2006) A scan of molecular variation leads to the narrow localization of a selective sweep affecting both Afrotropical and cosmopolitan populations of Drosophila melanogaster. Genetics 172:1093–1105. https://doi.org/10.1534/genetics.105.049973

202. Depaulis F, Brazier L, Veuille M (1999) Selective sweep at the Drosophila melanogaster suppressor of hairless locus and its association with the In(2L)t inversion polymorphism. Genetics 152:1017–1024. https://doi.org/10.1017/S0016672398003462

203. Beisswanger S, Stephan W, De Lorenzo D (2006) Evidence for a selective sweep in the wapl region of Drosophila melanogaster. Genetics 172:265–274. https://doi.org/10.1534/genetics.105.049346

204. Garud NR, Messer PW, Buzbas EO, Petrov DA (2015) Recent selective sweeps in North American Drosophila melanogaster show signatures of soft sweeps. PLoS Genet 11:e1005004. https://doi.org/10.1371/journal.pgen.1005004

205. Schlenke TA, Begun DJ (2004) Strong selective sweep associated with a transposon insertion in Drosophila simulans. Proc Natl Acad Sci 101:1626–1631. https://doi.org/10.1073/pnas.0303793101

206. Teshima KM, Coop G, Przeworski M (2006) How reliable are empirical genomic scans for selective sweeps? Genome Res 16:702–712

207. Thornton KR, Jensen JD (2007) Controlling the false-positive rate in multilocus genome scans for selection. Genetics 175:737–750

208. Pavlidis P, Jensen JD, Stephan W (2010) Searching for footprints of positive selection in whole-genome SNP data from nonequilibrium populations. Genetics 185:907–922. https://doi.org/10.1534/genetics.110.116459

209. Crisci JL, Poh YP, Mahajan S, Jensen JD (2013) The impact of equilibrium assumptions on tests of selection. Front Genet 4:235. https://doi.org/10.3389/fgene.2013.00235

210. Hermisson J, Pennings PS (2017) Soft sweeps and beyond: understanding the patterns and probabilities of selection footprints under rapid adaptation. Methods Ecol Evol 8:700–716. https://doi.org/10.1111/2041-210X.12808

211. Jensen JD (2014) On the unfounded enthusiasm for soft selective sweeps. Nat Commun 5:5281. https://doi.org/10.1038/ncomms6281

212. Kaplan NL, Hudson RR, Langley CH (1989) The "Hitchhiking Effect" revisited. Genetics 123:887–899

213. Casillas S, Barbadilla A (2017) Molecular population genetics. Genetics 205:1003–1035. https://doi.org/10.1534/genetics.116.196493

214. Sella G, Petrov DA, Przeworski M, Andolfatto P (2009) Pervasive natural selection in the Drosophila genome? PLoS Genet 5:e1000495. https://doi.org/10.1371/journal.pgen.1000495

215. Sawyer S, Kulathinal RJ, Bustamante CD, Hartl DL (2003) Bayesian analysis suggests that most amino acid replacements in Drosophila are driven by positive selection. J Mol Evol 57:S154–S164

216. Schneider A, Charlesworth B, Eyre-Walker A, Keightley PD (2011) A method for inferring the rate of occurrence and fitness effects of advantageous mutations. Genetics 189:1427–1437. https://doi.org/10.1534/genetics.111.131730

217. Keightley PD, Campos JL, Booker TR, Charlesworth B (2016) Inferring the frequency spectrum of derived variants to quantify adaptive molecular evolution in protein-coding genes of Drosophila melanogaster. Genetics 203:975–984. https://doi.org/10.1534/genetics.116.188102

218. Andolfatto P (2007) Hitchhiking effects of recurrent beneficial amino acid substitutions in the Drosophila melanogaster genome. Genome Res 17:1755–1762. https://doi.org/10.1101/gr.6691007

219. Jensen JD, Thornton KR, Andolfatto P (2008) An approximate Bayesian estimator suggests strong, recurrent selective sweeps in

Drosophila. PLoS Genet 4:e1000198. https://doi.org/10.1371/journal.pgen. 1000198

220. Macpherson JM, Sella G, Davis JC, Petrov DA (2007) Genomewide spatial correspondence between nonsynonymous divergence and neutral polymorphism reveals extensive adaptation in Drosophila. Genetics 177:2083–2099. https://doi.org/10.1534/genetics.107.080226

221. Stephan W (2010) Genetic hitchhiking versus background selection: the controversy and its implications. Philos Trans R Soc L B Biol Sci 365:1245–1253. https://doi.org/10.1098/rstb.2009.0278

222. Charlesworth B (2012) The role of background selection in shaping patterns of molecular evolution and variation: evidence from variability on the Drosophila X chromosome. Genetics 191:233–246

223. Hudson RR, Kaplan NL (1995) Deleterious background selection with recombination. Genetics 141:1605–1617

224. Comeron JM (2014) Background selection as baseline for nucleotide variation across the Drosophila genome. PLoS Genet 10: e1004434. https://doi.org/10.1371/journal.pgen.1004434

225. Elyashiv E, Sattath S, Hu TT et al (2016) A genomic map of the effects of linked selection in Drosophila. PLoS Genet 12:e1006130. https://doi.org/10.1371/journal.pgen. 1006130

226. Siepel A (2005) Evolutionarily conserved elements in vertebrate, insect, worm, and yeast genomes. Genome Res 15:1034–1050. https://doi.org/10.1101/gr.3715005

227. Bergman CM, Kreitman M (2001) Analysis of conserved noncoding DNA in Drosophila reveals similar constraints in intergenic and intronic sequences. Genome Res 11:1335–1345

228. Harmston N, Baresic A, Lenhard B (2013) The mystery of extreme non-coding conservation. Philos Trans R Soc Lond Ser B Biol Sci 368:20130021. https://doi.org/10.1098/rstb.2013.0021

229. Berr T, Peticca A, Haudry A (2019) Evidence for purifying selection on conserved noncoding elements in the genome of Drosophila melanogaster. bioRxiv 623744. https://doi.org/10.1101/623744

230. Brody T, Yavatkar AS, Kuzin A et al (2012) Use of a Drosophila genome-wide conserved sequence database to identify functionally related cis-regulatory enhancers. Dev Dyn 241:169–189. https://doi.org/10.1002/dvdy.22728

231. Enright AJ, John B, Gaul U et al (2003) MicroRNA targets in Drosophila. Genome Biol 5:R1. https://doi.org/10.1186/gb-2003-5-1-r1

232. Lai CK, Miller MC, Collins K (2003) Roles for RNA in telomerase nucleotide and repeat addition processivity. Mol Cell 11:1673–1683

233. Casillas S, Barbadilla A, Bergman CM (2007) Purifying selection maintains highly conserved noncoding sequences in Drosophila. Mol Biol Evol 24:2222–2234. https://doi.org/10.1093/molbev/msm150

234. Clark AG (2001) The search for meaning in noncoding DNA. Genome Res 11:1319–1320. https://doi.org/10.1101/gr.201601

235. Shields DC, Sharp PM, Higgins DG, Wright F (1988) "Silent" sites in Drosophila genes are not neutral: evidence of selection among synonymous codons. Mol Biol Evol 5:704–716. https://doi.org/10.1093/oxfordjournals.molbev.a040525

236. Hershberg R, Petrov DA (2008) Selection on codon bias. https://doi.org/10.1146/annurev.genet.42.110807.091442

237. Lawrie DS, Messer PW, Hershberg R, Petrov DA (2013) Strong purifying selection at synonymous sites in D. melanogaster. PLoS Genet 9:e1003527. https://doi.org/10.1371/journal.pgen.1003527

238. Machado HE, Lawrie DS, Petrov DA (2017) Strong selection at the level of codon usage bias: evidence against the Li-Bulmer model. bioRxiv. https://doi.org/10.1101/106476

239. Duret L, Mouchiroud D (1999) Expression pattern and, surprisingly, gene length shape codon usage in Caenorhabditis, Drosophila, and Arabidopsis. Evolution (N Y) 96:4482–4487

240. Powell JR, Moriyama EN (1997) Evolution of codon usage bias in Drosophila. Proc Natl Acad Sci U S A 94:7784–7790

241. Bierne N, Eyre-Walker A (2006) Variation in synonymous codon use and DNA polymorphism within the Drosophila genome. J Evol Biol 19:1–11

242. Vicario S, Moriyama EN, Powell JR (2007) Codon usage in twelve species of Drosophila. BMC Evol Biol 7:226. https://doi.org/10.1186/1471-2148-7-226

243. Zeng K, Charlesworth B (2009) Estimating selection intensity on synonymous codon usage in a nonequilibrium population.

Genetics 183:651–662, 1SI-23SI. https://doi.org/10.1534/genetics.109.101782

244. Andolfatto P, Wong KM, Bachtrog D (2011) Effective population size and the efficacy of selection on the X chromosomes of two closely related Drosophila species. Genome Biol Evol 3:114–128. https://doi.org/10.1093/gbe/evq086

245. Clemente F, Vogl C (2012) Evidence for complex selection on four-fold degenerate sites in Drosophila melanogaster. J Evol Biol 25:2582–2595. https://doi.org/10.1111/jeb.12003

246. Poh Y-P, Ting C-T, Fu H-W et al (2012) Population genomic analysis of base composition evolution in Drosophila melanogaster. Genome Biol Evol 4:1245–1255. https://doi.org/10.1093/gbe/evs097

247. Nielsen R, Bauer DuMont VL, Hubisz MJ, Aquadro CF (2006) Maximum likelihood estimation of ancestral codon usage bias parameters in Drosophila. Mol Biol Evol 24:228–235. https://doi.org/10.1093/molbev/msl146

248. Jackson BC, Campos JL, Haddrill PR et al (2017) Variation in the intensity of selection on codon bias over time causes contrasting patterns of base composition evolution in Drosophila. Genome Biol Evol 9:102–123. https://doi.org/10.1093/gbe/evw291

249. Kirkpatrick M, Kern A (2012) Where's the money? Inversions, genes, and the hunt for genomic targets of selection. Genetics 190:1153–1155. https://doi.org/10.1534/genetics.112.139899

250. Charlesworth B, Charlesworth D (1973) Selection of new inversions in multi-locus genetic systems. Genet Res (Camb) 21:167. https://doi.org/10.1017/S0016672300013343

251. Schaeffer SW, Miller JM, Anderson WW (2003) Evolutionary genomics of inversions in Drosophila pseudoobscura: evidence for epistasis. Proc Natl Acad Sci U S A 100:8319–8324. https://doi.org/10.1073/pnas.1432900100

252. Kirkpatrick M, Barton N (2006) Chromosome inversions, local adaptation and speciation. Genetics 173:419–434. https://doi.org/10.1534/genetics.105.047985

253. Charlesworth B, Barton NH (2018) The spread of an inversion with migration and selection. Genetics 208:377–382. https://doi.org/10.1534/genetics.117.300426

254. Bhutkar A, Schaeffer SW, Russo SM et al (2008) Chromosomal rearrangement inferred from comparisons of 12 Drosophila genomes. Genetics 179:1657–1680. https://doi.org/10.1534/genetics.107.086108

255. Stevison LS, Hoehn KB, Noor MAF (2011) Effects of inversions on within- and between-species recombination and divergence. Genome Biol Evol 3:830–841. https://doi.org/10.1093/gbe/evr081

256. McGaugh SE, Noor MAF (2012) Genomic impacts of chromosomal inversions in parapatric Drosophila species. Philos Trans R Soc Lond B Biol Sci 367:422–429. https://doi.org/10.1098/rstb.2011.0250

257. Lohse K, Clarke M, Ritchie MG, Etges WJ (2015) Genome-wide tests for introgression between cactophilic Drosophila implicate a role of inversions during speciation. Evolution (N Y) 69:1178–1190. https://doi.org/10.1111/evo.12650

258. Lee Y, Collier TC, Sanford MR et al (2013) Chromosome inversions, genomic differentiation and speciation in the African malaria mosquito Anopheles gambiae. PLoS One 8: e57887. https://doi.org/10.1371/journal.pone.0057887

259. Ayala D, Guerrero RF, Kirkpatrick M (2013) Reproductive isolation and local adaptation quantified for a chromosome inversion in a malaria mosquito. Evolution (N Y) 67:946–958. https://doi.org/10.1111/j.1558-5646.2012.01836.x

260. Rieseberg LH (2001) Chromosomal rearrangements and speciation. Trends Ecol Evol 16:351–358. https://doi.org/10.1016/S0169-5347(01)02187-5

261. Abbott JK, Nordén AK, Hansson B (2017) Sex chromosome evolution: historical insights and future perspectives. Proc R Soc B Biol Sci 284:20162806. https://doi.org/10.1098/rspb.2016.2806

262. Lemeunier F, Aulard S (1992) Inversion polymorphism in Drosophila melanogaster. In: Krimbas CB, Powell JR (eds) Drosophila inversion polymorphism. CRC Press, p 576

263. Rebollo R, Romanish MT, Mager DL (2012) Transposable elements: an abundant and natural source of regulatory sequences for host genes. Annu Rev Genet 46:21–42. https://doi.org/10.1146/annurev-genet-110711-155621

264. Casacuberta E, González J (2013) The impact of transposable elements in environmental adaptation. Mol Ecol 22:1503–1517. https://doi.org/10.1111/mec.12170

265. Villanueva-Cañas JL, Rech GE, de Cara MAR, González J (2017) Beyond SNPs: how to detect selection on transposable element insertions. Methods Ecol Evol

8:728–737. https://doi.org/10.1111/2041-210X.12781

266. González J, Karasov TL, Messer PW, Petrov DA (2010) Genome-wide patterns of adaptation to temperate environments associated with transposable elements in Drosophila. PLoS Genet 6:e1000905. https://doi.org/10.1371/journal.pgen.1000905

267. Merenciano M, Ullastres A, de Cara MAR et al (2016) Multiple independent Retroelement insertions in the promoter of a stress response gene have variable molecular and functional effects in Drosophila. PLoS Genet 12:e1006249. https://doi.org/10.1371/journal.pgen.1006249

268. González J, Lenkov K, Lipatov M et al (2008) High rate of recent transposable element–induced adaptation in Drosophila melanogaster. PLoS Biol 6:e251. https://doi.org/10.1371/journal.pbio.0060251

269. González J, Macpherson JM, Petrov DA (2009) A recent adaptive transposable element insertion near highly conserved developmental loci in Drosophila melanogaster. Mol Biol Evol 26:1949–1961. https://doi.org/10.1093/molbev/msp107

270. Guio L, Barrón MG, González J (2014) The transposable element Bari-Jheh mediates oxidative stress response in Drosophila. Mol Ecol 23:2020–2030. https://doi.org/10.1111/mec.12711

271. Ellison CE, Bachtrog D (2013) Dosage compensation via transposable element mediated rewiring of a regulatory network. Science 342:846–850. https://doi.org/10.1126/science.1239552

272. Sentmanat MF, Elgin SCR (2012) Ectopic assembly of heterochromatin in Drosophila melanogaster triggered by transposable elements. Proc Natl Acad Sci U S A 109:14104–14109. https://doi.org/10.1073/pnas.1207036109

273. Perrat PN, DasGupta S, Wang J et al (2013) Transposition-driven genomic heterogeneity in the Drosophila brain. Science 340:91–95. https://doi.org/10.1126/science.1231965

274. Charlesworth B, Coyne JA, Barton NH (1987) The relative rates of evolution of sex chromosomes and autosomes. Am Nat 130:113–146. https://doi.org/10.1086/284701

275. Charlesworth B, Campos JL, Jackson BC (2018) Faster-X evolution: theory and evidence from Drosophila. Mol Ecol 27:3753–3771. https://doi.org/10.1111/mec.14534

276. Betancourt AJ, Kim Y, Orr HA (2004) A pseudohitchhiking model of X vs. autosomal diversity. Genetics 168:2261–2269. https://doi.org/10.1534/genetics.104.030999

277. Campos JL, Halligan DL, Haddrill PR, Charlesworth B (2014) The relation between recombination rate and patterns of molecular evolution and variation in Drosophila melanogaster. Mol Biol Evol 31:1010–1028. https://doi.org/10.1093/molbev/msu056

278. Endler JA (1977) Geographic variation, speciation, and clines. Princeton University Press, Princeton, NJ

279. Huxley JS (1938) Clines: an auxiliary taxonomic principle. Nature 142:219–220. https://doi.org/10.1038/142219a0

280. de Jong G, Bochdanovits Z (2003) Latitudinal clines in Drosophila melanogaster: body size, allozyme frequencies, inversion frequencies, and the insulin-signalling pathway. J Genet 82:207–223. https://doi.org/10.1007/BF02715819

281. Adrion JR, Hahn MW, Cooper BS (2015) Revisiting classic clines in Drosophila melanogaster in the age of genomics. Trends Genet 31:434–444. https://doi.org/10.1016/j.tig.2015.05.006

282. Flatt T (2016) Genomics of clinal variation in Drosophila: disentangling the interactions of selection and demography. Mol Ecol 25:1023–1026. https://doi.org/10.1111/mec.13534

283. Rosenzweig ML (1968) The strategy of body size in mammalian carnivores. Am Midl Nat 80:299

284. Blanckenhorn WU, Demont M (2004) Bergmann and converse bergmann latitudinal clines in arthropods: two ends of a continuum? Integr Comp Biol 44:413–424. https://doi.org/10.1093/icb/44.6.413

285. Stillwell RC (2010) Are latitudinal clines in body size adaptive? Oikos 119:1387–1390. https://doi.org/10.1111/j.1600-0706.2010.18670.x

286. Kivelä SM, Välimäki P, Carrasco D et al (2011) Latitudinal insect body size clines revisited: a critical evaluation of the saw-tooth model. J Anim Ecol 80:1184–1195. https://doi.org/10.2307/41332025?ref=no-x-route:0c31195f82739061ed649686a0a9f292

287. Colombo PC, Remis MI (2015) Morphometric variation in chromosomally polymorphic grasshoppers (Orthoptera: Acrididae) from South America: Bergmann and Converse Bergmann patterns. Florida Entomol 98:570–574. https://doi.org/10.1653/024.098.0228

288. Schmidt PS, Paaby AB (2008) Reproductive diapause and life-history clines in North American populations of Drosophila melanogaster. Evolution (N Y) 62:1204–1215. https://doi.org/10.1111/j.1558-5646.2008.00351.x

289. Addo-Bediako A, Chown SL, Gaston KJ (2000) Thermal tolerance, climatic variability and latitude. Proc R Soc B Biol Sci 267:739–745. https://doi.org/10.1098/rspb.2000.1065

290. Hoffmann AA, Anderson A, Hallas R (2002) Opposing clines for high and low temperature resistance in Drosophila melanogaster. Ecol Lett 5:614–618. https://doi.org/10.1046/j.1461-0248.2002.00367.x

291. Castañeda LE, Lardies MA, Bozinovic F (2005) Interpopulational variation in recovery time from chill coma along a geographic gradient: a study in the common woodlouse, Porcellio laevis. J Insect Physiol 51:1346–1351

292. Gockel J, Kennington WJ, Hoffmann AA et al (2001) Nonclinality of molecular variation implicates selection in maintaining a morphological cline of Drosophila melanogaster. Genetics 158:319–323

293. Turner TL, Levine MT, Eckert ML, Begun DJ (2008) Genomic analysis of adaptive differentiation in Drosophila melanogaster. Genetics 179:455–473. https://doi.org/10.1534/genetics.107.083659

294. Behrman EL, Howick VM, Kapun M et al (2018) Rapid seasonal evolution in innate immunity of wild Drosophila melanogaster. Proc R Soc B Biol Sci 285:20172599. https://doi.org/10.1098/rspb.2017.2599

295. De Mita S, Thuillet A-C, Gay L et al (2013) Detecting selection along environmental gradients: analysis of eight methods and their effectiveness for outbreeding and selfing populations. Mol Ecol 22:1383–1399. https://doi.org/10.1111/mec.12182

296. Vasemägi A (2006) The adaptive hypothesis of clinal variation revisited: single-locus clines as a result of spatially restricted gene flow. Genetics 173:2411–2414. https://doi.org/10.1534/genetics.106.059881

297. Prevosti A, Serra L, Segarra C et al (1990) Clines of chromosomal arrangements of Drosophila subobscura in South America evolve closer to Old World patterns. Evolution (N Y) 44:218. https://doi.org/10.2307/2409539

298. Machado HE, Bergland AO, O'Brien KR et al (2016) Comparative population genomics of latitudinal variation in Drosophila simulans and Drosophila melanogaster. Mol Ecol 25:723–740. https://doi.org/10.1111/mec.13446

299. Barton NH (1999) Clines in polygenic traits. Genet Res 74:223–236

Chapter 16

Genomic Access to the Diversity of Fishes

Arne W. Nolte

Abstract

The number of fishes exceeds that of all other vertebrates both in terms of species numbers and in their morphological and phylogenetic diversity. They are an ecologically and economically important group and play an essential role as a resource for humans. This makes the genomic exploration of fishes an important area of research, both from an applied and a basic research perspective. Fish genomes can vary greatly in complexity, which is partially due to differences in size and content of repetitive DNA, a history of genome duplication events and because fishes may be polyploid, all of which complicate the assembly and analysis of genome sequences. However, the advent of modern sequencing techniques now facilitates access to genomic data that permit genome-wide exploration of genetic information even for previously unexplored species. The development of genomic resources for fishes is spearheaded by model organisms that have been subject to genetic analysis and genome sequencing projects for a long time. These offer a great potential for the exploration of new species through the transfer of genomic information in comparative analyses. A growing number of genome sequencing projects and the increasing availability of tools to assemble and access genomic information now move boundaries between model and nonmodel species and promises progress in many interesting but unexplored species that remain to be studied.

Key words Teleostei, Genomic makeup, Genome size, Ploidy, Nonmodel organisms

1 Diversity of Fishes

Fishes are the most diverse group of vertebrates on earth [1]. As of 2017, a total of 33,554 species of fishes have been described [2], and many more remain to be discovered. They have colonized marine and freshwater habitats alike and display tremendous anatomical and ecological diversity. The term fishes (Pisces) includes the most basal jawless fishes (lampreys and hagfishes) that live as parasites or scavengers and the lobe-finned fishes (lungfish and coelacanths) that gave rise to the tetrapods. The exclusion of the latter renders fishes as a whole a paraphyletic group. The ray-finned fishes comprise the basal bichirs (Polypteridae) and sturgeons (Ascipenderidae) as well as the holostei [Bowfins (Amiidae) and gars (Lepisosteidae)]. The most speciose group by far is the Teleosts that have undergone a spectacular diversification since the Cretaceous period

Julien Y. Dutheil (ed.), *Statistical Population Genomics*, Methods in Molecular Biology, vol. 2090,
https://doi.org/10.1007/978-1-0716-0199-0_16, © The Author(s) 2020

[3]. They are prevalent in many aquatic ecosystems and are of manifold importance to man. Fishes are found in the deepest ocean trenches and up to 5200 m elevation in the Himalayas [2]. They have colonized rivers, lakes, and oceans but also extreme habitats like caves where they live in constant darkness, the Arctic, or desert springs with high temperature and salt conditions. Some killifishes even survive dry periods by laying drought-resistant eggs, and these fishes are also among the most short-lived ones [4]. The oldest fishes to date may be Greenland sharks that have been estimated to be close to 400 years old [5]. The south-east Asian *Paedocypris* with a standard length of 7.9 mm possibly represents the world's smallest vertebrate [6] while the largest nonmammalian vertebrate is given by the whale shark that may reach a length of up to 13 m. The diversity of reproductive modes in fishes includes egg laying and oviposition or the birth of fully developed young. Egg laying fishes have evolved numerous modes of brood care including mouthbrooding, substrate brooding, nest building, pelvic fin brooding, or ventral pouch brooding whereby the parental care may be performed through the father, the mother or both parents. Eggs of fishes may be released into the pelagial zone in mass spawning events and left to themselves, deposited into caves, mussels, gravel rudds or nests out of plant matter or air bubbles [7]. All fishes have a direct development, but may go through extended and distinct larval periods including the blind and worm-like ammocoetes larvae of lampreys or the marine leptocephalus larvae of eels and tarpon as opposed to the fully developed offspring of fishes that give birth. The feeding types of fishes are equally diverse. While some feed on microscopic algae (Silver carp, *Hypophthalmichthys molitrix*), scrape algae of surfaces (Nase, *Chondrostoma nasus*) or feed on higher plants (grass carp, *Ctenoparyngodon idella*) the majority of fishes feed on animals. Again, there is a range from plankton feeders (herring, *Clupea harengus*), to piscivorous fishes (pike, *Esox lucius*) and top predators like sharks that may prey upon marine mammals. There a numerous highly specialized feeding strategies in which fishes specialize on detritus, decaying wood, snails, or mussels. They eat scales, skin, eyes or parasites of other fishes or specialize on crabs, shrimp, insects, coral, fruit, sponges, and many other food items that are taken on occasion. The exploitation of these different food sources is typically facilitated by evolutionary accommodation of the feeding apparatus, which constitutes a key element that has determined the impressive adaptive radiation of fishes. Another factor that has contributed to the diversity of fishes is the many means by which they use their body or fins to move. Fishes can swim, whereby different species use fins very differently to propel themselves and for fine maneuvering. Some are constant swimmers whereas others are sit-and-wait predators or almost sessile in very confined spaces where they tend to camouflage. Some eel-shaped species, puffer fish or flatfishes are

able to bury themselves into different substrates. Different fishes can use their modified mouth or fins to cling to hard substrates, enabling them to persist in strong currents or to climb steep waterfalls. Finally, mudskippers have even colonized the intertidal zone above the water level where they can move rapidly using body movements and modified fins. Fishes use the same senses to acquire signals from their environment like humans, including vision with highly developed eyes that enable them to see color, and sometimes ultraviolet or polarized light. They can hear sounds with the help of the Weberian apparatus and they have a sensory system equipped to feel pain. The smell or taste of fishes is well developed within the nose, but also through taste buds that are distributed across their body. Beyond these, fishes can detect currents or waves underwater through their lateral line and head canal system, and some are able to detect electrical fields of prey items or the earth's magnetic field for orientation. Although fishes are mostly harmless to humans, there are species that are highly toxic or venomous, and that are of medical importance. Tissues of the Japanese pufferfish (*Takifugu* sp.) contain tetrodotoxin that can kill humans if consumed and tropical marine predators like moray eels or barracuda may accumulate toxins that originate from dinoflagellate blooms. Finally, there are also venomous species, such as the stonefish (*Synanceia verrucosa*) or the related lionfish (*Pterois volitans*) that possess venom and inflict painful and life-threatening injuries when touched or unintentionally stepped on.

The diversity of fishes has permitted them to exploit niches in aquatic ecosystems in many specialized ways and to become dominant components in food webs. Fishes represent top predators that convert energy from lower trophic levels to biomass that is harvested by humans and other top predators. For this reason, fishes have been naturalized across the globe in hope to create prolific food resources for human consumption, including the release of carp, trout, salmon, *Tilapia*, Nile perch, eel, catfish, and many other species outside their native range. However, it is now clear that considerable detrimental side effects on local ecosystems are common whenever such introductions were successful. Humans also employ fishes in attempts to manipulate ecosystems as biological control agents, for example, the silver carp (*Hypophthalmichthys*) or grass carp (*Ctenopharyngodon*) to control algae, or aquatic weeds or mosquito fish (*Gambusia*) to control malaria vectors, again often accompanied by undesirable side effects.

Humans have long kept fishes as ornamental pets, and the history of domestication and aquaculture dates back a long time. Goldfish were already bred in China in 1000 AD [8]. Nowadays, they often are the first pets that children are acquainted with, and seed a positive image fishes have for humans. Additional species such as Koi Carp (*Cyprinus carpio*), Siamese fighting fish (*Betta splendens*), platyfish (*Xiphophorus maculatus*), zebrafish (*Danio rerio*), flowerhorn chichlids (*Amphilophus* hybrids), and many wild ornamental species are kept as pets. Other marine and freshwater including tuna, flatfishes, sea bass, seabream or freshwater fishes like trout, salmon, carp, *Tilapia*, catfishes, and sturgeon are targets of intensive aquaculture to meet the growing demands for food. Likewise, wild populations of fishes are managed and exploited as the most important food resource from aquatic environments. Together, aquaculture and fisheries provide food, income, and livelihoods of hundreds of millions of people and the world per capita fish supply reached a record high of 20 kg in 2014 [9]. Given the growing world population and the limited availability of space for agriculture, fish will play a central role in providing future generations with adequate nutrition. They not only play a direct role but are used to produce fish oil as a food complement, fish meal as food for other livestock and manure to fertilize fields. Accordingly, whole industries are built around fisheries, fish farming, and fish products. Fishes play an important socioeconomic role in recreational angling, and some can serve as flagship species to transport conservation issues into a broader public.

Due to their economic value and because of the essential role fishes play in ecosystems, they are subject to management, conservation efforts, and scientific studies. Fishes are targets of applied research that aims at improving harvests, but also out of broader interest in fish biology or because fishes can serve as model vertebrates in studies that aim at obtaining results of direct relevance to humans in fundamental medical research or ecotoxicology. Fishes are prime models in evolutionary studies. It is this prevalence of fishes and the diverse ways in which they are exploited by humans that makes them targets for genomic exploration.

2   The Genomic Makeup of Fishes

Compared to other vertebrates, fishes seem to have more plastic and variable genomes, which is associated with the fact that they display frequent polyploidization, have high speciation rates and carry a diversity of repetitive genetic elements [10, 11]. The majority of fishes that are intensively studied have relatively compact genomes, but fish genomes may vary in size between 0.35 and 133 Gb [12]. Among these, the teleosts have the most compact genomes ranging from 0.35 to 10 Gb, followed by Chondrichthyes (1.5–17.5 Gb) and finally the lobiform Dipnoi (80–132 Gb) [13]. The Japanese pufferfish *Takifugu rubripes* was targeted in one of the first fish genome sequencing projects because of its compact genome size of 0.39 Gb, which still marks the lower end of the spectrum of vertebrate genome sizes. The three-spined stickleback (*Gasterosteus aculeatus*) genome has a size of 0.46 Gb, the one of the Zebrafish *Danio rerio* has 1.67 Gb and the Japanese

Medaka 0.70 Gb. The genome of a basal "fishes" such as the sea lamprey *Petromyzon marinus* has a size of 0.65 Gb and the sarcopterygian *Latimeria chalumnae* has a genome size of 2.86 Gb that is only a little bit smaller than our own. The largest fish genome can be found in the marbled lungfish *Protopterus aethiopicus* (133 Gb), which represents the largest genome known from any metazoan. Fish genome size and diversity are affected by their variable content of repetitive DNA elements [14] that make a more important relative contribution in fish genomes than in mammals [12, 15, 16]. Besides affecting genome size and structure, repetitive genetic elements have been found to be involved in functional genetic divergence among fishes, such as the rapid evolution of new sex-determining loci or the emergence of barriers to reproduction that reduce the viability of hybrid offspring [11, 17, 18]. The diversity of repetitive genetic elements in fishes exceeds that in higher vertebrates and the relative contribution of repetitive genetic elements may vary from 6% in *Tetraodon* to 55% in *Danio*. The distribution of TE families across the phylogeny demonstrates that their presence and abundance may be highly lineage-specific, and that periods of TE diversification occur independently among different lineages of fishes. The *Sarcopterygii* have lost TE diversity, a trend that manifested even more in the notable reduction of TE diversity in birds and mammals [15]. While the diversity of TEs in fish genomes represents an important component of their between and within lineage genomic diversity, repetitive elements pose challenges for the assembly and thus the analysis of their genomes [19].

Genome duplication events have been postulated to represent major evolutionary events that have facilitated the extraordinary diversification of fishes [20]. There is evidence that rounds of genome duplications have occurred in the stem lineage of the vertebrates, and that an additional round of tetraploidization followed by rediploidization has occurred early in the evolution of the ray-finned fishes (*Actinopterygii*). This process has generated redundant gene copies that may have vanished but also taken up new functions [11, 21, 22]. The evolution of multigene families such as the Hox cluster has been explained through ancient gene duplications and adds another level of complexity to the genetic makeup of fishes [23]. Species of fishes that deviate notably toward larger genome sizes include a range of species that have undergone lineage-specific and more recent genome duplications. Examples include the ancient tetraploid Salmonidae (trout, salmon, whitefishes), but also taxa like the sturgeons (Acipenseridae) where ploidy ranges from diploid to octaploids and that may carry several hundred chromosomes [24]. Although the genome size increases, it is common that redundant gene copies are lost in a process of rediploidization that occurs even after multiple rounds of polyploidization [25, 26]. However, it is also possible that copies of duplicated genes diverge after polyploidization to acquire new functions. Paralogy relationships within genomes can still be tracked as genomes rediploidize, as in the Atlantic salmon genome [27].

While many genome duplication events may be quite ancient [24], there is a range of polyploid species of more recent origin. The range of modes of reproduction of these fishes often leads to patterns of inheritance that deviate from a classical Mendelian pattern. While this comprises interesting phenomena in itself, it poses challenges for the exploration of their genome content as a single organism may contain more than two alleles of a given gene and because divergence between gene copies will reach levels that would otherwise be encountered in separated species. These issues are typically not considered in default parameters of data analyses tools, which can introduce massive bias in attempts to identify orthologous and paralogous sequences and in all studies on genetic variation. Examples include the Eurasian diploid–polyploid species complexes of *Cobitis* loaches in which polyploid hybrids can carry sets of chromosomes originating from parental species that do not co-occur with the hybrid lineages any more [28, 29]. Comparable examples belong to the cypriniform fishes such as the Iberian cyprinid *Squalius alburnoides* or the North American Minnow *Phoxinus eos-neogaeus*. All of these taxa of recent polyploidy origin are allotetraploid, that is, they have arisen as hybrids between two divergent lineages that can apparently only continue to exist when species-specific sets of chromosomes are inherited as a whole. They may use different reproductive modes to pass their genetic material on to the offspring including normal meiosis, asexual reproduction through gynogenesis, where male sperm initiates development but genetic material is excluded, and hybridogenesis [30–33]. An example from central America and the first vertebrate in which unisexuality was discovered [34] includes the Amazon molly, a species of hybrid origin, in which gynogenetic females mate with males of another species to initiate development but exclude the male genetic material from the developing zygote [35, 36]. Experimental studies [37] suggest that fishes are flexible and actively choose their mode of diploid – polyploid reproduction depending on the genotype of the parents, which explains the diversity and success of such lineages in nature. There is one species of fish, the North- and Central-American mangrove killifish *Kryptolebias marmoratus* that exhibits true hermaphroditism and must have existed as a self-fertilizing lineage for a long time [38, 39].

Although Teleostei are at the base of the evolution of the vertebrates, the explosive diversification that has resulted in most of today's diversity of ray-finned fishes (Actinopterygii) has taken place between the late Mesozoic and early Cenozoic [40]. Gene sequences and the order of genes within the genome (synteny) have been conserved. This now permits a transfer of positional genomic

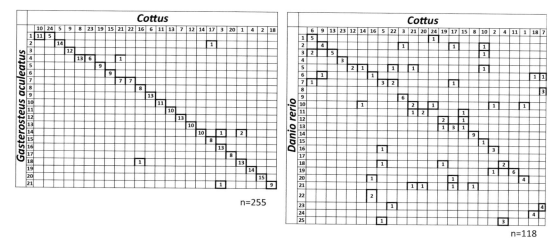

Fig. 1 Oxford grids exploring synteny relationships between *Cottus* ssp. (European sculpin) linkage groups (*x*-axis) and chromosomes of model organisms such as the Stickleback (*Gasterosteus aculeatus*) or the zebrafish (*Danio rerio*) (*y*-axis). For this purpose, genomic fragments for which the position in *Cottus* was genetically mapped were BLAST searched against fully sequenced genomes of the other species. Numbers in fat squares indicate shared markers among known chromosomes and *Cottus* linkage groups. The synteny of *Cottus* and the stickleback is well conserved while the zebrafish genome is more divergent both in terms of chromosome organization and in the low number of markers that could be mapped overall (*n* given at bottom right of each graph). This highlights the utility of the stickleback genome as a genomic reference for the exploration of *Cottus*. (Modified from [42])

information between fully sequenced genomes of model organisms and the wealth of emerging model systems [41, 42]. Conservation of synteny can be visualized by means of oxford grids (Fig. 1) or more elaborate circle graphs (*see* Fig. 2 in [27]) all of which illustrate which regions of the genome contain homologous sequences that are arranged in the same order.

Such inference can infer homology among chromosomes, chromosome fissions and fusion and remnants of duplicated chromosomes. A related issue is that conserved synteny can support inference about homology when gene annotations are transferred between species. Finally, knowledge about syntenic relationships between two genomes lets one predict which genetic elements can be found near a given marker, even if that part of the genome of one of the species is not fully sequenced or assembled. Together with the rapidly growing number of fully sequences fish genomes that sample the fish phylogeny more and more densely, these inferences contribute greatly to the exploration of as yet unexplored species [23]. Even when genomes are not fully sequenced, the conservation of synteny can be exploited to validate newly generated genetic maps [43] or to explore the most likely gene content of QTL regions that have not been fully sequenced in the target species [42, 44]. The number of fully sequenced and annotated fish

Fig. 2 The exploration of fishes like alpine char (*Salvelinus umbla*, top) or grayling (*Thymallus thymallus*, middle) is facilitated through advances in genomics in the closely related Atlantic salmon (*Salmo salar*). The former can be referred to as satellite species of the latter as a transfer of genomic information is very promising. This in turn supports studies on their own biology in manifold ways. Less well-known fishes like loaches (*Cobitis* spp., bottom) have been difficult to study because of their hybrid origin and polyploid genomes. Long read sequencing and continuing development of approaches to assemble genomes now enable better access to such fascinating systems in fundamental research. (All pictures A. Hartl)

genomes that are made available through databases such as Ensembl [45] is currently rapidly growing due to the development of sophisticated assembly strategies and the rise of long read sequencing that spans genomic fragments that are difficult to assemble.

3   Genomics in Studies on the Biology of Fishes

Fishes are studied genetically to infer basic biological and evolutionary processes. The details of such studies have often relied on population genetic approaches in which descriptors of population structure were inferred. This included studies on the distribution of lineages across their ranges [46] and the outcome of secondary contact when such lineages hybridize [47]. Studies often aimed to infer population structure from an evolutionary perspective [48] but also with the goal to improve stock management [49]. While such studies have been extremely successful in identifying units of biodiversity and evolutionary patterns, they have often relied on information from anonymous, neutrally evolving genetic markers and they applied the neutral evolutionary theory. However, there is a deep interest in identifying the loci that drive evolutionary processes and that determine the phenotypic properties of organisms. The latter aspects require that additional information be integrated with population genetic patterns observed at a given marker. First, anonymous markers need to be assigned to genome positions to infer whether they are associated with genetic elements and their functions. Moreover, a dense sampling of markers ordered along the genome permits powerful statistical analyses as patterns of individual markers can be combined in sliding window analyses that test for shared signals. Such data is useful to detect genetic signatures of selection that are expected when adaptive evolutionary change takes place. Moreover, genetic elements themselves have to be cataloged and functionally studied to understand their molecular functions and the higher-level phenotypes they affect. While such analyses have been conducted in the field of developmental biology and quantitative genetics the data that becomes available now permits the inference of populations genetic differentiation in conjunction with likely causative genetic variants in genome-wide studies of the association of phenotypes with genetic variation. A hallmark example in fishes is given by [50] who have studied genome-wide genetic variation in sticklebacks to identify genetic loci involved in the phenotypic and ecological diversity. Other intensively studied groups of fishes have been subject to intense genomic exploration as well as shedding light on study systems that have intrigued biologists for a long time [51]. Population genetics in fishes will doubtlessly move forward toward integrative analyses [52] that rely heavily on the interpretation of evolutionary or ecological patterns in nature in the light of detailed genomic and functional genetic information.

Fish genomics is driven by the progress that has been made in intensively studied species such as the zebrafish (*Danio rerio*) or medaka (*Oryzias latipes*). These species have long been favorite ornamental fishes and make excellent laboratory animals because of their short generation times and the ease of their care and breeding. Fundamental biological processes were uncovered in these fishes and could then be explored, generalized and extended to other species. The transfer of knowledge from model organisms has already decidedly influenced areas of applied research such as medical studies, ecotoxicology, environmental sensing systems, and sustainable aquaculture strategies [53]. The integration of knowledge from model organisms with the so-called satellite species that are related closely enough to facilitate the transfer of genomic information paves the way to study ecologically relevant taxa, and more broadly the evolution and diversity of all known fishes and other species [54]. The applicability of this approach will tremendously increase as the progress in next-generation sequencing fully includes nonmodel organisms and more and more fish genomes and biological knowledge about different species accumulates. The wave of next-generation sequencing has turned fishes into a highly informative group, a "new model army," in which long-standing questions on the evolution of their biodiversity can be addressed [23] (Fig. 2).

The zebrafish and the medaka were the first fish model systems that were intensively studied genetically and for which methods to conduct mutagenesis screens were established [55, 56]. Studies on these fishes were at the forefront of developmental biology, biomedical, and genomic studies. Moreover, they complemented each other in that they differ notably in their phylogenetic position and properties, which provided insights into the possibilities and power of comparative analyses between these models [57]. Since then, a large community of researchers has exploited the zebrafish system to pioneer many fields of fish genetics. Sophisticated methods that have been first developed in model systems [58] are now becoming applicable in other species. The wealth of knowledge that is available for the zebrafish has been collected in the form of a dedicated book [59] and is accessible online in the ZFIN database [60]. Likewise, there are comprehensive books [61, 62] and a website [63] for the medaka. Beyond the fully sequenced and annotated genomes, these resources (1) provide information on the laboratory use (protocols) of zebrafish and medaka, (2) summarize information on known mutants and transgenic strains as well as wild type strains, (3) provide access to genetic, genomic and developmental information, (4) aid in the transfer of information between species and databases, and (5) facilitate the use of fishes as a model for human focused medical research. Finally, they (6) serve as a general platform for researchers, and a collection of husbandry and laboratory protocols are provided. Only a few other model organisms parallel this rich set of genomic resources and genomic information is increasingly added to and curated in public open access databases [45].

A growing number of additional fish genomes have been fully sequenced and extend the genomic exploration of fishes. Each was initially planned with a different biological emphasis and each has distinct advantages related to the biology of the species or its use from a human perspective. Disadvantages relative to zebrafish and medaka vary and may be relatively longer generation times, more demanding husbandry and difficulties in breeding them. Pufferfish genomes such as the ones from *Takifugu rubipes* and *Tetraodon nigroviridis* were initially targeted because of their compact genome sizes [64, 65]. These studies revealed that Pufferfish, nonetheless, carries a number of genes that is comparable to the human genome and targeted a species that are of fundamental biology questions and economic importance. Other species were targeted with a more focused view on phenomena that are of medical importance. The genome of the African Killifish *Nothobranchius furzeri* was sequenced to gain access to a species that served as a model to study senescence [66]. This species is extremely short lived and can thus serve to genetically map traits related to ageing in relatively short experimental timescales. A species that has received interest from the field of developmental biology is the Mexican cave tetra *Astyanax mexicanus* [67] that lives in subterranean caves and is distinguished from its surface dwelling relatives by a number of reduced traits such as the loss of vision but also by the gain of other sensory abilities. Its genome has enabled mapping of the genetic basis of these traits and added great detail to our understanding of the genetic changes that cause phenotypic evolution. The platyfish (*Xiphophorus maculatus*) was sequenced as a model to study the development of skin melanoma, and to study the genetics of live-bearing and sex determination [68]. A growing number of fishes is targeted for their economic importance and with the goal to study genomic resources that may be relevant to improve aquaculture. However, studies on Atlantic salmon [27], turbot [69], the European sea bass [70], and tilapia [71] illustrate that their genomes also gave rich insights into questions related to environmental adaptation, development, and genome evolution. Other fish species have been specifically targeted with the aim to develop model systems to study evolutionary processes that have given rise to the diversity of fishes. As a prime example, the stickleback has been dubbed a supermodel that is amenable for the full integration of behavioral, developmental, ecological, and genetic data [72]. Its genome has been sequenced and served in hallmark studies in the field of ecological genomics that illustrated the power of genomics to unravel evolutionary processes and to link genotype with phenotype information [50, 73]. These studies, among many others, have carried a species that has received long-standing interest of researchers as a model in behavioral studies into the genomic era. Likewise, cichlids have been favorite study systems to understand the explosive diversification that must have occurred in the east African lakes of the rift valley where hundredth of species have evolved within each of the separated lakes. These systems have received interest to study the process of speciation, functional morphology of the feeding apparatus, and color polymorphism and its role in mate choice. Progress in these fields as well as in aspects of the molecular evolution of this group of fishes has been greatly facilitated by several cichlid genomes [51].

Clearly, the previous trend to sequence genomes only for particularly well-studied species for which a wealth of information is available will not be the only path for future research. Numerous genome sequencing projects that are not mentioned here and their number is growing exponentially. The resulting sequences and the tools to assemble and access the information make genome sequencing project feasible for more and more species that are interesting for a smaller community or single researchers. This trend clearly moves boundaries between model species and non-model species and promises progress in many exceptional species that remain to be studied.

References

1. Helfman G, Collette BB, Facey DE, Bowen BW (2009) The diversity of fishes: biology, evolution, and ecology, 2nd edn. Wiley-Blackwell, Hoboken. 736 pages, ISBN: 978-1-405-12494-2

2. Froese R, Pauly D (2018) FishBase (version Jun 2017). In: Roskov Y, Abucay L, Orrell T, Nicolson D, Bailly N, Kirk PM, Bourgoin T, DeWalt RE, Decock W, De Wever A, van Nieukerken E, Zarucchi J, Penev L (eds) Species 2000 & ITIS catalogue of life. Species 2000, Naturalis, Leiden. ISSN 2405-8858. , 30th January 2018. Digital resource at www.catalogueoflife.org/col.

3. Betancur R, Broughton RE, Wiley EO et al (2013) The tree of life and a new classification of bony fishes. PLOS Curr Tree Life. [last modified: 2013 Apr 23]

4. Valenzano DR, Benayoun BA, Singh PP et al (2015) The African turquoise killifish genome provides insights into evolution and genetic architecture of lifespan. Cell 163 (6):1539–1554

5. Nielsen J, Hedeholm RB, Heinemeier J, Bushnell PG, Christiansen JS, Olsen J, Ramsey CB, Brill RW, Simon M, Steffensen KF, Steffensen JF (2016) Eye lens radiocarbon reveals centuries of longevity in the Greenland shark (Somniosus microcephalus). Science 353 (6300):702–704

6. Kottelat M, Britz R, Tan HH, Witte KE (2005) Paedocypris, a new genus of Southeast Asian cyprinid fish with a remarkable sexual dimorphism, comprises the world's smallest vertebrate. Proc R Soc B 273:895–899

7. Breder DM, Rosen DE (1966) Modes of reproduction in fishes, American Museum of Natural History. Natural History Press, Garden City, NY. 941 pp

8. Komiyama T, Kobayashi H, Tateno Y, Inoko H, Gojobori T, Ikeo K (2009) An evolutionary origin and selection process of goldfish. Gene 430(1–2):5–11

9. FAO (2016) The State of World Fisheries and Aquaculture 2016. Contributing to food security and nutrition for all. FAO, Rome. 200 pp. isbn:978-92-5-109185-2

10. Venkatesh B (2003) Evolution and diversity of fish genomes. Curr Opin Genet Dev 13:588–592

11. Volff JN (2005) Genome evolution and biodiversity in teleost fish. Heredity 94:280–294

12. Kapusta A, Suh A, Feschotte C (2017) Dynamics of genome size evolution in birds and mammals. Proc Natl Acad Sci U S A 114: E1460–E1469

13. Gregory, T.R. (2005). Animal genome size database. http://www.genomesize.com

14. Shao F, Wang J, Xu H, Peng Z (2018) Fish-TEDB: a collective database of transposable elements identified in the complete genomes of fish. Database 2018

15. Chalopin D, Naville M, Plard F, Galiana D, Volff JN (2015) Comparative analysis of transposable elements highlights mobilome diversity and evolution in vertebrates. Genome Biol Evol 7:567–580

16. Gao B et al (2016) The contribution of transposable elements to size variations between four teleost genomes. Mob DNA 7:4

17. Cioffi MB, Bertollo LAC (2012) Chromosomal distribution and evolution of repetitive DNAs in fish. In: Garrido-Ramos MA (ed) Repetitive DNA, vol 7. Genome Dyn, Basel, Karger, pp 197–221

18. Schartl M et al (2018) Sox5 is involved in germ-cell regulation and sex determination in medaka following co-option of nested transposable elements. BMC Biol 16:16

19. Sotero-Caio CG, Platt RN, Suh A, Ray DA (2017) Evolution and diversity of transposable elements in vertebrate genomes. Genome Biol Evol 9:161–177

20. Taylor JS, Van de Peer Y, Braasch I, Meyer A (2001) Comparative genomics provides evidence for an ancient genome duplication event in fish. Philos Trans R Soc Lond Ser B Biol Sci 356:1661–1679

21. Jaillon O et al (2004) Genome duplication in the teleost fish Tetraodon nigroviridis reveals the early vertebrate proto-karyotype. Nature 431:946–957

22. Meyer A, Van de Peer Y (2005) From 2R to 3R: evidence for a fish-specific genome duplication (FSGD). BioEssays 27:937–945

23. Braasch I, Peterson SM, Desvignes T, McCluskey BM, Batzel P, Postlethwait JH (2015) A new model army: emerging fish models to study the genomics of vertebrate Evo-Devo. J Exp Zool Mol Dev Evol 324:316–341

24. Leggatt RA, Iwama GK (2003) Occurrence of polyploidy in the fishes. Rev Fish Biol Fish 13:237–246

25. Ludwig A, Belfiore NM, Pitra C, Svirsky V, Jenneckens I (2001) Genome duplication events and functional reduction of ploidy levels in sturgeon (Acipenser, Huso and Scaphirhynchus). Genetics 158:1203–1215

26. Rajkov J, Zhaojun S, Berrebi P (2014) Evolution of polyploidy and functional diploidization in sturgeons: microsatellite analysis in 10 sturgeon species. J Hered 105:521–531. https://doi.org/10.1093/jhered/esu027

27. Lien S, Ben F, Koop BF et al (2016) The Atlantic salmon genome provides insights into rediploidization. Nature 533:200–205

28. Saitoh K, Chen W-J, Mayden RL (2010) Extensive hybridization and tetrapolyploidy in spined loach fish. Mol Phylogenet Evol 56:1001–1010

29. Vasil'ev VP, Lebedeva EB, Vasil'eva ED (2011) Evolutionary ecology of clonal-bisexual complexes in spined loaches from genus Cobitis (Pisces, Cobitidae). J Ichthyol 51:932–940

30. Bohlen J, Ritterbusch D (2000) Which factors affect sex ratio of spined loach (genus Cobitis) in lake Müggelsee? Environ Biol Fish 59:347–352

31. Goddard KA, Megwinoff O, Wessner LL, Giaimo F (1998) Confirmation of gynogenesis in Phoxinus eos-neogaeus (Pisces: Cyprinidae). J Hered 89:151–157

32. Collares-Pereira MJ, Coelho MM (2010) Reconfirming the hybrid origin and generic status of the Iberian cyprinid complex Squalius alburnoides. J Fish Biol 76(3):707–715. https://doi.org/10.1111/j.1095-8649.2009.02460.x

33. Cunha C et al (2011) The evolutionary history of the allopolyploid Squalius alburnoides (Cyprinidae) complex in the northern Iberian Peninsula. Heredity 106:100–112. https://doi.org/10.1038/hdy.2010.7

34. Hubbs CL, Hubbs LC (1932) Apparent parthenogenesis in nature, in a form of fish of hybrid origin. Science 76(1983):628–630

35. Schlupp I (2005) The evolutionary ecology of gynogenesis. Annu Rev Ecol Evol Syst 36:399–417

36. Lampert KP, Schartl M (2008) The origin and evolution of a unisexual hybrid: Poecilia formosa. Philos T R Soc B 363:2901–2909

37. Zhang J, Sun M, Zhou L, Li Z, Liu Z, Li XY, Liu XL, Wei Liu W, Gui JF (2015) Meiosis completion and various sperm responses lead to unisexual and sexual reproduction modes in one clone of polyploid Carassius gibelio. Sci Report 5:10898

38. Harrington RW (1961) Oviparous hermaphroditic fish with internal self-fertilization. Science 134:1749–1750

39. Tatarenkov A, Lima SMQ, Taylor DS, Avise JC (2009) Long-term retention of self-fertilization in a fish clade. PNAS 106(34):14456–14459

40. Near TJ, Eytan RI, Dornburg A, Kuhn KL, Moore JA, Davis MP, Wainwright PC, Friedman M, Smith WL (2012) Resolution of ray-finned fish phylogeny and timing of diversification. PNAS 109:13698–13703

41. Sarropoulou E, Fernandes JMO (2011) Comparative genomics in teleost species: knowledge transfer by linking the genomes of model and non-model fish species. Compar Biochem Physiol 6:92–102

42. Cheng J, Czypionka T, Nolte AW (2013) The genomics of incompatibility factors and sex determination in hybridizing species of Cottus (Pisces). Heredity 111:520–529

43. Rexroad CE, Palti Y, Gahr SA, Vallejo RL (2008) A second generation genetic map for rainbow trout (Oncorhynchus mykiss). BMC Genet 9:74

44. Cheng J, Sedlazeck F, Altmüller J, Nolte AW (2015) Ectodysplasin signalling genes and phenotypic evolution in sculpins (Cottus). Proc Royal Soc Ser B 282(1815):20150746

45. Zerbino DR, Achuthan P, Wasiu Akanni W et al (2018) Ensembl 2018. Nucleic Acids Res 46 (D1):D754–D761

46. Bernatchez L, Wilson CC (1998) Comparative phylogeography of nearctic and palearctic fishes. Mol Ecol 7:431–452

47. Nolte AW, Freyhof J, Tautz D (2006) When invaders meet locally adapted types: rapid moulding of hybrid zones between two species of sculpins (Cottus, pisces). Mol Ecol 15:1983–1993

48. Barluenga M, Stölting KN, Salzburger W, Muschick M, Axel Meyer A (2006) Sympatric speciation in Nicaraguan crater lake cichlid fish. Nature 439:719–723

49. Bradbury IR, Hamilton LC, Sheehan TF, Chaput G, Robertson MJ, Dempson JB, Reddin D, Morris V, King T, Louis Bernatchez L (2016) Genetic mixed-stock analysis disentangles spatial and temporal variation in composition of the West Greenland Atlantic Salmon fishery. ICES J Marine Sci 73 (9):2311–2321

50. Jones FC, Grabherr MG et al (2012) The genomic basis of adaptive evolution in threespine sticklebacks. Nature 484:55–61

51. Brawand D, Wagner CE et al (2014) The genomic substrate for adaptive radiation in African cichlid fish. Nature 513:375–381

52. Bernatchez L, Renaut S, Whiteley AR, Campbell D, Derome N, Jeukens J, Landry L, Lu G, Nolte AW, Østbye K, Rogers SM, St-Cyr J (2010) On the origins of species: insights from the ecological genomics of whitefish. Philos Trans R Soc Lond 365:1783–1800

53. Spaink HP, Jansen HJ, Dirks RP (2013) Advances in genomics of bony fish. Brief Funct Genomics 13:144–156

54. Tagu D, Colbourne JK, Nègre N (2014) Genomic data integration for ecological and evolutionary traits in non-model organisms. BMC Genomics 15(1):490

55. Shima A, Shimada A (1991) Development of a possible nonmammalian test system for radiation-induced germ-cell mutagenesis using a fish, the Japanese medaka (Oryzias latipes). Proc Natl Acad Sci U S A 88:2545–2549

56. Mullins MC, Hammerschmidt M, Haffter P, Nüsslein-Volhard C (1994) Large-scale mutagenesis in the zebrafish: in search of genes controlling development in a vertebrate. Curr Biol 4(3):189–202

57. Wittbrodt J, Shima A, Schartl M (2002) Medaka—a model organism from the far east. Nat Rev Genet 3(1):53–64

58. Kirchmaier S, Naruse K, Wittbrodt J, Felix Loosli F (2015) The genomic and genetic toolbox of the teleost medaka (Oryzias latipes). Genetics 199(4):905–918

59. Westerfield M (2007) The zebrafish book, 5th ed; a guide for the laboratory use of zebrafish (Danio rerio). University of Oregon Press, Eugene

60. Howe DG, Bradford YM, Conlin T, Eagle AE, Fashena D, Frazer K, Knight J, Mani P, Martin R, Moxon SA, Paddock H, Pich C, Ramachandran S, Ruef BJ, Ruzicka L, Schaper K, Shao X, Singer A, Sprunger B, Van Slyke CE, Westerfield M (2013) ZFIN, the Zebrafish Model Organism Database: increased support for mutants and transgenics. Nucleic Acids Res 41(Database issue): D854–D860

61. Kinoshita M, Murata K, Naruse K, Tanaka M (2012) Medaka: biology, management, and experimental protocols. Wiley, New York. 9780813808710

62. Naruse K, Tanaka M, Takeda H (eds) (2011) Medaka: a model for organogenesis, human disease, and evolution. Springer, New York. ISBN 978-4-431-92690-0

63. NBRP Medaka. https://shigen.nig.ac.jp/medaka/

64. Aparicio S, Chapman J, Stupka E et al (2002) Whole-genome shotgun assembly and analysis of the genome of Fugu rubripes. Science 297:1301–1310

65. Jaillon O, Aury JM et al (2004) Genome duplication in the teleost fish Tetraodon nigroviridis reveals the early vertebrate proto-karyotype. Nature 431:946–957

66. Valenzano DR, Benayoun BA et al (2015) The African turquoise killifish genome provides insights into evolution and genetic architecture of lifespan. Cell 163:1539–1554

67. McGaugh SE, Gross JB et al (2014) The cavefish genome reveals candidate genes for eye loss. Nat Commun 5:5307

68. Schartl M, Walter RB, Shen Y et al (2013) The genome of the platyfish, Xiphophorus maculatus, provides insights into evolutionary

adaptation and several complex traits. Nat Genet 45:567–572

69. Figueras A, Robledo D et al (2016) Whole genome sequencing of turbot (Scophthalmus maximus; Pleuronectiformes): a fish adapted to demersal life. DNA Res 23:181–192

70. Tine M, Kuhl H et al (2014) European sea bass genome and its variation provide insights into adaptation to euryhalinity and speciation. Nat Commun 5:5770

71. Conte MA, Gammerdinger WJ et al (2017) A high quality assembly of the Nile Tilapia (Oreochromis niloticus) genome reveals the structure of two sex determination regions. BMC Genomics 18:341

72. Gibson G (2005) The synthesis and evolution of a super-model. Science 307:1890–1891

73. Colosimo PF, Hosemann KE et al (2005) Widespread parallel evolution in sticklebacks by repeated fixation of Ectodysplasin alleles. Science 307(5717):1928–1933

Chapter 17

Avian Population Genomics Taking Off: Latest Findings and Future Prospects

Kira E. Delmore and Miriam Liedvogel

Abstract

Birds are one of the most recognizable and diverse groups of organisms on earth. This group has played an important role in many fields, including the development of methods in behavioral ecology and evolutionary theory. The use of population genomics took off following the advent of high-throughput sequencing in various taxa. Several features of avian genomes make them particularly amenable for work in this field, including their nucleated red blood cells permitting easy DNA extraction and small, compact genomes. We review the latest findings in the population genomics of birds here, emphasizing questions related to behavior, ecology, evolution, and conservation. Additionally, we include insights in trait mapping and the ability to obtain accurate estimates of important summary statistics for conservation (e.g., genetic diversity and inbreeding). We highlight roadblocks that will need to be overcome in order to advance work on the population genomics of birds and prospects for future work. Roadblocks include the assembly of more contiguous reference genomes using long-reads and optical mapping. Prospects include the integration of population genomics with additional fields (e.g., landscape genetics, phylogeography, and genomic mapping) along with studies beyond genetic variants (e.g., epigenetics).

Key words Population genomics, Bird, Avian, Demography, Conservation, Speciation, Ecology

---

1 Introduction

Birds have played a central role in our understanding of many research fields. Notable examples include (1) the development of methods essential for behavior and ecology by Margaret Nice [1] using populations of song sparrows (*Melospiza melodia*), and (2) the definition of species as groups of populations reproductively isolated from one another by Ernst Mayr [2], inspired by the geographic distribution of birds and galvanizing the field of evolution. Current declines in natural populations of birds worldwide may cause work by researchers like Nice and Mayr to be overshadowed by efforts in conservation. Concern about the loss of birds to the millinery trade triggered Harriet Hemenway and Mina Hall to establish the National Audubon Society in 1886. This

Julien Y. Dutheil (ed.), *Statistical Population Genomics*, Methods in Molecular Biology, vol. 2090,
https://doi.org/10.1007/978-1-0716-0199-0_17, © The Author(s) 2020

society is one of the first nonprofit environmental organizations and generates datasets essential for population monitoring and forecasting to this day (e.g., annual Christmas and Great Backyard Bird Counts to census birds worldwide and eBird, an online database of bird observations).

The influential role of birds across varied research fields continued with the development of population genetics. This field emerged in the 1980s following the advent of sequencing technologies to quantify marker based genetic variation, including sequence variation in mitochondrial DNA (mtDNA) and length polymorphisms in microsatellites. Tools from population genetics can be used to evaluate the role mutation, genetic drift, selection and gene flow play in generating variation within and between populations [3, 4], with relevance to behavior, ecology, evolution, and conservation. Avian blood has nucleated red blood cells, making it ideally suited for DNA extraction and subsequent population genetic analyses, early examples include the use of mtDNA to identify taxonomic units for conservation (e.g., dusky seaside sparrows [*Ammodramus maritimus nigrescens*], [5, 6]).

The last 10 years has seen a change in both the scale and depth of genetic analyses, with the transition from the use of one or a few genetic markers to tens of thousands of markers genome-wide marking the development of population genomics. This transition was stimulated by de novo assembly of reference genomes along with the advent of *high-throughput sequencing* (HTS). HTS is discussed in detail elsewhere (e.g., [7, 8]), but briefly, is a set of platforms that sequence DNA from multiple genomic regions and individuals in parallel. HTS has increased the proportion of the genome that can be sampled and decreased the time and cost of sequencing, allowing its use on most organisms of interest. Again, birds are well-suited for this extension. Not only can good amounts of DNA be generated easily, but they also have small (mean ~1.45 billion base pairs; [9]), compact (e.g., fewer transposable elements [TEs] and repetitive regions; [10]) genomes, allowing for relatively easy genome assembly and mapping at high coverages. The first avian reference genomes were the chicken (*Gallus gallus*, [11]) and zebra finch (*Taeniopygia guttata*, [12]), assembled using Sanger sequencing and followed by an explosion of bird genomes assembled using HTS [13].

We review the population genomics in birds here, emphasizing the role this group has played in both the development of this field and its application to questions in behavior, ecology, evolution, and conservation. Population genomics can be used to answer questions at both the genome and locus levels. Work at the genome level informs our understanding of population processes (e.g., demography and population structure), while work at the locus level helps identify genomic regions affected by mutation, drift, selection and/or gene flow [3]. We follow this division in this chapter, introducing some of the latest findings from birds, highlighting the benefits of applying population genomics tools to these questions (vs. traditional population genetic techniques with fewer markers), and finishing by outlining future prospects along this trajectory.

2   Latest Findings

2.1   Relevance of Genomic Insight for Evolution

Demography is the study of changes in effective population size (N_e) through time, gene flow and divergence. Information on these dynamics is essential for understanding the evolution of species, populations and traits and important for setting baselines beyond which evolutionary processes can be examined. This is especially true in the current literature, where genome scans are being used to identify loci associated with phenotypic traits and/or involved in adaptation and speciation (*see* below; e.g., [14–17]). Early demographic analyses were based on the coalescent (or divergence), assuming that pairwise sequence divergence is proportional to the time of the coalescent, and relying on nonrecombining loci (e.g., mtDNA, Fig. 1a left, [18]). These analyses were expanded to include nuclear loci and permit estimates of changes in population size and gene flow, but remained limited to a small number of demographic scenarios and markers and were computationally expensive [19]. The availability of genome-wide data and new tools for their analysis has revolutionized this field and its progression is demonstrated well by a series of studies on collared and pied flycatchers (*Ficedula albicollis* and *F. hypoleuca*).

Early demographic work with collared and pied flycatchers suggested these species diverged during the Pleistocene, expanded their ranges following the last glacial maximum, and came into secondary contact in Central Europe where introgression from pied into collared flycatchers is greater [20, 21]. Nadachowska-Brzyska et al. [22] used *whole genome resequencing* (WGS) data to expand on these findings, comparing 15 demographic models using an Approximate Bayesian Computation (ABC) approach. A model with recent divergence time (230,000–240,000 years ago [ya]), unidirectional gene flow (0.16–0.36 pied individuals per generation) and ancient reductions in population size (e.g., 600,000–80,000 in collared) provided the best fit to the data (Fig. 1 right). This work represents one of the first times an ABC approach with HTS was applied to study demography in a non-model system, providing a level of detail unattainable with earlier methods [22]. Flycatchers exhibit strong reproductive isolation, with evidence for both pre- and postzygotic barriers to gene flow (e.g., nearly complete female sterility, [23]). Accordingly, these results suggest speciation in birds can occur quite rapidly.

Nadachowska-Brzyska et al. [24] validated findings from the ABC analysis using the Pairwise Sequentially Markovian Coalescent

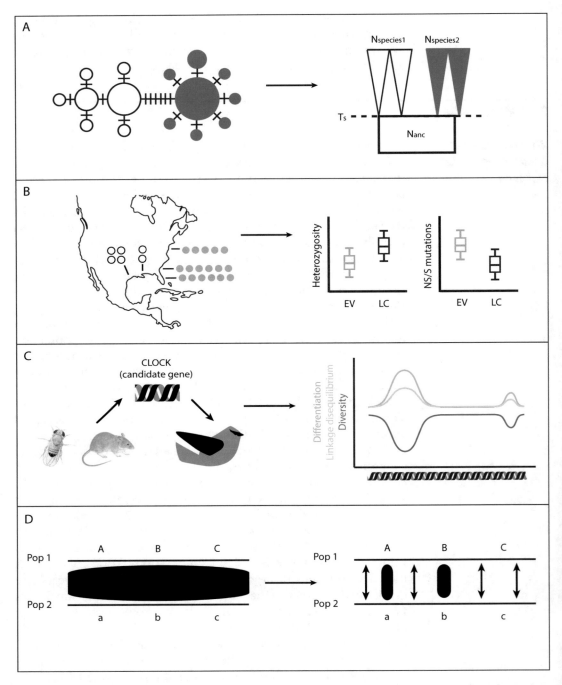

Fig. 1 Conceptualized figure showing examples of questions that were answered using genetic data and how they have been expanded with the transition from population genetics (left) to genomics (right). Panel **a** is relevant to *demography* and show a minimum spanning network where circles represent haplotypes, sizes of circles reflect abundance, and bars across branches represent single nucleotide changes that can be used to estimate divergence time (yellow), and more detailed analyses using genomic data where a demographic scenario with recent migration and population size changes derived from recent ABC analyses (N—effective

(PSMC, Li and Durbin [25]; similar to the model described in Chapter 7) and WGS data. PSMC estimated a slightly later divergence time between the species (386,000–888,000 ya) and provided additional information on changes in population size over time, with collared flycatchers undergoing an expansion 100,000–200,000 ya and subsequent decline, likely corresponding with glaciation period marine isotope stage (MIS) 6 (191,000–130,000 ya). Pied flycatchers did not exhibit the same decline, perhaps because they can tolerate glacial climates more effectively than collared flycatchers that occur at lower latitudes and altitudes. Instead, pied flycatchers appear to be increasing in population size, suggesting they are outcompeting collared flycatchers in the present day, which falls in line with behavioral observations in the system [23]. Results from this study on flycatchers support earlier work using data from 38 bird species [26], documenting similar variability in estimates of population size over time. Together these results provide important inferences about the population dynamics of temperate species that have experienced glacial cycles throughout their history and caution against assuming simple demographic history (e.g., constant population sizes, a single expansion, and/or similar trends for populations).

Of perhaps less evolutionary interest, but equal importance, Nadachowska-Brzyska et al. [24] conducted a thorough analysis of the impact of sequence coverage and missing data on the accuracy of demographic inference with PSMC. They conclude at least $18\times$ mean coverage is needed and no more than 25% missing data can be permitted. These kinds of considerations are important when implementing analyses with genomic data as biases resulting from poor filtering can have a significant impact on results. In the case of PSMC, low filtering cutoffs lead to homozygous sites being considered heterozygous, affecting the size of recombination blocks and estimates of N_e (both magnitude and the shape of curves). It is important to note that newer versions of PSMC exist and provide many benefits but are not easily applied to birds. For example,

Fig. 1 (continued) population size [anc—ancestral population], T_s—divergence time). Panel **b** is relevant to *conservation* and shows the geographic distribution of individuals differentiated at a small number of markers and potentially requiring separate management status, and more recent comparisons of heterozygosity and the proportion of nonsynonymous to synonymous mutations between an endangered/vulnerable species and one of least concern providing more detail when establishing management plans. Panel **c** is relevant to the *genetic basis of phenotypes* and shows a candidate gene identified in *Drosophila* and *Mus musculus* and surveyed in populations of birds as a potential regulator of migration and more recent uses of genomic data to identify selective sweeps related to a phenotype in an unbiased way (based on [30]). Panel **d** is relevant to the *genetics of speciation* and shows how speciation was originally conceived as complete loss of gene flow between population, and more recent ideas where speciation can still occur if regions of the genome continue to exchange genes, with black blocks representing areas where gene flow is no longer occurring and fixed alleles between population arise (based on [40])

multiple sequentially Markovian coalescent (MSMC, *see* Chapter 7) can provide information on more recent population dynamics but requires accurate *phasing*, which can be difficult to obtain for birds as reference panels of known haplotypes are often used and derive from trios (adult and both parents) or pedigreed populations which are also difficult to obtain for birds.

2.2 Relevance of Genomic Insight for Conservation

Demographic analyses provide us with information on population dynamics mostly in the past. Population genomics can also be used to understand dynamics in present day populations and are especially relevant for conservation (e.g., helping inform the management of threated species). Of prime importance is the application of these tools to study genetic factors that can compound reductions in population size already experienced by threatened species and transfer of findings from common species to those that are under threat. Here the availability of HTS is allowing researchers to obtain far more accurate and precise estimates not only of population structure and dynamics, but also loss of genetic diversity and inbreeding (Fig. 1b). Using a comparative framework and genome sequences spanning nearly the full phylogenetic spectrum of birds Li et al. [27] highlighted the potential of these data for conservation. We will discuss this work below but want to emphasize the importance of the application of population genomics to these questions for birds where 1375/10,000 species (13%) are threatened with extinction (IUCN https://www.iucn.org/). Habitat loss is among the main threats and the ramifications of extinction in birds will be far reaching, as they are essential for ecosystem functioning (e.g., as seed dispersers) and serve as important indicators of ecosystem health (e.g., tracking changes in habitat, water and climate).

Li et al. [27] is one of 27 papers that were released by the Avian Phylogenomics Project in December 2014 based on 48 genomes assembled using HTS. These authors classified each species as endangered/vulnerable (EV) or of no conservation concern and observed that EV species exhibited lower levels of heterozygosity and more nonsynonymous (and potentially deleterious) mutations than species of no conservation concern. Nonsynonymous mutations were associated with increased *linkage disequilibrium* (*see* also Chapter 1) across the genome. Combined, these findings suggest EV species may be at risk of *inbreeding depression*. Estimates of inbreeding can also be obtained using pedigrees in the form of inbreeding coefficients (the probability of a locus being identical by decent). Nevertheless, as noted already pedigrees are rare in wild bird populations so the application of HTS in the framework used by Li et al. [27] will be imperative for understanding risks to natural populations of birds [28].

Li et al. [27] took their work one step further, obtaining WGS data from eight individuals of the critically endangered crested ibis

(*Nipponia nippon*). These data were used to document changes in population size similar to work described by Nadachowska-Brzyska et al. [24] with flycatchers. However, of potentially greater importance for conservation, Li et al. [27] also used these data to develop a set of genetic loci to track individuals of the species. 166,000 degenerate STR (short tandem repeats) loci were identified. Among these loci, 23 were informative and will have many applications, including the estimation of sex and paternity along with the reconstruction of pedigrees to identify birds for breeding programs. Similar applications are being promoted by many authors as a way of implementing rapid biodiversity screening for risk assessment, including the quantification of genetic diversity and population structure in natural populations. This work will allow researchers to identify taxonomic units quickly, permitting the delineation of geographic areas for conservation and management.

2.3 Locus-Level Work to Examine the Genetic Basis of Phenotypic Traits

The application of population genomics to questions of demography and conservation are examples of genome-level analyses. Population genomics can also be used at the locus-level and include bottom-up approaches to examining the genetic basis of phenotypic traits, including morphological and behavioral traits. These analyses involve the identification of populations that differ in a trait of interest with a strong genetic component and the estimation of summary statistics along the genomes of these populations to detect *selective sweeps* (*see* also Chapter 1). Selective sweeps can derive from positive or divergent selection and evidence for these events include reductions in nucleotide diversity and increased linkage disequilibrium within populations. Elevated differentiation between populations also provides evidence for selective sweeps [29, 30]. The number of studies examining the genetic basis of phenotypic traits in birds is increasing and one pattern that is emerging is the importance of inversions for the control of phenotypic traits. Inversions are a form of rearrangement where portions of a chromosome are flipped, disrupting chromosome pairing during meiosis and preventing recombination from occurring. These regions should be inherited largely as single units, explaining how loci involved in the expression of phenotypic traits can work in concert.

One well-known example of an inversion associated with a phenotypic trait in birds comes from the ruff (*Philomachus pugnax*) where an inversion controls the expressive of different reproductive morphs. Additional examples are beginning to accumulate and include the willow warbler (*Phylloscopus trochilus*) and white-crowned sparrow (*Zonotrichia leucophrys gambelii*). Two subspecies of willow warblers form a ring around the Baltic Sea and differ in migratory orientation, forming a migratory divide in central Scandinavia. Using genome-wide SNP data, Lundberg et al. [31] identified three regions of the genome that exhibit extremely high differentiation compared to the rest of the genome and form distinct haplotype clusters. These clusters suggest recombination is rare in these regions and could be explained by inversion polymorphisms. Haplotypes at one region correlated with environmental features (altitude and latitude) while the other two correlated with migratory orientation. Tuttle et al. [32] used WGS to identify a series of potential inversions on avian chromosome 2 between two color morphs of white-crowned sparrows that also differ in reproductive behavior (promiscuous vs. monogamous). This region spans ~100 Mb and includes 1100 genes. F_{ST} between morphs is elevated in this region and linkage disequilibrium is high. One morph is homozygous for alleles at this inversion and the other heterozygous. These authors used a phylogenetic analysis to show the inversion evolved before sparrows diverged from their most closely related relative.

There are several benefits of using the bottom-up approach of population genomics to study the genetic basis of phenotypic traits. Early work on this topic was limited to a set of candidate genes that were often identified in model organisms that were distantly related from the focal species (e.g., [33]). Work with HTS allows researchers to study all genes in the genome, permitting unbiased assays of genomic variation and allowing for the de novo identification of candidate genes and new biological processes underlying traits (Fig. 1c). This genome-wide perspective also provides a broader understanding of how phenotypic traits are controlled, including the number, size and distribution of loci that underlie these traits. The bottom-up approach also has considerably more power than other methods. For example, genome-wide association studies (GWAS) can be used, but often require data from hundreds of individuals as they are conducted in single populations that vary in a trait of interest and have low levels of linkage disequilibrium. Bottom-up approaches can use of as few as 10 individuals/population which can be important for nonmodel organisms like birds where large numbers of individuals may be hard to sample. Nevertheless, there are still some problems associated with the bottom-up approach, including the fact that processes other than positive or divergent selection can generate signals similar to selective sweeps (e.g., *background selection, see* also Chapter 1). We discuss these problems and potential solutions below.

2.4 Locus-Level Work to Understand the Genetics of Adaptation and Speciation

Similar to work focused on identifying the genetic basis of phenotypic traits, the estimation of summary statistics along the genome can be used to study the processes of adaptation and speciation more broadly. In this case, studies normally compare closely related populations from the same or related species and focus on estimates of genomic differentiation like F_{ST}. One of the chief findings from this work is that differentiation can be highly variable across the genome, with areas of elevated F_{ST} interspersed with areas of reduced F_{ST} (e.g., [34–39]). An important inference drawn from these findings is that speciation can proceed through a few focal changes and does not require divergence across the entire genome (Fig. 1d, [40]). While this conclusion does not seem to be controversial, the processes that generate variable patterns of differentiation have received considerable interest and include two main models, divergence-with-gene-flow and selection-in-allopatry.

The divergence-with-gene-flow model posits that divergent selection at loci involved in speciation must be protecting some regions of the genome from gene flow, elevating an otherwise homogenized (or low) landscape of differentiation [41, 42]. This model received considerable enthusiasm when it was first developed as it suggests researchers can identify loci involved with speciation relatively easily, by looking for areas of elevated differentiation between closely related populations. Recent work has encouraged caution with this approach and work with birds has been at the forefront of this wave, promoting a second model to explain variation in F_{ST}, selection-in-allopatry. The selection-in-allopatry model posits that variation in the strength of selection can explain variation in differentiation on its own [43]. This model derives from the observation that F_{ST} is a relative measure of differentiation that includes a term for within population variation. As a result, it can be elevated by reductions in within population variation alone. These reductions can derive from any form of linked selection, including *genetic hitchhiking* and background selection that is unrelated to speciation ([43, 44]; also increases in neutral variance generated by population structure [45]) and mean that gene flow is not necessarily needed to explain variation in genomic differentiation.

Contrasts between windowed-estimates of F_{ST} and d_{XY} have been used to support the selection-in-allopatry model. d_{XY} is an absolute measure of differentiation that does not include a term for within population variation. Accordingly, it should be unaffected by reductions in within population variation and show limited associations with estimates of F_{ST} across the genome. Work with birds supports this prediction. Burri et al. [46] estimated F_{ST} and d_{XY} between collared and pied flycatchers and noted that d_{XY} was not elevated where F_{ST} was elevated. In fact, d_{XY} seemed to show the opposite pattern of F_{ST}, being reduced where F_{ST} was elevated. Burri et al. [46] argued that recurrent linked selection in regions of reduced recombination in ancestral populations were responsible for this pattern. This form of selection would reduce d_{XY} to zero prior to population splitting. Similar findings have been documented between Swainson's thrushes [35], greenish warblers (*Phylloscopus trochiloides*, [36]), stonechats (*Saxicola rubicola*, [37]), and Darwin's finches (*Geospiza fortis*, [38]).

Burri et al. [46] added an additional dimension to the selection-in-allopatry model. These authors documented an association between F_{ST} and recombination rates in flycatchers, suggesting genomic features like reduced recombination that extend the effects of linked selection by preventing linked neutral sites from recombining off their shared background could also play a role in generating variation in F_{ST}. Delmore et al. [47] evaluated this idea further using a comparative analysis, estimating genomic differentiation (F_{ST} and d_{XY}) between eight population pairs of birds that span a broad taxonomic scale (sharing a common ancestor ~52 million ya). Features of the local genomic landscape are highly conserved across birds, including chromosome number, recombination rate and synteny [48–54]. Accordingly, Delmore et al. [47] predicted that if genomic features are generating variation in differentiation across genomes they should generate correlated patterns of differentiation across population pairs of birds. In support of this prediction, a significant proportion of variation in windowed-estimates of F_{ST} and d_{XY} could be explained by correlations across pairs (up to 3% for F_{ST} and 26% for d_{XY}). In addition, genomic regions showing high repeatability across pairs were correlated with several genomic features (e.g., reduced recombination rates [approximated using GC content], elevated gene densities and chromosome size [higher on micro- vs. macrochromosomes]).

As a final note on the genetics of adaptation and speciation, support for the divergence-with-gene-flow model versus the selection-in-allopatry model will likely depend on the geographic context in which speciation occurs. Much of the work on speciation genomics in birds focuses on species in the temperate region that have experienced periods of allopatry and sympatry [55]. Accordingly, a model including allopatric periods (selection-in-allopatry) will likely be more relevant. In addition, there are variants to the selection-in-allopatry model. For example, Delmore et al. [35] and Irwin et al. [36] describe a scenario where selective sweeps upon secondary contact could also reduce d_{XY} in regions of elevated F_{ST}, with globally adaptive alleles evolving during allopatric periods and sweeping across both populations when they come into secondary contact. These alternatives are not mutually exclusive.

3   Roadblock: Genome Assemblies, Novel Genes and Structural Variants

Much of the work in population genomics makes use of reference genomes to place (or map) resequencing reads from individuals or populations. Reference genomes for birds are of variable quality [56]. As noted earlier, the first references were assembled using Sanger sequencing. They represent the most complete reference genomes for birds, but are not perfect. For example, annotations are still lacking for some of the microchromosomes in these references [57] and most chromosomes still include random sequences that cannot be placed. Each reference also includes unassigned scaffolds with unknown chromosome location. More recent genomes assembled using HTS often integrate data from short insert libraries and mate pairs. Data from small insert libraries are used to construct *contigs* and contigs are combined into *scaffolds* using mate pair libraries. Mate pair libraries help bridge complex regions of the genome that cannot be assembled (e.g., highly repetitive or heterozygous regions) but resulting genomes remain quite incontinguous, consisting of thousands of scaffolds for genome that consist of only 33 chromosomes (on average). In addition, these scaffolds are often assigned to chromosomes using synteny with the chicken or zebra finch genome. While karyotype and synteny are conserved across birds, intrachromosomal rearrangements are quite common [58] and it is possible that regions of the genome containing genes relevant to your focal species will not be present in these highly domesticated species. It is also highly likely that structural variations (e.g., inversions and duplications) controlling phenotypic traits and involved in adaptation and speciation are missing or misassembled.

One of the next steps in the population genomics of birds will be to improve these genome assemblies. Linkage maps would help join scaffolds but are hard to generate in birds as for several crosses cannot easily be made in the lab and a limited number of pedigreed populations in the wild exist. Nevertheless, alternative methods to join scaffolds are being developed and include both long-read technologies (e.g., Nanopore and 10× genomics) and optical mapping with BioNano technology (Fig. 2a). Optical maps are generated by shearing DNA into large molecules (>250 kb), linearizing them in nanochannels, and barcoding them with restriction enzymes. These maps are visualized with fluorescence microscopy and generated for each fragment before being combined into a consensus map. Nick sites from restriction enzymes are used to order and orient HTS scaffolds generated using traditional techniques and estimate gap size between them, ultimately producing hybrid assemblies with an increased N50 (defined as the minimum threshold of sequence length above which all scaffolds cover at least 50% of the total assembly size) and less, longer scaffolds for each genome. One example of a hybrid genome assembly generated using optical mapping comes from the ostrich (*Struthio camelus*). The original Illumina-based assembly had an N50 of 3.59 Mb and 414 scaffolds. Using optical mapping the N50 was increased to 17.71 and the number of scaffolds was reduced to 75 [59].

The desired quality of reference genomes will depend on the objectives of each study. If for example, researchers are interested in specific genes associated with a trait of interest, they may be missing from annotation. If structural variants are of interest (and results from birds thus far suggest they may be), more contiguous genomes will be needed to identify them. On the other hand, many studies, especially those interested in genome-level processes like demography and population structure, may not need a high quality

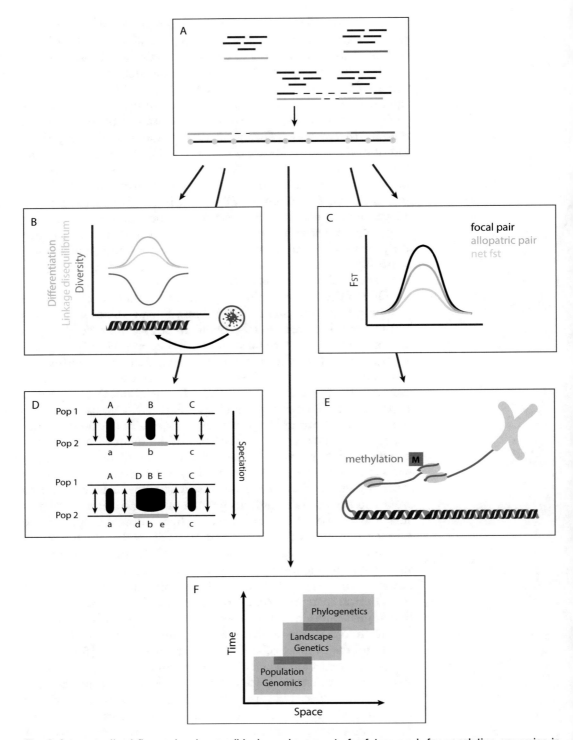

Fig. 2 Conceptualized figure showing roadblocks and prospects for future work for population genomics in birds. Panel **a** shows a roadblock that will need to be overcome (incomplete *genome assemblies*), using optical mapping (yellow) to extend traditional genomes based on contigs (blue and pink) and scaffolds (orange)

reference genome as specific loci or variants are not of interest. The same considerations apply to choosing sequence technologies for resequencing individuals or populations. If locus-level questions are of interest, whole genome resequencing data will likely be needed as linkage disequilibrium can drop off quite quickly in birds [35, 60]. If specific loci are not important (e.g., for analyses of population structure or preliminary work characterizing genomic regions), restriction site associated DNA sequencing (RAD) or Genotyping-by-sequencing (GBS), where DNA is cut with restriction enzymes and sequencing limited to those cut sites may be sufficient [61]. Target capture approaches can also be used if specific loci are not important, including enrichment of ultraconserved elements and their flanking regions [62].

4   Prospects

4.1 Continued Application of Population Genomics for Conservation

The use of HTS data can provide more accurate estimates of genetic diversity in threatened populations and allow for the development of reference platforms to quickly survey populations. Another major goal of conservation genetics is to identify population structure and delineate species boundaries. A few studies have begun doing this (e.g., Bell's vireo [*Vireo bellii*, [62]]; Mottled duck [*Anas fulvigula*, [63]]), but caution will be needed as the increased resolution provided by HTS may permit the identification of fine scale population structure that does not warrant separate taxonomic status. Work to identify adaptive variants important for population survival and maintenance will also be important (Fig. 2b). To the best of our knowledge this has not yet been done in birds, but there are examples from fisheries; for example, two ecotypes of kokanee salmon (*Oncorhynchus nerka*) that differ in their reproductive strategies (stream vs. shore-spawning) exist that are panmictic outside the breeding season and morphologically indistinguishable. These ecotypes are currently managed as separate

Fig. 2 (continued) from short read HTS assemblies. The remaining panels show prospects, including (**b**) the potential to identify *selective sweeps in populations of conservation* concern associated with traits of importance (here a region protecting the population from pathogen infection), (**c**) controlling for the effects *processes unrelated to adaptation and speciation* have on differentiation—estimating F_{ST} between a focal pair and allopatric populations that do not differ in the trait of interest, subtracting allopatric F_{ST} from focal F_{ST} to obtain net F_{ST}, (**d**) extension of Fig. 1d (right) with genomic features (shown in blue, e.g., reduced recombination) that extend the effects of linked selection causing alleles nearby and those under selection to fix at later *stages of speciation*, (**e**) complete reference genomes with high quality annotation allow for analyses of general *regulatory and epigenetic mechanisms*, such as characterization of open chromatin regions, histones, and mapping of chemical marks (e.g., methylation), and (**f**) *integrating* three fields together that differ in their temporal and geographic scales to gain a broader understanding of behavior, ecology, evolution, and conservation (based on [80])

stocks and marker based genetic characterization supports this separation [64, 65]. These authors used RAD sequencing to show these ecotypes are genetically differentiated from one another and identified 12 contigs using transcriptome sequencing that matched pathogens known to reduce the fitness of salmonids. These contigs were limited to stream-spawners, suggesting this ecotype has evolved a way to reduce pathogen load. Combined these results suggest different management strategies for both ecotypes. *See* Chapter 14 for more information on the genomics of fishes.

4.2 Control for Alternative Processes in Genome Scans and Expand Studies to Focus on the Process of Speciation

The estimation of genomic differentiation between closely related populations had led to the observation that levels of differentiation are highly variable across the genome. There is interest in using these differentiation patterns to identify loci involved in adaptation and speciation but processes other than positive or divergent selection can elevate differentiation and include background selection that is not related to adaptation or speciation. Accordingly, future work on this topic must control for regions affected by these processes and the framework outlined by Vijay et al. [66] using the crow species complex (genus *Corvus*) provides one approach. These authors obtained WGS data from populations at different stages of differentiation and from different color morphs (all-black or pied). One of the objectives of their study was to identify genomic regions associated with this color difference using populations in three hybrid zones between the morphs. They estimated F_{ST} between allopatric populations that did not differ in color and used these values as a null beyond which F_{ST} in hybrid zones had to exceed to provide evidence of positive selection relevant to color (Fig. 2c). Interestingly, these hybrid zones did not share the same peaks of differentiation, suggesting different genes were involved in each color transition. Instead, each contact zone had at least one gene in a differentiated region that was associated with the Wnt signalling component of the melanogenesis pathway suggesting the pathways but not necessarily the same genes to be important.

Scans of genomic differentiation across genomes can also be expanded to study the *process* of speciation, rather than a snapshot in what is a continuous process from initial population divergence to complete reproductive isolation (Fig. 2d). Speciation is also not a deterministic process; populations can go extinct of fuse back into a single unit before reaching speciation. Results from Delmore et al. [47] provide one example. Recall, these authors documented consistent patterns of differentiation across eight species pairs of birds implicating linked selection at genomic features in generating variation in genomic differentiation. Linked selection at genomic features is expected to have a greater effect at later stages of speciation when enough mutations have accumulated in populations to reflect processes occurring at these features [67]. Under this prediction, patterns of species pairs at later stages of speciation should show increasing levels of consistency. The species pairs included in Delmore et al. [47] span the full speciation continuum and an analysis comparing correlation coefficients between population pairs in windowed-estimates of F_{ST} support this prediction, with pairs later in the continuum (e.g., with more narrow hybrid zones) exhibiting more consistent patterns. This pattern was not observed using d_{XY} and likely relates to the fact that d_{XY} reflects processes that have accumulated over multiple speciation events. Accordingly, it will not matter when d_{XY} is estimated in the process of speciation, it will always reflect linked selection at genomic features.

4.3 Expand Beyond Studies of Genetic Variation Alone

Thus far we have focused mainly on genetic variants, SNPs or structural features like inversions. As noted, more direct evaluations of structural variants are needed but additional expansions would also profit research in birds. For example, it is possible that in addition to genetic sequence variation, expression dynamics and/or epigenetic mechanisms often play an important role in the expression of complex traits (Fig. 2e). These traits are typically controlled by many loci of small effect [68, 69] that may depend on the regulation of gene expression to produce the downstream phenotype ([70] but *see* [71–73]). Changes in gene expression can derive from chemical and molecular modifications, including methylation, phosphorylation, acetylation, accessibility of DNA in chromatin, and occupancy of regulatory sequences by transcription factors.

Epigenetic studies are beginning to accumulate in birds and include investigations of vocalization (zebra finch [74]); learning and cognition (great tit [*Parus major*, [75]]), and beak size and shape (Darwin's finches, [76, 77]). It is likely that these studies will increase in prevalence as the field moves from studying relatively simple traits (e.g., color) to more complex traits (e.g., behavior, such as migration). Along with this expansion, greater precision will be needed, and study designs require a carefully controlled experimental framework, making sure hindering noise (diet has a huge effect on expression profiles, profiles vary significantly across sexes, developmental stages) is kept to a minimum. Even within tissues, especially within brain, epigenetic makeup depends on the exact location of brain area. In birds this has been studied for bird song: behaviorally regulated gene expression profiles vary depending on time and between anatomical structures. It was hypothesized that drivers of this variability are signalling cascades modulate by transcription factors, cis and trans regulatory regions and epigenetic chromatin states [74]. This study focused on identifying transcriptionally active chromatin regions for song nuclei involved in their focal trait (singing). Their results suggest the presence of epigenetic modifications that prime gene regulation differ between brain regions, so that specific target regions are in a chromatin state that allows to immediately modulate transcription of behavior specific genes once the behavior kicks in (upon neuronal firing).

4.4 Integrate Population Genomics with Additional Fields

We will finish our review by highlighting the importance of integrating population genetics with additional fields to answer broader questions relevant to ecology, behavior, conservation, and evolution. Specifically, the fields of landscape genetics and phylogeography were originally developed as bridges to population genetics, to study the interaction between landscape features and micro- or macroevolutionary processes, respectively. Both involve geographic sampling beyond what is done for population genetics. The advent of HTS has blurred the boundaries between these three fields, with principles from landscape genetics and phylogeography being used to (1) identify loci under selection, (2) identify correlations between genomic data and the environment, and (3) reconstruct their history of divergence within species [60, 80]. One example comes from [78] who focused on populations of white-breasted nuthatches (*Sitta carolinensis*) that occupy the sky islands of Arizona. RAD sequencing showed that genetic differentiation between these islands was mediated largely by ecological distance rather than geographic distance, identifying eight loci associated with ecological distance (e.g., strong associations with minimum precipitation of driest month and maximum temperature of the hottest month). This integration provides a more complete picture of adaptation and speciation (Fig. 2f) and could be especially important for conservation, allowing researchers to understand how organisms relate to their environment and will handle future changes to their landscape.

Similar integrations with genetic mapping have the potential of being quite powerful as well, using the top-down approach of genetic mapping to identify loci associated with a trait of interest and the bottom-up approach of population genomics to examine the selective context underlying these traits. Work by Delmore et al. [79] demonstrates the utility of this integration. These researchers focused on a hybrid zone between two subspecies groups of Swainson's thrushes (*Catharus ustulatus*) in western North America that differ in migratory orientation. Hybrids between these groups take intermediate and inferior routes, likely helping reduce gene flow between subspecies (i.e., maintain subspecies boundaries). Delmore et al. [79] focused on hybrids, tracking these birds on migration with light-level geolocators and genotyping them using GBS. *Admixture mapping* using these data identified a region on chromosome 4 associated with migratory orientation. This region includes 60 genes, many involved in cell signalling, the nervous system and the circadian clock and had been identified in smaller-scale studies of migration in other animal groups supporting the idea there may be a common genetic basis to migration in animals. Delmore et al. [79] took traditional genetic mapping one step further by obtaining resequencing data form pure populations and showing this region is strongly differentiated between the subspecies, suggesting it is under divergent selection, supporting behavioral observations of inferior hybrid behavior showing differences in migration could be helping maintain subspecies boundaries.

5   Conclusion

We look forward to both participating and watching how the field of population genomics evolves for birds in the coming years. As sketched in this chapter, birds have played an important role in the development and application of population genomics, with work on demography and genetic variation at the genome level informing evolution and conservation and work at the locus level providing information on the genetic basis of phenotypic traits, adaptation and speciation. The improvement of reference genomes, use of long-read mapping for studying structural variation and expansion beyond genetic variation to epigenetics will take us a long way to expanding population genomics in birds. Integrations with other fields and beyond landscape genetics and phylogenetics (e.g., inclusion of functional genomics and genome editing) are also on the horizon for some species and will undoubtedly help to unravel many mysteries concerning the behavior, ecology, evolution, and conservation of birds.

Glossary

Admixture mapping: A genetic mapping approach that makes use of recombination and backcrossing in natural hybrid zones instead of producing crosses (Quantitative trait locus [QTL] mapping) and/or examining variation within a single population (Genome wide association studies [GWAS]).

Background selection: Reduction in diversity at sites linked to a variant under purifying selection, removing deleterious mutations from the populations.

Contigs: Continuous sequences derived from overlapping reads.

Genetic hitchhiking: Reduction in diversity at sites linked to variant under positive selection.

High-throughput sequencing: Set of platforms that sequence DNA from multiple genomic regions and individuals in parallel.

Inbreeding depression: Reduced fitness of a population resulting from inbreeding, the breeding of related individuals.

Linkage disequilibrium: Nonrandom association of loci in the genome, causing specific loci to be inherited together more often than expected by chance.

Phasing: Estimation of haplotypes from genotype data, of alleles on a particular chromosome that are inherited together.

Scaffolds: Series of contigs linked together. Noncontinuous—contigs separated by gaps of known length.

Selective sweeps: The reduction or loss of variation at loci linked to one experiencing positive or divergent selection.

Whole genome resequencing (WGS): Form of HTS that theoretically obtains data from the entire genome. When Illumina technology is used, this results in short reads (no more than 250 bp) with insert size of ~300 kb that are aligned/mapped to a reference genome and used to call variants, often SNPs.

References

1. Nice M (1939) The watcher at the nest. Macmillan, New York
2. Mayr E (1942) Systematics and the origin of species, from the viewpoint of a zoologist. Harvard University Press, Michigan
3. Luikart G, England PR, Tallmon D et al (2003) The power and promise of population genomics: from genotyping to genome typing. Nat Rev Genet 4:981–994
4. Begun DJ, Holloway AK, Stevens K et al (2007) Population genomics: whole-genome analysis of polymorphism and divergence in *Drosophila simulans*. PLoS Biol 5:e310
5. Avise JC, Nelson WS (1989) Molecular genetic relationships of the extinct dusky seaside sparrow. Science 243:646–648
6. Zink RM, Kale HW (1995) Conservation genetics of the extinct dusky seaside sparrow *Ammodramus maritimus nigrescens*. Biol Conserv 74:69–71
7. Stapley J, Reger J, Feulner PGD et al (2010) Adaptation genomics: the next generation. Trends Ecol Evol 25:705–712
8. Davey JW, Hohenlohe PA, Etter PD et al (2011) Genome-wide genetic marker discovery and genotyping using next-generation sequencing. Nat Rev Genet 12:499–510
9. Gregory TR (2005) Synergy between sequence and size in large-scale genomics. Nat Rev Genet 6:699–708
10. Organ CL, Shedlock AM, Meade A et al (2007) Origin of avian genome size and structure in non-avian dinosaurs. Nature 446:180–184
11. International Chicken Genome Sequencing Consortium (2004) Sequence and comparative analysis of the chicken genome provide unique perspectives on vertebrate evolution. Nature 432:695–716
12. Warren WC, Clayton DF, Ellegren H et al (2010) The genome of a songbird. Nature 464:757–762
13. Ellegren H (2014) Genome sequencing and population genomics in non-model organisms. Trends Ecol Evol 29:51–63
14. Eyre-Walker A, Keightley PD (2009) Estimating the rate of adaptive molecular evolution in the presence of slightly deleterious mutations and population size change. Mol Biol Evol 26:2097–2108
15. Zeng K (2013) A coalescent model of background selection with recombination, demography and variation in selection coefficients. Heredity 110:363
16. Bank C, Ewing GB, Ferrer-Admettla A et al (2014) Thinking too positive? Revisiting current methods of population genetic selection inference. Trends Genet 30:540–546
17. Beissinger TM, Wang L, Crosby K et al (2016) Recent demography drives changes in linked selection across the maize genome. Nat Plants 2:16084
18. Weir JT, Schluter D (2004) Ice sheets promote speciation in boreal birds. Proc R Soc Lond B Biol Sci 271:1881–1887

19. Pinho C, Hey J (2010) Divergence with gene flow: models and data. Annu Rev Ecol Evol Syst 41:215–230

20. Borge T, Lindroos K, Nádvorník P et al (2005) Amount of introgression in flycatcher hybrid zones reflects regional differences in pre and post-zygotic barriers to gene exchange. J Evol Biol 18:1416–1424

21. Saetre G-P, Borge T, Lindell J et al (2001) Speciation, introgressive hybridization and nonlinear rate of molecular evolution in flycatchers. Mol Ecol 10:737–749

22. Nadachowska-Brzyska K, Burri R, Olason PI et al (2013) Demographic divergence history of pied flycatcher and collared flycatcher inferred from whole-genome re-sequencing data. PLoS Genet 9:e1003942

23. Saetre G-P, Saether SA (2010) Ecology and genetics of speciation in *Ficedula* flycatchers. Mol Ecol 19:1091–1106

24. Nadachowska-Brzyska K, Burri R, Smeds L, Ellegren H (2016) PSMC analysis of effective population sizes in molecular ecology and its application to black-and-white Ficedula flycatchers. Mol Ecol 25:1058–1072

25. Li H, Durbin R (2011) Inference of human population history from individual whole-genome sequences. Nature 475:493–496. https://doi.org/10.1038/nature10231

26. Nadachowska-Brzyska K, Li C, Smeds L et al (2015) Temporal dynamics of avian populations during Pleistocene revealed by whole-genome sequences. Curr Biol 25:1375–1380

27. Li S, Li B, Cheng C et al (2014) Genomic signatures of near-extinction and rebirth of the crested ibis and other endangered bird species. Genome Biol 15:557

28. Kardos M, Taylor HR, Ellegren H et al (2016) Genomics advances the study of inbreeding depression in the wild. Evol Appl 9:1205–1218

29. Siol M, Wright SI, Barrett SC (2010) The population genomics of plant adaptation. New Phytol 188:313–332

30. Nielsen R, Williamson S, Kim Y et al (2005) Genomic scans for selective sweeps using SNP data. Genome Res 15:1566–1575

31. Lundberg M, Liedvogel M, Larson K et al (2017) Genetic differences between willow warbler migratory phenotypes are few and cluster in large haplotype blocks. Evol Lett 1:155–168

32. Tuttle EM, Bergland AO, Korody ML et al (2016) Divergence and functional degradation of a sex chromosome-like supergene. Curr Biol 26:344–350

33. Delmore KE, Liedvogel M (2016) Investigating factors that generate and maintain variation in migratory orientation: a primer for recent and future work. Front Behav Neurosci 10:3

34. Ellegren H, Smeds L, Burri R et al (2012) The genomic landscape of species divergence in *Ficedula* flycatchers. Nature 491:756–760

35. Delmore KE, Hübner S, Kane NC et al (2015) Genomic analysis of a migratory divide reveals candidate genes for migration and implicates selective sweeps in generating islands of differentiation. Mol Ecol 24:1873–1888

36. Irwin DE, Alcaide M, Delmore KE et al (2016) Recurrent selection explains parallel evolution of genomic regions of high relative but low absolute differentiation in a ring species. Mol Ecol 25:4488–4507

37. Van Doren BA, Campagna L, Helm B et al (2017) Correlated patterns of genetic diversity and differentiation across an avian family. Mol Ecol 26:3982–3997

38. Han F, Lamichhaney S, Grant BR et al (2017) Gene flow, ancient polymorphism, and ecological adaptation shape the genomic landscape of divergence among Darwin's finches. Genome Res 27:1004–1015

39. Poelstra JW, Vijay N, Bossu CM et al (2014) The genomic landscape underlying phenotypic integrity in the face of gene flow in crows. Science 344:1410–1414

40. Wu C-I (2001) Genes and speciation. J Evol Biol 14:889–891

41. Nosil P (2012) Ecological speciation (Oxford series in ecology and evolution). Oxford University Press, Oxford

42. Nosil P, Funk DJ, Ortiz-Barrientos D (2009) Divergent selection and heterogeneous genomic divergence. Mol Ecol 18:375–402

43. Cruickshank TE, Hahn MW (2014) Reanalysis suggests that genomic islands of speciation are due to reduced diversity, not reduced gene flow. Mol Ecol 23:3133–3157

44. Noor MA, Bennett SM (2009) Islands of speciation or mirages in the desert? Examining the role of restricted recombination in maintaining species. Heredity 103:439–444

45. Bierne N, Roze D, Welch JJ (2013) Pervasive selection or is it...? Why are FST outliers sometimes so frequent? Mol Ecol 22:2061–2064

46. Burri R, Nater A, Kawakami T et al (2015) Linked selection and recombination rate variation drive the evolution of the genomic landscape of differentiation across the speciation continuum of *Ficedula* flycatchers. Genome Res 25:1656–1665

47. Delmore KD, Lugo JS, Van Doren BM et al (2018) Comparative analysis examining patterns of genomic differentiation across multiple

episodes of population divergence in birds. Evol Lett 2:76–87

48. Dawson DA, Akesson M, Burke T et al (2007) Gene order and recombination rate in homologous chromosome regions of the chicken and a passerine bird. Mol Biol Evol 24:1537–1552

49. Griffin DK, Robertson LBW, Tempest HG, Skinner BM (2007) The evolution of the avian genome as revealed by comparative molecular cytogenetics. Cytogenet Genome Res 117:64–77

50. Backström N, Qvarnström A, Gustafsson L, Ellegren H (2006) Levels of linkage disequilibrium in a wild bird population. Biol Lett 2:435–438

51. Stapley J, Birkhead TR, Burke T, Slate J (2008) A linkage map of the zebra finch Taeniopygia guttata provides new insights into avian genome evolution. Genetics 179:651–667

52. Kawakami T, Smeds L, Backström N et al (2014) A high-density linkage map enables a second-generation collared flycatcher genome assembly and reveals the patterns of avian recombination rate variation and chromosomal evolution. Mol Ecol 23:4035–4058

53. Zhang G, Li C, Gilbert M et al (2014) Comparative genomic data of the avian phylogenomics project. GigaSci 3:26

54. Kawakami T, Mugal CF, Suh A et al (2017) Whole-genome patterns of linkage disequilibrium across flycatcher populations clarify the causes and consequences of fine-scale recombination rate variation in birds. Mol Ecol 26:4158–4172

55. Hewitt G (2000) The genetic legacy of the quaternary ice ages. Nature 405:907–913

56. Botero-Castro F, Figuet E, Tilak M-K et al (2017) Avian genomes revisited: hidden genes uncovered and the rates versus traits paradox in birds. Mol Biol Evol 34:3123–3131

57. Miller MM, Robinson CM, Abernathy J et al (2013) Mapping genes to chicken microchromosome 16 and discovery of olfactory and scavenger receptor genes near the major histocompatibility complex. J Hered 105:203–215

58. Skinner BM, Griffin DK (2012) Intrachromosomal rearrangements in avian genome evolution: evidence for regions prone to breakpoints. Heredity 108:37–41

59. Zhang J, Li C, Zhou Q, Zhang G (2015) Improving the ostrich genome assembly using optical mapping data. GigaSci 4:24

60. Edwards SV, Shultz AJ, Campbell-Staton S (2015) Next-generation sequencing and the expanding domain of phylogeography. Folia Zool 64:187–206

61. Puritz JB, Addison JA, Toonen RJ (2012) Next-generation phylogeography: a targeted approach for multilocus sequencing of non-model organisms. PLoS One 7:e34241

62. Faircloth BC, McCormack JE, Crawford NG et al (2012) Ultraconserved elements anchor thousands of genetic markers spanning multiple evolutionary timescales. Syst Biol 61:717–726

63. Klicka LB, Kus BE, Burns KJ (2016) Conservation genomics reveals multiple evolutionary units within Bell's Vireo (*Vireo bellii*). Conserv Genet 17:455–471

64. Lemay MA, Donnelly DJ, Russello MA (2013) Transcriptome-wide comparison of sequence variation in divergent ecotypes of kokanee salmon. BMC Genomics 14:308

65. Lemay MA, Russello MA (2015) Genetic evidence for ecological divergence in kokanee salmon. Mol Ecol 24:798–811

66. Vijay N, Bossu CM, Poelstra JW et al (2016) Evolution of heterogeneous genome differentiation across multiple contact zones in a crow species complex. Nat Commun 7:13195

67. Burri R (2017) Interpreting differentiation landscapes in the light of long-term linked selection. Evol Lett 1:118–131

68. Ellegren H, Sheldon BC (2008) Genetic basis of fitness differences in natural populations. Nature 452:169–175

69. Mackay TF, Stone EA, Ayroles JF (2009) The genetics of quantitative traits: challenges and prospects. Nat Rev Genet 10:565–577

70. Carroll SB (2000) Endless forms: the evolution of gene regulation and morphological diversity. Cell 101:577–580

71. Hoekstra HE, Coyne JA (2007) The locus of evolution: evo devo and the genetics of adaptation. Evolution 61:995–1016

72. Wray GA (2007) The evolutionary significance of cis-regulatory mutations. Nat Rev Genet 8:206–216

73. Albert FW, Kruglyak L (2015) The role of regulatory variation in complex traits and disease. Nat Rev Genet 16:197–212

74. Whitney O, Pfenning AR, Howard JT et al (2014) Core and region-enriched networks of behaviorally regulated genes and the singing genome. Science 346:1256780

75. Laine VN, Gossmann TI, Schachtschneider KM et al (2016) Evolutionary signals of selection on cognition from the great tit genome and methylome. Nat Commun 7:10474

76. Skinner MK, Gurerrero-Bosagna C, Haque MM et al (2014) Epigenetics and the evolution

of Darwin's finches. Genome Biol Evol 6:1972–1989

77. McNew SM, Beck D, Sadler-Riggleman I et al (2017) Epigenetic variation between urban and rural populations of Darwin's finches. BMC Evol Biol 17:183

78. Manthey JD, Moyle RG (2015) Isolation by environment in white-breasted Nuthatches (Sitta carolinensis) of the Madrean Archipelago

sky islands: a landscape genomics approach. Mol Ecol 24:3628–3638

79. Delmore KE, Toews DPL, Germain RR et al (2016) The genetics of seasonal migration and plumage color. Curr Biol 26:2167–2173

80. Edwards SV, Potter S, Schmitt CJ et al (2016) Reticulation, divergence, and the phylogeography–phylogenetics continuum. Proc Natl Acad Sci 113:8025–8032

Further Reading

Alkan C et al (2011) Genome structural variation discovery and genotyping. Nat Rev Genet 12:363–376

Carstens A et al (2013) How to fail at species delimitation. Mol Ecol 22:4369–4383

Edwards SV et al (2015) Next-generation sequencing and the expanding domain of phylogeography. Folia Zool 64:187–206

Feltus FA (2014) Systems genetics: a paradigm to improve discovery of candidate genes and

mechanisms underlying complex traits. Plant Sci 223:45–48

Manel N et al (2010) Perspectives on the use of landscape genetics to determine genetic adaptive variation in the field. Mol Ecol 19:3760–3772

Wolf JB, Ellegren H (2017) Making sense of genomic islands of differentiation in light of speciation. Nat Rev Genet 18:87–100

Chapter 18

Population Genomics of the House Mouse and the Brown Rat

Kristian K. Ullrich and Diethard Tautz

Abstract

Mice (*Mus musculus*) and rats (*Rattus norvegicus*) have long served as model systems for biomedical research. However, they are also excellent models for studying the evolution of populations, subspecies, and species. Within the past million years, they have spread in various waves across large parts of the globe, with the most recent spread in the wake of human civilization. They have developed into commensal species, but have also been able to colonize extreme environments on islands free of human civilization. Given that ample genomic and genetic resources are available for these species, they have thus also become ideal mammalian systems for evolutionary studies on adaptation and speciation, particularly in the combination with the rapid developments in population genomics. The chapter provides an overview of the systems and their history, as well as of available resources.

Key words House mouse, Population genomics, Evolution, Rodents, Adaptation

1 Introduction

Population genomics can address very different biological questions related to speciation, divergence of closely related species, within species population structure or within population evolutionary processes that affect adaptation. In the era of next-generation sequencing (NGS) with increasing taxonomic sampling, the crucial factor to apply population genomics is not any longer the number of genetic markers (quantity) but it is quality and complexity of the massive amount of available information that needs to be integrated and interpreted.

In this chapter, we focus on studies of population genomics in rodents and in particular on the Murinae. Murinae as a subfamily of rodents comprises more than one hundred genera and it is among mammals one of the largest subfamilies with species native to most continents. Murinae includes the house mouse (*Mus musculus*) and the brown rat (*Rattus norvegicus*) of which laboratory strains have been used since decades for biomedical research, as well as to serve

Julien Y. Dutheil (ed.), *Statistical Population Genomics*, Methods in Molecular Biology, vol. 2090,
https://doi.org/10.1007/978-1-0716-0199-0_18, © The Author(s) 2020

as models to study human diseases. Further, as human commensal species, both harbor also vectors for spreading infectious diseases that makes the wild living animals and populations of special interest. But also their evolutionary histories make them perfect models for studying general evolutionary processes, such as speciation, rapid adaptation and behavioral changes.

1.1 History of the House Mouse

A recent book, "Evolution of the House Mouse" [1], provides a broad overview on a variety of evolutionary aspects for the house mouse. Other general reviews can be found in [2, 3]. Here, we provide a short summary.

Mice consist of four major clades (*Coelomys, Mus, Nannomys,* and *Pyromys*), of which the subgenus *Mus* harbors the species *Mus musculus*, the house mouse. House mouse genetics began early in the twentieth century based on the first inbred strains from wild derived animals to study modes of inheritance [4, 5]. The worldwide distribution range of the house mouse is depicted in Fig. 1. It shows three main subspecies, the southeastern Asian house mouse (*Mus musculus castaneus*), the eastern European house mouse (*Mus musculus musculus*) and the western European house mouse (*Mus musculus domesticus*). Next to these main subspecies, there exist other subspecies (e.g., *M. m. molossinus*, a presumptive hybrid species between *M. m. castaneus* and *M. m. musculus;* [6], *M. m. gentilulus* [7, 8], *M. m. homoulus* [9], and further recently diverged ones like *M. m. helgolandicus* [10]). Most inbred strains and the reference genome sequence are derived from *M. m. domesticus*. The mouse genome was the first sequenced mammalian genome published in 2002 (Mouse Genome Sequencing Consortium, 2002) [8]. The genome consists of 19 autosomes and 2 sex chromosomes (X and Y) with a total length of 2.7 Gbp (currently with 22,612 coding and 15,402 noncoding genes annotated). The mouse ENCODE [11] consortium and genome assemblies of wild-derived inbred strains of the main subspecies have further enhanced the available genomic information [12–14], complemented by detailed recombination maps [15–17]. Genomic and transcriptomic data from wild derived populations of the subspecies and the sister species *Mus spretus* were reported in [14].

As one of the prominent human commensals, the dispersal and phylogenetic history of the house mouse were intensively studied. The ancestor of all subspecies within *Mus musculus* was initially thought to have lived in India [1, 18], but a broader sampling has shown that the Iranian plateau shows the highest diversity of lineages, including some as yet unnamed lineages [10]. The main subspecies started to diverge ~350–500 thousand years ago. As recently diverged species, one finds frequently phylogenetic discordance at different loci, whereby the statistical analysis of discordance patterns shows a strong deviation from a neutral model of pure lineage sorting [19]. Based on population data, it was shown

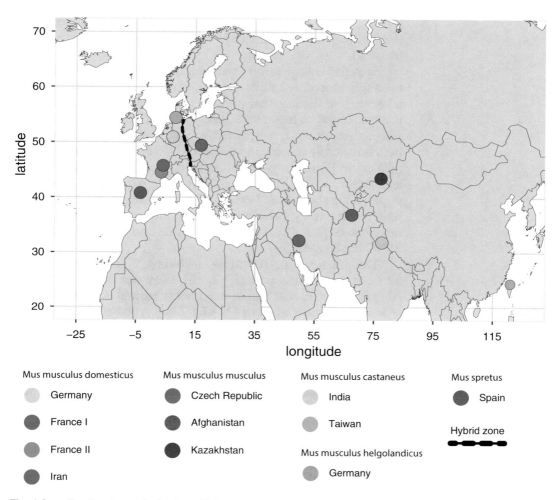

Fig. 1 Sampling locations of mice for which public population scale WGS data exist. Population scale sampling locations of house mouse and close relatives

that this is most likely due to secondary adaptive introgression, even across large geographic distances [20, 21]. The overall phylogenomic analysis suggests that *M. m. musculus* and *M. m. castaneus* are sister groups and that *M. m. domesticus* is more basal [12, 19].

The subspecies meet in several zones of secondary contact, where they form hybrid zones [2, 18, 22]. Fertility of offspring is impaired across these hybrid zones, and this serves as a general model to study the genetic basis of hybrid sterility as part of speciation processes (e.g., [22–26]).

Studies on house mouse phylogeography showed that the spread of the populations, especially those of *M. m. domesticus*, reflects human colonization and settlement history. For example, by looking into mtDNA haplotypes of worldwide distributed mouse samples, some historical human movements, such as following the seafarer routes of Vikings [27–29] or the colonization history of sub-Antarctic islands, could be reconstructed [30, 31].

Systematic population-level sampling of mouse populations has been introduced by Ihle et al. [32], where the sampling regime has taken care of the fact that mice tend to show inbreeding in family groups. Initial microsatellite based scans of populations that were sampled in this way suggested a high rate of positive selection between closely related populations [33]. The colonization history of Western European populations was traced by fossil evidence [34] and shown to be less than 3000 years ago. Nonetheless, these populations show clear genomic differentiation [20, 32, 33], differences in gene expression [35, 36], ultrasonic vocalization and mate choice [37, 38]. They harbor also a number of deme-specific MHC haplotypes [39].

Despite genomic resources, including a variant database of 17 laboratory inbred strains [12, 40], there was the need to derive laboratory strains that harbor most of the natural variation found in wild-derived populations [41, 42]. Genotype arrays were established that were constructed to maximize variant information at low sequencing costs [43]. The still commonly used genotyping arrays are MegaMUGA with a set of 77,808 SNP markers and GigaMUGA with a set of 143,259 SNP markers [44], which only represent a fraction of variants found between any sequenced inbred strain and the reference genome (~4 to 5 million SNPs; [12, 45]). However, researchers started to complement their analyses with NGS based datasets and genomic resources for wild populations of the house mouse are now common ground for subsequent analysis [14].

1.2 Brown Rat History

Mice and rats approximately diverged 7–12 million years ago [46]. Similar to house mice, brown rats (*Rattus norvegicus*) have been used for more than two centuries for biomedical studies to learn about the basis of human diseases and to deal with human pest management [47, 48]. The genome of the brown rat was published in 2004 [49] and consists of 20 autosomes and 2 sex chromosomes (X and Y) with a total length of 2.8 Gb (currently with 22,250 coding and 8934 noncoding genes annotated). The house mouse genome and the brown rat genome show a high number of shared syntenic homologous blocks with different levels of recombination [50]. Approximately 30% of the rat genome aligns only with the mouse genome, which might correspond to rodent-specific repeats [49]. A syntenic view of both genomes is given in Fig. 2 to illustrate the pairwise chromosome assignment obtained from the Synteny Portal (*see* Table 1 for web page URL link; [51]).

The origin of the laboratory brown rat (*Rattus norvegicus*) and the black rat (*Rattus rattus*) most likely lies in central Asia [52]. Spatial population genomics studies were conducted on brown rats living in New York City [53] and, like in mice studies,

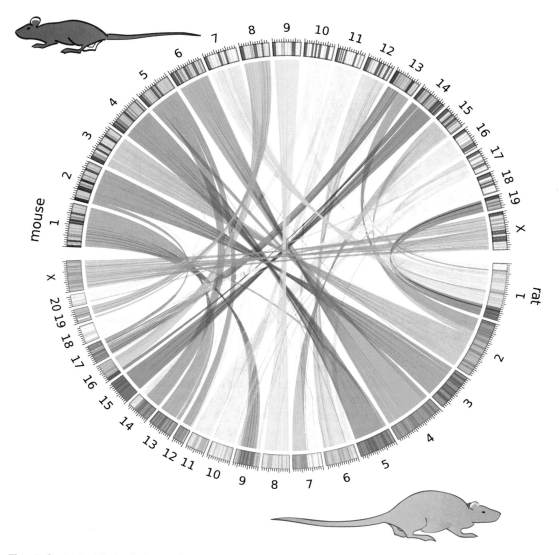

Fig. 2 Syntenic blocks between house mouse and rat. House mouse (GRCm38/mm10) and brown rat (Rnor_6.0/rn6) chromosome-wide syntenic blocks obtained via the web-based Synteny Portal [51]

mtDNA haplotype data could disentangle the phylogeography of brown rats in the countries surrounding the South Atlantic Ocean [54]. While the phylogeography of black rats, like the phylogeography of house mice, reflects human colonization and settlement history [53, 55, 56, 57], brown rats did not appear in Europe until the sixteenth century. Their dispersal routes from Asia to Europe are still under debate [57]. For example, one route is thought to lead via northeast China and Siberia, while another route inferred on whole-genome sequencing may represent an expansion via a Southern East Asia route [58]. Figure 3 illustrates the sampling distribution of *Rattus norvegicus* from publicly available whole-genome data sets.

Table 1
Useful public URL links for house mouse resources

Database	URL
SyntenyPortal	http://bioinfo.konkuk.ac.kr/synteny_portal/
Mouse ENCODE consortium	http://www.mouseencode.org/
Ensembl Mouse strains	https://www.ensembl.org/Mus_musculus/Info/Strains
UCSC wildmouse tracks	https://genome.ucsc.edu/cgi-bin/hgTracks?hgS_doOtherUser=submit&hgS_otherUserName=dtautz&hgS_otherUserSessionName=wildmouse
Colloborative Cross	http://csbio.unc.edu/CCstatus/CCGenomes
Inbred Strain Variant Database	http://isvdb.unc.edu/

2   Population Genomics

As mammal species expand, they are faced with new abiotic and biotic factors, such as different climatic conditions, different food or new pathogens, prey and/or predators, which potentially lead to adaptation and contributes to shaping the genome over time. Evolutionary changes in the genome can result from mutation, gene flow, random genetic drift, recombination and selection. Genome-wide scans for deviation from modelled neutrality aim at revealing such evolutionary processes. Genome-wide scans can help to identify genotypic and phenotypic variation, and by taking demographic events into account, they can even detect genes under recent positive selection [59]. Negative selection leads to sequence conservation by removing disadvantageous alleles. Positive selection can yield to an excess of nonsynonymous fixed differences or lead to an altered allele-frequency spectrum (AFS). Multiple approaches exist to detect adaptation, each with its own caveats. For example, dN/dS ratios can be used in comparative studies to detect selection on genes. But this analysis is limited to species that represent a certain evolutionary distance to allow a sufficient number of substitutions to have occurred [60]. When samples are drawn from different populations of the same species, it is necessary to study frequency changes of polymorphisms instead of substitutions. As compared to studies with a limited number of neutral markers, population genomics uses high marker density to robustly infer genome-wide effects, usually as signals of departure from expectations of the neutral theory of molecular evolution (*see* Chapter 5 for a detailed description how to detect positive selection).

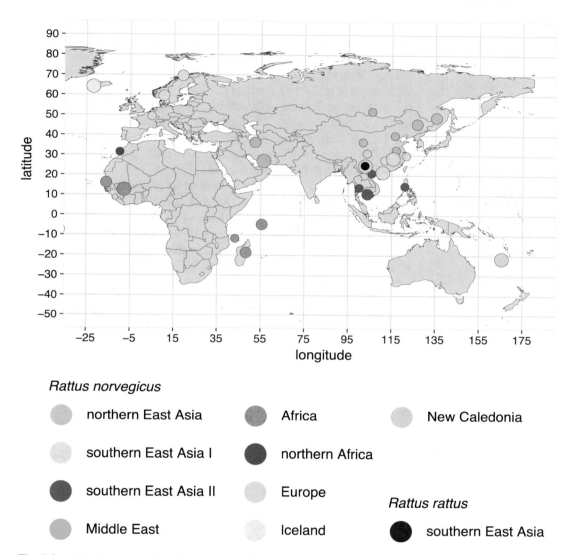

Fig. 3 Sampling locations of rats for which public population scale WGS data exist. Population scale sampling locations of brown rats obtained from [58]

2.1 House Mouse Genetic Variation

Population genetic studies revealed a fairly large effective population size (N_e) for wild natural populations of mice in the order of $N_e = 5 \times 10^5$ to 2×10^6 [61, 62] with two to three generations per year. Based on a genotyping array, the effective population sizes for the subspecies were estimated to range between $N_e = 0.25 \times 10^5$ to 1.2×10^5 for *M. m. musculus*, $N_e = 0.58 \times 10^5$ to 2×10^5 for *M. m. domesticus* and $N_e = 2 \times 10^5$ to 7×10^5 for *M. m. castaneus* [63]. This assumption was validated recently by a population genomic study on nucleotide diversity within the subspecies of *M. m. castaneus* [64]. In the same study an excess of adaptive substitutions in protein-coding genes, UTRs and conserved noncoding elements (CNE) were observed [64]. A follow-up study based on the same data recently inferred the recombination landscape within the same subspecies and revealed that genetic diversity is positively correlated with the rate of recombination [17] (*see* ref. 13 for the recombination landscape in the collaborative cross [41] and *see* ref. 65 for mouse inbred strains). The frequency-weighted mean estimate of the recombination rate was inferred from a broad-scaled map to $4N_e r/\mathrm{bp} = 0.0092$ for autosomes per bp and to $4N_e r/\mathrm{bp} = 0.0026$ for the X chromosome [17].

One candidate gene that is known to influence recombination break points in mammals is PRDM9 [66–69]. PRDM9 is highly polymorphic in natural populations of the house mouse [70, 71] and it was recently shown that some alleles are preferred over others in hybrid mice [72]. What is remarkable in the study of Booker et al. [17] is the high level of variability of recombination hot spots within one population and between wild-derived and classical inbred strains, which is worth further consideration. For example, phasing approaches should depend on an accurate recombination map and the question arises whether global heterogeneous recombination rates provide sufficient information for fine-scaled phasing inference.

Researchers need to rely on high-quality genome information to perform reference-based whole-genome analysis to retain variant information for the populations under study. However, in some cases the sequence divergence of the analyzed population and the reference is high and might produce mapping artefacts [73]. To cope with such situations Sarver et al. [74] performed a pseudo-reference based approach using exome data to infer the phylogenetic relationship and gene tree incongruence of the *Mus* clade. While Sarver et al. [74] used the D-statistic [75] to detect introgression between *M. m. musculus* and *M. m. domesticus*, other methods have been recently applied to infer introgression signals [8, 20, 21, 76, 77].

In their genomic comparison, Harr et al. [14] incorporated the two other house mouse subspecies *M. m. domesticus* and *M. m. musculus* together with the *M. m. castaneus* samples. In total this study covers a divergence time of roughly two million years by complementing the data with samples from the sister species *M. spretus* and the recently diverged species *M. m. helgolandicus* [14]; *see* Fig. 1. In combination with the short generation time of mice, this constitutes a substantial molecular divergence, which is, for example, larger than the divergence between humans and Hominidae across the same time scale. Figure 4 represents the inferred population sizes for the subspecies *M. m. domesticus* and the diverged species *M. m. helgolandicus*, this data set was analyzed with the smc++ software setting the mutation rate to $\mu = 5 \times 10^{-9}$ per base pair per generation [78].

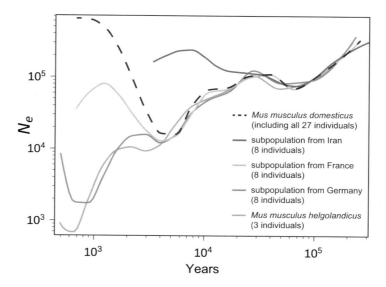

Fig. 4 Inferred population history for subspecies of the house mouse. Effective population size inference across populations of the house mouse subspecies *M. m. domesticus* and *Mus musculus helgolandicus*. SNP data from [14] was filtered to only retain intergenic regions without any feature annotation. For each population a separate smc++ [78] model was created setting the per generation mutation rate to 5×10^{-9} (*see* **Note 1** for a detailed method description)

Population genetic variation in segmental duplications (copy number variation) was systematically studied by Pezer et al. [79]. They found among the most copy-number variable genes three highly conserved genes that encode the splicing factor CWC22, the spindle protein SFI1, and the Holliday junction recognition protein HJURP. These genes showed population-specific expansion patterns that suggested an involvement in local adaptations. Other variable genes were found to encode proteins that are relevant for environmental and behavioral interactions, such as vomeronasal and olfactory receptors, as well as major urinary proteins. In a follow-up study, it was suggested that duplications in the Androgen-binding protein gene region might specifically have contributed to species diversification [80].

Another study also identified the CWC22 region as a region which shows major segmental duplication in the house mouse. It received the genetic name R2d and it was shown that the structural mutation rate appears to depend on the diploid configuration at that locus [81]. By reconstructing the origin and history of copy-number variants (CNVs), the study of Morgan et al. [81] is a nice example how important refined analyses are to disentangle complex genome structures. This is particularly true for genomic regions that are duplicated and are absent from the reference genome, which the author termed the "missing genome" [81].

The sequence and structural diversity of Y chromosomes in natural populations was studied in [82]. The mouse Y chromosome is in comparison to other mammals larger and harbors more annotated genes. The authors could show that CNV on the long arm of both sex chromosomes is highly variable, but sequence diversity as compared to autosomes is low in nonrepetitive regions.

The autosomal AFS of neutral intergenic regions was used to infer demography of all subspecies with the software "∂a ∂i" [83]. All simple models applied predicted effective population sizes that fall inside the range mentioned above (*M. m. domesticus*: $N_e = 1.6 \times 10^5$, *M. m. musculus*: $N_e = 1.6 \times 10^5$, *M. m. domesticus*: $N_e = 4.2 \times 10^5$; [82] but could not explain the reduction of sex chromosome diversity. Important findings are for instance that there is a moderately strong selective sweep on the Y chromosome in the *M. m. domesticus* population and that positive selection of genes expressed in the male germline might shape the sex chromosomes.

2.2 Brown Rat Genetic Variation

Rats and in particular the species *Rattus norvegicus* have an effective population size comparable to that of the mice subspecies *M. m. domesticus* and *M. m. musculus*. Denium et al. [84] estimated the effective population size to be $N_e = 1.24 \times 10^5$, based on silent mutations of 12 wild-derived animals. The authors highlight a recent bottleneck in rats (20,000 years ago) based on a 'PSMC' [85] analysis (*see* Chapter 7 for a discussion of MSMC and MSMC2). This bottleneck might be the cause of negative estimates of the rate of adaptive evolution in proteins and noncoding elements. Compared to mice, rats show a larger proportion of mildly deleterious mutations and concordantly a lower rate of highly deleterious mutations [84]. However, the reduction in diversity around exons is comparable to values obtained for mice [64]. Considering the different N_e of mice and rats, Denium et al. [84] estimated linkage disequilibrium (LD) decay to be six to seven times faster in mice than in rats.

As for mice, researchers looked into speciation and introgression events using population genomics. Teng et al. [86] used the Himalayan field rat (*Rattus nitidius*) as an outgroup, which is geographically restricted to Southeast Asia, to investigate introgression in brown rats sampled in China. With whole-genome data from 44 individuals, the N_e for brown rats and Himalayan field rats was estimated to $N_{e \text{ brown rats}} = 2.53 \times 10^5$ and $N_{e \text{ Himalayan field rats}} = 5.18 \times 10^5$, which reflects a difference of similar order to that of the house mice subspecies *M. m. musculus* and *M. m. castaneus*. According to the "PSMC" analysis the sibling species *R. norvegicus* and *R. nitidius* diverged ~650 thousand years ago, that is, within a time frame where the mouse divergence is suggested to be at the level of subspecies. The proportion of admixed fragments was estimated to 1.59% with admixture block sizes from 100 kbp to

1.42 Mbp [86]. Among the 346 introgressed regions detected, 92 loci were classified as adaptive. The strongest candidate is located on chromosome 1 overlapping with the "vomeronasal 1 receptor cluster," a chemical communication protein. As in mice [20], the regions were enriched in biological terms like "chemosensory perception" and "immune response." Next to regions showing signals of introgression, 352 regions were identified as having undergone a selective sweep based on allele frequency differentiation between populations "XP-CLR" [87] and cross population extended haplotype homozygosity calculations "XP-EHH" [88] which, like introgressed regions, are enriched in proteins involved in immune-response and metabolism.

Zeng et al. [58] extended the publicly available whole-genome sample set of brown rats to a world-wide distribution. With more than 100 individuals the authors investigated the geographic origin and migration paths. In contrast to previous hypothesis that *Rattus norvegicus* dispersed from northern Asia to Europe, their data supports the southern East Asian dispersal route to Europe [58]. Similar to Teng et al. [86], Zeng et al. [58] consistently identified candidate genes with signatures of positive selection that are associated with the immune-response by comparing European and Chinese populations.

3   Examples of Genes Under Positive Selection

In this section, we discuss three of several examples of genes that have been shown to be involved in adaptation in mice and rats. One prominent example is the evolution of the resistance against warfarin, a rodent pest management poison.

3.1   Rodent Resistance to Anticoagulants: Vkorc1

As vectors for human diseases, rodents have been reduced over half a century by rodenticides. Common compounds of rodenticides target the blood coagulation (e.g., warfarin) and target the vitamin K reductase reaction [89]. Several mutations have been found in house mice and brown rats within the *Vkorc1* gene that confer resistance against warfarin [90]. Song et al. [76] suggested that an allele introgressed from the Algerian mouse (*Mus spretus*) into *M. m. domesticus* led to anticoagulant resistance. Both species live today in sympatry in south-western Europe. *Vkorc1* was subject to adaptive protein evolution in *M. spretus* since it separated from other *Mus* lineages and four introgressed polymorphisms could be linked to a strong resistance phenotype [76, 91]. Based on whole-genome data [14], this region shows negative Tajima's D values within western European mouse populations in contrast to a population from Iran (*see* Fig. 5a), compatible with recent positive selection acting on it.

(A)

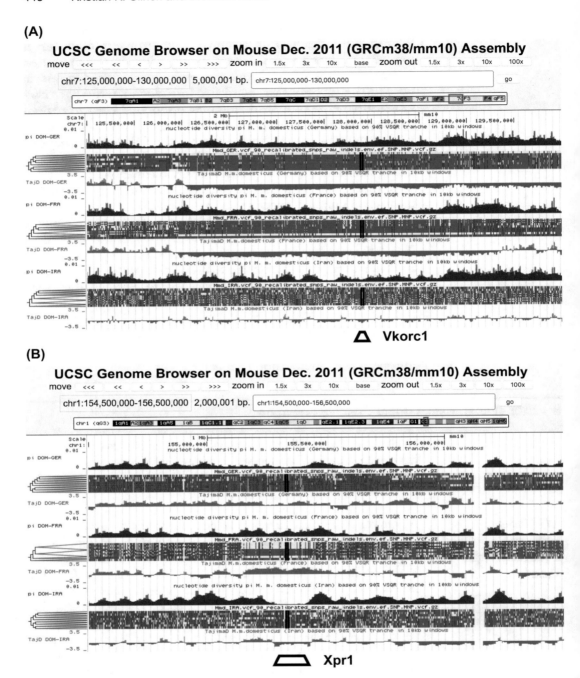

Fig. 5 Views from the UCSC genome browser showing haplotypes, nucleotide diversity and Tajima's D values for *M. m. domesticus* subpopulations. UCSC tracks are shown for (**a**) *Vkorc1* region on chromosome 7 and (**b**) *Xpr1* region on chromosome 1. Tracks were obtained from data published in [14] via a "public track hub" showing haplotypes from SNPs, nucleotide diversity (pi) and Tajima's D values for the subpopulations from France (DOM-FRA), Germany (DOM-GER) and Iran (DOM-IRA). pi and Tajima's D was calculated on 10 kbp windows (*see* Table 1 for web page URL link)

3.2 Pathogen Related Resistance: Xpr1

Next to artificial human-made selection pressure, there exists natural selection caused by pathogens. Hasenkamp et al. [92] have studied the gene *Xpr1*, coding for the receptor of murine leukemia virus (MLV) They found that the gene has been subject to a recent selective sweep in the population from Iran and that the selected haplotype has adaptively introgressed into a population from France, where it has mixed with existing haplotypes and thus creates a higher average population diversity than in the nonintrogressed population from Germany (*see* Fig. 5b). It seems that the *Xpr1* gene itself is under frequent positive selection and that alleles coping with new virus variants can rather quickly spread into other subpopulations if these are actively dealing with infectious cycles of that virus variant [92].

3.3 Segmental Duplications and Selective Sweeps: R2D2

As mentioned above, R2d is a CNV region on chromosome 2 that was found to cause nonrandom segregation [93]. Didion et al. [93] showed that signatures of selective sweeps obtained via genome-wide scans can be mimicked by "selfish" alleles. Within the 127 kbp genome region of R2d there is one annotated gene, namely *Cwc22*, which is a spliceosomal protein. Based on haplotype sharing, analysis of almost 400 individuals sampled across Europe revealed that all individuals with an extreme excess of shared identity showed a high copy number of R2d. If only one subpopulation was analyzed, the haplotype sharing methods failed to detect this "selfish" sweep. However, if individuals from different geographically locations were included in the analysis, R2d was identified as a selective sweep. Morgan et al. [81] showed for the same locus that an initial duplication event ~3.5 million years ago led to R2d1 and R2d2 and, therefore, mouse strains containing a single copy must have lost the second one. The authors identified nonallelic gene conversion in R2d1, which were transferred from R2d2 and caused the appearance of deep coalescence among R2d1 sequences [81]. Given both the patterns of concerted evolution, as well as the evolutionary dynamics of the selfish alleles, this could be a case of evolution through "molecular drive" [94].

3.4 The t-Haplotype as Meiotic Drive Element

Meiotic drive elements, or segregation distorters, transmit themselves to over 50% of the progeny of heterozygous individuals. The mouse t-haplotype, located within several inversions on chromosome 17, is a classic example of such a meiotic drive element [1]. Despite a strong driving capacity, t-haplotypes remain at relatively low frequency in natural populations, since homozygous individuals have strongly reduced viability [95]. The population genomics of the t-haplotype was studied in [96] based on the data provided in [14]. They found evidence for an accumulation of nonsynonymous substitutions within the inversions, but also signatures of recombination events that appear to have regenerated coding sequences that had accumulated deleterious mutations.

Based on the corresponding transcriptome data in [14] they could show that individuals carrying a t-haplotype display also a change in the testis expression of genes outside of the t-complex.

4 Conclusion

Per sample cost reduction for sequencing has led to an exponential increase in available whole-genome data for model and nonmodel organisms. Being among the longest studied mammals, both house mouse and brown rat have proven to serve as models for studying the processes that shape genome evolution in natural populations, including introgression and positive selection. However, while the public domain is steadily filled with population genomic usable datasets, there is still a gap between studies that predict candidates and studies that functionally validate them. As a consequence, functional studies to prove that genes have a direct impact on fitness in a certain species should be extended. The experimental set up to measure fitness will always depend on the species level and should be imbedded in an environmental context.

5 Note

1. SMC++ [78] analysis is based on 24 *Mus musculus domesticus* and 3 *Mus musculus helgolandicus* individuals described earlier [14]. SMC++ version 1.12.1 was used to infer population history for *Mus musculus domesticus* subpopulations based on a Variant Call Format (VCF) file obtained via the following URL: *http:// wwwuser.gwdg.de/~evolbio/evolgen/wildmouse/vcf/AllMouse.vcf_ 90_recalibrated_snps_raw_indels_reheader_PopSorted.vcf.gz*. First, bcftools version 1.3.1 [97] was used to filter SNP positions (*bcftools filter*) aside indel regions (*--SnpGap 3*), setting genotypes of failed samples to missing values (*--set-GTs.*) and excluding all sites with either low coverage or low genotype quality (*FORMAT/DP<5 | FORMAT/GQ<30*). Further, bcftools was used to retain only biallelic SNPs (*bcftools view -m2 -M2 -v snps*) and SMC++ was used to convert the VCF file to SMC++ format. Only subpopulations indicated above were retained from the input VCF file and only autosomes were extracted individually by additionally masking all exons, regulatory features, simple repeats, and missing sites from the reference mm10 (exons URL: *ftp://ftp.ensembl.org/pub/release-90/gtf/mus_musculus/ Mus_musculus.GRCm38.90.chr.gtf*; regulatory features URL: *ftp://ftp.ensembl.org/pub/release-90/regulation/mus_musculus/ mus_musculus.GRCm38.Regulatory_Build.regulatory_features. 20161111.gff*;

simple repeats URL: *http://hgdownload.soe.ucsc.edu/goldenPath/mm10/database/simpleRepeat.txt*). The per generation mutation rate was set to 5×10^{-9} to fit a size history for each subpopulation based on the extracted autosome data (*scm ++ estimate 5e-9 chr*.smc.gz*) and plotted with SMC++ (*smc++ plot*) as shown in Fig. 4.

References

1. Macholán M, Baird SJ, Munclinger P (eds) (2012) Evolution of the house mouse. Cambridge University Press, Cambridge
2. Guénet J-L, Bonhomme F (2003) Wild mice: an ever-increasing contribution to a popular mammalian model. Trends Genet 19:24
3. Phifer-Rixey M, Nachman MW (2015) The Natural History of Model Organisms: insights into mammalian biology from the wild house mouse Mus musculus. eLife Sci 4:e05959
4. Russell ES (1978) Origins and history of mouse inbred strains: contributions of Clarence Cook Little. Academic, New York
5. Silver LM (1995) Mouse genetics: concepts and applications. Oxford University Press, Oxford
6. Yonekawa H, Moriwaki K, Gotoh O et al (1988) Hybrid origin of Japanese mice "Mus musculus molossinus": evidence from restriction analysis of mitochondrial DNA. Mol Biol Evol 5:63
7. Prager EM, Orrego C, Sage RD (1998) Genetic variation and phylogeography of central Asian and other house mice, including a major new mitochondrial lineage in Yemen. Genetics 150:835
8. Yang H, Wang JR, Didion JP et al (2011) Subspecific origin and haplotype diversity in the laboratory mouse. Nat Genet 43:648
9. Terashima M, Suyanto A, Tsuchiya K et al (2003) Geographic variation of Mus caroli from East and Southeast Asia based on mitochondrial cytochrome b gene sequences. Mammal Study 28:67
10. Hardouin EA, Orth A, Teschke M et al (2015) Eurasian house mouse (Mus musculus L.) differentiation at microsatellite loci identifies the Iranian plateau as a phylogeographic hotspot. BMC Evol Biol 15:26
11. Yue F, Cheng Y, Breschi A et al (2014) A comparative encyclopedia of DNA elements in the mouse genome. Nature 515:355
12. Keane TM, Goodstadt L, Danecek P et al (2011) Mouse genomic variation and its effect on phenotypes and gene regulation. Nature 477:289
13. Morgan AP, Gatti DM, Najarian ML et al (2017) Structural variation shapes the landscape of recombination in mouse. Genetics 206:603
14. Harr B, Karakoc E, Neme R et al (2016) Genomic resources for wild populations of the house mouse, Mus musculus and its close relative Mus spretus. Sci Data 3:160075
15. Cox A, Ackert-Bicknell CL, Dumont BL et al (2009) A new standard genetic map for the laboratory mouse. Genetics 182:1335
16. Liu EY, Morgan AP, Chesler EJ et al (2014) High-resolution sex-specific linkage maps of the mouse reveal polarized distribution of crossovers in male germline. Genetics 197:91
17. Booker TR, Ness RW, Keightley PD (2017) The recombination landscape in wild house mice inferred using population genomic data. Genetics 207:297
18. Boursot P, Auffray J-C, Britton-Davidian J et al (1993) The evolution of house mice. Annu Rev Ecol Syst 24:119
19. White MA, Ané C, Dewey CN et al (2009) Fine-scale phylogenetic discordance across the house mouse genome. PLoS Gen 5:e1000729
20. Staubach F, Lorenc A, Messer PW et al (2012) Genome patterns of selection and introgression of haplotypes in natural populations of the house mouse (Mus musculus). PLoS Gen 8:e1002891
21. Liu KJ, Steinberg E, Yozzo A et al (2015) Interspecific introgressive origin of genomic diversity in the house mouse. PNAS 112:196
22. Teeter KC, Payseur BA, Harris LW et al (2008) Genome-wide patterns of gene flow across a house mouse hybrid zone. Gen Res 18:67
23. Forejt J (1996) Hybrid sterility in the mouse. Trends Genet 12:412
24. Turner LM, Schwahn DJ, Harr B (2012) Reduced male fertility is common but highly variable in form and severity in a natural house mouse hybrid zone. Evolution 66:443
25. Turner LM, Harr B (2014) Genome-wide mapping in a house mouse hybrid zone reveals hybrid sterility loci and Dobzhansky-Muller interactions. eLife Sci 3:e02504

26. Dumont BL (2017) X-chromosome control of genome-scale recombination rates in house mice. Genetics 205:1649

27. Searle JB, Jones CS, Gündüz İ et al (2009) Of mice and (Viking?) men: phylogeography of British and Irish house mice. Proc R Soc B 276:201

28. Jones EP, van der Kooij J, Solheim R et al (2010) Colonization and interactions of two subspecies of house mouse (Mus musculus) in Norway. Mol Ecol 19:5252

29. Jones EP, Skirnisson K, McGovern TH et al (2012) Fellow travellers: a concordance of colonization patterns between mice and men in the North Atlantic region. BMC Evol Biol 12:35

30. Hardouin EA, Chapuis J-L, Stevens MI et al (2010) House mouse colonization patterns on the sub-Antarctic Kerguelen Archipelago suggest singular primary invasions and resilience against re-invasion. BMC Evol Biol 10:325

31. Gray MM, Wegmann D, Haasl RJ et al (2014) Demographic history of a recent invasion of house mice on the isolated Island of Gough. Mol Ecol 23:1923

32. Ihle S, Ravaoarimanana I, Thomas M et al (2006) An analysis of signatures of selective sweeps in natural populations of the house mouse. Mol Biol Evol 23:790

33. Teschke M, Mukabayire O, Wiehe T et al (2008) Identification of selective sweeps in closely related populations of the house mouse based on microsatellite scans. Genetics 180:1537

34. Cucchi T, Vigne J-D, Auffray J-C (2005) First occurrence of the house mouse (Mus musculus domesticus Schwarz & Schwarz, 1943) in the Western Mediterranean: a zooarchaeological revision of subfossil occurrences. Biol J Linnean Soc 84:429

35. Bryk J, Somel M, Lorenc A et al (2013) Early gene expression divergence between allopatric populations of the house mouse (Mus musculus domesticus). Ecol Evol 3:558

36. Lorenc A, Linnenbrink M, Montero I et al (2014) Genetic differentiation of hypothalamus parentally biased transcripts in populations of the house mouse implicate the Prader–Willi syndrome imprinted region as a possible source of behavioral divergence. Mol Biol Evol 31:3240

37. von Merten S, Hoier S, Pfeifle C et al (2014) A role for ultrasonic vocalisation in social communication and divergence of natural populations of the house mouse (Mus musculus domesticus). PLoS One 9:e97244

38. Montero I, Teschke M, Tautz D (2013) Paternal imprinting of mating preferences between natural populations of house mice (Mus musculus domesticus). Mol Ecol 22:2549

39. Linnenbrink M, Teschke M, Montero I et al (2018) Meta-populational demes constitute a reservoir for large MHC allele diversity in wild house mice (Mus musculus). Front Zool 15:15

40. Morgan AP, Didion JP, Doran AG et al (2016) Whole genome sequence of two wild-derived Mus musculus domesticus inbred strains, LEWES/EiJ and ZALENDE/EiJ, with different diploid numbers. G3 6:4211

41. Churchill GA, Airey DC, Allayee H et al (2004) The Collaborative Cross, a community resource for the genetic analysis of complex traits. Nat Genet 36:1133

42. de Koning D-J, McIntyre LM (2017) Back to the future: multiparent populations provide the key to unlocking the genetic basis of complex traits. Genetics 206:527

43. Yang H, Ding Y, Hutchins LN et al (2009) A customized and versatile high-density genotyping array for the mouse. Nat Metods 6:663

44. Morgan AP, Fu C-P, Kao C-Y et al (2015) The mouse universal genotyping array: from substrains to subspecies. G3 6:263

45. Doran AG, Wong K, Flint J et al (2016) Deep genome sequencing and variation analysis of 13 inbred mouse strains defines candidate phenotypic alleles, private variation and homozygous truncating mutations. Gen Biol 17:167

46. Wu S, Wu W, Zhang F et al (2012) Molecular and paleontological evidence for a post-cretaceous origin of rodents. PLoS ONE 7 (10): e46445

47. Jacob HJ (1999) Functional genomics and rat models. Gen Res 9:1013

48. Aitman TJ, Critser JK, Cuppen E et al (2008) Progress and prospects in rat genetics: a community view. Nat Genet 40:516

49. Consortium RGSP (2004) Genome sequence of the Brown Norway rat yields insights into mammalian evolution. Nature 428:493

50. Jensen-Seaman MI, Furey TS, Payseur BA et al (2004) Comparative recombination rates in the rat, mouse, and human genomes. Gen Res 14:528

51. Lee J, Hong W, Cho M et al (2016) Synteny portal: a web-based application portal for synteny block analysis. Nucleic Acids Res 44:W35

52. Aplin KP, Suzuki H, Chinen AA et al (2011) Multiple geographic origins of commensalism and complex dispersal history of black rats. PLoS One 6:e26357

53. Combs M, Puckett EE, Richardson J et al (2018) Spatial population genomics of the brown rat (Rattus norvegicus) in New York City. Mol Ecol 27:83

54. Hingston M, Poncet S, Passfield K et al (2016) Phylogeography of Rattus norvegicus in the South Atlantic Ocean. Diversity 8:32

55. Matisoo-Smith E, Robins JH (2004) Origins and dispersals of Pacific peoples: evidence from mtDNA phylogenies of the Pacific rat. PNAS 101:9167

56. Tollenaere C, Brouat C, Duplantier JM et al (2010) Phylogeography of the introduced species Rattus rattus in the western Indian Ocean, with special emphasis on the colonization history of Madagascar. J Biogeogr 37:398

57. Puckett EE, Park J, Combs M et al (2016) Global population divergence and admixture of the brown rat (Rattus norvegicus). Proc R Soc B 283:20161762

58. Zeng L, Ming C, Li Y et al (2018) Out of Southern East Asia of the brown rat revealed by large-scale genome sequencing. Mol Biol Evol 35:149

59. Ometto L, Glinka S, De Lorenzo D et al (2005) Inferring the effects of demography and selection on Drosophila melanogaster populations from a chromosome-wide scan of DNA variation. Mol Biol Evol 22:2119

60. Pond SK, Muse SV (2005) Site-to-site variation of synonymous substitution rates. Mol Biol Evol 22:2375

61. Baines JF, Harr B (2007) Reduced X-linked diversity in derived populations of house mice. Genetics 175:1911

62. Halligan DL, Oliver F, Eyre-Walker A et al (2010) Evidence for pervasive adaptive protein evolution in wild mice. PLoS Genet 6: e1000825

63. Phifer-Rixey M, Bonhomme F, Boursot P et al (2012) Adaptive evolution and effective population size in wild house mice. Mol Biol Evol 29:2949

64. Halligan DL, Kousathanas A, Ness RW et al (2013) Contributions of protein-coding and regulatory change to adaptive molecular evolution in murid rodents. PLoS Genet 9: e1003995

65. Simecek P, Forejt J, Williams RW et al (2017) High-resolution maps of mouse reference populations. G3 7:3427

66. Brunschwig H, Levi L, Ben-David E et al (2012) Fine-scale maps of recombination rates and hotspots in the mouse genome. Genetics 191:757

67. Parvanov ED, Petkov PM, Paigen K (2010) Prdm9 controls activation of mammalian recombination hotspots. Science 327:835

68. Baudat F, Buard J, Grey C et al (2010) PRDM9 is a major determinant of meiotic recombination hotspots in humans and mice. Science 327:836

69. Berg IL, Neumann R, Lam K-WG et al (2010) PRDM9 variation strongly influences recombination hot-spot activity and meiotic instability in humans. Nat Genet 42:859

70. Buard J, Rivals E, de Segonzac DD et al (2014) Diversity of Prdm9 zinc finger array in wild mice unravels new facets of the evolutionary turnover of this coding minisatellite. PLoS One 9:e85021

71. Kono H, Tamura M, Osada N et al (2014) Prdm9 polymorphism unveils mouse evolutionary tracks. DNA Res 21:315

72. Smagulova F, Brick K, Pu Y et al (2016) The evolutionary turnover of recombination hot spots contributes to speciation in mice. Genes Dev 30:266

73. Liu Q, Guo Y, Li J et al (2012) Steps to ensure accuracy in genotype and SNP calling from Illumina sequencing data. BMC Genomics 13:S8

74. Sarver BAJ, Keeble S, Cosart T et al (2017) Phylogenomic insights into mouse evolution using a pseudoreference approach. Gen Biol Evol 9:726

75. Durand EY, Patterson N, Reich D et al (2011) Testing for ancient admixture between closely related populations. Mol Biol Evol 28:2239

76. Song Y, Endepols S, Klemann N et al (2011) Adaptive introgression of anticoagulant rodent poison resistance by hybridization between Old World mice. Curr Biol 21:1296

77. Rosenzweig BK, Pease JB, Besansky NJ et al (2016) Powerful methods for detecting introgressed regions from population genomic data. Mol Ecol 25:2387

78. Terhorst J, Kamm JA, Song YS (2017) Robust and scalable inference of population history from hundreds of unphased whole-genomes. Nat Genet 49:303

79. Pezer Ž, Harr B, Teschke M et al (2015) Divergence patterns of genic copy number variation in natural populations of the house mouse (Mus musculus domesticus) reveal three conserved genes with major population-specific expansions. Gen Res 25:1114

80. Pezer Ž, Chung AG, Karn RC et al (2017) Analysis of copy number variation in the Abp gene regions of two house mouse subspecies suggests divergence during the gene family expansions. Gen Biol Evol 9:1393

81. Morgan AP, Holt JM, McMullan RC et al (2016) The evolutionary fates of a large segmental duplication in mouse. Genetics 204:267

82. Morgan AP, Pardo-Manuel de Villena F (2017) Sequence and structural diversity of mouse Y chromosomes. Mol Biol Evol 34:3186

83. Gutenkunst RN, Hernandez RD, Williamson SH et al (2009) Inferring the joint demographic history of multiple populations from multidimensional SNP frequency data. PLoS Genet 5:e1000695

84. Deinum EE, Halligan DL, Ness RW et al (2015) Recent evolution in Rattus norvegicus is shaped by declining effective population size. Mol Biol Evol 32:2547

85. Li H, Durbin R (2011) Inference of human population history from individual whole-genome sequences. Nature 475:493

86. Teng H, Zhang Y, Shi C et al (2017) Population genomics reveals speciation and introgression between brown norway rats and their sibling species. Mol Biol Evol 34:2214

87. Chen H, Patterson N, Reich D (2010) Population differentiation as a test for selective sweeps. Gen Res 20:393

88. Sabeti PC, Varilly P, Fry B et al (2007) Genome-wide detection and characterization of positive selection in human populations. Nature 449:913

89. Pelz H-J, Rost S, Hünerberg M et al (2005) The genetic basis of resistance to anticoagulants in rodents. Genetics 170:1839

90. Rost S, Pelz H-J, Menzel S et al (2009) Novel mutations in the VKORC1 gene of wild rats and mice—a response to 50 years of selection pressure by warfarin? BMC Genet 10:4

91. Goulois J, Hascoët C, Dorani K et al (2017) Study of the efficiency of anticoagulant rodenticides to control Mus musculus domesticus introgressed with Mus spretus Vkorc1. Pest Manag Sci 73:325

92. Hasenkamp N, Solomon T, Tautz D (2015) Selective sweeps versus introgression-population genetic dynamics of the murine leukemia virus receptor Xpr1 in wild populations of the house mouse (Mus musculus). BMC Evol Biol 15:248

93. Didion JP, Morgan AP, Yadgary L et al (2016) R2d2 drives selfish sweeps in the house mouse. Mol Biol Evol 33:1381

94. Dover G (1982) Molecular drive: a cohesive mode of species evolution. Nature 299:111

95. Lindholm AK, Dyer KA, Firman RC et al (2016) The ecology and evolutionary dynamics of meiotic drive. Trends Ecol Evol 31:315

96. Kelemen RK, Vicoso B (2018) Complex history and differentiation patterns of the t-haplotype, a mouse meiotic driver. Genetics 208:365

97. Li H, Handsaker B, Wysoker A et al (2009) The sequence alignment/map format and SAMtools. Bioinformatics 25:2078–2079

Chapter 19

Population Genomics in the Great Apes

David Castellano and Kasper Munch

Abstract

The great apes play an important role as model organisms. They are our closest living relatives, allowing us to identify the genetic basis of phenotypic traits that we think of as characteristically human. However, the most significant asset of great apes as model organisms is that they share with humans most of their genetic makeup. This means that we can extend our vast knowledge of the human genome, its genes, and the associated phenotypes to these species. Comparative genomic studies of humans and apes thus reveal how very similar genomes react when exposed to different population genetic regimes. In this way, each species represents a natural experiment, where a genome highly similar to the human one, is differently exposed to the evolutionary forces of demography, population structure, selection, recombination, and admixture/hybridization. The initial sequencing of reference genomes for chimpanzee, orangutan, gorilla, the bonobo, each provided new insights and a second generation of sequencing projects has provided diversity data for all the great apes. In this chapter, we will outline some of the findings that population genomic analysis of great apes has provided, and how comparative studies have helped us understand how the fundamental forces in evolution have contributed to shaping the genomes and the genetic diversity of the great apes.

Key words Population genomics, Great apes, Incomplete lineage sorting, Demography, Distribution of fitness effects, Recombination, X chromosome, Selective sweeps

1 Species Trees and Incomplete Lineage Sorting

The sequencing of all the great ape genomes [1–6] has allowed us to paint a detailed picture of the species relationship between humans and their closest relatives. By joint analysis of full genomes from pairs of species, coalescent hidden Markov models (CoalHMM) (*see* Chapter 8) can efficiently model both sequence divergence and recombination by approximating the full ancestral recombination graph as a Markov process along the genome. The states in these hidden Markov models represent different gene trees separated by recombination events. Such models can jointly estimate both the time of reproductive isolation (the time of speciation) and the size of the ancestral population that gave rise to the two species. Figure 1 provides an overview of the estimated split

Julien Y. Dutheil (ed.), *Statistical Population Genomics*, Methods in Molecular Biology, vol. 2090,
https://doi.org/10.1007/978-1-0716-0199-0_19, © The Author(s) 2020

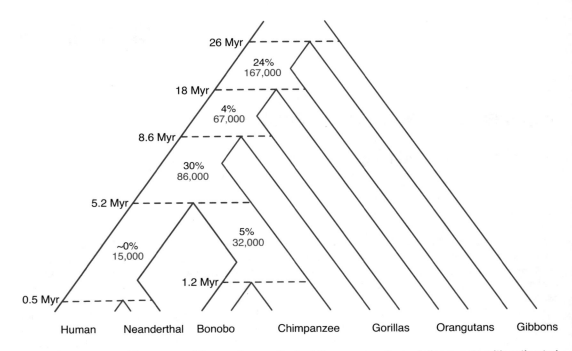

Fig. 1 Species tree of humans and the great apes. Dashed lines represent speciation events with estimated dates. Gray numbers show estimated sizes of ancestral populations, and percentages show the estimated amount of incomplete lineage sorting (*see* below) between two descendant species and their immediate outgroup (e.g., 30% ILS between human, chimpanzee, and gorilla). (The figure is adapted from Mailund et al. [7])

times and ancestral population sizes. The estimated split times are computed assuming that the mutation rate across the species tree has remained constant at the rate of 0.6×10^{-9} per year as observed in humans. However, speciation dates produced using a constant mutation rate does not square with the physical dating of ancestral fossil species. To reconcile DNA and fossil evidence, it has been proposed that the yearly mutation rate has slowed down across in the great ape lineage [8], possibly resulting from the development of larger body sizes and longer generation times.

The time to the most recent common ancestor of sequences sampled from two species lies much further into the past than the time when the species split apart. For this reason, a common ancestor of two lineages from separate species may not be found in the population ancestral to the two species, but even further into the past, in a population ancestral to additional species. When sampling more than two species, this allows for the possibility that lineages from other than the most closely related species find a common ancestor before those most closely related. This is especially true for the relationship between human, chimpanzee, and gorilla. Between the speciation events separating human and chimpanzee and that separating human and gorilla, a lot of ancestral polymorphism was conserved in the large ancestral population. The more rapid the succession of speciation events, and the larger the ancestral population between them is, the more ancestral polymorphism will be conserved. The implication is that individual gene trees along the alignment of these three species will not always group the same two species as the species tree does. The phenomenon is called incomplete lineage sorting (ILS) because the lineages of individual gene trees are not completely sorted according to species (*see* Chapter 1 for further details).

One coalescent hidden Markov model compares three closely related species and exploits information from sequence divergence and ILS to estimate the time of the two speciation events as well as the size of the ancestral population [9, 10]. From this model, it is also possible to extract the proportion of discordant gene trees with a topology different from the species tree. Applying this method to the human, chimpanzee, and gorilla showed that for ~15% percent of the genome, humans are more closely related to gorillas than chimpanzees, and for another ~15%, chimpanzees and gorillas are more closely related to each other than to humans [3] (*see* Fig. 1). The same model has been applied to alignments of bonobo, chimpanzee, and human, and showed that ~5% of the genome is subject to ILS [4]. Because the proportion of ILS is determined by ancestral population size and the time between speciation events, we can compute the estimated proportions of ILS for trios of species where these parameters have been estimated by other means. Between human, gorilla, and orangutan it is expected to be ~4%, and for human, orangutan, and gibbon it is expected to be ~24% [7]. The great apes thus also showcase how misleading phylogenies built from individual genes may be since a phylogeny built from long regions of a recombining sequence will not represent the population genetic processes that distribute individual lineages among species.

2   Gene Flow and Demography

Most coalescent hidden Markov models assume that speciation is instantaneous and that the initial split of two populations is not followed by gene flow between the diverging populations. Other coalescent hidden Markov models account for the possibility that such gene flow has occurred [11]. Among species splits in great ape evolution, most have involved a period of gene flow before consolidation of the populations as separate species [11]. The divergence of the orangutan from the human–chimp–bonobo–gorilla ancestor involved several hundred thousand years of gene flow. The speciation of humans and chimpanzees-bonobo most likely also included an extended period of gene flow. Only the speciation separating the bonobo from the chimpanzees seem to be a clear example of an abrupt and permanent split, possibly produced as the Congo River provided a physical barrier between the populations.

An alternative way to estimate gene flow between separating populations is using methods such as MSMC [12] (*see* Chapter 7) that can estimate the relative rate of cross-coalescence between populations from the present and into the past. This approach measures the proportion of gene pairs that find common ancestry between two sampled populations rather than within them. Inspecting a curve of this relative cross-coalescence rate can help identify both the time of speciation and whether this was a clean split rather than a protracted period of reduced gene flow. MSMC as well as a similar method modeling only one diploid sample (PSMC [13]) also estimate the historical effective population size of a species. Such methods have been used to identify how the great apes have responded to environmental changes and show that great apes have experienced a decline in their effective population size across the last few hundred thousand years [5]. Comparing the curves of historical effective population sizes may also reveal when the species split apart. Across the time in the past where two species share an ancestor their historical population sizes will be the same, but at the time this ancestral population split into two, the size of these two populations will be free to follow different trajectories through time and will reveal the species split as a separation of the curves of historical population sizes. Along with methods such as approximate Bayesian computation, PSMC has helped describe the relationship between chimpanzee subspecies [5] showing that eastern and central chimpanzees are most closely related, forming a group separate from Nigeria–Cameroon and western chimpanzees.

3 Selection

One of the most intriguing questions in great ape evolution is how the adaptive evolution of particular genes has contributed to shaping phenotypes in present-day species. A study comparing a large number of orthologous genes addressed adaptive evolution along the branches of humans and chimpanzees by comparing the rate of evolution at synonymous sites (sites where a mutation will not change the encoded protein) with nonsynonymous sites (sites where mutation replaces an amino acid) [14]. Many of the identified genes were involved in sensory perception and immune defenses, but the genes showing the strongest evidence of positive selection were genes involved in tumor suppression and apoptosis, and genes involved in spermatogenesis [15]. Another way to identify selection in primate genomes is by measuring the patterns of genetic diversity along the genomes. Slightly deleterious variants will reduce genetic diversity in a genomic region around the deleterious variant, a process called background selection (*see*

Charlesworth [16] for a review), in effect reducing the local effective population size. Positive selection also removes variation in a region around it, leaving a signature in local genetic variation that can be distinguished from that of background selection if the positive selection is strong enough and occurred recently. When a new variant is subject to strong positive selection, variation in the flanking regions is depleted because linked variants are carried to fixation along with the selected variant. This is called a selective sweep because it sweeps variation in a region around the selected variant and produces a wide genomic region where all individuals from a species share a recent common ancestor [17]. The size of the swept region depends on the strength of selection, the size of the population and the rate of genetic recombination. Several methods have been developed to detect sweeps from information in population samples such as the site frequency spectrum, linkage disequilibrium and population differentiation [18] (*see* Chapter 5). Due to the relatively small sample sizes available in great apes, no striking examples of recent sweeps on great ape autosomes have been reported (but see Sect. 5 for strong selective sweeps on the X chromosome). However, thanks to the McDonald and Kreitman test framework there are many estimates of the proportion of beneficial nonsynonymous substitutions (α) across primates (*see* Chapter 1 for a formal definition of α). Genome-wide estimates in humans and nonhuman primates are very low, $\alpha < 10$–20% [1, 19–23], but α can be as high as 50% for some particular genes like immune genes, testis genes, or virus interacting protein genes [24, 25].

It is still debated if positive or negative selection is more prominent in shaping diversity along great ape genomes, and we are still trying to figure out whether selective sweeps are mainly due to new mutations [17] or selection on standing variation [26], and which are more important for adaptation and the surrounding patterns of DNA diversity. One argument to suggest that sweeps from new mutations contribute significantly to variation in diversity is that great apes with larger population sizes show more dramatic reductions in diversity near genes [27]. This dependence of population size is consistent with the action of positive selection rather than negative selection and suggests that new beneficial mutations leading to sweeps arise more often in species with a larger number of individuals subject to mutation. Identification of selective sweeps, from depressions in diversity or distortions of the site frequency spectrum, is limited to the recent past, where a sample of individuals is expected to be represented by many ancestors. An alternative method to quantify the impact of sweeps on longer timescales is to identify extended regions devoid of incomplete lineage sorting. A sweep in an ancestral species will induce common ancestry for all lineages in a wide region around the selected variant and thus precludes the possibility of incomplete lineage sorting in the region. By identifying and comparing such regions in both the ancestor to human and chimpanzee and the ancestor to human and orangutan, it was possible to show that the human–chimpanzee ancestor experienced a higher frequency of strong sweeps than the human–orangutan ancestor [28].

Addressing the forces of positive and negative selection in the great apes, we need to know what proportion of new mutations are advantageous, neutral, or deleterious and whether these proportions differ across these species. The distribution of fitness effects (DFE) describes the proportions of new mutations that are effectively neutral and new mutations that are under selection [29, 30] (*see* Chapter 1). The DFE further distinguishes between advantageous mutations, which increase the fitness of the organism, and deleterious mutations, which impair survival or fertility. Several methods are available to infer this continuum of selective effects from DNA sequence data [19–21, 31–34]. Initial studies in humans with modest sample sizes found ~25% of effectively neutral nonsynonymous mutations ($-1 > 2Ns < 1$), ~15% of weakly deleterious nonsynonymous mutations ($-10 > 2Ns \leq -1$) and ~60% of moderately to strongly deleterious nonsynonymous mutations ($2Ns \leq -10$) [19–21, 31–34]. A recent study with a large sample size was able to further refine the estimate of new nonsynonymous mutations which are strongly deleterious ($2Ns \leq -100$) to 14–22% and the proportion of weakly deleterious mutations ($-10 > 2Ns \leq -1$) to 25–33% [32]. The DFE for new nonsynonymous mutations is quite similar across great apes despite the differences in the species long-term Ne [22, 35]. This similarity may be explained by the highly leptokurtic DFE of these species, which predicts that substantial changes in Ne will only have a modest impact on the selective effects of mutations. Nonetheless, very different methods and assumptions have been invoked to estimate the DFE across species, and even the shape of the DFE is still a contentious issue. There is very limited knowledge about the DFE of new noncoding mutations, and all we know relies on measures of DNA conservation across mammals and primates. Thus, for noncoding DNA we are only able to say which proportion of new mutations are effectively neutral ($-1 > 2Ns < 1$) and effectively selected against ($2Ns \leq -1$). These rough conservation scores show that only 2–5% of point mutations at noncoding sites might be under purifying selection in humans and the rest of primates [36–40].

Balancing selection is another mode of selection that differs from directional selection in that it does not drive selected variants toward fixation or extinction (*see* Chapter 1 for a definition of balancing selection). Instead, it maintains genetic variation by stabilizing alleles at intermediate frequencies. There are several methods to detect loci under recent and/or long-term balancing selection [22, 41, 42]. A recent study in great apes has confirmed that immune genes are enriched in signals of balancing selection, and it has found that genes involved in the formation of the skin are also under balancing selection [22]. Some of these polymorphisms maintained by balancing selection are even shared between humans and chimpanzees; the most prominent example is the major histocompatibility complex (MHC).

4 Recombination

The rate of recombination varies along the genome. The local recombination rate in each part of the genome can be estimated from patterns of linkage disequilibrium (LD) [43]. It can also be inferred from individually called recombination events by comparing many parent and offspring genomes [44] or by examining genomes of individuals with mixed ancestry [45]. The landscape of varying recombination rate across the genome is referred to as a recombination map. For humans, recombination maps have been produced by all three approaches. Among the great apes, detailed recombination maps only exist for bonobo, chimpanzee, and gorilla. These are produced using the same LD-based method used in humans, allowing direct comparison of recombination maps across species. In all four species, recombination rate varies on a large scale (millions of bases), and this variation is associated with the size of chromosomes, the chromosomal position, the sequence GC content, the gene density, and several other factors [46]. At the fine scale (thousands of bases) recombination rate is determined by the location of the so-called recombination hotspots where about 60% of recombinations occur despite that these hotspots constitute only ~6% of the genome [47]. The location of hotspots is determined by the affinity of the PRDM9 protein for certain DNA motifs present at hotspots. This affinity is encoded in a zinc-finger array whose DNA contacting residues are under strong positive selection. It is now clear that biased gene conversion favors alleles that disrupt hotspots. This depletion of hotspot motifs may result in selection for PRMD9 variants recognizing alternative motifs, producing a turnover of hotspot locations [48]. A comparative analysis of recombination maps of the four species [49, 50] showed that recombination rate on a megabase scale is highly conserved across species, but that the location recombination hotspots are completely different. Only a few hotspots are shared even between chimpanzee subspecies, revealing that turnover of hotspot locations commence at short evolutionary timescales [50].

Comparative studies of recombination have less power than comparative studies of genome sequences: whereas sequence change can be assigned to individual species branches using standard models of molecular evolution, change in recombination rate has so far only been observed as differences between pairs of species. This is because the differences between two species cannot be resolved into the change that occurred in each species without knowledge of recombination rates in the species common ancestor. Fortunately, it is now also possible to construct recombination maps for ancestral species if enough incomplete lineage sorting is present [2]. This approach takes advantage of the fact that gene trees with different topologies must be separated by a recombination event. When sequences are sampled from three different species, the majority of recombination events separating gene trees with different topology will occur in the species ancestral to the two most closely related species. This approach has been used to produce a recombination map of the ancestor of human and chimpanzee [3]. By resolving the differences between humans and chimpanzees into the changes that occurred in each species since their divergence it was shown that recombination rate had evolved more rapidly in humans than in chimpanzees and that striking changes in recombination rate had resulted from a genomic inversion and a chromosome fusion in the human lineage.

5   The X Chromosomes of Great Apes

The unique mode of inheritance of X chromosomes exposes them to population genetic process that differs from that of the autosomes. In a simple population genetic model, the effective population size of the X chromosome will be 3/4 that of the autosomal one. However, this ratio is influenced by many factors such as a difference in generation time and reproductive variance between the sexes, or a stronger propensity of one sex to migrate between subpopulations. More recently it has been suggested that linked selection on the X chromosome in the form of selective sweeps may contribute significantly to a reduced X–autosome ratio. Analysis of diversity along the X chromosomes of the great apes identified extreme selective sweeps in the form of wide regions with strongly reduced diversity and a higher proportion of singleton polymorphisms [51]. The swept regions overlap partially between species, suggesting some amount of recurrent positive selection on the same genes. A separate study exploiting patterns of ILS to measure the cumulative effect of sweeps in the human–chimpanzee ancestor, identified a set of wider regions, spanning the regions identified in extant great ape species [52]. This suggests that regions of the X chromosome are subject to recurrent very strong positive selection. Since these extreme sweeps are only observed on the X chromosomes, it is possible that this is the result of selection of "selfish genes." Such selfish genes, catering only for the preferential transmission of X or Y chromosomes into viable sperm are potentially subject to a particular kind of positive selection called meiotic drive. Even modest transmission distortions will provide selective advantages strong enough to explain the magnitude of these sweeps.

6   Conclusion

The examples of insights provided above represent only the first glimpses of the evolutionary history we share with the great apes as well as the evolution that is private to each species. As genetic diversity across the ranges of each great ape is assayed in more detail, we will get a much deeper understanding of how diverse population genetic processes have shaped genomes very similar to our own.

References

1. Chimpanzee Sequencing and Analysis Consortium (2005) Initial sequence of the chimpanzee genome and comparison with the human genome. Nature 437:69–87

2. Locke DP et al (2011) Comparative and demographic analysis of orang-utan genomes. Nature 469:529–533

3. Scally A et al (2012) Insights into hominid evolution from the gorilla genome sequence. Nature 483:169–175

4. Prüfer K et al (2012) The bonobo genome compared with the chimpanzee and human genomes. Nature 486:527–531

5. Prado-Martinez J et al (2013) Great ape genetic diversity and population history. Nature 499:471–475

6. Kronenberg ZN et al (2018) High-resolution comparative analysis of great ape genomes. Science 360:pii: eaar6343

7. Mailund T, Munch K, Schierup MH (2014) Lineage sorting in apes. Annu Rev Genet 48:519–535

8. Scally A, Durbin R (2012) Revising the human mutation rate: implications for understanding human evolution. Nat Rev Genet 13:745–753

9. Dutheil JY et al (2009) Ancestral population genomics: the coalescent hidden Markov model approach. Genetics 183:259–274

10. Hobolth A, Christensen OF, Mailund T, Schierup MH (2007) Genomic relationships and speciation times of human, chimpanzee, and gorilla inferred from a coalescent hidden Markov model. PLoS Genet 3:e7

11. Mailund T et al (2012) A new isolation with migration model along complete genomes infers very different divergence processes among closely related great ape species. PLoS Genet 8:e1003125

12. Schiffels S, Durbin R (2014) Inferring human population size and separation history from multiple genome sequences. Nat Genet 46:919–925

13. Li H, Durbin R (2011) Inference of human population history from individual whole-genome sequences. Nature 475:493–496

14. Kosiol C et al (2008) Patterns of positive selection in six Mammalian genomes. PLoS Genet 4:e1000144

15. Nielsen R et al (2005) A scan for positively selected genes in the genomes of humans and chimpanzees. PLoS Biol 3:e170

16. Charlesworth B (2013) Background selection 20 years on: the Wilhelmine E. Key 2012 invitational lecture. J Hered 104:161–171

17. Smith JM, Haigh J (1974) The hitch-hiking effect of a favourable gene. Genet Res 23:23–35

18. Nielsen R (2005) Molecular signatures of natural selection. Annu Rev Genet 39:197–218

19. Boyko AR et al (2008) Assessing the evolutionary impact of amino acid mutations in the human genome. PLoS Genet 4:e1000083

20. Eyre-Walker A, Keightley PD (2009) Estimating the rate of adaptive molecular evolution in the presence of slightly deleterious mutations and population size change. Mol Biol Evol 26:2097–2108

21. Galtier N (2016) Adaptive protein evolution in animals and the effective population size hypothesis. PLoS Genet 12:e1005774

22. Cagan A et al (2016) Natural selection in the great apes. Mol Biol Evol 33:3268–3283

23. McManus KF et al (2015) Inference of gorilla demographic and selective history from whole-genome sequence data. Mol Biol Evol 32:600–612

24. Enard D, Cai L, Gwennap C, Petrov DA (2016) Viruses are a dominant driver of protein adaptation in mammals. elife 5:pii: e12469

25. Enard D, Messer PW, Petrov DA (2014) Genome-wide signals of positive selection in human evolution. Genome Res 24:885–895

26. Przeworski M, Coop G, Wall JD (2005) The signature of positive selection on standing genetic variation. Evolution 59:2312–2323

27. Nam K et al (2017) Evidence that the rate of strong selective sweeps increases with population size in the great apes. Proc Natl Acad Sci U S A 114:1613–1618

28. Munch K, Nam K, Schierup MH, Mailund T (2016) Selective sweeps across twenty millions years of primate evolution. Mol Biol Evol 33:3065–3074

29. Eyre-Walker A, Keightley PD (2007) The distribution of fitness effects of new mutations. Nat Rev Genet 8:610–618

30. Keightley PD, Eyre-Walker A (2010) What can we learn about the distribution of fitness effects of new mutations from DNA sequence data? Philos Trans R Soc Lond Ser B Biol Sci 365:1187–1193

31. Keightley PD, Eyre-Walker A (2007) Joint inference of the distribution of fitness effects of deleterious mutations and population demography based on nucleotide polymorphism frequencies. Genetics 177:2251–2261

32. Kim BY, Huber CD, Lohmueller KE (2017) Inference of the distribution of selection coefficients for new nonsynonymous mutations using large samples. Genetics 206:345–361

33. Schneider A, Charlesworth B, Eyre-Walker A, Keightley PD (2011) A method for inferring the rate of occurrence and fitness effects of advantageous mutations. Genetics 189:1427–1437

34. Tataru P, Mollion M, Glémin S, Bataillon T (2017) Inference of distribution of fitness effects and proportion of adaptive substitutions from polymorphism data. Genetics 207:1103–1119

35. Bataillon T et al (2015) Inference of purifying and positive selection in three subspecies of chimpanzees (Pan troglodytes) from exome sequencing. Genome Biol Evol 7:1122–1132

36. Cooper GM et al (2005) Distribution and intensity of constraint in mammalian genomic sequence. Genome Res 15:901–913

37. Davydov EV et al (2010) Identifying a high fraction of the human genome to be under selective constraint using GERP++. PLoS Comput Biol 6:e1001025

38. Pollard KS, Hubisz MJ, Rosenbloom KR, Siepel A (2010) Detection of nonneutral substitution rates on mammalian phylogenies. Genome Res 20:110–121

39. Siepel A et al (2005) Evolutionarily conserved elements in vertebrate, insect, worm, and yeast genomes. Genome Res 15:1034–1050

40. Lindblad-Toh K et al (2011) A high-resolution map of human evolutionary constraint using 29 mammals. Nature 478:476–482

41. Andrés AM et al (2009) Targets of balancing selection in the human genome. Mol Biol Evol 26:2755–2764

42. Leffler EM et al (2013) Multiple instances of ancient balancing selection shared between humans and chimpanzees. Science 339:1578–1582

43. McVean GAT et al (2004) The fine-scale structure of recombination rate variation in the human genome. Science 304:581–584

44. Kong A et al (2002) A high-resolution recombination map of the human genome. Nat Genet 31:241–247

45. Hinch AG et al (2011) The landscape of recombination in African Americans. Nature 476:170–175

46. Spencer CCA et al (2006) The influence of recombination on human genetic diversity. PLoS Genet 2:e148

47. Frazer KA et al (2007) A second generation human haplotype map of over 3.1 million SNPs. Nature 449:851–861

48. Coop G, Myers SR (2007) Live hot, die young: transmission distortion in recombination hotspots. PLoS Genet 3:e35

49. Auton A et al (2012) A fine-scale chimpanzee genetic map from population sequencing. Science 336:193–198

50. Stevison LS et al (2016) The time scale of recombination rate evolution in great apes. Mol Biol Evol 33:928–945

51. Nam K et al (2015) Extreme selective sweeps independently targeted the X chromosomes of the great apes. Proc Natl Acad Sci U S A 112:6413–6418

52. Dutheil JY, Munch K, Nam K, Mailund T, Schierup MH (2015) Strong selective sweeps on the X chromosome in the human-chimpanzee ancestor explain its low divergence. PLoS Genet 11:e1005451

Correction to: Statistical Population Genomics

Julien Y. Dutheil

Correction to:
Chapter 2 in: Julien Y. Dutheil (ed.), *Statistical Population Genomics*,
Methods in Molecular Biology, vol. 2090,
https://doi.org/10.1007/978-1-0716-0199-0_2

Chapter 2, "Processing and Analyzing Multiple Genomes Alignments with MafFilter," was previously published without including the Electronic Supplementary Material. This has now been included in the revised version of this book.

Correction to:
Chapter 3,4,6,7,8,9 and 10 in: Julien Y. Dutheil (ed.), *Statistical Population Genomics*,
Methods in Molecular Biology, vol. 2090,
https://doi.org/10.1007/978-1-0716-0199-0

Chapter 3, 4, 6, 7, 8, 9, and 10 was previously published without including the Extra Supplementary Files. This has now been included in the revised version of this book.

The updated online version of this chapter can be found at
https://doi.org/10.1007/978-1-0716-0199-0_2
https://doi.org/10.1007/978-1-0716-0199-0_3
https://doi.org/10.1007/978-1-0716-0199-0_4
https://doi.org/10.1007/978-1-0716-0199-0_6
https://doi.org/10.1007/978-1-0716-0199-0_7
https://doi.org/10.1007/978-1-0716-0199-0_8
https://doi.org/10.1007/978-1-0716-0199-0_9
https://doi.org/10.1007/978-1-0716-0199-0_10

Julien Y. Dutheil (ed.), *Statistical Population Genomics*, Methods in Molecular Biology, vol. 2090,
https://doi.org/10.1007/978-1-0716-0199-0_20, © The Author(s) 2021

INDEX

Printed in the United States
by Baker & Taylor Publisher Services